Wood & Forest Based Industries of India

(An Inter-State and Inter-Region Comparison)

The Author

Dr L M S Baghel, M.A., M.Phil., Ph.D. presently working as an Associate Professor in the Department of Economics' at Jawaharlal Nehru Memorial Post-Graduate College, Barabanki (UP). He was a meritorious student and obtained M.A., M. Phil. degree by securing second position in University and completed Doctoral degree (Ph.D.) from the Rani Durgavati Vishwavidyalaya, Jabalpur. He has published more than 140 Research papers/articles in various reputed National and International Journals and also presented research papers in National and International Conferences/Seminars. His two Research/Reference Books entitled –"Technological Change and Productivity Growth in Total Manufacturing Sector of India" has been published in 2008 and "Terrorism, Naxalism, Corruption & Regionalism (Challenges & Threats to Development)" published in 2014. Not only this, he has also completed major and minor research projects during the Doctoral degree and Post Doctoral Research Fellowship. He is also contributing to research and development in the field of Economics and Higher Education as a whole as Editor & Publisher of an Multi Disciplinary National & International Research Journals viz. "Indian Journal of New Dimensions" since 2011. Moreover, he has completed a Major Research Project of UGC; New Delhi as a Principal Investigator in year 2015-16. He has been honored by the ADRRI Research Journal, Ghana, Africa as a Reviewer and Member of Editorial Board. Presently, he is holding various academic and non academic positions in College such as member of Proctorial Board & Admission Committee, Career Counselor. At last but not the least, he has also contributed in the field of Astrology and more than 2 dozen predictions on celebrity/politicians and analytical articles too as well has been published in various Astros Magazines and News Papers.

Wood & Forest Based Industries of India

(An Inter-State and Inter-Region Comparison)

Lal Mrigendra Singh Baghel,
M.Phil., Ph.D.
Principal Investigator- UGC MRP
Associate Professor (Economics)
Editor- *Indian Journal of New Dimensions*
Department of Economics
JNMPG College, BARABANKI
(UP)- 225 001, India

2019

Daya Publishing House®

A Division of

Astral International Pvt. Ltd.

New Delhi – 110 002

ISBN: 9789390384747 (HB)

Publisher's Note:

Every possible effort has been made to ensure that the information contained in this book is accurate at the time of going to press, and the publisher and author cannot accept responsibility for any errors or omissions, however caused. No responsibility for loss or damage occasioned to any person acting, or refraining from action, as a result of the material in this publication can be accepted by the editor, the publisher or the author. The Publisher is not associated with any product or vendor mentioned in the book. The contents of this work are intended to further general scientific research, understanding and discussion only. Readers should consult with a specialist where appropriate.

Every effort has been made to trace the owners of copyright material used in this book, if any. The author and the publisher will be grateful for any omission brought to their notice for acknowledgement in the future editions of the book.

Published by : **Daya Publishing House®**
A Division of
Astral International Pvt. Ltd.
– ISO 9001:2015 Certified Company –
4736/23, Ansari Road, Darya Ganj
New Delhi-110 002
Ph. 011-43549197, 23278134
E-mail: info@astralint.com
Website: www.astralint.com

Dedicated to:

My father late Lal Brijendra Singh
Village & Post - Singodi,
Tahsil - Vijayraghavgarh,
Distt - Katni (MP)

Preface

Since time immemorial, Forests have been one of the most valuable natural resources of India. They serve the people in many ways. The economic value of forest increases as Science and technology find new ways to utilize their products. To an agricultural country like India, their importance can hardly be exaggerated as they have been regarded as "green-gold" and the greatest "renewable natural resources". After gaining independence from British Rule in 1947, India followed a path of planned economic development and simultaneously the government of India recognized the role and importance of rural industries but neglible importance has been given to Wood and forest-based industries in the economic upliftment of the rural millions. The rural industries sector has received much emphasis in most of the major policy pronouncements on development in India but this sector has not received proper attention towards its development. While Wood and forest based industries linkages are the best instruments to achieve balanced development of the country/states. As well as to minimize regional inequalities in industrial disbursement they can play a positive role. The importance of forests in national economy requires no explanation. It can be also added that forests and the forest & wood-based industry sector are very important from an ecological, economic and social perspective, especially when they form an industry cluster in a certain region/state of country like India. Considering increasing challenges for the forest sector, a study on industry-wise and inter–state, inter-region perspective can be very helpful for showing the complex structure and the various interactions between the different industry and along the productivity growth, technological change and production function approach. Therefore, keeping these views in mind, this study has been conducted with the help of financial assistance provided by UGC, New Delhi.

This Major Research Project conducted and completed by me and final draft submitted to UGC, New Delhi in the year 2016. Title of the project was –"Wood & Forest Based Industries of India: An Inter-State and Inter-Region Comparison…." {(F.No. 5-183/2013(HRP)}. The main objective of this study was devoted to the productivity trends and statistical estimation of production functions and technical change in the wood and forest-based industries for Indian states and India as a whole. With the overall objective of the proposed study was to analyze the growth of technological change and productivity growth in wood and forest-based industries of Indian states as well as India as a whole.

This study has been conducted with the aims at finding out the extent of technological change in wood and Forest Industries and productivity growth. The SFP ratios have enabled us to quantify the unit factor requirements on saving in the use of factor of production. Thus, inter-state difference in SFP found useful in explaining the gap between developed and less developed states. The TFP have also throws light on overall efficiency, which would be helpful for the policy makers in framing policy for efficiency improvement in less developed states. The estimates of production function are giving an idea of marginal product of factor, returns to scale elasticity of substitution among the regions of India and India as a whole as well. The variation in all the measured parameters among the states and regions has given the idea of extent of disparity existing among the states and regions. Thus, finally it can be said that all the objectives of the study have been covered and analyzed in proper and prospective manner and set goal have been achieved which objectively and hypothetically has been set. I am fully satisfied with this achievement because, No any study of this nature on WFBI previously has been conducted by and academicians/researchers or institutions.

The review of literature serves as a background for any research and scientific investigation which helps in understanding it in proper perspective. An exhaustive review of literature help any researcher to identify variables relevant for research, decide tools and techniques to be adopted, delineate a new area of study and relate the present study with the previous ones. Keeping these views in mind, all the available literatures concerning the problem has been reviewed under the 4 heads and in their various sub-heads. The main purpose of this study is devoted to the productivity trends and statistical estimation of production functions and technical change in the wood and forest-based industries for Indian states and India as a whole. With the overall objective of the study was to analyze the growth of technological change and productivity growth in wood and forest-based industries of Indian states as well as India as a whole. For the purpose of analysis, this study covers 20 industries group of wood & forest based industries and total as well; 16 States, and 5 regions along with All India as whole. The data used for the analytical purpose are gathered through secondary sources for the period 1973-74 to 2012-13. (Last year of study depends upon the availability of the data).

This study has cover trends in SAGR, 7 types of partial factor productivity ratios (SFPs) and 3 types of total factor productivity growth (TFPG) indices, and estimate parameters of Cobb-Douglas, CES, and VES production function. At, last, on the basis of overall analysis and in-depth elaboration of issues taken here, we once again certainly want to explore that Wood and forest based industries can play a vital role in the development process of a country like India because most of the rural population living in and around the forests. Rural population and their dependency on agriculture and forest are still heavy. Economic prosperity of agrarian economy will not only depend upon the integration of agriculture with industry but also upon the integration of forest with industry. Hence, for the development of rural areas, Wood and forest based industries have a special significance. In the midst of all these problems a special importance can be attached to the question of balance between forest and industry in development.

Finally, it can be said that this study will prove a mile stone in developmental literature of this country. By 2030, India is likely to have GDP of USD 4 trillion dollars and a population of 1.5billions. To achieve sustainable growth of our economy, it is essential to develop forest sector and its related activities like forest based industrial development, to achieve double digit per annum growth of GDP. The study like this has not been done earlier in India by any academicians, researchers, and also not by any institutions too. Therefore, there is an essential need to work out on the result finally interpreted here in this work for the economic growth and sustainable development of manufacturing sector like Wood and Forest based Industries.

At last, I am very much grateful to the University Grants Commission, New Delhi for providing financial assistance to conduct such a vast study. I am also thankful to the Principal, JNMPG College, Barabanki for the encouragement and cooperation which I have been fortunate to receive all along not only in our project tenure but also in our academic pursuits continuously going on. In the preparation of the present study, I have made wide use of library facilities at the various institutions such as CSO, New Delhi; Ministry of Environment & Forests, New Delhi; CSO Kolkata; ASI Data Library, Greater Noida; MOSPI, R.K. Puram, New Delhi; NITI AAYOG, New Delhi; IIPA, New Delhi; TFRI, Jabalpur; FRI, Kanpur; Centre for Social Forestry and Eco-Rehabilitation, Allahabad; Library Lucknow University; and Dept. of Economics, RDVV, Jabalpur. I am extremely thankful to all the aforesaid institutions and their Staff members for proving facilities, materials and supports. I would also like to express my thanks to Prof. N.G. Pendse, Dept. of Econ. RDVV, Jabalpur, for their constant encouragement. I am grateful to my better half Mrs. Laxmi Singh who associated with me from the beginning in compiling and for assisting in typing and correction. Finally, I must express my gratitude to the Publisher, Astral International, New Delhi and their dedicated staff for taking the initiative of bringing out this publication in a short time. May God always bless them who should get the credit for any merit the book may yield. Last but not the least, I would like to dedicate this work to my parents and heartfelt thanks to my Ultimate Almighty who are living in inner core of my heart for inspiring me always invisibly in each and every steps of my life.

LMS Baghel, M.Phil., Ph.D.,

Principal Investigator- UGC- MRP

Associate Professor (Economics)

Editor-*Indian Journal of New Dimensions.*

Dept. of Economics, JNMPG College,

Barabanki (UP) – 225 001, INDIA.

Mob.: 9451397832, 9554162122.

Email: lmsbaghel@rediffmail.com, journalijnd@gmail.com

Contents

1

Introduction

1.1: Background

The dominant economic theme in human-forest Relationships since the creation of human being and somehow especially over the past two millennia, but more especially during the last two centuries, has been one of value abstraction and increasingly intensive management. Forests have been an integral part of human eco-system. Humans have depended on forests for a remarkable variety of products, services, and benefits again since the beginning of human came into existence in earth. With rapid economic growth and technological changes, they have refined and added to the values forests make available. Even as deforestation has accompanied early stages of economic growth and economic development of advanced levels are closely associated with forest transitions and increased forest cover. The economic contributions from forests, the state of the literature enable us to know how forests contribute to human well-being through the many different goods and services they provide, the ways in which they are managed and governed, and some directions for more effective protection of the heritage they constitute and enhancing of the benefits they provide.

Historically, forests have played a major role to influence Patterns of economic development of an economy, supporting livelihoods not only for human but also for the habitat living in and around the forest, helping structure economic change, and promoting sustainable growth. For millennia before the industrial revolution, forests, woodland, and trees were the source of land for cultivation and settlement, of construction materials, of fuel and energy, and indeed of food and nutrition as well (Williams 2002).

Historical evidence also indicates that the early societies comprising mostly hunting-gathering households depended on forests for nearly all of their livelihood needs. The agricultural transition translated into deforestation for cultivable land at a scale that was limited only by available technologies, by the labor available to cultivate land and grow crops and by the presence or absence of markets. Forest provided fodder, firewood, and subsistence timber–goods for which they are still the major source for most poor households in the developing world. The extended use and exploitation of forest resources even before the industrial revolution had led to efforts to conserve forested areas and plant new trees in specific regions of the world. In Europe, France and Germany were leaders in developing policies in the 17th and 18th centuries to regulate the use of and to protect forests. The emergence of forestry as a science with its focus on sustainable timber production was also a hallmark of colonial forest departments founded all over the developing world by European colonizers (Barton 2001, Peluso & Vandergeest 2001, Troup 1940).

The foundational justification for many forestry departments all over the world was to improve management and enhance the public benefits of forests in terms of soil conservation, watershed protection and flood control. But most forestry departments were also under substantial pressure to generate revenues and often sought to protect forests for commercial exploitation. The consequence was tensions between government forest agencies and the poor populations that depended on forests for their livelihoods (Guha, 2000). Forests continue today to provide the high levels of commercial benefits to households, companies, and governments that formed the initial impetus for protective statutes and policies.

The forest sector consists of a wide range of industry branches, ranging from forestry over wood processing industries (*e.g.* sawmilling & wood-based panel industry), wood manufacturing industries (*e.g.* furniture & wood construction industry) and other wood-based industries (*e.g.* pulp & paper industry) to non-timber forest products and services (*e.g.* forest-based tourism). Forest & wood-based industries can show spatial agglomerations. Forests and the forest and wood-based industry sector are very important from an ecological, economic and social perspective, especially when they form an industry cluster in a certain region or country (Thorsten M. *et al.* 2010). However, from a policy and socio-economic perspective, mostly administrative units are applied for describing the forest sector at different spatial scales (International, National, Sub-national, Regional, and Local). The forest sector is also characterized by a complex policy arena with a large number of different stakeholder groups representing the different industry branches and the different spatial scales (Krott M., *et al.* 2005).

An estimates of FAO highlight the facts that forest industries contribute more than US$ 450 billion to national incomes, contributing nearly one percent of the global GDP in 2008 and providing formal employment to 0.4% of the global labor force (FAO 2012). Forests also provide other sources of incomes and subsistence benefits, generate informal work opportunities, and constitute reservoirs of economic values that help ameliorate shocks to household incomes –particularly in rural areas in poor countries (Chomitz and Kumari 1998). But systematically collected data to permit aggregate estimates of non-industrial economic contributions of forests are simply not available either at the global or at the national level. The individual country level estimates that are available demonstrate that such benefits are substantial, in many cases 3-10 times higher than those for which systematic national and global data are collected.

Since last sixty years, changing views about the importance of forests and their relationship to humans have been associated with corresponding changes in emphases in how forests are managed. Earlier efforts to manage forests for sustained yields of commercial timber were in many countries replaced in the 1970s and 1980s by the watchwords of sustainable forest management and ecosystem management as the multi-functional nature of forests and woodlands became obvious. Increasing levels of deforestation despite efforts to reduce and reverse loss forests led many governments and international agencies to attempt policy experiments that would have a positive influence on net deforestation. So at last, for this introductory section of the final drafting of UGC sponsored major research project, it can be said that Forests perform a number of productive, protective and regulative functions. They are important components of the economic and ecological systems. The role of forestry is to regulate the hydrological and atmospheric cycles, to provide a habitat for biodiversity and to prevent soil erosion, siltation and flooding. The economic returns are derived from the large variety of forest products. Forests also provide recreational, aesthetic and cultural functions and are an important source of livelihood for local communities.

1.2: Historical Aspect of Indian Forest

In Indian history, there is enough evidence to show that dense forests once covered India. The changing forest composition and cover can be closely linked to the growth and change of civilizations. Over the years, as man progressed and entered in civil society, the forest began gradually depleting. The growing population, industrialization, urbanization and man's dependence on the forest have been mainly responsible for this. All ancient texts have some mention of the forest and the activities that were performed in these areas. Forests were revered by the people and a large number of religious ceremonies centered on trees and plants. The principles of forest conservation and sustainable management were well entrenched in the pre-historic India (Kumar, 2008).

Starting with the days of Vegas and extending into *post-Vedic* and *Puranic times* (200 BC to 100 AD) *(http://www.ejvs.layrasjanacademy.com/ejvs1006/1006.txt)* environment consciousness, besides natural resource and biodiversity conservation, were intrinsic features of Hindu religions rituals and practices. In Vedic Forestry, both protective and productive aspects of forest vegetation were emphasized a lot. In *Shruti Vedas* such as *Rig, Sama, Yajur and Atharvaveda* much attention were given on forests and contain several discriptions on the uses, protection and management of forests. Over centuries, *Aranyakas Samhita* came into existence by giving much more attention on forests and uses of forest products such as herbs, shrubs, timber and non timber forest products. Of these, *Aryanyakas* or the 'forests' works (*'Aranya'* is Sanskrit means *'Forest'* and *'Aranyaka'* means *'in the forest'*) and the Upanishad *'Brhadaranyaka'* (translated as 'Great Forest Text') are particularly important from the all perspective of forestry traditions (Pande, 1994, Keith 2005, Witzel, 2005).

Traditions in Vedic age affirm that every village will be complete only when certain categories of forest vegetation or tress (*e.g.* Mahavan, Shrivan and Tapovan) are preserved in and around its territory. Of these, Mahavan or the great natural forest is perhaps equivalent to the protected area of today. According to Banwari (2002) in Prime, R. 2002, once some of the original forest was cleared, the Vedic culture also necessitated that another kind of forest be established in its place which is equivalent to today's *"production forests"*, and it provides the essential goods and services to humans and livestock (*e.g.* fresh air, shelter, herbs, shrubs, timber, fuelwood, roots, fodder besides maintaining soil fertility quality water). Traditionally, these are called *Shrivan* or the *'forestry of prosperity'* or *'forest of wealth'*. The third category of forests is *Tapovan* or the *'forest of religion'* where on animal, or trees could be harmed. This type of forest is natural and untended. The Agni Purana, written about 4000 years ago, stated that man should protect trees to have material gains and religious blessings.

Vedic literatures of Charaka, Susruta, Dhanwantari, Nagarjuna, Parashara, Valmiki and various others seers, exemplify that Himalaya and surrounding areas of it have been regarded as a vast repository of medicinal plants. During Emperor of King Ashoka's days, planting of medicinal herbs, plants and trees besides shade trees along the road and fruit plants on the wastelands was mandatory analogous to the social and agro-forestry programmes of today, in some way. In Ancient Bharat and up to some extent in this era of modern India, several trees and shrubs were regarded as sacred because of their medicinal/aesthetic/natural qualities as well as some because of their proximity to a particular deity. The Indian religious literature is replete with ideas of forest conservation, utilization, and regeneration. The traditional conservation ethos is also reflected biodiversity conservation in natural ecosystems and helps reduce the harvest pressure (Kumar, 2008).

Around 2500 years ago, Mahatma Buddha preached that man should plant a tree every five years. In his period, around the Bauddha Temple, Aashram were marked where certain rules and regulations applied. Chandra Gupta Maurya (about 300 BC) also realized the importance of the forests and they had appointed a high level authority to look after the forests. While King Ashoka stated that wild animals and forests should be preserved and protected and also launched various programmes to plant trees on a large scale. Rules of the Ashoka period were continued even during the Gupta period. During the early period of Muslim rules, invades a large number of people had to flee from the attacks and take refuge in the forests. This was the beginning of a phase of migration to the forest. In fact, deforestation of forest in large scale has been started during Muslim rules and they cleared vast areas of forests to make way for settlements. Their arm forces were all keen hunters and therefore had to have patches of forests where they could go hunting. This ensured that the trees in these areas were not felled, and the forest ecology was not tampered with. Ruler of this period showed more interest in gardens and their development. Akbar ordered the planting of trees in various parts of his kingdom. Jahangir was well known for laying out beautiful gardens and planting trees.

During the beginning of British rule, trees were felled without any positive thought and management. In British period, large numbers of trees mostly the Sal, Teak, and Sandalwood were cut for export. The history of modern Indian forestry was a process by which the British gradually appropriated forest resources for revenue generation. Trees could not be felled without prior permission and knowledge of the British authority. With keeping sole ownership of forest, such step was taken to ensure that they were the only sole users of the forest trees. But after some time, somehow they realized the importance of forest and then the

British began to regulate and conserve the Indian forests. Such realization was seen when in 1800 a commissioner was appointed to look into the availability of teak in the Malabar forests. In 1806, the Madras government appointed Capt. Watson as the commissioner of forests for organizing the production of teak and other timber suitable for the building of ships. Later on in 1855, Dalhousie framed regulations for conservation of forest in the entire country.

During World War I, forest resources were severely depleted as large quantities of timber were removed to build ships and railway sleepers and to pay for Britain's war efforts. The great advancements in scientific management of the forests were made between the two World Wars, with many areas undergoing regeneration and sustained harvest plans being drawn up. World War II made even greater demand on the forest than World War I had done. Sadly, it has been seen that emphasis was still not on protection and regeneration but on gaining maximum revenue from the forests. After independence in 1947, a great upheaval in forestry organization occurred. The princely states were managed variably, giving more concessions to the local populations. The transfer of these states to the government led to deforestation in these areas. A lot of forest had existed and has been lost since the government took over these states. The new Forest Policy of 1952 recognized the protective functions of the forest and aimed at maintaining one-third of India's land area under forest. Certain activities were banned and grazing restricted. Much of the original British policy was kept in place, such as the classification of forest land into two types.

Thus, past history clear throw a picture that India had rich forest and forest resources in the past due to favorable climatic conditions, extending over major part of the country. Clearing of forests for cultivation, food, shelter and pasture and selective removal of valuable timber trees adversely affected the resources. The owners of the forests were interested only in collection of revenue. In the early days of establishment of British rule in 1800s, the accessible forests suffered due to large scale felling of valuable timber trees. The forests of western coastal tract were heavily exploited for teak and other timber required for the Navy. Later, Britishers began to realize that forests were not so inexhaustible and there was need for conservancy, leading to introduction of systematic working and regeneration measures.

It is to be noted here that during the pre- independence era, the forests in India were owned by the public, princely states and individuals. Only those forests which were owned by the public were put under working plans for management. The recorded total area of forests available with British India was only 34.76 million ha, of which about 74 per cent area was covered under working plans by 1947. The various working plans helped in managing the forests at local and divisional level effectively. The knowledge of the forest resource at the national or sub-national level was however lacking except for the area under forests. After political integration and abolition of proprietary rights in forests of feudal states especially of Saurashtra, Mysore, Madhya Bharat, Kashmir and Vindhya Pradesh, a big addition to the forest areas took place. The recorded forest area almost doubled to 71.8 million ha in 1950- 51, and up to 74.8 million ha in 1970-71. Most of the forests under feudal states had only primitive forms of working, with practically no working plans to regulate their management. On that time, it was therefore necessary to do the spade work first such as survey, demarcation, mapping and giving legal status to these forests. Thus, besides revising the existing working plans, preparation of working plans for the 'newly acquired' forests became an important responsibility of the provincial governments.

Thus, long history of India has witnessed colonial exploitation which has resulted minimization of forest covered area and in far below per capita income generation from forests as compared to developed countries and some of the developing countries too as well.

1.3: Forest Policy of India and Legal Framework

In India, forests and environment has always been low in the priority of the Governments. This is reflected in the negligible share of plan budget being allocated for environmental management. Even in the era of global environmental disturbances, many State Boards receive almost no financial support from the State Governments and are totally dependent on cess. Though environmental concerns are being given greater importance at the national level, commensurate investment is hardly evident. The total annual plan of the MoEF is currently of the order of Rs 2000 crores, which is about 0.25% of GoI's plan budget. The expenditure on environment in India is a very small fraction of the GDP, much below most developed and emerging economies. Currently, the percentage of GDP spent on environment in India is 0.012, where as it is 1.0 in Japan and 0.4 in USA.

The concept of environmental protection and resource conservation has been an integral part of our tradition and way of life. However, changing lifestyles, increased urbanisation, industrialisation, infrastructure development and population growth have put enormous pressure on environment. At the same time, there is heightened public awareness about the perils of unbridled industrial expansion and increased public pressure to strike a harmonious balance between rapid economic growth and protection of environment. This provides an excellent opportunity to the Government to mainstream sustainability concerns in the development projects and improve compliance with environmental laws and standards.

The scale of the environmental challenge is set to increase in magnitude as well as complexity. The following are the major threats: (i) Imperative of maintaining high economic growth; (ii) Increasing Urbanization, Population Growth and Industrialization; (iii) Unmet basic needs; (iv) Life style changes; (v) Huge Biotic pressure. The foremost challenge is to ensure that environment is not degraded with rapid economic growth. There are huge unmet basic demands in terms of electricity, drinking water and sanitation. Urban population is rapidly growing. There will be corresponding increase in the levels of effluent and waste generation. These trends are bound to put enormous pressure on the natural resources and the environment. Thus, the broad vision of the Ministry with sustainable development as the core underlying theme is enshrined in a number of policy documents brought out from time to time. These policy statements provide the conceptual and ideological underpinnings for the regulatory and institutional framework and the plans and schemes of the MoEF. The formal Forest Policy and legal framework for protection, conservation and management of forests have been in effect since 1894 and 1865 respectively. Early forest policies tended to consider timber production as the primary function of the forest. In today's context, a multiplicity of interests competes for forest outputs and correspondingly forest policies have become increasingly complex. Thus, the important policy statements can be outlined as below:

- Before the advent of colonial rule in India, there was no unified formal forest policy, various princely states having different approaches towards the forestry resources available in their areas. With the advent of British in India, forest management started in the mid-eighteenth century and the first policy statement was announced in 1894. However, there was no mention of any percentage of the land area that should be under the forest cover in India, probably because no need was felt at that time.
- NFP 1894, or Circular F 22 of 1894, was the first formal forest policy statement of India, and was based on the Voelcker Report in 1893 on Improvement of Indian Agriculture. The main stated objective of this policy was to manage the State Forests for public benefit. However, the policy also provided for regulation of rights and restriction of privileges of users in the forest area.
- The Forest Policy of 1952 was among the notable initiatives taken by the Government in post-independent India. It enunciated a target of maintaining one third of the total land area under forests. In 1976, The National Commission on Agriculture fully acknowledged the role to be played by forestry in the development of the country and recommended large scale plantations on degraded forest areas and social forestry in community and private lands to reduce the growing gap in timber and firewood requirements. It also suggested the formation of Forest Corporations to raise plantations in degraded forestlands. This led to the initiation of large scale social forestry projects from 1980 onwards with international assistance and as part of the rural development programme since the 7th Five Year Plan.
- The Forest Policy 1952 recognized the protective role of forests and discarded the notion that forestry has no intrinsic right to land. It stipulated that the country should aim at having at least one third of its total land area under forests. Wildlife conservation became prominent in 1972 with the promulgation of the Wildlife (Protection) Act. The legal regime for protection of forests is mainly limited to the Indian Forest Act 1927.
- Most of the Indian States have promulgated separate legislations to meet the specificities of the respective states. The Constitution of India assigns fundamental duties to the citizens of the country and directs States for conservation and protection of forest, wildlife and the environment.
- The country shifted forest from the state list to the concurrent list in the Constitution of India due to emerging ecological needs in the mid-1970s.

India has shown political commitment for the conservation of forests with the initiative to enact the Forest (Conservation) Act 1980. It is regulatory Act and maintains balance between development and conservation. It was a milestone step in the direction of forest conservation in the country.

- The National Afforestation and Eco-development Board (NAEB) has the mandate to regenerate degraded forests in the country with the active involvement of the public and stakeholders. The National Wasteland Development Board (NWDB) focuses on improving land capabilities. The subject 'Forest' has been in the concurrent list of the Indian Constitution since 1976. The Forest (Conservation) Act 1980 (amended in1988, 2003) empowers the central Government (Ministry of Environment and Forests) to guide states in matters related to the diversion of forestland for non-forestry purposes, conversion of natural forests into plantations and even priorities of forest management in line with the National Forest Policy.
- India had taken a revolutionary shift in forest management from a regulatory to a participatory approach with the promulgation of the National Forest Policy (NFP) 1988. This Forest Policy came four years before the Earth Summit which embodies all elements of SFM and India's forests are treated as social and environmental resources. Ecological security became the prime objective and focus was given for providing livelihoods for forest dependent communities. India initiated the implementation of this policy in a major way to involve local communities in the conservation, protection and management of forests through JFM institutions in 1990 and expanded this programme to more than 22 million hectares of forests with the involvement of approximately 21 million people. India has shown sensitivity towards the recognition of tenurial rights of the tribals on forestland with the issuance of guidelines to state governments in 1990.
- The main aims of NFP 1988 is to increase the area under forests in the country to 33% through massive afforestation programs and protection of the existing forest cover, regulating management of forest based industries and building on the symbiotic relationship between forests and the forest dwellers.
- The Environment Protection Act was enacted in 1986 for improving the environment of the country. India has shown remarkable progress during the last 15 years for enhancing the contribution of forests towards poverty alleviation through empowering people with the ownership of NWFPs.
- India has provided a legal regime for biodiversity conservation with the promulgation of national level legislation i.e. The Biological Diversity Act 2002.
- Another milest one was achieved in 2006 with the enactment of a national level legislation for assigning habitation and occupation rights on forests along with the responsibility for conservation of biological resources and maintenance of the ecological balance to communities. Fine tuning of the

Indian Forest Act 1927, the State Forest Acts and the Wildlife Protection Act is needed in future.

- Furthermore, forests in India are treated primarily as social and environmental resource and only secondarily as commercial resources. The country largely depends upon the forest for its requirement of wood and wood products from private lands. For making relevant policy interventions to provide an enabling environment for the private sector to grow more trees, the policy and legal regime has been reviewed by an independent body, i.e. the National Forest Commission (NFC) which submitted its report in 2006.
- The view of the NFC was to continue with the National Forest Policy 1988. The forestry sector is impacted directly by the policies of other sectors such as Agriculture, Rural Development, Panchayati Raj, Education, Energy, and Water Resources and indirectly by the policies of industry and commerce.
- India is looking for certain policy interventions to motivate people to grow more trees in future and also thinking about establish a mechanism to involve the private sector in investing financial resources in degraded forests. Our mandate is to achieve 33% forest and tree cover in the country and also to contribute towards achieving global objectives for the sustainable development of all types of forests. The policy and legal regime in the forestry sector will focus on poverty alleviation through forestry, increasing productivity, enabling the private sector to grow more trees, ecological security, empowerment of communities along with their capacity building and biodiversity conservation in 2020.
- Traditionally, Forests have been the habitats of tribal communities with a variety of lifestyles ranging from nomads, hunters, wild food gatherers to agrarians. Thus, the traditional lifestyles of tribes and their recorded rights have been respected and embedded in forest management practices as well as in subsequent policies. A law was enacted in 2006 recognizing the rights of occupation of forests by tribes and forest dwellers and empowering them for the management of forests (used by them) as common property resources. It is estimated that about 20% of the government controlled and managed forestland will come under the occupational titles recognized under this law. The recognition of the right of common use conforms to the policy prescription of participatory forest management and also accepted principles of biodiversity conservation as well as community involvement in conservation.
- Within the ambit of the national policy and legislation, states can promulgate legal instruments and undertake suitable measures to facilitate smooth functioning of the sector. A number of state laws have been passed to regulate forest resource use, including timber and NWFPs.

The review of the NFP by Indian Institute of Forest Management (IIFM, 2001) has suggested to resolve, protect and improve the environment and forests of the country by initiating key programs including forest protection and afforestation, JFM, forest fire control measures, treatment of drought prone areas, strengthening of infrastructure, wildlife conservation, pollution control measures and implementation of environment law. But much of these activities are not justified or well integrated within the forest policy cycle till date. Thus, there is an urgent need for a new or revised NFP proposal considering new and healthy approaches for NFP implementation for a long-term strategy on the Concept of Forestry Development.

1.4: Definition of Forests, Forestry Sector & Forest based Industries

The word 'Forests' has been derived from the latin words 'Foris' means 'outside', which referred to a village boundary of fence. In technical terms, forest is defined as an area set aside or maintained under vegetation for any indirect benefits, namely climatic, productive or environmental and/or for production of wood and non-wood products.

According to Allen, S.W. (1950), "A *forest* is a community of living trees and associated organisms, covering a considerable area , utilizing sunshine, air, water and earthly materials to attain maturity and to reproduce itself, and capable of furnishing mankind with indispensable products and services".

While in legal terms, a forest can be defined as an area of land notified to be a forest under a forest law. Thus, a forest is the most complex, terrestrial, biotic portion of the ecosystem and is characterized by a predominance of trees. Forestry is the profession that embraces the science, art and business of managing the forested portion of the ecosystem in a manner that assures the maintenance of biological, genetic, and functional diversity and productivity for the perpetual production of amenities, services and goods for human use.

In the available literature, there is no commonly agreed definition of the forestry sector. Ideally, the sector should include all economic activities that mostly depend on the production of goods and services from forests. This would include commercial activities that depend on the production of wood fibre (i.e., production of industrial round wood, fuel wood & charcoal; sawn wood & wood-based panels; pulp & paper; and wooden furniture). It would also include activities such as the commercial production and processing of non-wood forest products (NWFPs) and the subsistence use of forest products. It could even include economic activities related to provision of forest services.

FSI defines forests as "all the lands, more than one hectare in area, with a tree canopy density of more than 10%." Champion & Seth (1968), classified India's forests into four major ecosystems groups, namely, tropical, subtropical, temperate, and alpine. These major groups are further divided into 16 types. Of the 16 forests types, tropical dry deciduous forests form the major percentage that is 38% of the forest cover in India (Sharma & Chaudhry, 2013). With the fast-growing economy in India, the growing pressure for environmental protection and increasing environmental awareness require deep understanding of the interactions between forest and other sectors concerning economic development, forest resources consumption and forest environmental degradation (MoSPI, 2013). Forests contribute directly to welfare through the provision of amenity values, which may not satisfy the SNA's definition of 'production'. They also provide other industries with services, such as watershed protection, whose value the SNA records as part of the operating surplus of recipient industries instead of as services furnished by forests. For these reasons, the SNA likely understates the economic contribution of forests (Vincent & Hartwick, 1997).

A forest dependent industry or forest based industries is any industry that uses raw materials from the forested portion of the ecosystem, including amenities and services, such as oxygen, water, electricity, recreation, and migratory animals. A forest based industry also includes any industry that uses extractive good such as minerals wood fiber, forage for livestock, resident fish and game animals, and pelts from fur-bearing mammals. Forest dependent industries based on amenities and services are extractive if the products either enter and/or leave the forest under their own volition. Such industries includes both the sport and commercial fisher who catches migratory salmon and steel hood, the farmer who uses water to irrigate crops, the person who markets those crops, the electrical company that uses water converted to electricity and the municipal water company itself.

Forest based industries that are based on extractive products, on the other hand, physically remove raw materials from the forest and, for the most part, make them available for further refinement and use that goes beyond the initial extraction. Such industries include timber companies that cut trees, ranches who graze their livestock in forested allotments, miners who extract ore, and hunters, fishers and trappers who kill and remove forest-dwelling wildlife. Forest dependent industries that refine the extracted products include jewelers, Carpenters, boat builders, artisan woodworkers and furriers. Finally, these forest-based industries are all interwoven because each industry uses one or more of the other's produced, such as water, electricity, wood fiber, red meat, vegetables, and so on.

Wood and forest-based industries are essentially those industries which depend on raw materials drawn from forests; save for some exceptional cases. Development of forest based industries is the story of development of civilization. Today our access to major forest products is through industry only. Special features of forest-based industries become active when they furnish a very wide range of products, both consumable goods and intermediate goods flowing into many sectors of the economy. In the context of country like India, the forest-based industries do play a significant role in the economic development, putting forest products to more meaningful use.

Therefore, wood & forest-based Industry contributes significantly to a nation's economic development for following reasons: (1) Wood & forest based industries are a primary method of transforming raw –material of forests and/or products into finished products for various uses of human life, (2) Wood and forest based industries often constitute the majority of a developing nation's manufacturing sector, (3) Wood and Forest based industrial products are the major exports from a developing nation; and (4) The Wood and forest based industries provide diverse products, employment and income.

1.5: Contributions of Forestry Sector in India's Economy

Since time immemorial, Forests have been one of the most valuable natural resources of India. They serve the people in many ways. The economic value of forest increases as Science and technology find new ways to utilize their products. To an agricultural country like India, their importance can hardly be exaggerated as they have been regarded as ***"green-gold"*** and the greatest "renewable natural resources". The forests constitute a perennial economy unlike minerals whose exploitation is a rubber economy.

At present, total land area of the world is 13,077 million hectares of which 32 percent is covered under forests. Of these, India occupies about 2.4 percent of world's land area and only 1.8 percent of its forest covers. Regions having forest cover more than World average (32%) are Latin America (46%), former USSR (41%), North America (37%) and Europe (37%). Worldwide, more than 430 million hectares of forests, or 11% of the total forest estate, are officially community owned or managed and another 13.8% are owned by smallholders and firms. Local rights are more marked in developing countries, where, in 2008, 27% of all forest lands were community owned or administered. Tenure shapes a country's forest industry and economy. There is ample evidence in some developed forested countries—*e.g.* the United States, Sweden, and Finland—and developing countries—such as Mexico and China—that the recognition of local rights has a profound effect on the structure of industry and increases the potential for forestry to generate jobs and economic growth and contribute to good governance. Small-scale and community initiatives around forests can also provide invaluable ecosystem services, including climate change mitigation—given the necessary tenure reform (Augusta Molnar, *et al.*, 2011).

In India, forestry has been considered as a significant rural industry and a major environmental resource for the masses. India is one of the ten most forest-rich countries of the world along with the Russian Federation, Brazil, Canada, United States of America, China, Democratic Republic of the Congo, Australia, Indonesia and Sudan. Together, India and these countries account for 67 percent of total forest area of the world (Global Forest Resources Assessment 2010). India's forest cover grew at 0.22% annually over 1990-2000, and has grown at the rate of 0.46% per year over 2000-2010 (Global Forest Resources Assessment 2010), after decades where forest degradation was a matter of serious concern (FAO, 2002). As of 2010, the Food and Agriculture Organization of the United Nations estimates India's forest cover to be about 68 million hectares, or 24% of the country's area. The 2013 Forest Survey of India states its forest cover increased to 69.8 million hectares by 2012, per satellite measurements; this represents an increase of 5,871 square kilometers of forest cover in 2 years (State of Forest Report, 2013). However, the gains were primarily in northern, central and southern Indian states, while northeastern states witnessed a net loss in forest cover over 2010 to 2012.

India is a densely populated country with high pressure on land and forest resources. The total geographic area of the country is 32,87,240 km^2 (327.8 million ha) and the population is 1,210 million (Census 2011) which gives population density of 382 persons per km^2 and 0.06 ha per capita forest. It is the 7th largest country of the World's geographic area, and 1.8 per cent of its forests, but supports almost 18 percent of its population. At present, India's population is almost equal to be combined population of USA, Brazil, Indonesia, Pakistan, Bangladesh and Japan. Comparing with USA, India's geographic area is almost one-third of the area of USA whereas the population is 4 times, thus population density is 12 times of that of USA. As per Census 2011, about 68.8 percent of India's population (833 million) lives in rural areas, and most of them have land based economy which uses forest resources one way or the other. It is estimated that about 200 million people live in and around forests, and fully depend for their livelihood on forest resources. In addition, the livestock population of India is about 530 million as per the 18th Livestock Census, 2007 (MoA 2010) and has registered a growth of 9.2 per cent during 2003 to 2007. About 38% of livestock depend on fodder derived from forests by direct grazing or by harvesting, causing additional burden on the forests, and is responsible for forest degradation. With only 2.4% of the land area, India accounts for 7 to 8 percent of the recorded species of the world (Gokhale, 2010). This biodiversity is of immense economic, ecological, social, and cultural value. Approximately 275 million people in India (27% of the total population) are known to live in the forest fringes and earn bulk of their livelihood from forests (World Bank, 2001; 2006; Poffenberger, 2000, Sinha *et al.*, 2010).

Though India's major area falls in the tropical zone, due to large altitudinal variations and presence of high mountains, there exist almost all the climatic conditions, from hot to cold deserts and from arid to wet areas. If we consider by Topographically, the country is divided into 4 broad regions, these are: (i) The *Himalayan Mountains* in the north contain the cold deserts and fertile valleys, (ii) The vast *Indo-Gangetic Plains* formed by the basin of three distinct river systems - the Indus, the Ganges and the Brahmaputra, the flat, fertile and most densely populated area, (iii) The region comprising the *Thar Desert* in the West comprising major part of Rajasthan and lower regions of Punjab and Haryana and Rann of Kutch in Gujarat and (iv) The *Southern (Deccan) Peninsula* bounded by the Western and Eastern Ghats and, the Coastal Plains and Islands which are also densely populated.

Of the total 327.8 million ha geographic area of the country, the land-use statistics of 23 million ha are not available, as approximately 12 million ha is under the occupation of Pakistan and China and about 11 million ha is under permanent snow or in inaccessible region. Further, about 42 million ha area has been used for constructing roads and railway tracks, stations and in habitation under cities and villages. The detailed land-use is given in Table 1.1. It can be seen that in addition to 69 million ha area under forest land-use, only about 183.5 million ha is cultivable. Though, net sown area under agriculture is only 141 million ha, other cultivable lands are also in one or other type of cultivation and encroachment (ICFRE Report 2010).

Table 1.1: Land-use Statistics of India in 2009

Classification of Area	Area in (million ha)
Geographical area of India	328.7
Reported area*	305.8 (100)
Forests	69.8 (22.83)
Not available for cultivation	42.5 (13.9)
Permanent pastures	10.5 (3.43)
Misc. tree crops and groves	3.4 (1.11)
Cultivable wasteland	13.2 (4.79)
Current fallow	14.8 (23.22)
Other fallow	11.3 (3.69)
Net sown area	140.9 (46.07)

**(excludes area under Pakistan & Chinese occupation and snow covered and inaccessible areas).*

Note: *Figures in bracket are the percentage of reported Area.*

Source: *Forest Sector Report India 2010, ICFRE, Dehra Dun.*

In 2002, forestry industry contributed 1.7% to India's GDP (FAO, 2002). While in 2010, the contribution to GDP dropped to 0.9%, largely because of rapid growth of the economy in other sectors and the government's decision to reform and reduces import tariffs to let imports satisfy the growing Indian demand for wood products. India produces a range of processed forest (wood and non-wood) products ranging from wood panel products and wood pulp to make bronze, rattazikistan ware and pern resin. India's paper industry produces over 3,000 metric tonnes (MT) annually from more than 400 mills. The furniture and craft industry is another consumer of wood. India's wood-based processing industries consumed about 30 million cubic metres of industrial wood in 2002. (FAO, 2002) India annually consumes an additional 270 million tonnes of fuel wood, 2800 million tonnes of fodder, and about 102 million cubic meter of forest products - valued at about 27500 crore (US$4.2 billion) a year.

India is the world's largest consumer of fuel-wood. India's consumption of fuel wood is about five times higher than what can be sustainably removed from forests. However, a large percentage of this fuel wood is grown as biomass remaining from agriculture, and is managed outside forests. Fuel wood meets about 40% of the energy needs of the country. Around 80% of rural people and 48% of urban people use fuel-wood (FAO, 2002). Unless India makes major, rapid and sustained effort to expand electricity generation and power plants, the rural and urban poor in India will continue to meet their energy needs through unsustainable destruction of forests and fuel wood consumption. India's dependence on fuel-wood and forestry products as a primary energy source is not only environmentally unsustainable; it is a primary cause of India's near-permanent haze and air pollution (United Nations Environmental Programme, 2002 and WHO and UNEP, 2011).

Forestry in India is more than just about wood and fuel. India has a thriving non-wood forest products industry, which produces latex, gums, resins, essential oils, flavours, fragrances and aroma chemicals, incense sticks, handicrafts, thatching materials and medicinal plants. About 60% of non-wood forest products production is consumed locally. About 50% of the total revenue from the forestry industry in India is in non-wood forest products category. In 2002, non-wood forest products were a source of significant supplemental income to over 400 million people in India, mostly rural (FAO 2002).

1.5.1(A): Contribution of Forestry Sector in GDP

Forests play an important role in India's economic development in terms of their contribution to gross domestic products (GDP), employment, and livelihoods of millions of poor people, who are mainly dependent on forests. Besides, they are also the main source of meeting food, fuel, fodder and timber requirements of the forest dwellers. India has 16 different types of forests, with wide range of flora and fauna. These forests meet the basic requirements of millions of forest-dwelling communities. They contribute significantly to the economy, conserve biodiversity, moderate the environment and maintain the resilience of the eco-system (Maler 2008).

The System of National Accounts (SNA) forms the basis of national income accounts in India. The Central Statistical Organisation (CSO), Ministry of Statistics and Programme Implementation (MoSPI), Government of India (GoI) prepares the National Accounts Statistics regularly for the country of which GDP is one aggregate. In this role, the CSO releases annual national accounts statistics. Officially first estimate of national income was prepared by CSO in 1953 with the base year 1948-49 for estimates at constant and current prices. With gradual improvement in the availability of basic data over the years, comprehensive review of methodology for National Accounts Statistics (NAS) has been regularly undertaken with a view to updating the database and shifting the base year to the more recent year. The reason for periodically changing the base year of the national accounts is to take into account the structural changes which have taken place in the economy and to depict the true picture of economy through macro aggregates like GDP, consumption expenditure, capital formation, etc.

The entire economy is divided into various sectors and sub-sectors. The 'Agriculture, Forestry and Fishing' sector concerns forests and includes sub-sectors: (i) Agriculture, (ii) Forestry and Logging, and (iii) Fishing. In GDP estimation, the economic activities of 'forestry and logging' sub-sector include (i) Forestry (*e.g.* planting and conservation of forests, gathering of forest products, charcoal burning carried out in the forests), (ii) Logging (*e.g.*, felling and rough cutting of trees, hewing or rough shaping of poles, blocks etc.) and transportation of forest products to the sale depots/assembly centres, and (iii) Farmyard wood (industrial wood and fuel wood collected by the primary producers from trees outside regular forests). The forest products are classified into two broad groups viz., (a) Major products comprising industrial wood (i.e. timber, round wood, match and pulpwood) and fuel wood (i.e. firewood and charcoal wood) and (b) Minor products comprising a large number of wild growing forest material such as bamboo, fodder, lac, sandalwood, honey, resin, gum, tendu leaves, cork, balsams, vegetable fibre, eelgrass, acorns, horse chestnuts, mosses, lichens etc. Production of field crops (Jhum cultivation etc) and extraction of minor and major minerals from forests are included in agriculture and mining sectors respectively

(CSO 2007). Thus value of industrial wood, fuel wood and minor forest produce are used for income estimation of forestry sector.

A large part of the forest production consisting of fuel, fodder, medicine and food are removed without payment and without any record by the rural and tribal people. According to Ahmed (1997), the total annual value of India's harvest of all forest produce is estimated to be Rs. 300,000 millions (compared to the investment of Rs. 8000 in the sector). The low estimate of contribution to the GDP resulted in low priority for forestry investments in five-year plans. Efforts are needed for monitoring the services provided by the forests so as to appreciate their contribution to human well being. Over 50% of the revenue earned by the forest departments comes from NWFPs. Their growth is generally 40% higher than timber (MoEF 2000).

The SNA (1993) forms the basis of national income accounts in India, as in a large number of other countries. The SNA defines production as "an activity in which an enterprise uses inputs to produce outputs". However, the SNA covers all such activities carried out by an institutional unit. An institutional unit further is defined as "an economic entity that is capable in its own right, of owning assets, incurring liabilities and engaging in economic activities and in transactions with other entities". Institutional units are either (i) persons or groups of persons in the form of households or (ii) legal and social entities whose existence is recognized by law or society independent of persons or other entities who may own or control them.

The principle ways in which forest-economy interacts and which need to be provided for in a satellite account are shown here as under (Kanchan Chopra, 2006):

1. Forests as a source of timber, renewable in the main but potentially depletable. It is usually harvested by government corporations or private loggers and used as an input in wood-based industries.
2. Forests as a source of tangible non-timber forest products collected and consumed by households (*e.g.* fuel wood, resin, fruit, leaves, game, etc) but not always bought and sold in markets.
3. Forests as a source of less tangible forest amenities consumed directly either in the present or future (bio-diversity related benefits).
4. Forests as a source of environmental services that benefit other productive sectors (*e.g.* watershed protection for downstream agriculture, forest based recreation and tourism).
5. Forests as a disposal site for air pollutants that may be damaging to forest health (acid disposition).
6. Forests as a sink and a source of carbon dioxide which potentially damages other sectors through global climate change (carbon sequestration).
7. Through deforestation, forests as a source of land for other sectors, in particular agriculture and urban construction.

8. Forest management as an activity of the governmental and private sectors involving the use of variable inputs (labour and materials) and human capital.

Using the above comprehensive list of forest-economy interactions and defining forest related production in the context of output in the rest of the economy, **Vincent** (1999) defines the adjustments required to be made in NDP, conventional GDP and in the level of GDP. While, the details of the theory are not reproduced here, we give below the kinds of adjustments, he suggests are to be included:

Adjusted NDP = Conventional GDP + Non-market Values to be added to GDP- Depreciation of human-made capital + Net accumulation of Natural Capital.

Further, changes in the value added by industry and agriculture are to be made to allow for the contribution of the forestry sector to them. This is in the form of the contribution of forests to pollution disposal and carbon sequestration services, should they accrue to these sectors. Some carbon sequestration services would flow to the rest of the world as well.

On the other side, the income from forest resources is aggregated at the national and state levels under the head "Income from Agriculture, Forestry and Fisheries" and the sub-head of "Forestry and Logging" which includes income accruing from industrial wood, fire wood and Minor Forest Produce (MFP) (**CSO**, 2002). At the State level, the State Directorate of Economics and Statistics (DES) prepared estimates of State Domestic Product (SDP) and Net State Domestic Product (NSDP). The estimates are prepared at the state level first, which are then consolidated to obtain estimates at the national level (**CSO** 2007). Gross Domestic Product (GDP) and Net Domestic Product (NDP) from the forestry sector are computed as follows:

GDP = Value of Output- Repairs, Maintenance and other Operational Costs

GDP = GDP - Consumption of Fixed Capital (whereby Consumption of Fixed Capital is Depreciation of Fixed Assets)

The GDP from forestry sector can be estimated by following either the production approach through Gross Value Added (GVA) like timber or the consumption approach (*e.g.* Fuel wood). It aims at estimating the value of output at factor cost in the first instance and then deducting the value of various inputs at purchasers' prices (Mali *et al.*, 2011).

The data on production and prices of industrial wood and that on Minor Forest Products (MFPs) are provided by the State Forest Departments (SFDs). The data on production of industrial wood generally relates to the quantities sold/ auctioned or given on royalty by the SFDs. In the case of minor forest products some information on production is obtained from Forest Development Corporations. The value of minor forest products could be often the economic value derived from the royalty figures which are much lower than the market value. The production and price of fuelwood are prepared from the side of consumption of firewood by the households, on the basis of quinquennial survey on consumer expenditure of

NSSO (Singh, 2011). In addition to total consumption of firewood, the survey also records consumption from market purchases and own sources.

The economic and environmental benefits of the extensive biodiversity-rich forest areas of India are immense and are depended upon by the people for their livelihoods. However, for historical, topographical, logistic and complex sociological reasons, quantitative data on most of the economic and environmental components of forests are lacking. Quantification of both tangible and intangible forest benefits separately remains a neglected research area. But we provide some available data on the forest contribution. The share of forestry sector in the GDP during different period of time is given in Table 1.2.

Table 1.2: Share of Forestry Sector in GDP Over Years

Year	Forestry Sector Share in India's GDP (%)
1950-51	2.6 %
1960-61	1.9 %
1970-71	1.8 %
1980-81	2.2 %
1990-91	1.6 %
2000-01	1.0 %
2005-06	0.7 %
2007-08	0.67%
2010-11	0.9%
2014-15	1.0% (Approx.)

In India, forestry is a major land use next to agriculture. However, it is almost a neglected sector in the structure of Indian economy. It is consistently and seriously under-valued in economic and social terms. Forestry has to be treated as one of the few core sectors of Indian economy (K.C. Raut, 2004). Despite making considerable contribution to the ecology, economic and sociocultural development of the country, the forests hardly get due recognition of their contribution in national income (i.e. GNP) The value of forest reflected in the System of National Accounts (SNA) represents less than 10% of the real value. In 200203 forests contributed Rs. 270,130 million to India's GDP at the current prices, which was 1.2% of the total GDP. The contribution of forest to India's GDP has varied from 1.01.5% over the period from 199394 to 20022003. The GDP estimates for the forestry sector was Rs. 29,069 crores for the year 2007-08, which was 0.67% of the total GDP of the country. Similarly the contribution of forestry and logging to India's Net Domestic Product (NDP) also varied from 1.6–1.3% during the same period (Ram Prasad, 2006). On the other side, rapid decline in share of forestry sector in GDP has been due to higher growth in other sectors. The average annual growth in GDP of forestry sector during 1950-2011 was 0.9% against growth in overall GDP of 4.6%.

Forestry & logging contributes to 1.2% of India's GDP (Economic Survey, Ministry of Finance, 2011). The Indian forest products industry had total revenue of $65,844.6 million in 2011, representing a compound annual growth rate (CAGR) of 5.5% between 2007 and 2011. Industry consumption volumes increased with a CAGR of 0.2% between 2007-2011, to reach a total of 355.4 million cubic meters in 2011. The performance of the industry is forecast to accelerate, with an anticipated

CAGR of 7.7% for the five-year period 2011-2016, which is expected to drive the industry to a value of $95,467 million by the end of 2016 (*http://store.marketline.com*).

The main share (about 83%) was attributed to fuel wood which was properly estimated by NSSO for household sector, industrial wood accounted for about 9% and NTFPs about 8%. While estimating GDP for 2008-09, the CSO included the estimated production of timber from 'Trees Outside Forests' (TOF) as estimated by the Forest Survey of India. The value of the timber was determined after getting the price data from SFDs. The contribution of TOF was about Rs. 34,000 crores. Further, CSO also included the 'input for livestock feed' derived from forests on the basis of study report of FSI. It was estimated by FSI that 15.5% of adult cattle units of the country are totally dependent on forest for their feed. CSO estimated the fodder value from forests and added to GDP of forests under MFP which was about Rs. 12,500 crores. These two additions made a sharp rise in GDP estimates of the sector at 'current prices' as Rs. 88,000 crores for the year 2008-09 which was 1.70% of the total GDP of the country.

In forestry sector there are still many products specially the MFPs which are undervalued or not accounted for inclusion into the sector's contribution to GPD. For example lot of bamboo is given to right holders/local people either free of costs or on nominal royalty basis. Rationalization of the price of such forest products will help rationalize the contribution of forestry in the GDP.

Now we take a fresh look on index of industrial production (IIP) which is a measure of industrial performance complied and released every month by CSO. In fact, IIP is a fixed weight, fixed base index. CSO revised the base year of IIP from 1993-94 to 2004-05 and the present series were launched in June, 2011. This IIP series has an enlarged and more representative basket of the Indian Industrial sector. On the one side, IIP comprises three components of industry, viz., Mining, manufacturing and electricity, and on the other side it has also categorized by Use Based Classification.

Table 1.3: IIP Growth Rate-by Wood & Forest based Industry Groups

Industry Group	Weight	Growth Rate (%)						
		2008-09	2009-10	2010-11	2011-12	2012-13	2013-14	2014-15 (Apr-Jun)
Tobacco products	15.7	4.4	-0.6	2.1	5.4	-0.4	0.8	14.8
Wood & Wood Products	10.51	4.9	3.1	-2.2	1.8	-7.1	-2.2	-0.7
Paper & Paper Products	9.99	4.8	2.6	8.5	5.0	0.5	-0.1	2.1
General Index: All Manufacturing sector	1000	2.5	5.3	8.2	2.9	1.1	-0.1	3.9

Source: *Compiled from Central Statistics Office, New Delhi.*

The Table 1.3 above shows the IIP growth rate of WFBI for the year of 2008-09 to 2014-15. Table reveals that negative growth rate in the tobacco products for the 2009-10 and 2012-13, in wood & wood products for the year 2010-11 and 2012-13 to 2014-15, while the same declining growth rate was observed in paper and paper products for only the year 2013-14. Recovery is visible in manufacturing sector on the whole in the current year registering a growth of 3.9% during the first quarter of year 2014-15 over the corresponding periods of 2011-12 to 2013-14.

Natural Resource Accounting is a revaluation of the National Income Accounts of a country, adjusting for the values of natural resources used in various economic activities during the past fiscal year. Natural resources, as they appear in nature, get degraded in quality and depleted in stock due to economic and human activities. They also go through natural decay and regeneration. They may also have been enhanced due to planned intervention. There is a need to understand the extent of such losses or gains and to work out their "use values" and generate accounts of such resource deletions and additions. Natural resources being a part of the wealth of a nation, initiatives have to be taken to integrate the natural resource accounting along with the System of National Accounts (SNA).

India's economy is agriculture based with major per cent of people living in rural areas, and many millions of them are land-less. The distribution of size of land holdings is highly skewed. Though the average land holding per house hold (HH) is 1.33 ha, about 62 per cent house hold have less than 1 ha with an average of 0.40 ha land comprising only 18 per cent of the total operational holdings. On the other hand 18 per cent house hold own 63 per cent operational lands.

For the sustainable development, Infrastructure development is the main focus for sustaining the economic growth rates projected for the coming years. India has committed to developing state-of-art infrastructure for facilitating growth and improving quality of life. Communication, transport, housing, energy, healthcare and knowledge infrastructure are the priorities. Where, Land is the first and foremost requirement. The safeguards adopted in India in terms of making environmental impact assessments mandatory for development projects and also for decision making on diversion of forests at the highest level to ensure objective considerations of balanced growth. The diversion rate of forests for development has been inevitable in recent years.

Now, we will take a fresh look on WPI based Inflation in Forest products related manufacturing sector, and it is presented in Table 1.4 (A) and Table 1.4 (B). The current series of index numbers of wholesale prices (base year 2004-05=100) was introduced from Sept., 2010. The inflation rate per WPI for overall manufacturing sector has shown decrease from 5.4% in 2012-13 to 3.0% in 2013-14. Similarly tobacco products and wood & wood products have also shown a decrease in inflation in 2013-14 as compared to previous year. The WFBI which have shown an increase in the inflation in 2013-14 is paper & paper products.

Table 1.4 (A): Index Numbers of Wholesale Prices for Wood & Forest based Industry Groups

Industry Group	Weight	Financial Year Averages (Base Year: 2004-05=100)						
		2006-07	2008-09	2009-10	2010-11	2011-12	2012-13	2013-14
Tobacco products	1.7625	110.0	128.3	136.2	146.2	163.3	175.3	186.0
Wood & Wood Products	0.5874	111.9	130.7	143.3	149.0	161.0	171.0	179.1
Paper & Paper Products	2.0335	108.4	116.3	118.9	125.2	131.9	136.6	143.0
All Commodities	100.0	111.4	126.0	130.8	143.3	156.1	167.6	177.6

Source: *Compiled from Central Statistics Office, New Delhi.*

Table 1.4 (B): Rate of Inflation in terms of WPI for Wood & Forest based Industry Groups

Industry Group	Weight	Financial Year Averages (Base Year: 2004-05=100)						
		2006-07	2008-09	2009-10	2010-11	2011-12	2012-13	2013-14
Tobacco products	1.7625	5.1	9.5	6.1	7.4	11.7	7.4	6.1
Wood & Wood Products	0.5874	5.8	9.5	9.6	4.0	8.1	6.2	4.7
Paper & Paper Products	2.0335	4.6	4.2	2.2	5.3	5.4	3.5	4.6
All Manufactured Products	64.97	5.7	6.2	2.2	5.7	7.3	5.4	3.0

Source: *Compiled from Central Statistics Office, New Delhi.*

1.5.1(B): Contribution of Forestry Sector in Employment

India has a huge population living close to the forest with their livelihoods critically linked to the forest ecosystem. There are around 1.73 lakh villages out of 6.41 lakh villages (Total Number of Villages as per census 2011) located in and around forests (MoEF, 2006). Though there is no official census figures for the forest dependent population in the country, different estimates put the figures from 275 million (World Bank, 2006) to 350- 400 million (MoEF, 2009). People living in these forest fringe villages depend upon forest for a variety of goods and services. These includes collection of edible fruits, flowers, tubers, roots and leaves for food and medicines; firewood for cooking (some also sale in the market); materials for agricultural implements, house construction and fencing; fodder (grass and leave) for livestock and grazing of livestock in forest; and collection of arrange of marketable non-timber forest products.

Of the total wage employment in the forestry sector, NWFPs account for more than 70% of the opportunities for self-employment for the forest dwellers as farm mechanization has not developed well in India. According to an ILO estimate, one hectare of forest plantation creates nearly 630 mandays from the raising of nurseries to the harvesting stage. 70% of the budget al.ocated to plantations or afforestation is spent on providing direct wages to the workers and only 30% goes towards purchase of seeds, planting materials, equipment etc. It would not be out of place to mention that 50% of the workforces on forest plantations are women and tribal. Rural women use 70-80% of the mandays in collection of NWFPs, fuel and fodder.

Activities related to NWFPs provide employment during slack periods and a buffer against risk and household emergencies. In the remotest areas, sometimes the forest is the only source of employment and income. Nearly 350 million people living in and around forests in India depend on NWFPs for their sustenance and supplemental income which is worth Rs.400 billion annually (**Tewari** 1994). Studies in Orissa, Madhya Pradesh, Himachal Pradesh and Bihar have indicated that over 80% of forest dwellers depend entirely on NWFPs. Similarly 17% landless depend on daily wages related to the collection of NWFPs. 39% people are, however, involved in NWFPs collection as a subsidiary occupation.

Moreover, so far as the employment analysis based on NSSO and CSO taking into consideration, the available workforce data from the Census of India indicates a much lower number of household industry workers in both the year i.e. 2000-01 and 2010-11. The employment shares of household enterprises (see **Table 1.5**) have declined in all WFBI which is the matter of serious concern.

Table 1.5: Share of Household Employment by Wood & Forest based related Industries

Industry*	2000-01	2010-11	Change in percentage
Tobacco Products	89.1	84.8	-4.3
Wood & Wood Products	86.0	77.1	-8.8
Paper & Paper Products	49.9	21.7	-28.2
All Manufacturing	54.5	46.6	-8.0

**National Industrial Classification 2004.*

***Sources:** NSSO Enterprise Surveys and ASI Summary results of respective years.*

Recently a study conducted by KPMG (2013) on '*Human Resource Skill Requirements in the Furniture and Furnishing Sector (2013-17, 2017-22)*' have reported that as per an ASSOCHAM report, there are total 1,419 registered furniture factories in India. However, only 1,157 factories were under operation as of 2011-12 i.e. about 20 percent of registered furniture factories in India were non-operational due to labor dogs & lack of modernization. Demand for furniture in India surged at 12% annual rate over 2007-2012, and in 2013 it increased at a rate of 15%. This KPMG report has also studied demographic features of India's States workforce and the same presented in Table 1.6.

Table 1.6: Demographic Characteristics of Workforce: 2013.

States	Furniture Employment	Proportion of Employment
West Bengal	355,124	16.58%
Uttar Pradesh	238,544	11.14%
Maharashtra	203,124	9.49%
Bihar	150,996	7.05%
Gujarat	138,213	6.45%
Kerala	137,156	6.41%
Odisha	122,181	5.71%
Andhra Pradesh	117,997	5.51%
Assam	104,121	4.86%
Tamil Nadu	94,117	4.40%
Rajasthan	83,652	3.91%
Punjab	82,592	3.86%
Karnataka	73,238	3.42%
Madhya Pradesh	55,643	2.60%
Rest of India	184,663	8.62%
Total	**2141362**	**100%**

Source: *NSSO 68th Round, KPMG in India Analysis, ASSOCHAM*

NSSO employment data gives the state-wise total employment figures for the furniture manufacturing sector and thus, the numbers are inclusive of the unorganized sector employment figures as well. According to its 68th Round, it has been observed that 50% of the total employment in furniture manufacturing is concentrated in the five states of West Bengal, Uttar Pradesh, Maharashtra, Bihar and Gujarat. West Bengal, UP and Maharashtra alone account for nearly 38% of the work force. Interestingly, it has been mentioned in said report that about 97% of the workforce involved in furniture sector is school drop outs. Of this, 88% have an education qualification of secondary education or less. In the organized furniture segment, workers account for more than half of the total workforce. Moreover, managerial and supervisors account for 10% of the total workforce and contractor workers account for nearly 25 per cent of the total workforce.

So far as the employment in Indian manufacturing sector are concerned, estimates of National Sample Employment-Unemployment Survey indicate that between 1999-2000 and 2011-12, manufacturing employment increased by 17 million from 42.8 million to 59.8 million. This amounts to an increase of approximately 1.4 million per annum. Results from ASI which report formal employment in this sector reveal that for the same period employment in the organized manufacturing sector increased from 7.9 million to 12.2 million in 2011-12 (an increase of about 4.6% per annum).

While on the other side, so far as the employment scenario at global perspective are concerned, it has been observed that over the last decade, total employment in the (formal forestry sector decreased by about six percent, from 14.0 million in 2000 to 13.2 million in 2011. Most of the decline came from the forestry sub-sector (a 21% decrease over 2000-2011) and the developed regions (i.e. Northern America, Western Europe and the Developed Asia-Pacific region). These losses were partly offset by an increase in employment in the forestry sector in the developing regions.

At the global level, the forestry sector employed about 0.4 percent of the labour force, contributed about 1.0% to GDP and accounted for about 2.4% of global merchandise trade in 2011. However, the contribution of the forestry sector to the total economy has steadily declined in recent years, due to faster growth in other sectors. In general, value-added in the forestry sector has not increased rapidly except in a few countries where development of the sector has been a specific national development priority. Very few countries have focused on the development of the forestry sector, preferring instead to promote the development of other sectors. Thus, it is generally the case that the forestry sector has been left behind, particularly in rapidly growing economies. Furthermore, this suggests that the forestry sector is not a major driving force for economic growth and development except in specific circumstances.

1.6: Wood & Forest Based Industries in India

After gaining independence from British Rule in 1947, India followed a path of planned economic development and simultaneously the government of India recognized the role and importance of rural industries but neglible importance has been given to Wood and forest-based industries in the economic upliftment of the rural millions. The rural industries sector has received much emphasis in most of the major policy pronouncements on development in India. In fact, the government has made rural industrialization as one of the major planks of Indian development strategy ever since the I^st^ Five Year Plan. Promotional measures taken under the 3rd FYP were directed to achieve the main objectives of the programme:

(i) Improvement of skills and productivity,

(ii) Progressive reduction in the role of subsidies,

(iii) Sales rebates and sheltered markets,

(iv) Promotion of growth of industries in rural areas and small towns,

(v) Development of small industries as ancillaries to large industries, and

(vi) Organization of artisans and craftsmen on cooperative lines.

Wood and forest based industries linkages are the best instrument to achieve balanced development of the country/states. As well as to minimize regional inequalities in industrial disbursement they can play a positive role. The importance of forests in national economy requires no explanation. Therefore, it can be said that '*wood and forest based industries can be considered as an agent of growth in new economic environments*'.

Wood and Forest based Industry play a very vital role in shaping the robust growth of the Indian Economy. This industry has true potential to grow manifolds from the existing levels and is poised for a sustainable growth annually. Since last two decades, it has been seen that India becomes fast growing manufacturing hub for the global markets. The Wood, Plywood & Allied Products Industry is one of the key sectors having immense potential for gaining from these developments as India is one of the major wood-users in the Asia pacific region. Asia Pacific region has over 4500 varieties of wood-yielding species & has some of the best known and most highly prized tropical hardwoods.

Wood-based industries are key industrial contributors to income and employment generation, particularly in the rural and underdeveloped areas in India. Wood-based industries in India produce a range of processed wood and non-wood products including sawn wood, composite panel products and pulp & paper. Broadly, there are three categories of wood based industries, i.e., the saw mills, paper mills, and plywood & panel industries and almost all are in private sector. The wood supply to these industries is from government forests, trees growing in private and non forest lands and from import. The market for wood and wood products in India is predominantly domestic in nature. The export of wood – logs, timber, stumps, roots, barks, chips, powder, flakes, dust, and charcoal – has been totally banned and that of the wood products restricted. Thus, this industrial sector includes all stakeholders with major interests in forestry, forest-based materials and products. It provides essential products and services for a more sustainable society.

India's forest-based secondary industry encompasses a wide range of small, medium and large scale firms that process wood into a variety of products for the domestic market. With the improvement of productivity, net supply of wood products may improve. However, the principles of SFM and fair trade practices including eco-labeling of wood products might limit the wood supply from natural forests. In the circumstances, wood from trees outside forests, mostly agro-forestry will become the main source of raw material for industries. It is also likely that wood processing units will modernize and improve their efficiency to meet the growing demands of the house construction and furniture industry, infrastructure and other industrial requirements (India Forestry Outlook Study, 2009).

Indian plywood industry is as big as Rs.5,000 Crore equivalent to 1 billion US Dollar. The industry is growing at rapid pace of 10-20% per annum. Approximately 600 units are currently functioning all over the country. There is tremendous growth potential as the players are yet to penetrate majority of the international market. Panel and plywood products are the main wood products in India. Product categories include veneer sheets, particle board (composite wood core with plastic laminate finish), panel products (fiber board), plywood made from both hard and softwood (veneered panels and laminated woods), and medium density fiber board. Indian particle board and plywood industry accounts for 15% of the total production, producing, more than 30 million square meters of plywood and block boards.

The Indian wood & furniture sector in present is predominantly in the hands of unorganized small units. After Liberalization, fortunately, some large corporate houses have shown their interest in production of modern furniture. The furniture market in India is the second largest wood processing segment after timber & logs, making India a fast emerging market for high-end, value-added imported products. The manufacture of pre-fabricated doors and windows is relatively new and the current market is growing at 10-15% per annum. The total annual market for timber & furniture in India is estimated to be US$ 1.25 billion about 90% of which is for wooden products. The branded (higher quality) wooden furniture industry is growing at 15% annually.

In spite of all above little bit positive and attracting circumstances, this sector is predominantly characterized by inefficiency and low productivity due to insufficient technical skills and know-how. Contribution of Forestry based industries in National income clearly indicate that even in the era of modern technology, this sector still suffers from poor product quality and outdated product designs that severely limit access to world markets. To improve the efficiency of forest based industries, the following initiatives should be taken:

- Wood & Forest-based industries are not efficient due to technological obsolescence, inappropriate machinery and its maintenance, unskilled manpower and poor quality of products. Such industries need to be modernized in proper manner. Initiatives for the modernization of technology, reduction and recycling of waste, and regulations regarding the use of seasoned and treated material, promotion of standards and codes for wood products, etc. will be encouraged and initiated like USA, Canada, China, Russia, UK.
- The use of fuel wood in open hearths is inefficient because it leads to considerable heat loss and causes health hazards. Improved and modified methods of cooking can reduce fuel wood needs and improve hygiene.
- With the relaxation of trade barriers and liberalized imports, customs duty on logs and wood chips was substantially reduced. Though, on the one hand, the liberalized import reduced the demand on our natural forests, it also acted as a deterrent to the growth of indigenous production and forest based industries.
- Furniture to bamboo-based boards, numerous industrial bamboo products is now being exported in considerable quantities by the countries. An increasingly consequences should be adopted so that development of the generation of employment opportunities in remote rural areas can be enhanced.
- While maintenance of ecological balance remains the pre-eminent objective of forest management, contributions of forests to the subsistence and livelihood needs of millions of rural poor, especially the tribal communities, should be one of the primary considerations and need to be improved.
- There is need for improved interactions forest dwellers and public servants.
- The rising price of forest produce, growing interests in medicinal and nutritional values of NWFPs and the emerging trade scenario under the WTO regime are also creating significant investment opportunities for private organisations. The debate on private participation has remained polarized on account of the appropriate role for the industry and control over public forestland. New approaches need to be defined to create a

win-win situation while addressing the concerns and apprehensions of stakeholders.

- India's forests will have to be managed holistically, taking into account tangible and intangible benefits and involving forest communities and other stakeholders who will participate directly as shareholders.
- Forest communities could be rewarded for conserving biodiversity, providing carbon sinks, protecting watersheds, and maintaining scenic beauty or recreational values. The healthy way this can be done if these services have real values attached to them. Payment for environmental services from forests is one of the phenomenons gaining ground worldwide, and so many countries are already on this path to conservation.

1.6.1: Various Aspects of Wood Industries & Their Articles: All around Globe

Various sources and available literature confined that the main importing countries in world for Wood and Wood articles are the USA (11%), China (10%), Japan (9.5%), Germany (6.1%), United Kingdom (4.6%), Italy (4.5%), France (4.2%), Netherlands (3.2%), Canada (2.8%), Belgium (2.6%), Netherlands (2.5%), Austria (2.4%) and Rep. of Korea (2.1%).

The major markets for Indian furniture's (i.e. Kitchen, Bedroom and Office furniture's etc.) in the world are USA, UK, France, Switzerland, Canada and Netherlands. China, USA, Japan, UK, Italy, Germany and France are the leading importers for Sawn wood. On the other side, France, USA, Spain, Italy, Germany, Russian Federation, Chile, China and Japan are the leading importers for Cork and Cork Products in the world.

It has also been seen that global imports of Wood, Plywood and Panel Industry reached to US $ 157.97 billion during 2010. The broad group of the products exported globally under Wood, Plywood and Panel Industry are Furniture and Parts (41%), Sawn Wood (19%), Plywood (7%), Builders' joinery and carpentary of wood (7%), Fibre board of Wood (5%), balance in Cork & Cork Products, Hard board of wood fibre, other articles of wood, Parquet Panel and Sandal wood Chips etc., as detailed given below in Table 1.7.

Table 1.7: Global Imports of Wood, Plywood and Panel Industry

Products	Value in 2010 (US$ Million)	% Share
Furniture and parts	63648	41
Sawn Wood	29761	19
Plywood	11343	7
Joinery & Carpentary of Wood	11140	7
Fibre board	8183	5
Others	33899	21
Total	**157973**	**100**

Source: *ITC Trade Map*

Furthermore, the exports of Wood and Wood products from India has reached to all time high of US $ 453 million during 2010-11 showing a growth of 20% compared to previous year. For the past 4 years, exports of timber products growing an average of Compounded Annual Growth Rate (CAGR) of over 7% from India. The major export destinations for Indian Plywood and Wood Products are USA (23%), Germany (10%), UK (9%), France (7%), UAE (7%), Italy (4%), Netherlands (3%), Australia (3%), Belgium (3%) and Spain (2%), as per the Table 1.8.

Table 1.8: Exports of Plywood and Wood Products from India (Value: Rs. Millions)

Products	2009- 2010	2010- 2011	% Change	Export Destinations
Wooden furniture	11220.45	16207.13	44.44	USA, Germany & France
Other articles of wood	3546.03	4560.69	28.61	USA, UK & Canada
Sawn timber	687.34	831.15	20.92	UAE, Italy & Oman
Veneer	925.47	717.52	-22.47	Turkey, UAE & USA
Other plywood	674.00	566.16	-16.00	Turkey, UAE & Netherlands
Hard board of wood fibre	583.46	486.17	-16.68	UAE, Saudi Arabia & Qatar
Cork and cork products	53.98	116.23	115.34	USA, UAE & Russia
Sandalwood chips	50.57	92.84	83.58	UAE, Malaysia & Saudi Arabia
Decorative plywood	152.92	33.50	-78.10	Nepal, Canada & Turkey
Tea chest panel	1.34	0.98	-26.51	Nepal, Germany & Japan
Grand Total	**17895.55**	**23612.36**	**31.95**	

Source: *MoC Export Import Data Bank*

1.6.2: India's Export Market Challenges

(A). Globalization

Like many other industries, the forest industry has undergone profound changes in recent years. With the globalization & opening of economy in the last decade of 20th century, there is entry of very well known multinational brands in India, many products and services are now delivered to consumers in a similar way across the world and consumers are now aware of trends, tastes and fashions as in other parts of the world. These developments present opportunities to increase efficiency in the delivery of products and services across a much larger global marketplace as well as they also enable firms to gain competitive advantage through overseas market knowledge, product differentiation and the development of local market niches. This globalization and liberalization is posing big challenges for exports of timber products from India to across the globe. Consumers expect high quality materials at par with international standards particularly with various certifications in place.

(B). Increasing competition from other Asian countries

Indian producers are facing increasingly higher amount of competition particularly from other Asian countries of China, Thailand and Vietnam. Since the manufacturing of plywood, furniture/veneers etc. are labour intensive processes in India and whereas a good number of jobs are being created for export of these products. Industries in India are still to be organized & need huge investment & technical support in the form of technology, plant & machineries, education & training. There is substantial quality difference in international products compared to domestic products. This is also a big challenge for Indian Exporters to improve as per international market demand & standards.

(C). Increasing costs and Automation

In India, both wood raw materials and skilled labour are now in short supply. Because labour was cheap in India the wood and wood industry developed using labour intensive production processes. Today however, the situation is different and because of low cost labour is no longer readily available many factories are running at around only 60% of their capacity. To overcome this persistent problem the industry now prefers to have as much automation as possible and is retooling production plant accordingly. This change is driving up demand for high tech wood processing equipment.

(D). Forest product certification

In India, Forest product certification was developed during the 1990s as a mechanism to identify forest products that come from sustainably managed forests. Main elements of the certification process are: (i) the development of agreed standards defining sustainable forest management; (ii) auditing of forest operations and issuance of certificates to companies that meet those standards; (iii) auditing of the chain-of-custody to ensure that a company's products come from certified forests; and (iv) the use of product labels so that certified products can be identified in the marketplace. In present, there are more than 50 certification programmes in different countries around the world, many of which fall under the two largest umbrella organizations, they are: (a) the Forest Stewardship Council (FSC) and (b) the Programme for the Endorsement of Forest Certification (PEFC).

The area of certified forests covered by the two main organizations has steadily increased since the 1990s to reaches about 350 million hectares in 2010. Globally a number of barriers to more widespread adoption of certification have been identified. Two of the most important of these are the costs of certification (especially for small forest owners) and the lack of a price premium for certified forest products in the marketplace. Although the latter has been noted in almost all developed country markets for forest products, one benefit of certification is that it facilitates entry to those markets, where prices generally may be higher than in countries where there is no demand for certified forest products.

Although forest certification seems to fail to stimulate widespread changes in forest management and harvesting practices in all parts of the world, it remains an important tool for companies in the forest industry to demonstrate their commitment to meeting high social and environmental performance standards.

Indeed, many of the largest forest products companies are certified and can use this to gain competitive advantage by differentiating their products and communicating their superior performance to consumers.

The major exporting countries of logs are still to get the forest certification & hence it is difficult to get certified logs. Due to all these reasons, Indian exporters cannot supply certified wood products. Further, this certification adds to the cost of production to a great extent which stands alone will not be competitive internationally and it has been become drawbacks for the establishment and further development of WBI.

(E). Understanding and meeting customers' needs

In the era of *'Make in India'*, making the wood and wood products more customer-focused and responsive to changing needs is one of the biggest challenges of India. The wood and wood industry remains very conservative with the old technology in comparison with many other consumer product markets and it's essential that manufacturers and merchants become more dynamic, innovative and attuned to market trends and customer requirements. Across the Global, the price-conscious, customers adopting a more cautious approach to committing to significant home improvement projects. Such type of shift in demand means that quality becomes even more important, as customers purchasing premium wood and wood products have higher expectations of design, finish and durability, as well as service and after-sales customer care support and services.

(F). Use of Alternative materials to Wood

Perhaps the biggest challenge for the wood and wood industry as a whole is to position the material as complementary to modern homes, were alternative materials such as metal and glass are now commonplace within shelves, storage units, tables, chairs and other furniture. Customers are including a whole range of different materials, finishes and textures within their homes, and wood and wooden manufacturers and merchants need to innovate to either integrate these materials within their ranges or design wood & wood products to complement them. The wood and wood products industry in India need to understand and respond towards changing needs, if it is to grow and compete with other consumer product suppliers.

(G). Social & Domestic Market: Trends & Scenario

Social trends changes public opinions, attitudes and lifestyles that occur when income rises. For example, it has been seen that as income increases in India, people move beyond basic needs and start to seek new products and services so that they can improve their quality of life, but is subject and accordance to their tastes and preferences. Other wealth or assets related factors also affect consumption, such as increases in home ownership (including second homes), trends towards larger homes and greater leisure time, as well as changes in the amount of time spent at home. This creates an additional demand for the domestic market and thus the surplus available for the export reduces. There are lots of pressures on forest land and resources as demand for construction for new homes increases.

Secondly, India is experiencing a rapid phase of urbanization with a change in lifestyles, a growing demand for engineered wood panel products, and a high infrastructure, industry sources expect positive growth for wood products such as plywood, particleboard, medium density fiberboard, oriented-strand board and laminated veneer lumber in near future. Therefore manufacturers are not really willing to export when they have readymade domestic market to cater. There is huge young population aging between 18 – 40 years in India adopting high quality of life standards with high income group & having more spending power. Therefore, the domestic market is quite strong and housing sector as well as other relevant industries is booming. This affects exporters to decide between domestic and export market.

(H). Lack of Market Information

In India, still there is no proper market information system to provide the prevailing price on wood on a day-to-day basis, nor is there any support price fixed by the Government as has been the case in agricultural products. Indian exporters need to participate in every possible wood fairs & exhibitions to get first hand information on market and recent trends. This also helps in upgrading plant & machinery & brings about latest technology in wood industry. The other major challenges for the Indian exporters can be summarized as under:

- It has been observed that there is a acute shortage of good raw material as wood. The customers demand for high quality of products at higher rate for which the exporters need to procure best quality of raw material to ensure quality specification and to meet customer's expectation at satisfactory level.
- Huge capital & resources are required for technology upgradation & modernization, which seems to be a difficult task for small & medium size exporters.
- Shipping has become critical aspect as most of the raw materials are imported from African countries where container facility is not available & the cargo has to come in break bulk which involves huge finance & forex risk & also increases transaction cost too as well.
- India is still lacking in developing proper marketing strategies to penetrate the international market & to sustain against cut throat competition, as exporters need to have united efforts to meet ever- changing market challenges.

At last it can be said that Asia is one of the most interesting dynamic and fastest growing markets in the industry, and China in particular, just because of to its incredible economic prosperity. The region is becoming one of the most important areas in the world for furniture manufacturing. The development of demanding, quality orientated industries such as furniture manufacturing and interior decoration will ensure a steady and promising flow of business among India and other Asian countries for years to come. It has been also observed that the forest based industries and its related companies would not have had any assets to go beyond Western industrialized markets without supplementary political and economical efforts and actions for to modernize the forest sector in the developing countries like India. In fact, targeted internationalization support for this industrial sector has been significant. In future, it can be expected that a well focused and professionally driven cluster initiative could help both the traditional industry like WFBI and the Indian economy as a whole.

1.7: The Globalization of the Wood Processing and Furniture Industry

The international wood industry has become more complex due to globalization, production assortment and the development of technologies. Environmental protection also plays a major role to the development of this sector (processed materials, nature preservation, etc). European, Australian and USA manufacturers compete with the imported Asian products that are cheaper. The industry has responded to these threats by enhancing productivity, targeting niche markets and improving quality, design and marketing. In many companies, the production process consists of the assembly and gluing of particleboard which is coated with a decorative covering such as a veneer. The level of skills involved is not as sophisticated as in craft work.

Nevertheless, the skills are important particularly in respect of the operation and maintenance of computer numerically controlled machines, finishing techniques and in the case of soft furniture, the sewing, cutting and pattern making. The method of working, however, is as important as the actual skills of the operatives. Specifically, 'working in cells' has been shown to be more efficient in reducing lead times and in optimising machine utilisation. These working arrangements, however, require a high degree of multi-skilling on the part of the production operatives. The development of craft skills is essential for the survival of the European furniture industry. Craft skills are necessary to produce high quality, intricate furniture items based on solid wood. The market for such products is less sensitive to price than the market for mass produced furniture constructed from wood composites and it is a market therefore which can be successfully exploited by the European companies. The increasing complexity of the industry means that it requires a cadre of university qualified professionals to provide many of the technical, supervisory and management functions. Design and marketing have become the twin pillars on which any successful development strategy for this industry must be based (Aleksandras Abišala, 2008).

The liberalization of international trade has had a dramatic impact on the international furniture and wood processing industry. It has resulted in a significant transfer of the manufacturing processes of many European and American companies to Asia – particularly to China, Indonesia, Malaysia, Thailand, Vietnam, Hong Kong, Philippines and India.

The processing of wood is one of the most labour intensive of all manufacturing activities. As such, labour costs constitute a relatively high component of the final retail cost of any wood-based product. The wages of factory operatives in Asia in wood processing and furniture companies are only a fraction of the corresponding rates in West European and American plants. This wage differential has given Asian companies, particularly in China, Indonesia Malaysia, Vietnam, Hong Kong and the Philippines a significant competitive advantage and has resulted in a major shift of output and employment from the United States and Western Europe to Asia. China has now passed Italy as the World's largest exporter of finished and semi-finished furniture products. In contrast, production is contracting dramatically in the United States and over 40% of American companies have closed since the beginning of the decade.

The major furniture exporters still include Canada, Germany and of course Italy but the Asian economies, particularly Malaysia and Indonesia – in addition to China - now figure prominently among the world's top exporters while Poland and to a lesser extent the Czech Republic are also major exporters of furniture. In Latin America, both Brazil and Mexico have emerged as major exporters of furniture. The industry in the United States and Western Europe has responded to the threats posed by Globalization primarily through outsourcing their manufacturing process to Asian economies. They have done this in one of two ways; either by purchasing companies in these economies or through entering into franchising or other contractual relationship with existing companies for the purpose of securing low cost supplies of semi-finished and finished wood products. This process of relocation, in many cases, has involved a large element of technological transfer. The level of technology in wood processing and furniture companies in Asia in general is quite low and the emergence of significant clusters of foreign-owned companies has resulted in a much greater level of diffusion of technological innovation.

The Swedish company IKEA is perhaps the best exponent of the success of the outsourcing model. It has over 13,000 suppliers located in over 55 different countries. It produces less than 10% of its final product in Sweden itself. While some of its producers are located in West European countries such as Italy, and also more recently in the United States, most of its producers and suppliers are based in Asia or Eastern Europe. The trends towards outsourcing creates opportunities for the industry in countries such as Lithuania because wage costs in these countries are considerably less than in Western Europe and also because the industry has a strong tradition of good quality and well-designed furniture.

However the wage costs in Eastern Europe are significantly higher than in Asia. Consequently, the industry in Eastern Europe must achieve higher levels of productivity to continue to compete in the international market place and to attract business from West European and American companies. The industry in Poland provides a good example of how companies located in Eastern Europe can achieve an international competitive position despite having a relatively higher cost base than the industry in Asia. The Polish wood processing and furniture industry is booming and it accounts for over 10% of all Polish exports. It provides about 12% of IKEA's global needs.

1.7.1: Status & Trends of Globalization, Investment and WFBI: Brief Review

The forest sector worldwide went through a strong restructuring process during the 1990s, influenced basically by globalization. The restructuring process was based mainly on the consolidation process through M&A operations. Despite the relative importance of DI in the forest sector, the data available on such investments are limited. An IMF article has made reference to the role of private investments in the forest sector in comparison with other sources. The article pointed out that the total amount invested in 1993 in the forest sector worldwide (US$21.5 billion) was concentrated in private (46.5%) and domestic public investments (46.5%) and that ODA was responsible for only 7% of the total (Tomaselli, 2006).

According to UNCTAD data, the worldwide FDI in agriculture, forestry, hunting, and fishing activities combined reached US$1.8 billion in 2001–2003, representing around 3.5% of the total FDI worldwide in the primary sector. The worldwide entry of FDI into the forest industry (wood and wood products manufacturing) reached US$2.3 billion in 2001–2003, which represented 4.5% of the FDI in the secondary sector and only 1.4% (2001–2003) of the total global FDI (primary, secondary, and tertiary sectors). Over the years, DIs have generally contributed more than 90% of the total value invested, amounting to around US$60 billion p.a. (Tomaselli, 2006).

Total direct private investment (in the forest sector) rose from US$30,000 million in 1992 to US$118,000 million in 1998 and then went down to an estimated US$98,000 million in 1999. More significant is the fact that in 1992 private direct investment accounted for only 19% of total net resource flows from OECD/DAC countries and multilaterals, while it had reached more than 50% by 1999. Crossley *et al.* (1996), as cited by Greig-Gran, suggest that overall capital flows to the forest-based sector in developing countries is in the billions of dollars. (Gregersen & Contreras, 2001).

Private FDI in the forest sector considerably exceeds public official development assistance (ODA). In recent years, forest financing has been characterized by an increase in FDI in developing countries to approximately US$8–10 billion a year, and a decline in ODA to about US$1.75 billion a year (PROFOR 2003).

Most of the future investments in the forest industry will continue to be concentrated in the pulp and paper segment. To a lesser extent, but also importantly, investments will continue to flow to the reconstituted wood panel segment, mainly for MDF and OSB production. (Tomaselli, 2006).

DI in the pulp and paper segment in the short and medium run will be driven to maintain the growth in production observed in the past few years. DI will be concentrated in countries with a low cost of wooden raw material and high market potential, for instance, Brazil, China, Russia, and some Eastern European countries. The perspective is that FDI flows predominantly from the USA toward Latin America and from Western Europe toward Eastern Europe and Russia. The expectation is that the international trade may grow strongly in the coming years for forest product exports from Eastern European countries, Russia, and Brazil (Tomaselli, 2006).

M&A in the forest industry will likely continue in the future, but at a slower pace than observed in the past, given the changes in the world economy and competition regulations. The latter will in the short term be a strong barrier for the M&A. As for newsprint, for instance, the top five world producers already account for 85% of the production capacity in Western Europe. In the case of magazine (i.e. couches) paper, the top two world producers control 50% of the European market. Within this context, it is important to consider that regulation of competition will certainly call the attention of European companies to investing outside the region, primarily in Asia and in South America (IADB, 2004).

In the United States, the industry has been divesting itself of its ownership of forestlands. In the past 25 years, industry lands have been reduced by 50%, with nearly half of that decline in the past decade. Simultaneously, the industry has increased its ownership of offshore forestlands (Bael & Sedjo, 2006).

In Russia, current investments are largely focused on the development of basic production capacity, such as production of logs and timber for export. Longer-term development is likely to turn to production of value-added products. A current effort to establish large-scale wood processing facilities along the Russian–Chinese border provides an indication of what the future may hold (Taylor, 2004).

Other countries that are currently building capacity in the forestry and wood products sector include several countries in the Asia and the P acific region, a number of countries in Eastern Europe, and several countries in Latin America, most notably Brazil and Chile. The key producing countries in the southern hemisphere (Argentina, Australia, Brazil, Chile, Indonesia, Malaysia, New Zealand, and South Africa) have slowly but steadily raised their contribution to global wood products exports over the past f our decades, from under 6% to more than 16% (Whiteman, 2003). In tropical countries there is a clear trend toward development of capacity for production of primary processed and secondary processed products, with most of the output intended for export markets (Johnson, Adams and Miyake, 2003; Bowyer 2004).

In the EU net flow in FDI was negative in 2003–2004, with about €1 billion inward investment per year in wood, publishing, and printing (EUROSTAT). FDI in pulp and paper products tends to be around 10 times higher than FDI in other wood working industries (*e.g.*, US FDI [year 2000] in wood products is US$1.5 billion compared to US$15 billion in paper products). In 2004, 68.2% of the investment in Stora Enso was of a transnational nature (US$15,467 million in foreign assets). The company is number 85 in the world's top 100 non-financial TNCs, ranked by foreign assets (UNCTAD, 2006).

Globalization has also taken many companies in the forest cluster overseas. They have internationalized at least as rapidly as companies in the paper field. Outward (outside EU) investments are expected to be in the billions of euros. For example, Finnish forest companies had already invested nearly €2 billion in China by 2005 (FFI, 2005). Inward investment particularly in United Kingdom publishing and printing in 2003–2004 (EUROSTAT), Metso, a Finnish supplier of paper machines, says that over one-third of its present order backlog is from China (FFI, 2005).

European forest-based-industry investments have been made and the capacity is expected to increase by some 10% for the pulp and paper industry, by some 13% for the panels industry, and by nearly 15% for energy plants within the next five years, representing an additional need for 30 million m3 of wood in that period. One reason for this growing capacity is the expected increase in the consumption of paper and board and other wood-based products, mainly because of opportunities offered by the accession of Central and Eastern European countries to the EU. Apart from this result an extra amount of some 80 millions m^3/year is estimated to be needed to meet EU commitments in the field of renewable energy sources by 2030. (CEPI study, Galembert, Brasov 2003).

Numbers on domestic direct private investment in forestry and related forest-based activities are unavailable on an aggregate basis. Grieg-Gran *et al.* (1998) indicate the same point related to portfolio investment in forest-based activities (Gregersen & Contreras, 2001).

According to estimates the amount of DI (direct investment) in the forest sector on a global scale exceeds US$60 billion a year, which represents about 1% of total DI in the world. In the forest sector, following the general trend, DDI (domestic direct investment) has a predominant share (Tomaselli, 2006).

Thus, it can be said that Globalization of the forest sector is happening rapidly and these forces will substantially change the shape of the Indian forest sector between now and 2020-25. The Indian forest industry is already today facing some of the highest wood and energy costs in the world and the costs are bound to increase due to escalating deficits. Globalization will probably drive the Indian forest industry to import its raw material and invest in plantations abroad. At last, it can be also added that forests and the forest & wood-based industry sector are very important from an ecological, economic and social perspective, especially when they form an industry cluster in a certain region/state of country like India. Considering increasing challenges for the forest sector, a study on industry-wise and inter–state, inter-region perspective can be very helpful for showing the complex structure and the various interactions between the different industry and along the productivity growth, technological change and production function approach.

References

Ahmed, M. (1997): "In Depth Country Study –India", Asia–Pacific Forestry Sector Outlook Study, Working paper No. APFSOS/WP.26. Food and Agriculture Organization of the United Nations (FAO). Rome/Bangkok, October 1997.

Aleksandras Abišala (2008): "Study of Wood Sector: Research report on skill needs", Methodological Centre for Vocational Education and Training, Vilnius.

Allen, Shirely, W. (1950): "An Introduction to American Forestry", Mc-Graw Hill Book Co., New York, 1950.

Asian Development Bank (2009): "Indigenous Peoples' Forest Tenure in India".

Augusta Molnar, *et al.* (2011): "Community-Based Forest Management: The Extent and Potential Scope of Community and Smallholder Forest Management and Enterprises", Rights and Resources Initiative Washington DC.

Beena, P.L. *et al.* (2004): "Foreign Direct Investment in India", In *Saul Estrin and Klaus E. Meyer (Eds.)* 'Investment Strategies in Emerging Markets', Edward Elgar, Cheltenham, 2004.

Bergen V, Löwenstein W, Olschewski R. (2002): "Forest Economy: An Introduction to Economics [in German]", Munich: Vahlen 2002.

Besley, Timothy & Robin Burgess (2004): "Can labor regulation hinder economic performance? Evidence from India", *Quarterly Journal of Economics,* 119(1), 91-134.

Bhagwati, Jagdish N. (1982): Directly unproductive, profit-seeking (DUP) activities, *Journal of Political Economy,* 90, 988-1002.

Bhagwati, Jagdish N. (1988): "Export-promoting trade strategies: Issues and evidence, *World Bank Research Observer,* 3(1), 27-57.

B.H. Patel (2012): "*Speech On 'Export Market Challenges For Indian Timber Products'*", 19th Illegal Logging Stakeholder Update, 10th February 2012, Chatham House, 10 St James's Square, London.

Bhaumik, Sumon K & Paramita Mukherjee (2002): "The Indian Banking Sector: A Commentary", In: Parthasarathi Banerjee & Frank-Jurgen Richter (Eds.) '*Economic Institutions in India: Sustainability under Liberalization and Globalization*', Palgrave Macmillan, London.

Bhaumik, Sumon K., Shubhashis Gangopadhyay & Shagun Krishnan (2006): "Reforms, Entry and Productivity: Some Evidence from the Indian Manufacturing Sector", Discussion paper No. 2086, IZA – Institute for Study of Labour, Bonn.

Champion, H. G. & Seth, V. K. (1968): "A Revised Survey of the Forest Types of India Government of India".

Chopra, K., Bhattacharya, K., & Kumar, P. (2002): "Contribution of Forestry Sector to Gross Domestic Product", Institute of Economic Growth, New Delhi, India.

CSO (2002): "National Accounts Statistics Sources and Methods 2002", New Delhi: Central Statistical Organisation, MoSPI, Government of India.

CSO (2007): "National Accounts Statistics, Sources and Methods", Central Statistical Organisation, MoSPI, GoI, New Delhi.

Curitiba, (2004), Tomaselli I. (2006): "Brief Study on Funding and Finance for Forestry and Forest-Based Sector", Final Report for the UNFF Secretariat, United Nations Forum on Forests, New York.

FAO (1997): "Decentralisation and devolution of forest management in Asia and Pacific" (PDF), FAO-United Nations, 1997.

FAO (2004a): "Trade and Sustainable Forest Management—Impacts and Interactions".

FAO (2011): "Global Forest Resources Assessment 2010", FAO Forestry Paper 163, Food and Agriculture Organization of the United Nations (2011).

FAO (2002): "Forests and the Forestry Sector: India", Food and Agriculture Organisation of the United Nations. 2002.

Freeman, C., ed. (1987): "Output Measurement in Science and Technology", North-Holland, Amsterdam.

FSI (2009): "India State of Forest Report, Forest Survey of India", Dehradun, Ministry of Environment and Forests, New Delhi.

Gangopadhyay, Shubhashis & John D. Knopf (1998): "Dividends and Conflicts between Equity Holders and Debt Holders with Weak Monitoring: The case of India", In: John Doukas, Victor Murinde & Clas Wihlborg (Eds.) '*Financial Sector Reform and Privatization in Transition Economies*', North-Holland, Amsterdam.

Ganguli, B.N. (2000): "Forest Management Options Beyond 2000-Issues and Opportunities for India", Paper for Commonwealth Forestry Association (India), Seminar on India's forests beyond 2000. New Delhi, April 19-21, 2000.

Gregersen H. & Contreras A. (2001): "Investing in the Future: The Private Sector and Sustainable Forestry Management. International Workshop of experts on financing sustainable forest management", Oslo, Norway, 22–25 January 2001.

IADB (2004): Inter-American Development Bank, "Estudio sobre inversión directa en negocios forestales sostenibles: Documento Conceptual". Project ATN/NP-8323-RS; STCP:

India State of Forest Report (2011): "Forest Survey of India (2011)", Ministry of Environment & Forests, Government of India, New Delhi.

India Forestry Outlook Study (2009): "APFSOS II: India", The Ministry of Environment and Forests Government of India, Food And Agriculture Organization Of The United Nations Regional Office For Asia And The Pacific Bangkok, 2009

Jhala Y.V., Gopal R. & Qureshi Q. (2008): "Status of Tigers, Co-predators and Prey in India", (ed), National Tiger Conservation Authority and Wildlife Institute of India, Ministry of Environment and Forests, New Delhi.

Kanchan Chopra (2006): "The Contribution of the Informal Sector: A Study of the Forestry Sector in India", Expert Group on Informal Sector Statistics (Delhi Group), Inst. of Economic Growth, New Delhi, 11-12 May, 2006.

K.C. Raut, (2004): "Development of Indian Forestry and Tasks Ahead", Technical Address delivered in 57th Annual Conference of the Indian Society of Agricultural Statistics held at GB Pant University of Agriculture & Technology, Pantnagar (Uttarakhand) on February 05, 2004.

Kaliappa, Kalirajan & Shashanka Bhide (2004): 'The post-reform performance of the manufacturing sector in India'. *Asian Economic Papers*, 3(2), 126-157.

Krishna, Pravin & Devashish Mitra (1998): "Trade Liberalisation, Market Discipline and Productivity Growth: New Evidence from India", *Jour. of Development Econ.*, 56, pp.447-462.

Krott M, von Paschen R. (2005): "Forest Policy Analysis", Dordrecht: Springer Netherlands 2005.

KPMG (2013): 'Human Resource Skill Requirements in the Furniture and Furnishing Sector (2013-17, 2017-22)', Vol. 11, N.S.D.C., Govt. Of India & KPMG Advisory Services Pvt. Ltd. India

Mali, K.P., Singh, K., Kotwal, P.C., & Omprakash, M.D. (2011): "Evaluation of Contribution of Forestry Sector to Economy: Application of Forest Resource Accounting in Gujarat, India", *Journal of Social and Economic Development*, Vol. 13, No. 2, pp.207–225.

Matej Jošt (2011): "Development trends in economic and management in wood processing and furniture manufacturing" International Scientific Conference Proceedings, International Association for Economics and Management in Wood Processing and Furniture Manufacturing Wood EMA, i.a University of Ljubljana, Biotechnical faculty Department of wood science and technology, Kozina, Slovenia, 2011.

MOEF. (2000): "Sustainable Development", Special paper In: Economic Survey, Economic Division, Ministry of Finance, New Delhi. NFRP, 2000.

MoEF. (2006): "Report of the National Forest Commission", Ministry of Environment and Forests, Government of India, New Delhi, pp.421.

MoEF (2009): "Asia-Pacific Forestry Sector Outlook Study II: India Country Report", Working Paper No. APFSOS II/WP/2009/06. Bangkok: FAO pp.78

MoSPI (2013): "Green National Accounts in India-A Framework; A Report by an Expert Group Convened by the National Statistical Organization", Ministry of Statistics and Programme Implementation, Government of India.

NFAP. (1999a): "National Forestry Action Programme", Vol. 1. *Status of Forestry in India.* Ministry of Environment and Forests, GOI, New Delhi.

NFAP. (1999b): "National Forestry Action Programme", Vol. 2. *Status of Forestry in India,* Min. of Environment & Forests, GOI. New Delhi

Schmidt R. (2003): "Financial investment in sustainable forest management - status and trends", A Background paper for the Global Project: Impact Assessment of Forest Products Trade in Promotion of Sustainable Forest Management GCP/INT/775/JPN. FAO, Rome

Stringer C. (2006): "Forest Certification and Changing Global Commodity Chains", *Journal of Economic Geography* 2006 6(5):701–722;

Taylor, R. (2004): "Policy Issues and Initiatives Related to the Four 'Wild Card' Supplying Regions". Presentation as part of UNECE Timber Committee and FAO European Forestry Commission Market Discussions. Geneva, Switzerland, 5 October.

Gokhale, Y. (2010): "Cover Story", Terragreen Magzine.

Poffenberger, M. (2000): Community and Forest Management in South Asia—A Regional Profile of the Working Group on Community Involvement in Forest Management. *Forest, people and policies*, IUCN, Switzerland.

Ram Prasad (2006): "Economic Value Of Forests/Green Accounting and Non Market Valuation Of Forests", in *'National Status Report on Forests and Forestry in India'*, (Survey and Utilisation Division) Ministry of Environment and Forests, Government of India, New Delhi, Sept. 2006

Rodrik, Dani & Arvind Subramanian (2004): "From 'Hindu Growth' to Productivity Surge: The Mystery of the Indian Growth Transition", IMF Staff Papers, 52(2), 193-228.

Schmithüsen F, Schmidhauser A, Mellinghoff S, Kammerhofer A, Kaiser B. (2008): "Entrepreneurship in Forest and Wood Industries: Basics of Business Administration and Management Processes [in German]", 2nd ed. Gernsbach: Deutscher Betriebswirte-Verlag 2008.

Sharma, V., & Chaudhry, S. (2013): "An Overview of Indian Forestry Sector with REDD+ Approach", *ISRN Forestry*, Vol. 2013, Article ID 298735, 10 pages, 2013.

Sinha, B., Kala, C. P., & Katiyar, A. S. (2010): "Enhancing Livelihoods of Forest Dependent Communities through Synergizing FDA Activities with Other Development Programmes", Indian Institute of Forest Management, Bhopal, India.

Singh N.K. (2011): "Contribution of Forestry Sector in GDP: A Case Study of Gujarat", Ph.D Thesis, FRI University, Dehradun.

Shruti Garg & Vandana Sharma (2015): "Forest Resource Accounting: An Overview and Indian Perspective", *Universal Journal of Environmental Research and Technology*, Vol. 5, Issue 1: 8-19.

State of Forest Report-2013, Ministry of Forests and Environment, Govt. of India (2014).

Tewari, D.N. (1994): "Tropical Forest Produce", International Book Distributors, Dehra Dun.

Thorsten Mrosek *et al.* (2010): "A Framework for Stakeholder Analysis of Forest and Wood-Based Industry Clusters – Case Study at the State of North Rhine-Westphalia, Germany", *The Open Forest Science Journal*, 3, 23-37.

United Nations Environmental Programme (2002): "The Asian Brown Cloud: Climate and Other Environmental Impacts" (PDF).

USDA (2009): "Wood and Wood Products in India" (PDF), United States Department of Agriculture, Foreign Agricultural Service, 2009.

Vincent, J.R., & Hartwick, J.M. (1997): "Accounting for the Benefits of Forest Resources: Concepts and Experience", FAO Forestry Department, Food and Agriculture Organisation, Rome.

Viitamo E. (2001): "Cluster Analysis and the Forest Sector–Where are We Now?", Interim Report IR-02-012. Laxenburg: International Institute for Applied Systems Analysis 2001.

White *et al.* (2006): "China and the Global Market for Forest Products: Transforming Trade to Benefit Forests and Livelihoods, Forest Trends, US".

WHO and UNEP (2011): "Indoor Air Pollution and Household Energy".

World Bank (2001): "New global poverty estimates in India," World Bank.

World Bank (2006): "An article of World Bank on India: Alleviating Poverty through Forest Development", World Bank. http://www.worldbank.org/ieg.

World Bank (2006): "India: Unlocking Opportunities for Forest Dependent People in India", Report No. 34481 - IN, World Bank: South Asia Region, p.85.

World Bank, "Making WAVES", Forest Accounting for Development: Capturing the Value of Forests Using Natural Capital Accounting. www.wavespartnership.org

2

A Conceptual Frame & Theoretical Background

2.1: Productivity Growth

2.1.1: Introduction

Understanding the trends and sources of economic growth and their relevance on an economy is one of the top research agenda that has long history in economics. By understanding through studying production functions that relate production level to input factors including unobserved productivity we can better attribute the economic growth to its sources and trends. In the modern literature it is well understood that production function estimation in micro-level is subject to simultaneity because the unobserved productivity affects not only the production level but also the levels of input such as labour, capital so many other influencing factors. Productivity growth has always the basis of efficient economic growth and it has been defined as the process of a sustained increase in the production of goods and services with the aim of making available a progressively diversified basket of consumption goods to population. Basically, economic growth has traditionally been associated with industrialization but industrialization in the initial stages has the effect of making resource scarcities more acute, making it all the more necessary that available resources are utilized more productively. Productivity has considered a key and major factor in the victory of any socio-economic system because of its direct relationship with economic welfare and also to achieve various goals of economic growth. The concept of productivity has come into greater prominence during the recent years and further it has assumed great importance and significance in the context of industrial development of an economy.

Scarcity of physical, financial and human resources has always been recognized as a limiting factor on the process of economic growth. While output expansion based on increased use of resources is feasible, it is not sustainable. Therefore, efficiency or productivity of resources becomes a critical factor in economic growth. These terms indicate ability to obtain a given amount of good or service by using a lesser amount of input. Productivity growth, therefore, is critical for ensuring sustained increase in the production of goods and services. As we know that economic development necessarily depends upon productivity of the factors of production while productivity is the result of a large number of social and managerial factors. Being essential social process of economic development, all factors must contribute at their best and the objective cannot be the responsibility alone of any one factor. Labour as a factor of production cannot be highly productive if they took with which the workers have to functioned are inefficient. Organization and managerial weakness would vitally impair the achievement of the industrial establishment. Researcher always considered that if productivity of all the factors is increasing over the period of time and total factor efficiency is improving then one can draw an inference that some sort of technological progress is taking place. Thus, it becomes essential to us being a investigator of proposed research work sponsored by UGC to discuss these phenomenon *viz.* productivity, technological change and production functions thoroughly and separately as a chepter.

Productivity trends and changes have attracted attention as clear indicators of economic efficiency, and they have been the subject of a considerable amount of research from both theoretical and empirical approaches, at both the microeconomic and macroeconomic levels, as a driver of economic growth and fluctuations in economic conditions. Moreover, role of productivity growth in the process of economic growth became clear when in the 1950's it was found that accumulation of productive factors (capital and labour) could explain only a fraction of actual expansion of output. Empirical work on the American economy by Tinbergen (1942), Schmookler (1952), Fabricant (1954), Abramovitz (1956), Kendrick (1957), Solow (1957) and Denison (1962) showed that between 80 to 90 percent of observed increase in output per head could not be explained by increase in capital per head and was attributed to productivity growth. Further, Terleckyi (1974), Scherer (1982, 1987) and Griliches (1984) showed that technological advancement was a major source of productivity improvement for the American industry {this point is more fully discussed in Goldar (1986)}.

It is also well acknowledged that economic growth depends both on the use of factors of production such as labour and capital, the efficiency in resource use and technical progress. This efficiency in resource use is often referred to as productivity. Some researchers note that growth in productivity is the only plausible route to increase the standard of living (see for example, Balakrishnan and Pushpangadan, 1998) and is therefore a measure of welfare (Krugman, 1990). The relevance of economic growth is less meaningful if it has not affected productivity growth and hence the standard of living. This increase in productivity or productivity growth can be caused by several factors including investment in human capital, infrastructure, R&D apart from healthy business environment.

The interest of economists in productivity growth is venerable. John Stuart Mill, like Karl Marx, was a growth theorist. Alfred Marshall was much interested in long-run economic change. In classical growth theory, as in neoclassical growth theory, firms were viewed as profit seeking and industries as competitive. But the connotation was more flexible than that of contemporary orthodox price theory.

In the verbal discussion, if not in the formal analysis growth was viewed as an evolutionary process. It is worth noting that during the early post-war era, the micro-economic conceptions underlying empirical analyses of productivity growth seems closer to the older theoretical tradition than to newer one. Moses Abramovitz stressed the links of the current empirical research to classical thinking and remarked upon the absence of many recent theoretical developments. Yet, despite the absence of modern framework for thinking about productivity growth the papers by Jacob Schmooker (1952), Theodore Schultz (1953), Solomon Fabricant (1954), John Kendrick (1956), and Abramovitz (1956) are remarkable foreshadowing the central conclusion of studies done somewhat later within the neo-classical framework that the growth of output experienced in the United States has been significantly greater than reasonably can be ascribed to input growth.

The concept of productivity has come into greater prominence during the recent years and has assumed great importance and significance in the context of industrial development. If the same production units can be used more efficiently, the net addition to the total national product will be much higher which will result in the process of industrial growth. In fact, higher production, greater consumption and better quality of life are possible only by raising the productivity. In this regard, Harrod has quite appropriately stated that *there is no shortcut to economic growth, except raising the productivity ratio.* Economists have advocated with evidences that higher levels of productivity contribute significantly to the eventual attainment of a self-sustaining economic growth resulting in continuous increase in investments, employment and income. The size of national income depends on the volume of output. It has to be enhanced not only by use of the employed resources to attain rapid and genuine economic growth. 'Productivity' refers to measurable relationship between the production results and the relative production agents, in both the financial and physical terms, in relation to given time and conditions. In broad and fundamental sense the problem of increasing productivity may be said to be the problem of making more efficient use of all types of resources in employment of using them to produce as many goods and services as possible at the lowest possible real cost.

Productivity in *general sense* usually denotes the possession or use of power to create or to make. To many, it also can note the quality or state of being productive. Such knowledge, as possession of power to bring forth goods and services; or the quality or state of being productive, do not precisely explain what productivity in economic terms actually means. Nor are the figures of volume of output alone helpful in explaining the term 'productivity'. Solomon Fabricant, therefore, writes "of the several sense of 'power to produce' then, it is to the comparison of output with input particularly, the ratio of one to other – that the term, productivity, is ordinarily attached". Also, Davis, has observed that as common application of the term involves a notion of the rate or degree with which the power to create or make it is utilized, "*the meaning of productivity in economic field may be stated as the degree to which the power to make or provide goods or services having exchange value is utilized as measured by the output obtained for the resources expended*". Thus, in the *widest sense,* productivity is the measurement of economic soundness and potentialities of means.

Productivity increase is, thus, an indispensable and powerful stimulus to and the end-result of a complex socio-economic process of economic development. Since the objective purpose both of productivity increment and economic growth is to satisfy the material needs of individual members as well as of society to the fullest possible extent, economic growth is positively correlated with productivity increase. Even economic growth is measured in terms both of absolute aggregate supply of goods and services and the increase in productivity. That is why in the current concept of 'welfare function' of economic growth, which implies the widening of the human choice and lays stress on attainment of higher standards of life and improved social well being, productivity growth is not only the target but also the crux of the whole problem. In this sense productivity increase, *viz.*, economic growth and development, is now a universal ambition.

However Productivity cannot be increased in any country or under any social system by simple decree. What is required is some sort of action, which would not have occurred otherwise. Benham by giving remark on that *'Manna does not fall from Heaven'* has also suggested that the real root of productivity growth lie in the general conditions and ways of economic development. But the real problem with which this generation is faced is how to mobilize and channelize the 'psychological resources' of a nation into right and purposive directions, so that its new productive power and diverse resources may be used for individual advancement in particular and the economic development of the country in general. This means that a developing society will be successful in raising productivity per man or man-hours for maintaining the upward momentum of its economy only when it is able to discover itself, to alert its people, to remove most of the social, political or economic handicaps and to be determined to push ahead. Until certain changes such as social, political, cultural and economic are introduced in the society to provide incentives and opportunities to its people for their individual advancement, and for work, towards group and society and towards welfare and progress will not change and sustained in the long run.

Thus, the economic development of any nation depends upon two distinct but related factors. One should be considered as the rate of industrialization. The establishment of a large number of new firms or industries at a fairly rapid rate leads to increased employment on the one hand and the production of much-needed goods on the other. The Efficiency and competitive capacity of industry or the quality of the goods produced, these can only be achieved by introducing innovations and technological changes as soon as they become available, by making the best possible available resources such as equipment's, materials, and man-power, by the introduction of productivity services. It means only a high level of productivity in existing as well as newly established industries ensures an improvement in the standard of living in long run. This has been considered as the second factor of economic development. It will generally be found that the higher the level of industrialization, the greater the productivity of industries as well as individuals working in industries. This is only to be expected since a high the level of industrialization produces the necessary industrial climate, the necessary services such as productive and progressive education, innovative training and upgraded research and creates auxiliary industries as well as markets all of which become closely interlinked. Once full employment as classical economists described in their theories, is achieved through industrialization, any further increase in the standard of living can only be achieved through increased productivity.

Now taking one indicator of productivity *i.e.* Total Factor Productivity (TFP) in particular, it has been observed that sometimes it plays a major role in the business cycle. For the first time Keynes (1936) in their General Theory focused attention on the forces that determine employment policy followed in industrialization. Researcher propounded the theory that entrepreneurs will offer the amount of employment which maximizes their output and profit and stressed the productivity of labour as the determining factor of the level of employment. There is a positive relationship among productivity of labour, output and employment. Lewis (1954) on the other side has strongly advocated the application of labour intensive techniques of production to have a steady and smooth economic growth. Analysis of total factor productivity (TFP) measures the increase in total output which is not accounted for by increases in total inputs. The level of TFP can be measured by dividing total output by total inputs. The TFP index is computed as the ratio of an index of aggregate output to an index of aggregate inputs. Growth in TFP is therefore the growth rate in total output less the growth rate in total inputs. In other words,

Productivity growth in the manufacturing sector in general has the effect of moderating the growth. The degree of this moderation of course depends on magnitude and the nature of technological change. If technological change is neutral, in the sense that it affects all inputs equally, the degree of moderation will depend on the overall growth of technological progress. If it has a bias approach, there will be significant degree of moderation. If on the other hand, technological change has a bias approach, the economy is likely to experience a rapid increase, requiring explicit policy initiatives (Puran Mongia, 1998).

Technology is considered the major cause of economic growth. Social scientists agree on the role of ideas and innovations in acting as adeus ex machine in somehow increasing total factor productivity (Mokyr, 2005), which in turn raises world income per capita (Maddison, 2005), transforms the production processes and modifies the way to run a business. Several compelling examples of the solid link between new technologies and growth can be found in the history: since the 18th century, with the Industrial Revolution, the introduction of new General Purpose Technologies (GPTs), such as steam engines, electricity, automobiles and telephones, has exponentially increased the standard of living. Furthermore, in the second half of the 1990s, for some countries, for example, the US, the investment in new Information and Communication Technologies (ICT) implied a radical changes in the underlying structure of its economy, which experienced, after an extended and unexpected stagnation during the 70s and the 80s, high levels of output growth associated with a strong, across-the-board productivity boom.

In the present era of increasing globalization and easy access to modern communication technology, firm in all countries are under pressure to improve their production efficiency in order to survive and thrive in the face of competition from newly emerging domestic firms on the one hand and foreign competition on the other. This is especially true in the case of India, which initiated major economic reform in 1991 in an attempt to make a systematic shift toward an open economy along with privatization of a large segment of its economy. These reforms were phased in over a number of years and many aspects of the reforms are only recently being implemented Ahluwalia. The removal of barriers to entry has opened up the economy to international market forces which, coupled with rising economic and social aspirations of the population, has led to the rapid emergence of a highly competitive environment, especially in the industrial sector. This has emphasized the importance of continuous improvement in productivity and efficiency, particularly if India is to achieve the high growth rate target set for the industrial sector.

As found in most of the empirical works, productivity growth plays a major role in enhancing economic growth as well as the standard of living. Efficiency gains and productivity improvements in the industrial sector are among the most important factors for successful economic reforms. At firms, and industrial level, productivity positively influences profitability, reduction of cost and price, which ultimately strengthens the competitiveness of firms and industries in the global competitive world. It is observed that cross-country disparities in real incomes, growth rates and standard of living have been attributed to differences in productivity performance. Therefore, total factor productivity growth is considered as an indicator of sustained economic growth and improvement in the standard of living. It is a well recognized fact that countries with poor performance in factor productivity growth would lose international competitiveness and thus, do not fully benefit from globalization. On the contrary, a higher level of productivity growth may result in lower product prices, better remuneration and working conditions to the employees, better returns to investors, adequate surplus available for expansion and modernization, besides giving a positive edge to these firms in the international market.

Although liberalization has been implemented effectively at the center, there has been considerable variation in the speed and extent of implementation of the reform measures across the different states, which has exacerbated regional imbalances. For the overall industrial sector to surge ahead it is very important to address the regional problems. In view of this, an analysis of the regional variation in productivity growth across the states is important. Total Factor Productivity (TFP) is the measure of performance of a productive unit. TFP may be defined as an index of change in output net of changes in inputs over the same period. A series of TFP growth is calculated over a period of time and a suitable index is constructed to get the total factor productivity of a production unit. Productivity has been one of the most popular areas of applied economic research as it is based on the well-defined analytical framework of the standard economic theory of production.

2.1.2: Productivity and Productivity Growth

In general term, Productivity is defined as the ratio of output to input(s). In other words, Productivity is the ratio between output and input (land, labour, capital, raw material, *etc.*). Depending on what is taken as the measure of output and which input is considered, a number of measure productivity may be considered. Productivity is a relationship between production and the means of production or a relation of proportionality between the output of a good or service and inputs which are used to generate that output. This relationship is articulated through the given technology of production.

The meaning of productivity as enunciated by ILO (1951) is generally accepted. According to ILO – "ratio between output and one of factors of input is generally known as productivity of factor considered". Thus, productivity means ratio between output and any factors of production land, labour, capital, and organization.

Ewan Clague says that productivity is a word, which we use broadly to express the overall efficiency with which our industries perform. Here that productivity is spoken of in terms of industrial efficiency.

According to Fabricant (1959) –"Productivity is a subject surrounded by considerable confusion. People employ the same term and mean different things". Productivity stands for efficiency in all activities. It is elimination of waste in all forms. Productivity is the function of providing more and more of everything for more and more people less and less consumption of real resources.

According to Mukherji (1962) productivity in its broadest sense is an indicator of the utilization of resources. It is the measurement of the economic soundness of the means. It is the combine effect of a number of factors like scientific management, technological development, proper allocation and scientific utilisation of available resources, human or labour incentive *etc.* For the practical purposes, productivity is measured by ratio obtained by dividing the output of goods and services produced by one or all factors of production necessary to achieve the result.

Russel W. Fenske (1968) defined productivity in ways, they are: (a) productivity is a form of efficiency; (b) productivity is the utilization of resources or effectiveness of utilization of resources; (c) productivity is a ratio; (d) Productivity is a measure of some kind; and (e) productivity is a rate of return (primarily in monetary terms).

In the words of Sreeniwasan productivity is usually defined as the ratio between output of product or services and input or resources in the form of manpower, machinery, materials, *etc.* In technical term it can be used as production per factor unit.

In the Dictionary of Management the concept of productivity has been defined as 'output per unit of input'. Thus the term 'productivity' is simply used to refer to the ratio between input and output and to measure the economic soundness of the means.

Peter Drucker's definition of productivity as 'that balance between all factors of production that will give the greatest output for smallest effort' falls under this category. According to him, 'without productivity objectives, a business does not have direction without productivity it does not have control'.

According to Risk (1970), 'productivity' is a physical ratio; it relates to the quality of goods produced or services given in comparison with quality of resources consumed.

Dhillon have defined productivity as the contribution of all the inputs, they being combined in some composite fashion. His definition of productivity can be considered as related to the efficiency in general, on the one hand and or it can be considered as a relation between output and input single or a combination of few or all, broadly, on the other hand.

Kumar definition of productivity is 'the term productivity is used by everyone, even in common parlance. However, its exact impact is rarely understood. The various terms like industrial production per worker, agricultural production per unit of capital, *etc.* is used as a proxy of productivity. In general, the productivity can be considered as related to the efficiency. Broadly, it is a relation between output and input single or a combination of few or all.

In the works of Sapro (1982) productivity may be defined as 'optimizing/ maximizing' the economic utilization of all available resources and investigating and utilizing the best known resources, as also creating new resources for different activities, be it industrial, commercial, agricultural, services or any economic activity in our day-to-day life.

In terms of Economics, thus 'productivity' can be expressed as the ratio of input in the form of production resources (*i.e.*, labour, materials, capital, power, floor space, physical conditions of work and living conditions) to output in form of production of goods and services obtained from consumption of such resources. Industrial productivity, in the aforesaid broad and fundamental economic sense, has also come to mean a ratio between output of wealth and the input of resources used up in the process of production. This definition of productivity is broader in the sense that it does not constitute the ratio of output to man-hours of efforts or labour expended.

In a modern manufacturing industry, the effectiveness of the expenditure of all other resources such as materials, capital, power, floor space, and machines are also equally important, because the concept of productive efficiency is a function of many independent and inter-dependent variables. Industrial productivity has, therefore, been defined by Davis (1951) as the degree to which the power to make or provide goods or services having exchange value is utilized, as measured by the output obtained for the resources applied. It implies the utmost utilization of all available resources of men, machinery, material, money-power, and land *etc.*, the end in view being the highest possible production at the lowest possible cost within the shortest period of time. In the words of Bell (1960) also, "broadly and basically defined, productivity is, of course, the relation between output and input a measurement of the efficiency with which resources are transformed". This concept also emphasis the fact that productivity is the result not of one factor alone but of many and to use productivity measurements that are made without reference to planning, engineering and research, and managerial competence is to provide a base that is narrow, inaccurate, and misleading.

Moreover, in economic literature, the term productivity is preceded by a number of objectives, some of which are the following:

(I) Actual Productivity: This is the current level of productivity that may be higher or lower than the desired rate of productivity.

(II) Potential Productivity: This is the rate of efficiency (productivity) of productive enterprises that would like to achieve in order to have a self-sustained rate of economic growth.

(III) Volume Productivity: It refers to cost concept. When output is undertaken on a large scale, unit cost (*i.e.* fixed cost) tends to decline. Contrariwise, when output diminishes, unit cost may be expected to go up. A question might be raised. Does productivity rise with large scale output simply because costs happen to be low? Not necessarily. Productivity might remain unaltered in the long run because low cost does not mean higher productivity.

(IV) Real Productivity: It has to do with a basic and permanent change in the value of output, which can be obtained by use of a given amount of labour. Changes in the technique and organization of production affect real productivity.

Thus shorn of all technical jargon, the concept of productivity implies the rate of proper and efficient utilization of power of all the associated factors of production. The object is to minimize losses and waste of every type and in every sphere, so that goods and services of kind and quality, most wanted by consumers, may be produced at lower and lower costs. If productivity is increasing in an economy, it means that its factors of production and commodity inputs are manifesting increases in their output efficiency. The notion of higher productivity is, therefore, consequently related to the optimum utilization of the inputs of all the available potential production factors so as to increase output, both qualitatively and quantitatively. The concept of productivity in this sense, cannot the class of conceivable measure denoting output per unit of associated input in a sequence of compared periods, signifies measurable input/output relationships, that is relationships between the result achieved and means employed to achieve the result and elaborates only definable relationships between an output and the relative factors of production. Thus, productivity may be stated as the ratio of output to input of only one factor or two more factors production expended for the production of goods and services in a given period of time.

The two most commonly used measures of productivity are single factor productivity (SFP) and multifactor or total factor productivity (TFP). When multiple inputs of heterogeneous nature are used in the production process, aggregation of these inputs requires use of price indices. This implies that productivity can be affected by both changes in relative prices of inputs and input requirements per unit of output. Thus, the main kinds of productivity can be ascertained, such as:

2.1.3: Single Factor Productivity (SFP) and Total Factor Productivity (TFP)

Productivity can be measured with respect to a single input or a combination of inputs. The partial or single factor productivity (or SFP) is defined as the ratio of the volume of output (or value added) to the quantity of the factor of production for which productivity is to be estimated (*e.g.*, labour productivity or capital productivity). If only one production factor is required in a production process, the productivity is usually called simple productivity. 'Single Factor' or Partial Factor Productivity: When a number of factors are involved in a production process, but the output is related to any single factor unit, productivity, thus measured, is called single factor or partial factor productivity, such as labour productivity, capital productivity, raw- materials productivity, machine productivity, capital intensity or productivity of organization, *etc*. Thus, ratio of any factor of production to output is known as partial factor productivity. When the proportion in which the factors of production are combined (*e.g.*, labour and capital) undergoes a change, partial measures of productivity provide a distorted view of the contribution made by these factors in changing the level of production. In a situation where capital-labour ratio follows an increasing trend, productivity of labour is overestimated and that of capital, underestimated. For instance, capital deepening (shifts in technique of production) can lead to a rise in labour productivity and fall in capital productivity over time. In this case, a change in labour productivity is merely a reflection of substituting one factor by another (Majumdar, 2004). Similarly, improvements in labour productivity could also be due to changes in scale economies (Mahadevan, 2004).

In short, the partial measure does not provide overall changes in productive capacity since it is affected by changes in the composition of inputs. Despite the limitation, estimation of productivity of labour is regarded crucial from the welfare point of view. This is because it measures production per unit of labour employed and a country's ability to improve its standard of living over time depends on its ability to raise its output per worker. Chen (1979) using Singapore and Hong Kong as examples has shown that in the long run, it is the growth of labour productivity that is more important than TFP growth (Mahadevan, 2003: 366). Kendrick (1991) argues that labour productivity measure is useful in showing the savings achieved over time in the use of the input per unit of output. Sargent and Rodriguez (2000) have advocated the use of labour productivity to examine the trends over a period that is less than a decade given the biases in estimating capital stock to obtain TFP growth. Balakrishnan (2004) argues that labour productivity merit attention in its own right and serves a different purpose for which the TFP is not a substitute. He contends that labour productivity is a measure of potential consumption and a steady rise in the productivity of labour is necessary for a sustained increase in the standard of living of a population. Typically, labour productivity moves in the same direction as TFP but grows at a somewhat faster rate reflecting the influence of capital deepening (Mahadevan, 2004).

The concept of total factor productivity (TFP) tries to circumvent the problem encountered in the interpretation of SFP estimates in the event of changing factor intensities. TFP is defined as the ratio of output (or value added) to a weighted sum of the inputs used in the production process. It is an indicator of the change in efficiency in factor use in estimated simultaneously with the returns to scale which is another parameter which has a bearing on efficiency. In fact, total productivity measures only that component of technological change, which increases the overall efficiency of factor. Thus, when the output is related to the entire input complex (labour, capital, raw- materials, power, organization, *etc.* combine together), the relation between output and input is multifactor (or overall, or composite, or total) productivity. TFP is deemed to be the broadest measure of productivity and efficiency in resource use. It aims at decomposing changes in production due to changes in quantity of inputs used and changes in all the residual factors such as change in technology, capacity utilization, quality of factors of production, learning by doing, *etc.* An increase in TFP, therefore, implies a decrease in unit cost of production.

2.1.4: Total Productivity (TP) versus Total Factor Productivity (TFP)

At this stage, choice exists in regards to the specification of output as value added (V) as in equation Vt = f(Kt ,Lt) {Where, Vt = level of net output (value added). Kt = capital input (or service of factor capital) Lt = labour input, t = time} above or gross value of output (Y). In the latter case, material and energy inputs are explicitly accounted for in both the left and the right hand sides in the production function. This would give rise to the following general functional form which in recent years has come to be known as KLEM type production function.

$$Yt = g(Kt, Lt, Et, Mt, t)$$

{Where, Yt = level of gross output per unit of time, Kt = capital input (or service of factor capital) Lt = labour input Et = input of energy, Mt = material inputs. t = time}

The choice between one forms of the other depends on what one believes to be the correct measure of output. It also depends on whether one believes the production function to be separable in factor and material inputs or not. The above functional forms give rise to alternative concepts of productivity. One can define the productivity measure associated with the value added (V) production function as total factor productivity (TFP) and that associated with gross output (Y) production function as total productivity (TP). {This usage may not be universal. It was recently suggested by Rao (1996)}.

In the available literature on the issue, it has been seen that the majority of studies have been conducted using production functions with value added as output and with K and L as inputs. It is only recently that studies on production functions in India have been using gross output and K, L, E and M as inputs (Puran Mongia, 1998).

The concept of TFP growth and what it constitutes has been a subject of debate since the time the term was first introduced in 1940s (Mahadevan, 2003: 365). This has led to several definitions of TFP growth. The term has been used interchangeably with technological change/progress, embodied and/or disembodied technical change. Following are some of the definitions of TFP growth:

TFP Growth = Output Growth – Input Growth
= Technical/Technological Change/Progress
= Embodied (or endogenous) Technical Change + Disembodied (or exogenous) Technical Change
= Changes in Technical efficiency + Technological Progress

Of all these definitions, the first one is the most commonly used. As per the definition, TFP growth incorporates all the residual factors after accounting for input growth, and has also been hailed as an 'index of ignorance' (Abramovitz, 1956). Jorgenson and Griliches (1967) argue that if we measure all the inputs carefully, this residual might disappear. The last two definitions are conceptually identical as the change in technical efficiency essentially indicates embodied technical change and technological progress constitutes the disembodied technical change (Mahadevan, 2003). The Embodied technical change results from the efficient use of new and better types of capital so as to move towards the frontier. Disembodied technical change, on the other hand, results in the expansion of production boundaries itself due to increase in knowledge.

2.1.5: TFP Measurements: An Overview

(Approaches to the Measurement of Productivity Growth)

Measurement of technical change (TC) and total factor productivity (TFP) growth has been the subject of investigations in many empirical studies on industrial productivity (for example, Jorgenson, 1995). These studies have followed several well-known directions. The various approaches used in the literature have been classified by Diewert (1981) into: parametric estimation of production and cost functions, non-parametric indices, exact index numbers, and non-parametric methods using linear programming. In the non-parametric approach, the Divisia index has been widely used as a convenient measure of TFP growth over time and space as well. An important feature of the Divisia measure of TFP growth is that it coincides with the technical change when the underlying technology is homogeneous of degree one. However, empirical studies on production functions based on panel data do not support constant returns to scale technology (see Atkinson and Cornwell, 1994a, 1994b; and Biorn and Klette, 1996). If this property does not hold TFP growth becomes a mixture of technical change and scale effects. In the case of non- constant returns to scale technology, decomposition of TFP growth into its sources requires knowledge of scale effects, which require econometric estimation of parametric functions.

While on the other side Puran Mongia & Jayant Sathaye (1998) were the view that there are three principal approaches to measurement of productivity growth. These are: (i) The index number approach, (ii) parametric approach and (iii) non-parametric approach. In their survey, they had focus primarily on studies which have estimated productivity growth using the first approach. Wherever appropriate, the results from the estimation of cost and production functions have been mentioned in support of as alternative explanations to the results of the first approach. But the non-parametric approach which is based on linear programming models of relative efficiency is not reviewed in their survey. In Index Number Approach, the observed growth in output is sought to be explained in terms of growth in factor inputs. The unexplained part or the residual is attributed to growth in productivity of factors. It consists in assuming a certain functional form for the producers' production function and then deriving an index number formula that is consistent (exact) with the assumed functional form. Preferred functional forms are the flexible ones. These indices differ from each other on the basis of underlying production function or the aggregation scheme assumed. Kendrick Index and Solow Index are the most commonly used indexes in economic research. Last two approaches in details has been discuss in following paras.

Very few other issues in Indian economic development has generated so much debate than the measurement of total factor productivity (TFP) growth in Indian manufacturing (Balakrishnan and Pushpangadan, 1994; Dholakia and Dholakia, 1994; Goldar, 2002). This debate has intensified following the major economic reforms in 1991. Estimates of TFPG do not provide a clear picture on what has happened to the rate of productivity growth in manufacturing after the 1991 reforms. Studies by Krishna and Mitra (1998), Unel (2003) and Tata Services Ltd. (2003) find an acceleration in TFPG in the 1990s, whereas studies by Baghel & Pendse (1996) Trivedi *et al.* (2000), Srivastava (2000), Balakrishnan *et al.* (2000), Ray (2002), Goldar and Kumari (2003), Goldar (2004), Baghel & Gupta (2002), Goldar (2006), Das (2004), Kumar (2004) and RBI (2004) find a deceleration in TFPG in the 1990s.

Since the 1991 reforms were specifically targeted to the manufacturing sector due to the realization that the sector offered much greater prospects for capital accumulation, technical change and linkages and hence job creation, it is important to understand what has been the behavior of TFPG in the post-reform period. It should be noted that the trend in TFPG prior to the 1991 reforms has also been contested. A number of studies (see for example, Brahmananda, 1982; Ahluwalia, 1991; Dholakia and Dholakia, 1994; Majumdar, 1996; Rao, 1996a; Pradhan and Barik, 1998; Trivedi *et al.*, 2000 among others) have suggested a decline in the TFPG till 1970s with a turnaround taking place in mid-1980s in line with the more open trade and industrial policies. However, Balakrishnan and Puspangadan (1994) argue that the TFPG growth during the 1980s is the arte-fact of using single digit deflation method. The turnaround vanishes if double deflation approach is adopted.) However, given the different methods used in these studies along with differences in variable construction and data used does not allow for one to infer whether the large differences in estimates (not just in magnitude but also the direction of change) are due to the use of different methods or due to different approaches to variable construction or the use of different data-sets. In addition, most of these studies do not include the informal manufacturing sector in their estimation of TFPG, which is a significant omission, given the importance of this sector in total employment in Indian manufacturing {Balakrishnan & Pushpangadan (1998)}.

Measuring TFP growth is a key element not only for quantifying the impact of new technologies but also for understanding why an economic unit is richer than another one or whether the advances in technological goods can fragment the production processes and have stark and different effects on the employment composition, with large positive shifts in demand of skilled workers. In economics, management and operations research, it is possible to choose from several sets of parametric and non-parametric procedures for estimating technological change. Diewert (1981) divided these techniques into different groups: growth accounting procedures, mostly based on index, estimations of cost function, estimation of production function, and nonparametric methods, also known in the literature as data envelopment analysis (DEA). In most of these frameworks, technological growth is derived as a difference between the output produced and the inputs used. Even if these procedures are in general quite simple and straight forward to implement, accurate measurement of TFP growth represents one of the most challenging task in macroeconomics for several reasons. First of all, while output and employment are directly measurable in the production process, capital is not observable and should be constructed considering a number of assumptions regarding investment, the depreciation rate and capacity utilization. In addition, increasing returns to scale or spillovers can bias upward the TFP growth results. Finally, once all the inputs are correctly measured, technological progress can be influenced by other factors, such as culture, institutions, climate conditions and initial endowments.

Now on the basis of above discussions, we are considering main two techniques to measure TFP growth *i.e.* frontier and non-frontier approaches. These approaches are further divided into parametric and non-parametric techniques. Most studies in the Indian context and elsewhere have used non-frontier techniques with recent emphasis being on parametric estimations. The crucial distinction between frontier and non-frontier approaches lies in the definition of frontier. In frontier approach aim is to find the bounding function *i.e.*, the best obtainable positions given the inputs or the prices. A 'cost frontier' traces the minimum attainable cost given input prices and output and a 'production frontier' traces the set of maximum obtainable output for a given set of inputs and technology. This is different from the average function which is often estimated by the ordinary least square regression as a line of best fit through the sample data.

Apart from this, the frontier approach identifies the role of technical efficiency in overall firm performance, whereas the non-frontier approach assumes that firms are technically efficient. This difference results in different interpretation for TFP growth for the two approaches. The TFP growth as obtained from frontier approach consist of two components - outward shifts of the production function resulting from technological progress, and technical efficiency related to the movements towards the production frontier. On the other hand, the non-frontier approach considers technological progress as a measure of TFP growth.

Since in frontier approach, benchmarking is done where a firm's actual performance is compared with its own maximum potential performance, the approach is more suited to describe industry or firm's behaviour (Mahadevan, 2003: 373). Benchmarking has little room in the non- frontier approach. Earlier studies used non-frontier approach to compute estimates of TFP growth at the economy level and later on only with availability of more disaggregated data, the approach has been used for sectoral or industry level analysis. The parametric non-frontier approach is statistical in nature and evaluates firms relative to an average producer. A feature common to both the frontier and non-frontier approach is that both can be estimated using parametric and non-parametric methods.

In parametric method, an explicit functional form is specified for the frontier and the parameters are estimated econometrically using sample data for inputs and output. This implies that the accuracy of the derived estimates is sensitive to the functional form specified. Simultaneity bias is one ignored problem in parametric estimations of the production function. The problem arises because the right-hand sides variables containing the inputs are chosen in some optimal way by producers themselves, thus are not exogenous. Adopting Zellner *et al.* (1966) argument that producers maximize expected profit or assume profit maximization ex ante or ex post reduces the problem. Using instrument variable estimation is one way of correcting the problem econometrically. The difficulty in choosing the right instruments however makes the implementation difficult.

However, parametric approach requires a mathematical function that represents production structure of an industry. The function can be a production function that provides a description of production technology, or a dual representation in terms of cost function, revenue function or profit function. Estimation of cost function is often preferred since cost share equations can be obtained from the function, and then estimated together, enlarging the degree of freedom. Also, elasticities of substitution and technical change can be estimated more convenient according to the cost function. Many researchers have conducted parametric methods to evaluate the productivity in forest products industries. Most of them measured the indicators like returns to scale, productivity growth, and factors substitution.

In the parametric specification of technology using production/cost/profit functions, a widely used practice has been to use quadratic function of time trend to represent technical change. Notwithstanding its widespread use, the use of time trend is a mere reflection of our ignorance. Baltagi & Griffin (1988) has shown that if a panel data set is available, we could estimate a time specific parameter referring to the state of technology (general index of technical change) instead of using time trend. The method applied to analysis of manufacturing industry performance has shown evidence of erratic patterns of technical change which limits its usefulness in capturing technical change (see Kumbhakar & Heshmati 1996; and Kumbhakar, Nakamura & Heshmati, 2000). Different generalizations of time trend (TT) and general index (GI) models of technical change have been developed and their performance and sensitivity using different datasets evaluated (see Heshmati & Nafar, 1998; Kumbhakar, Heshmati & Hjalmarsson, 1999; Kumbhakar, 2000; and Oh, Heshmati & Loof, 2009).

Econometric approach where technical change has been represented by a simple time trend or time dummies still found dominating in the empirical research. The popularity of time trend model comes from the fact that it is good in revealing long-run trends in technical change while general index model is good in capturing year to year variations (which may be caused by economy wide, sector-specific or firm-specific product or process innovations and demand or supply shocks). Despite of this popularity, the time trend model has been criticized because it reflects only our ignorance about the process. The general index overcomes the trend limitation by not imposing any systematic structure on the behaviour of technical change, but it is by no means any better in explaining technical change. In both approaches TC is modelled entirely in terms of time and they fail to account for determinants of technological change and productivity growth. If two firms have the same inputs then their TC will also be the same. In the general index model determinants of TC are not directly used in the model. These are used in a second stage regression, therefore fails to take into account their direct or interactive effects with the traditional inputs.

The main advantage of the non-parametric method like mathematical programming approach or the Data Envelopment Approach (DEA) is that it is parameter free and does not assume any functional form. The major drawback is that no direct statistical tests can be carried out to validate the estimates. The parametric approach employs econometric technique and in this approach, the deviation of actual output from the maximum output is decomposed into two parts, *viz.*, the statistical noise and inefficiency. The various alternatives within the parametric approach are as follows: (a) econometric frontier approach; (b) thick frontier approach; and, (c) distribution free approach. Each of these approaches involves arbitrary assumptions regarding the distribution of the noise and inefficiency components. The prime difficulty in using the econometric approach lies in separating the noise from the inefficiency. Of late newer techniques have been developed that take care some of the problems of statistical testing *etc.* These include stochastic DEA, Bayesian approach, testing for statistical properties of DEA estimates using jackknifing and bootstrapping. Incidentally, many of these are being used in operations research rather than their use in computing productivity growth. Even among the commonly used methods available to compute TFP growth, the literature is inconclusive on the best method to estimate TFP growth (Mahadevan, 2003). Before embarking upon the proposed estimation of TFP growth we discuss issues in measurement of TFP and TFP growth.

2.1.6: Issues in Measurement of Total Factor Productivity:

The measurement of TFP involves several issues. Some of the concerns have arisen because the data for the estimation of TFP and TFP growth are not available in the required form. For every variable there are different possible ways to adapt the available data and each of these is liable for criticism (Srivastava & Dasgupta, 2000). Measurement error in the original data further complicates the issue. Apart from these, there is a choice between using the traditional growth accounting approach (also known as deterministic approach) and obtaining econometric (or stochastic) estimates. Perhaps that may be the reason that the empirical evidence has given conflicting results sensitive to the choice of method, data used and the manner in which variables have been measured. The first issue relates to the measurement of the output to estimate productivity. The issue is choosing between 'output (O)' and 'value-added (VA)'. If VA is chosen, then the related second issue is between the use of 'single-deflation' and 'double-deflation' methods for its measurement and separatibility of the production function. Once decided about the output measure, the third issue is about biases in input measurements. The fourth issue is choosing between different methods for estimating TFPG.

2.1.6 (A): Measurement of Output

Gross Output versus Value-added Two types of output measures can be used to calculate TFP and TFP growth: value added and gross output. The separability of the production function is the de rigueur for the legitimation of the use of real value added (Balakrishnan & Pushpangadan, 1998). The literature has exhibited strong preference for using value-added as the measure of production. See for example, studies by Goldar (1986), Ahluwalia (1991), Balakrishnan & Pushpangadan (1994, 1998) among others. Norsworthy & Jang (1992) attribute this to the fact that the concept of value-added is useful in national income accounting as it avoids double counting of intermediate inputs. Diewert (2000) argues that value added scores over gross output in inter- industry level studies because the latter includes cost of intermediate inputs which may vary greatly across industries. According to Griliches & Ringsted (1971), use of value added allows comparison between the firms that are using heterogeneous raw materials, and it also takes into account differences and changes in the quality of inputs (Salim & Kalirajan, 1999).

The use of gross output that demands the inclusion of raw material as an input variable in the model might diminish the role of capital and labour in productivity growth (Hossain & Karunakara, 2004). However, the use of value-added provides a distorted view of technology because the effect of changes in prices of purchased raw-material inputs is removed from the costs of production and technology. According to Norsworthy & Jang (1992), it is the aftermath of the energy crisis that has revealed the shortcoming of using real value-added vis-a-vis gross real output for productivity estimation. In contrast, some studies have employed gross output function framework by rejecting the 'implicitly maintained hypothesis of separability of intermediate inputs like materials and fuel from labour and capital inputs (Rao, 1996a; Pradhan & Barik, 1998; Ray, 2002; Trivedi, 2004; Mukherjee & Ray, 2004). They have argued that a production function relating labor and capital is meaningful only when material inputs are separable from the primary inputs. Often TFP growth based on value added measure is greater than that of output measure due to the upward bias created by the omission of intermediate goods and services. This bias, however, can be corrected if the ratio of inputs to gross output remained constant (Star, 1974).

2.1.6 (B): Measurement of Value Added

Single versus Double-Deflation Methods If value-added is used as a measure of output, nominal value-added needs to be converted into real value-added. This conversion can be done with either single deflation (SD) or double deflation (DD) method. In the case of the former, nominal value-added is deflated by the output price index, *i.e.*, both nominal output and nominal material inputs are deflated by the output price index. This is referred to as the SD method. The other alternative is to deflate the nominal output by output price index and the nominal material inputs by the input price index, *i.e.*, the DD method. If both the output and input prices change in the same proportion, then the ratio of input-output prices remains constant and in such a situation, the estimates of TFP growth obtained by both SD and DD methods will coincide. During the periods when the input price index increases at a faster rate than the output price index, the estimate of real value-added obtained by using SD method will be lower than that obtained by using DD method and vice versa. Bruno (1984) has highlighted the role of increasing relative price of raw materials to output in explaining the productivity slowdown in USA and has argued that its effect on the estimation of productivity is analogous to that of Hicks-neutral technological regress (Chen, E.K.Y. 1979).

Goldar (1986) states that the use of SD method based on product prices for estimation of real value- added may not be appropriate but due to the difficulty of compiling a materials price index required for DD method, most of the studies including his has used SD method. Ahluwalia (1991) has also expressed the problems associated with the use of the SD approach in the context of measurement of productivity for petroleum and coal industries with the caveat that in the absence of official estimate of value-added in these sectors by the DD method, productivity estimates for these industries need to be interpreted with caution.

The study by Balakrishnan & Pushpangadan (1994) for Indian manufacturing sector was the first of its kind to use the DD method and to highlight the importance of changing relative prices in estimation of growth of TFP. They pointed out that deflating value added by a single deflator (as had been done by Ahluwalia, 1991 & Goldar, 1986) would be valid if the price of material inputs did not change relative to the price of the output, which in ordinary circumstances would not be valid. Their study at the aggregate level for the manufacturing sector then refutes the claim made by Ahluwalia (1991) that there was a positive turnaround in TFPG in the Indian manufacturing sector in the 1980-81. It attributed this result to overestimation of productivity by the use of SD method in the event of declining relative prices in the early 1980s.

2.1.6 (C): Measurement of Output–other Issues

Apart from the issue of choosing value added or output and going for single or double deflation method, there are other concerns pertaining to the output. These include how to account for newer outputs and the problem of product mix. In reality, very few firms produce single homogenous product. Firms not only produce differentiated products, but also produce variety of products and often change their product mix over time. As a result of these changes in output, the input mix also changes (Mahadevan, 2003: 370). Any index of real output also has to account for quality. Market prices in the base period are often taken to reflect relative values that capture quality differences, but when quality changes are not associated with increases in production costs (and hence market prices) productivity is underestimated. The problem of considering quality changes is more pronounced in service output. For instance, how to account for improved communication system, faster transport and increased array of financial services? Since these are difficult to capture, any estimate of TFP using either VA or output would yield biased results.

2.1.6(D): Measurement of Inputs

(i) Labour

The customary way of measuring labour input is either to use the number of hours worked or the number of workers employed. A large number of studies have used the former as it accounts more accurately for part- and full-time employees in terms of actual hours worked. Still the measure suffers from a limitation if a mix of skilled and unskilled workers is employed. This is because the contribution of skilled workers to production is much higher than that of unskilled workers. Thus, appropriate labour measure would require incorporating the quality of the labour inputs accounting for the sex, education, employment status of the worker *etc.* (Mahadevan, 2003).

(ii) Capital

With respect to capital a number of issues exist. Irrespective of whether frontier or non- frontier approach is used in TFP growth estimation, the flow of output is linked to the flow of inputs' services. Since the data on the flow of capital services is not available, it is assumed that capital flows are proportional to net capital stock after depreciation. Moreover the depreciation rate is assumed specific to asset type instead of specific to industries, and also it does not change over time. Since the asset mix in a given industry might change significantly over time, a technology intensive asset mix would result in capital under-representation and vice versa. The capital also needs to be adjusted for utilization since the use of capital is subject to cyclical factors. For instance in recession capacity utilization is low. If excess capacity is understated, then the residual TFP growth would be understated. In a way, utilization rates are seen as a means of converting capital stock to flows. It is claimed that in the long run, cyclical fluctuations in the flow of services average out and one can take the ratio of capital services flow to the capital stock to be constant, which allows the use of the perpetual inventory equation to measure capital services (Mahadevan, 2003).

In practice, the measurement of capital input is the most complex of all input measurements. There is no universally accepted method for its measurement and, as a result, several methods have been employed to estimate capital stock. In many studies, the capital unit is treated as a stock measured by the book value of fixed assets. Some studies have employed the perpetual inventory method to construct capital stock series from annual investment data. In this case it is assumed that the flow of capital services is proportional to the stock of capital. However, it is essential to point out that each of these measures has drawbacks. The book value method has three limitations. First, the use of 'lumpy' capital data underestimates or overestimates the amount of capital expenditure. Second, the book value may not truly represent the physical stock of machinery and equipment used in the production. Third, it does not address the question of capacity utilization. Perpetual inventory method also does not address the question of capacity utilization. The flow measure is criticized on the ground that the depreciation charges in the financial accounts may be unrelated to the actual wear and tear of hardware.

2.1.7: Issues related to Method Selection

While discussing the pros and cons of the parametric and non-parametric methods, Lovell (1993) has concluded "...in my judgment neither approach strictly dominates the other, although not everyone agrees with this opinion, there still remains some true believers out there". From the above statement it is quite clear that no technique is perfect in TFP calculation. However, whether to use frontier or non-frontier approach depends on the question a researcher is addressing.

For instance, if the objective of the study is to assess the contribution made by each input to output growth or to estimate how much output, on average, has been obtained from a set of inputs, then non-frontier approach would be a better choice. On the other hand, to address the questions on maximum productive or best practice output levels, given the inputs and technology, the frontier approach would be the best method. Besides, to examine the sources of TFP growth, the frontier approach is more useful as it decomposes TFPG into various components. However, if the researcher wants to know if the output growth is due to TFP growth or input growth, then the non-frontier approach would serve the purpose.

If the preferred approach is the frontier approach, then the next question is whether to use SFA, which represent absolute frontier (maximality over all possible sample points) or DEA representing best practice frontier constructed from the given sample. The decision depends on many considerations/ factors. Table 1 gives a comparison of different methods based on seven key parameters.

If multiple inputs and outputs are involved in the production process, then DEA is the appropriate method. The DEA method can accommodate multiple inputs and multiple outputs simultaneously. One of the principal disadvantages of DEA is that it can be extremely sensitive to variable selection and data errors. For example the data collected from agricultural sector suffers from two errors - measurement error due to poor quality of data and weather playing a significant role. In this context, the parametric stochastic production frontier is highly recommended. However, DEA appears to be more appropriate when knowledge about underlying technologies is weak. Stated differently, if the employed functional form is close to the given underlying technology, SFA outperforms DEA. In any case, before deciding on the best approach, one should also collect additional information about the type of activity under study. For instance, information about scale and substitution possibilities is best handled with parametric approach. {For a more extended discussion on methods see Kathuria *et al.* (2011)}.

2.1.8: Empirical Work in India

A Review Depending on the coverage, studies on productivity can be classified into three major types, *viz.*, macro, meso and micro level studies (Wagner and Ark, 1996). Macro level studies deal with the entire economy, whereas, meso level studies pertain to a sector or an industry. Micro level studies are conducted at the firm level. Table 2 lists some of the important macro, meso and micro level studies carried out in the post-1980 period for India.

As is evident from Table 2, considerable research attention has been devoted to analyzing the various aspects of the formal manufacturing sector to the relative neglect of informal manufacturing. Recently only there have been some attempts to examine the productivity performance of the informal segment of manufacturing sector (Unni *et al.*, 2001; Marjit & Kar, 2009; Kathuria *et al.*, 2010; Raj, 2011; Raj & Babu, 2011). Studies by Marjit & Kar (2009), Raj (2011) and Raj & Babu (2011) that employed frontier approach to estimate TFPG report gain in TFPG in the informal sector in the 1990s and early 2000s. On the other hand, studies that employ non-frontier approach (Unni *et al.*, 2001; Kathuria *et al.*, 2010) find a deceleration in productivity following reforms.

Table 1: Comparison between Different Methods of Productivity Measurement

Problem	Semi- Parametric	Non-parametric		Parametric	
		GAA	DEA	Regression	SFA
Multiple inputs and outputs	Complex, rarely taken up	Simple	Simple	Complex, rarely taken up	Complex, rarely taken up
Specification of functional form	Required may be incorrect	Required	Not required	Required may be incorrect	Required may be incorrect
Outliers	Not as sensitive	Sensitive	Inaccurate efficiency assessment	Not as sensitive	Not as sensitive
Sample Size	Moderate sample size is required	Small sample size adequate	Small sample size can be adequate	Moderate sample size is required	Large Sample size is needed
Prevalence of high co-llinearity among inputs	Possible misleading interpretation of relationships	Possible misleading interpretation of relationships	Better discrimination	Possible misleading interpretation of relationships	Possible misleading interpretation of relationships
Noise, such as measurement error	Not specifically modeled, but assumptions required	Sensitive	Sensitive but highly	Affected but impact is less as compared to DEA	specifically modeled- strong distributional assumptions required
Statistical Testing	Straight forward statistical testing	Not possible	Sensitive analysis is possible but complex	Straight forward statistical testing	Straight forward statistical testing

Source: *Raj (2006), and Vinish Kathuria (2013)*

In brief, the table indicates that it is only in the last few years that the productivity studies in India have considered the use of gross output over the real value-added as a measure of production and a large number of studies have used GAA. Of late, application of frontier approach is found to be common among researchers for estimating TFPG in the Indian manufacturing sector.

Table 2: A Brief Review of Productivity Studies in India

Study/year	Period of Study	Sector/ Industry type	Measurement of output	Deflation Method	Estimation Approach	Functional form of PF	Index used in GA Approach
Brahmananda (1982)	1950-1981	Formal & Informal	NDP	SD	GAA	--	KI
Goldar (1986)	1951-1979	Formal	VSD	SD	GAA & PFA	CD & SMAC	TLI, KI, & SI
Ahluwalia (1991)	1959-1986	Formal	VSD	SD	GAA & PFA	CD, TL & CES	TLI
Mohanty (1992)	1970-1989		NDP	SD	PFA	CD	--
Balakrishnan and Pushpangadan (1994)	1970-1989	Formal	VDD	DD	GAA	--	TLI
ICICI Limited (1994)	1970-1992	Formal	VSD	SD	GAA	--	TLI, KI, & SI
Dholakia and Dholakia (1994)	1970-1989	Formal	VSD & VDD	SD & DD	GAA	--	TLI
Rao (1996a)	1973-1993	Formal	O	--	GAA	--	TLI
Rao (1996b)	1973-1993	Formal	O	--	GAA	--	TLI
Baghel & Pendse (1996)							
Krishna and Mitra (1998)	1986-1993	Formal	O	--	PFA	--	--
Pradhan and Barik (1998)	1963-1992	Formal	O	--	GAA	--	TLI
Hulten and Srinivasan (1999)	1973-92	Formal	O	DD	--	--	--
Balakrishnan *et al.* (2000)	1988-1998	Formal	O	--	PFA	--	--
Trivedi *et al.* (2000)	1973-1998	Formal	O	--	GAA	--	TLI
Unni et. al. (2001)	1978-1995	Formal and Informal	VSD	SD	GAA	--	SI

Baghel *et al.* (2002)		Formal	O				
Ray (2002)	1991-2001	Formal	O	--	DEA	--	MI
TSL (2003)	1981-2000	Formal	O	--	GAA	--	TLI
Das (2004)	1980-2000	Formal	O	--	GAA	--	SI
B. Unel (2003)	1979-1997	Formal	O	SD	--	--	--
Goldar and Kumari (2003)	1981-1998	Formal	O	--	GAA	--	TLI
Kumar (2004)	1982-2001	Formal	O	--	DEA	--	MI
Trivedi (2004)	1980-2001	Formal	O	--	GAA &PFA	CD	TLI
Goldar (2006)	1981-1998	Formal	O, VSD & VDD	SD & DD	GAA	--	SI
Banga and Goldar (2007)	1980-99	Formal	O	DD	--	--	--
Marjit and Kar (2009)	1989-2001	Formal and Informal	--	--	DEA	--	MI
Kathuria *et al.* (2010)	1994-2006	Formal and Informal	VSD	SD	PFA- LP	--	--
Adil Mohommad (2010)	1970-03	Formal	O	DD	--	--	--
Raj (2011)	1978-2001	Formal	VSD	SD	DEA	--	MI
Raj and Babu (2011)	1984-2006	Formal	VSD	SD	DEA	--	MI

Notes: *(a) VSD and VDD - single deflated and double deflated value added respectively; NDP is net domestic product; CD - Cobb-Douglas Production Function; O - gross output and; (b) KI – Kendrick Index, SI – Solow Index, TLI – Translog Index and MI – Malmquist Index; (c) GAA – growth accounting approach, PFA- production function approach, PFA-LP - Levinsohn and Petrin methodology (d) CD – Cobb-Douglas, TL – Translog, CES – constant elasticity of substitution and SMAC – Solow-Minhas-Arrow-Chenery function.*

Source: *Own compilations*

2.2: Technological Change

2.2.1: Introduction

Economic historians, researchers and students of technology agree that technological change is the major determinant of very long-term economic growth. Lipsey has argued this in several publications, for example, Lipsey (1992, 1993 and 1994). Of course, technological change and investment are interrelated, the latter being the main vehicle by which the former enters the production process. If we knew no more than the Vedic age of India, Mesopotamians, or Medieval Europeans, our living standards would not be far above theirs, a bit more, owing to things such as capital accumulation but not much. Yet over shorter periods of time, there is debate over what proportion of economic growth is due to technological change and what to other forces, such as the accumulation of physical and human capital. Such debates imply that we are able to separate the effects of technological change from those of the other determinants. Economic growth, productivity, competitiveness, innovation, environmental protection and development have been associated with technical change, and it is, not confined to the manufacturing industries, it is also relevant in the agriculture where it plays a special crucial role to meet future demands.

Adam Smith identified the division of labour, free markets, and technical change in form of new machines as the three important causes of increasing incomes. Subsequent classical economists also found it crucial to development. Malthus thought that production could increase through accumulation of capital, fertility of land and use of labour saving inventions. He has pointed out that, when the existing land had been saturated food shortages and starvation would limit further growth of the population. According to David Ricardo the most productive areas of land would be occupied first and the quality of the land to be occupied subsequently would gradually decline. As a consequence diminishing marginal productivity would be encountered in expanding the land under cultivation until no further increase in output could be obtained. Marx (1887) argued that the growth of the capitalist system would eventually come to a halt. A crisis in capitalist society would result from inevitable decline in the rate of profit, which would be brought about by production becoming more and more capital-intensive. The increasing ratio of constant capital (capital equipment, building, *etc.*) to variable capital (wages) c/v, which Marx called the organic composition of capital, would lead to a fall in the rate of profit since surplus can be extracted from labour power. Marx, on the other way, clearly recognized the importance of technical innovations in the development of capitalist system. Like Marx, Marshall also was very clear as far as the importance of technical progress is concerned. In his 'Principles of Economics' he conceived the importance of 'knowledge' as the chief engine of economic progress. In the same way, Schumpeter argued that the fundamental impulse that sets and keeps the capitalist engine in motion comes from the new consumers' goods, the new methods of production or transportation, the new market, the new form of industrial organization that capitalist enterprises creates. However, technological progress was brought to notice in early 20th century by the work of Schumpeter, who sought in it explanation of the short-term instability and long term dynamics of the capitalist system.

However, in the 20th century, technological change became largely recognized as the key factor of economic growth, but the economic theory firstly treated it as a mere 'residual', the unexplained part remaining after the contributions of an increased quantity and quality of capital, labor and natural resources in output growth have been accounted for. However, Technical Change or Technological change, in this sense, represents the impact of exogenous forces, whereas some economic researchers have absolutely consider this residual of the growth models to be a measure of economic ignorance. Recently, the theory of economic growth reconsidered the nature of technological change and the concept of knowledge. Since technical progress is closely associated with the knowledge emerging from research and development (R&D) activities, endogenous technical change is considered to be generated by formal R&D activities. Therefore, the new growth economic theory included R&D as a factor of influence in the macroeconomic models. The endogenous or exogenous nature of the technological change refers to its source: endogenous change is internal to the national economy, being created by domestic private or public enterprise, while exogenous change is external, originating from foreign sources (Gheorghe Z. & Zizi G., 2010).

In the growth theory, technological change is considered to express the impact of new knowledge on the production function of the firm or industry or sector of an economy or nation as a whole. This new phenomenon may be exogenous if generated outside the economic process or endogenous if it emerges from within the economic process. The hypothesis of the exogenous character of the technical progress in the neoclassical models of economic growth was challenged earlier by new theoretical and empirical developments. Some of the researchers have attempted to demonstrate the endogenous character of the technical change (the term was first introduced by Lucas in 1966) both from a theoretical perspective (Arrow, 1962; Kaldor and Mirrlees, 1962) and from an empirical one (Aghion and Howitt, 1992, 1998; Romer, 1990). The endogeneity or exogeneity of technological change refers to its source. The exogenous technological change is generated outside the economic process, falling like manna from heaven (Scherer 1971, p. 347), while the source of endogenous technological change is the economic process itself, in response to profit and loss. Investing resources in R&D and others types of knowledge, such as human capital, generates innovative opportunities that explain endogenous technological change. Griliches (1979) was the first to introduce in the production function the R&D stock, computed as accumulated value of R&D expenditure after depreciation, to be one of the variables in his so-called knowledge-capital model. Many empirical studies that followed estimated production functions of the C-D type including various input indicators, and, as a rule, the expenditures for R&D, either aggregated or decomposed into components such as fundamental and applicative research, or private or governmental research (Griliches, 1980; Mansfield, 1980; Nadiri, 1980; Scherer, 1982; Terleckyj, 1974; Griliches & Lichtenberg, 1984).

2.2.2: Concept & Definition of Technological Change

In the beginning, Classical and so many Neo-classical Economists have considered the effects of technological change on labor composition and wages. Malthus, Karl Marx, and Ricardo all expressed concern about the effects of innovation, especially in the form of new machinery, on the displacement of labor. Since the 1970s the role of technological change in economic development has become a focal point of debate. During the 1980s there was a series of International Conferences and Seminars on long waves and long term economic development. But it was the Schempeter (1939) who, more than any other undertook major efforts to explain the cyclical pattern of growth largely in terms of technological and organizational innovations. Thus, the modern theory of the process of technological change can be traced to the ideas of Josef Schumpeter (1942), who saw innovation as the hallmark of the modern capitalist system. Entrepreneurs, enticed by the vision of the temporary market power that a successful new product or process could offer, continually introduce such products. They may enjoy excess profits for some period of time, until they are displaced by sub-sequent successful innovators, in a continuing process that Schumpeter called "creative destruction".

Schumpeter distinguished three steps or stages in the process by which a new, superior technology permeates the marketplace. Invention constitutes the first development of a scientifically or technically new product or process. (The Schumpeterian "trichotomy" focuses on the commercial aspects of technological change). Inventions may be patented, though many are not. Either way, most inventions never actually develop into an innovation, which is accomplished only when the new product or process is commercialized, that is, made available on the market. (More precisely, an invention may form the basis of a technological innovation. Economically important innovations need not be based on new technology, but can be new organizational or managerial forms, new marketing methods, and so forth.) A firm can innovate without ever inventing, if it identifies a previously existing technical idea that was never commercialized, and brings a product or process based on that idea to market. The invention and innovation stages are carried out primarily in private firms through a process that is broadly characterized as "Research and Development (*i.e.* R&D)" (Bound *et al.*, 1984, National Science Board, 1998, and OECD, 2000). Finally, a successful innovation gradually comes to be widely available for use in relevant applications through adoption by firms or individuals, a process labeled diffusion. The cumulative economic or environmental impact of new technology results from all three of these stages, which we refer to collectively as the process of technological change.

The literature pertaining to the economics of technological change is quite large and diverse in nature. Major sub-areas include: the theory of incentives for research and development [Tirole (1988), Reinganum (1989), Geroski (1995)]; the measurement of innovative inputs and outputs [Griliches (1984, 1998)]; analysis and measurement of externalities resulting from the research process [Griliches (1992), Jaffe (1998a)]; the measurement and analysis of productivity growth [Jorgenson (1990), Griliches (1998), Jorgenson & Stiroh (2000)]; diffusion of new technology [Karshenas & Stoneman (1995), Geroski (2000)]; the effect of market structure on innovation [Scherer (1986), Sutton (1998)]; market failures related to innovation and appropriate policy responses [Martin and Scott (2000)]; the economic effects of publicly funded research [David, Hall & Toole (2000)]; the economic effects of the patent system [Jaffe (2000)]; and the role of technological change in endogenous macroeconomic growth [Romer (1994), Grossman & Helpman (1994)]. Furthermore, a lot of other studies have suggested that there is a close correlation between technological development and productivity, since technological change affects the use of inputs engaged in the production process. Innovative capacity is one of the main factors which determine the growth level of the economy of a country or a region, as well as the competitiveness ability level [Quah (1996), Fagerberg *et al.* (1997), Freeman & Soete (1997)] and technological variables are able to explain a significant part of the diverging trends in the economic growth [Fagerberg & Verspagen (1996)] and productivity [Abramovitz (1986), Fagerberg (1988, 1994)].

The terms technological change and technical change are used interchangeably in the literature under review, both being indicators of a shift in the production function. It would have been useful to reserve the latter term for indicating change in techniques or processes. The terms technological progress and technical progress are synonymous with technological change and technical change respectively, all change being considered as being for the better. Technological change can be understood in terms of technological evolution. As technological advancement has played a crucial role in industrial development, almost every nation is concerned with monitoring technological change. Technology policies such as research and development (R&D), technology planning, technology management, *etc.* are related to this concern. Thus, Technological advancement has been a major source to economic growth in the rapidly growing economies and the forces shaping technological progress are largely economic. The contribution of technical change to output growth has been obtained as residual. Over a substantial period of time since the path breaking article of R.M. Solow has appeared researchers in empirical economics treated technical change and total factor productivity alike. However, total factor productivity may be defined as the output per unit of combined inputs foregone during the course of production.

$$TEP = \frac{u}{F}$$ (Where TFP: Total factor productivity, u: Observed output, and F: Input index).

(Where TFP: Total factor productivity, u: Observed output, and F: Input index).

But, technical change refers to advances in knowledge in relation to art of production. Technological change reflects the ability to get more output from the same amount of inputs. It can be reflected by: (i) an increase in an index of quantity of output per unit of input, or (ii) a change in production function parameter. The chief tool to measure technical change is the production frontier, which relates inputs with outputs subject to best practice technology. Changes in technology are due course of time bring about changes in productivity. Due to technological progress, productivity rises and the produced output is generally in the shape of new products rather than old products (for instance, synthetic fabrics instead of cotton cloth). We may identify it also by one of the following three ways: (a)

Output increases for the same inputs. (b) Same output is realized with less input. (c) Output may not rise but quality and durability of good as well as working condition improve.

Thus, in economic term, Technical change or technological change or technological progress is a shorthand expression for any kind of upward shift in the production function indicating improvements. It is the pool of knowledge relating to art of production and signifies creation of new set of production alternatives. While in general, term used to describe incremental change in the quality and quantity of knowledge and ideas that are applied in the stream of activities to enhance the social and economic well being of the society. Due to the positive nature of the implied change, it is also referred to as technological progress. Scientifically it has been realized that Technological change occurs through the process of invention, innovation and diffusion (which has been described thoroughly later on in this section separately) that leads to the transformation of ideas and knowledge into tangible products that have high utility value to human needs. The effect of technological change propels economic transformation; a change in the structure of an economy over time from a lower, rudimentary and subsistence level to a higher and more sophisticated level of economic activities. Technological change has been defined directly or indirectly in terms of its effects on productivities of inputs in terms of Q/L and Q/K ratios, when technological progress takes place, both the ratios may increase and one of them relatively rises. Hicks (1932) classified technical change as a neutral and non-neutral. Technical change is neutral, if the marginal rate of substitution between inputs is not affected. Non-neutral technical change is generally described as either labour-saving (capital-using) or capital-saving (labour-using). Technical change is said to be labour saving if the marginal product of capital rises relative to marginal product of labour.

Solow defined technical change as a "catch-all "expression for any kind of shift in production function assuming returns to scale, homogeneous inputs and competitive equilibrium. According to him, any increase in output not explained by increase in capital and labour is assigned to technical change.

Ruttan (1960) stated that technological change has traditionally been defined in terms of changes in the parameters of a production function or the creation of new production function. The traditional procedure has been to use a partial productivity index (average output per unit of labour/capital) or total productivity index (output per unit of total input) as a measure of the impact of technological change.

Harrod (1961) defined technical change as capital-saving or labour-saving according to whether capital/output ratio decreases, remains unchanged or increases with a constant rate of interest.

Schmookler (1966) defined technological change to denote the art or producing new knowledge and technical change to incorporation of this knowledge in the production process of firms. In the literature no much distinction seems to be kept in view.

According to Srivastava *et al.* (1975) the technological innovations in agriculture can be divided into two broad types, *viz.*, Biological and Mechanical. Biological innovations refer mainly to inputs that increase the productivity of a given land base. High yielding plant varieties and fertilizer are the examples. Biological innovations are found to raise total farm cost. Mechanical innovations mainly are those that cause a reduction in total costs, while biological innovations are labour-saving. Green revolution is frequently described as a seed-fertilizer technology. In a sense it falls in the class of biological innovations.

Shaw (1979) defined a technological innovation as concrete identifiable new factor of production material as well as non-material to which the increase in production is attributed and which is not explained by the traditional factors of production. The discovery of high yielding variety of seeds, the package of practices for realizing their production potential, mechanization of agriculture are regarded as technological innovations in agriculture. There are technological innovations in farm mechanizations, post harvest technology, milk and poultry production, frozen semen technology.

Nair (1980) defined technical progress, as those changes in the production processes which reduce the marginal cost of output. This change can occur either employing the existing inputs but in different composition (a change in technique) or by introducing new factors of production either for replacing old ones or simply as additional inputs (technological innovation). Thus the technological change in either case is associated with a shift in the production function which describes the technical relation between output and inputs. Shifting production functions with rising marginal productivities are more appropriate in a dynamic situation. An increase in productivity may come through changes in the quality of inputs or via through changes like improvements in marketing, increased managerial efficiency *etc*. Shifts of production frontier are attributed to technical change. Thus, failure to accommodate new technology is reflected in the lack of frontier shift over time.

While Adam B. Jaffe *et al.* (2003) in their introductory chapter articulates that measurement of the rate and direction of technological change rests fundamentally on the concept of the transformation function,

$$T(Y, I, t) \leq 0$$

(Where, Y: represents a vector of outputs, I: represents a vector of inputs, and t is time).

Above Equation describes a production possibility frontier, that is, a set of combinations of inputs and outputs that are technically feasible at a point in time. Technological change is represented by movement of this frontier that makes it possible over time to use given input vectors to produce output vectors that were not previously feasible.

Lipsey & Carlaw define technological knowledge as the idea set that specifies all activities that create economic value. It comprises knowledge about product technologies, the specifications of everything that is produced, process technologies, the specifications of all processes by which goods and services are produced; and organizational technologies, the specification of how productive activity is organized. All these are often referred to as just technology, and they had followed that practice whenever there is no ambiguity. While, Heady and Denison claim that technological change accounts for all the growth in output that not accounted by growth in physical input use. They measure an exogenous shift of isoquants where input measures are unchanged. They include changes in quality of input as part of technological change and measure use of input in each single period.

Schultz-Jorgenson and Griliches claim that technological change reflects changes in production functions that cannot explained otherwise. Specially, technological change reflects changes that cannot be explained in terms of increased quality and/or quantity of inputs. Technological changes are measured in input {output ratios when quality changes are taken into account. The Schultz approach enlightens two manifestations of technological change: (i) Increase in knowledge of the production process; (ii) Improvement in input quality.

While Chappellin & Nijkamp (1990) define it as –"Technological change is, as a consequence, increasingly considered as a fundamental force in shaping the spatial patterns of transformation of the economy. Technological innovation does not, however, take place in uniformly on a nation-wide scale. Specific places provide a specific incubator function for knowledge generation and specific new activities.

Accordingly, Peterson and Hayami define technological change as "the phenomena of input quality improvements for an increase in knowledge leading to increase in output per unit of input." Considering Technological Change views expressed as above, they tries to generate a compromise in the embodied {disembodied controversy}. For years, economists have asked: Do technological changes embody themselves in new machinery or are they slow processes of change in production techniques? Again, do technological changes arise from better input or better management?

Mainsfield (1968) offers its direct definition as follows: Technological change is the advance of technology, such advance often taking the form of new methods of producing existing products, new designs which enable the production of products with important new characteristics and new techniques of organization, marketing and management.

However, technological change is often conceived through its indirect definition, Schumpeter, to whom technological change was synonymous with innovation, gives its indirect definition as: we will now define innovation more rigorously by means of the production function. This function describes the way in which quantity of products varies if quality of factors varies. If instead of quantities of factors we vary the form of the function, we have an innovation. Schumpeter laid a great stress on discontinuous nature of technical progress, a process characterized by major breaks, giant discontinuities that startle the society with appearance. Against this is a school of thought that lays emphasis on continuous nature of technical progress. A major proponent of this school is Usher, who feels that major inventions emerge from cumulative synthesis of relatively simple inventions, each of which requires an individual act of insight. Therefore, the continuously occurring large number of changes of small magnitude results in a single big change. But Schumpeter's emphasis on major breakthroughs had a great impact on the entire generation of economists concerned with productivity and technological change.

Testing empirically, with the tool of production, various economists found technological change to be the most important source of output growth. But what greatly proved to be decisive were the findings of Fabricant, Abramovitz and Solow. They find out that 80 to 90 percent of output growth per head in the U. S. economy could not be accounted for by increase in capital per head, thereby, giving this credit to technological change. Many other studies have also reached a similar conclusion.

Denison, with a slightly dissimilar approach, where he decomposed the sources of growth into a large number of elements, also reached the conclusion that technological change is very important factor of growth. However, in the study technological change contributed only about 40 percent (as against 80 – 90 per cent of above mentioned studies) of the total increase in per capita income. Apart from the studies on the United States economy, studies for Norway, Finland, and Great Britain experienced almost the similar conclusions.

However, it would be wrong to ascribe this large contribution to technological change alone, as there are some other factors also contribution to growth. These are technological progress in the narrow sense, economies of scale, external economies, improvement in health, education and the skill of the labour force, better management and the movement of labour force form low productivity to high productivity sectors. This is the reason why the names given to this group have ranged from output per unit of input, efficiency index, TFP, change in productive efficiency, technical change, and the measure of our ignorance. To emphasis the nature of this concept, it has also been called a residual. So far Domer has seen essentially four methods of expressing the Residual: (i) it is defined as the difference between the values of output and inputs in constant prices (by Davis 1955). (ii) it is ratio between arithmetic index of output and of input, in the works of Schmooker, Abramovitz and Kendrick; (iii) the residual is the ratio between an aggregate arithmetic index of output and inputs embodied in a linear homogeneous production function the work of Solow; (iv) finally, the Residual, or more correctly, its relative percentage rate of growth, is the weighted arithmetic average of relative changes in input coefficients between two points of time derived by Leontief from his input – output studies.

Therefore, attempts were made in the subsequent researches to firstly, disaggregate the residual factor measuring the factor inputs in conventional manner; and secondly, L and K inputs were adjusted for changes in quality and composition so that much more measured could be attributed to increase in factor inputs. Denison's study has been a major attempt on disaggregating the residual. He decomposed the sources of growth into a large number of elements. Apart from taking into account the contribution of factor input, factors affecting the quality of labour, such as, education shorter work days, *etc.*, were also incorporated. Among his most significant findings in estimating the components of residual were the importance of advance in knowledge and the role of economies of scale.

Making a distinction between the 'embodied' and 'disembodied' technical changes, between innovations that are incorporated in new capital goods and labour force and those that are not carried out the second line of thought. Earlier, in the works of Solow, Abramovitz and others, the technological change was taken as an unexplained residual meaning, thereby, that it is *disembodied*. But, further researches led by Solow attempted to search for a solution were technological change is embodied in the latest vintage of capital (*i.e.* the current year investment). Analogously, technological change can said to be embodied in the latest vintage of labour (*i.e.* improvements in the productive efficiency of new workers due to education, training, *etc.*).

Apart from making a distinction between embodied and disembodied technological changes, a distinction is also made between exogenous (or autonomous) and endogenous (or induced) innovations. The first view treated technical change as an unexplained phenomenon, an increase in production being simply attributed to the passage of time. In the endogenous technical change the expansion of technical possibilities is explained explicitly by one or more economic factors, such as, (a) long term changes in prices of the factors of production, (b) learning process concerning production, and (c) investment in education and research. People developed the notion of a "learning curve" where output/labor = f(time) or output/labor = f(output); f is found to be logarithmic. Arrow developed accordingly this theory of "learning by doing" in 1961.

The new technology is considered embodied in the machinery that need to be bought. Technological change reflects itself in the putty stage every year the possibilities in terms of machinery are better. Note however that these two theories {embodied vs. Disembodied are more complementary than contradictory. It is true that new machinery, embodying new knowledge, is introduced constantly, but use efficiency improves with experience, as the user acquires new managerial ability.

A second important issue relates to the direction or technological change. With technological change, is the neoclassical production function (the ex ante one, la Salter) changing in a neutral or a biased manner? Technological change can he described as a shift of the unitary isoquant toward the origin (one unit of output can be produced with less inputs), but the shift can be \parallel" (that is, the quantity of all inputs declines proportionally to their use) but can also be biased (some inputs reduce more than other). There are three measures of technological bias. All three measure what occurs to the ratio $F_K K/F_L L$ along a certain line and can be interpreted as measures or what happens to the (share of capital)/(share of labor) over time along that particular line:

5. Hicks measures bias along the fixed capital/labor ratio
6. Harrod measures bias along fixed capital/output ratio
7. Solow measures bias along fixed labor/output ratio.

New – Classical economists viewed technological progress as exogeneous variables treating it as *'manna from heaven'*. But it was subsequently that 'technical progress' does not occur by accident but through deliberate diversion of resources to activities, which generate progress in pursuit of fame and profit, or both. In the modern growth models, Kaldor and Mirless followed by others have tried to incorporate technological change as an endogeneous variable into the economic system. Kaldor and Mirless has related technical advancement to capital accumalation where knowledge grows with learning, which is related to investment, and thus, learning is made a function of proportionate rate of growth of investment. Arrow in his model, however, has made learning a function of past gross investments rather than the proportionate rate of growth of investment.

According to Salter, technological changes are neutral. He agrees that observed data indicate that K/L ratio is increasing over time, but he explains it as the consequence of the increase in the relative price of labor (the economy is moving along the isoquants towards a more intensive use of capital, but the shape of the isoquants remains unchanged). He states that manufacturers want to minimize cost and do not care if this is due to labour or capital saving. Therefore, they will encourage research possibilities that reduce both capital and labour requirements.

2.2.3: Factors Governing Changes in the number of Technology Elements

Evolutionary economists and modelers discuss a variety of factors or conditions that impact the number of inventions and innovations. These include, *e.g.*, (Fagerberg 2000):

- Entrepreneurial drive
- New innovations tend to facilitate or enhance others
- Market-driven environmental selection processes
- The persistence of technical paradigms and paradigm shifts
- Feedback loops between developers and between innovations
- Social, institutional, and political factors
- Firm behavior and routines
- Increasing returns to scale
- R&D funding and size of firms
- Degree of variation in technologies and knowledge
- Vintage and rate of capital investment
- Chance

A smaller set of factors that appear to have academic support in economic and modeling literature as having a role in new technology developments discussed here as below:

1. Quantity and quality of scientific and technical personnel and other resources available for development-
    - R&D funding and size of firms
2. Risk and rewards associated with development-
    - Social, institutional, and political factors
    - Increasing returns to scale Competition between developers

  - Entrepreneurial drive
  - Market-driven environmental selection processes
  - Firm behavior and routines
  - Vintage and rate of capital investment
3. Competition between developers
4. The current state of technology
  - New innovations tend to facilitate or enhance others
  - The persistence of technical paradigms and paradigm shifts
  - Feedback loops between developers and between innovations
  - Degree of variation in technologies and knowledge
  - Chance

2.2.4: Some Theories to Explain Technological Change

To understand and explain technical change, a combination of economic and sociological theory is necessary. We shall focus on a few recent and fruitful theories, which have the additional advantage of being relevant for the question of orienting technology. They contrast sharply with traditional economic theories.

Mainstream economics tends to treat technology as an exogenous variable, which does not have to be studied itself. When technological change is included in economic analysis, it is treated abstractly. Technical change may be treated as shorthand for any kind of shift in the production function. In other words, if economic growth cannot be explained by other economic variables, it is, by definition, the result of technical change. At the macro-level, technical progress then appears as the so-called residual.

Within mainstream economics there have been two attempts to indigenize technical change: the theory of induced innovation since the 1960s (Kamien & Schwartz 1968, Binswanger 1974) and new growth theory since the 1980s. In induced innovation models, technical change is assumed to respond to changes in relative prices and thus be directed toward economizing the use of a factor which has become relatively expensive. In endogenous growth models, technical change derives from research (leading to designs, blueprints, and general knowledge) and human capital accumulation, and is modeled as a stock variable, with spillovers to other factors of production (Romer 1986, Lucas 1988; for surveys, Verspagen 1992, Schneider & Ziesemer 1995). Compared with the features of technical change discussed in the preceding subsections, these models only explain incentives for firms to do research under different conditions.

Nelson & Winter (1982) and other evolutionary economists have struck out in a different direction to indigenize technical change. Their theories are an alternative to neoclassical economics (See Dosi *et al.* 1988). Nelson & Winter developed a dynamic picture of firms and an evolutionary theory of economic change in which an evolutionary theory of technical change was embedded.

Thus, they combine the focus on firms with a perspective on technology. The starting point of the theory is uncertainty (Nelson & Winter 1977). Firms do not know beforehand which technology will be successful; they even lack the possibility to check all technological alternatives and, as a result, their behavior should not be understood as maximizing. Instead, firms have heuristic search routines to which they hold for while. In later studies, the further point was made that, even within firms, search processes are also informed by technological paradigms (Dosi 1982), or technological guideposts (Sahal 1981), which are available at the level of the sector or of technological communities. This indicates that sociological explanations have to be added to the economic theories.

In the quasi-evolutionary approach (Van den Belt & Rip 1987, Rip 1992, Schot 1992), heuristic search practices leading to technological options, artifacts, or transferable skills (embodied knowledge) relate to shared repertoires embedded in an organization, in a community of technical practitioners, or in an inter-organizational network. The variation is not random, but guided by heuristics and by other promises of success. Hughes (1983, 1987), in his study of electric power networks has shown how network builders bring together social as well as technical elements in order to make the environment part of the system. An interesting finding is that the dedicated network (or system) builders are different in different phases: inventor-entrepreneurs, engineer-entrepreneurs, and financier-entrepreneurs. In parallel, an inherent logic of the system develops, with a momentum, a drive toward expansion (*e.g.*, because of load factor requirements), and the need to overcome obstacles in the expansion.

The socio technical approach has been used to address contemporary issues of large technical systems (Mayntz & Hughes 1988) and to understand telecommunication networks (*e.g.*, Schmid & Werle 1992). Stankiewicz (1992) argues that a qualitative change occurred in the 1980s from technology as local concrete systems to technology as a global socio cognitive system in itself. Technological activity is increasingly self-referential; the list of priorities is more and more derived from the needs of the global system than from the local systems. A similar interest in socio technical dynamics of heterogeneous actors, that is, without a limitation to firms, is visible in so-called actor-network theory. Case studies focus on the interactions between the actors and evolving techno-economic networks (for case studies, see Callon 1986a, b, Law 1987, Law & Callon 1988, Mangematin & Callon 1995, Latour 1996). Recently, the approach has been used to develop instruments for strategic analysis and policymaking (Callon *et al.* 1992; OECD 1992).

Thus, the advantage of the evolutionary, quasi-evolutionary, and socio technical explanations is that, in contrast to the neoclassical explanation, they also cover public technologies and the increasingly important public and private settings of technical change. Socio technical theories are especially flexible, because the context within which novelty creation occurs is not specified beforehand. This flexibility is also a disadvantage, however. Without any specific theoretical structure, researchers fall back on case studies, and, as it turns out, on heroic storylines, where novelty is attempted against overwhelming odds. In evolutionary and quasi-evolutionary theories, no such storyline can be followed. Dynamics and outcomes cannot be separated. The emergence of irreversibility and the relative non malleability of technology are outcomes, but at each moment they are also slices of dynamics. However, managing technology requires an understanding of the relation between actions (induced dynamics) and their outcomes. A theory of prospective technology dynamics is needed, even if the contingencies involved reduce any hope of arriving at determinants or factors of successful direction. To arrive at such a theory is methodologically complex because of the retrospective bias involved in using history to explain the present. Historians have highlighted the shifts in actual developments. Sociologists and political scientists have pointed at factors modifying or overriding immediate economic considerations.

2.2.5: The Classification of Technological Change

While discussing technological change, generally it is assumed that there is an aggregate economic system and are only two types of inputs L and K which are homogeneous within themselves. Analogous to the concept of unitary elasticity of demand, economists make use of the notion of "neutrality" to serve as a criterion to measure bias of technological progress. Neutral technical change means, those innovations which or neither capital, neither saving nor labour-saving *i.e.*, they are neutral in their effect in the use of the two factors at the margin. They raise the productivities of both L & K (MP_L, MP_K) proportionately. When neutral technical change occurs, it so happens that the rise in K/L is just equal to the rise in O/L so that Q/K ratio remains constant. However, in reality at firm or industry level, such proportionate increase is rare and hence technical change is biased *i.e.* it favours the use of relatively more of capital or labour.

It is posited that the firm can choose between 'neutral' technological changes and 'non-neutral' or 'biased' technological changes. If, as is supposed throughout the subsequent discussion, labour and capital are the sole factors of production, then those changes in the production function which leave the optimal capital labour ratio corresponding to each factor price combination unaltered will be referred to as "neutral" technical changes. "Non-neutral" or "biased" technical change would be either "labour saving" or "capital-saving" depending on whether the optimal capital-labour ratio corresponding to each factor price combination is increased or decreased.

However, the interests in classifying technological change as labour-saving, capital-saving and neutral stems historically from a concern for its effect upon the distribution of income. The terms neutral, capital – saving and labour– saving technological change were introduced by Pigou. However, as a matter of fact, these terms as defined by Hicks, Harrod and Solow respectively, are more frequently used in the literature. Hicks introduced the classification of technical progress associated with his name in his "*Theory of Wages*" as: "We can classify inventions according as their initial effects are to increase, leave unchanged or diminish the ratio of the marginal products of capital to that of labour." We may call these inventions labour–saving, 'neutral' and 'capital-saving', respectively. Technological change is Hicks-neutral when the relationship between the marginal rate of substitution and factor proportion is unchanged. By contrast, technical change is called Harrod-neutral when the relationship between the capital output ratio and the interest rate does not change, and Solow-neutral, when the relationship between output per worker and the wage rate is invariant.

The Hicks classification is biased upon the comparison of points at which the K/L ratio is constant. Technical progress is said to be Hicks-neutral if, at any constant value of the K/L ratio, the ratio of relative shares remains constant. Technical progress is said to be labour-saving in Hick's sense if, any constant value of the K/L ratio, the ratio of relative shares is increasing (*i.e.* the rate of change of relative shares is positive) While technical progress is said to be capital-saving in Hick's sense if, at any constant value of the K/L ratio, the ratio of relative shares is decreasing. Hicks classification of technological change is based on short-run where input supply is fixed, if we concentrate on the immediate effects of technological change, but Harrods classification of technological change is related to long run, where supplies are flexible in the long-run impact of technical change.

Harrod classification of technical progress associated with his article "*Towards a Dynamic Economic*". His concept of neutrality pertains to long-run equilibrium, which has relevance in the recent theories of golden age (steady growth). The long run equilibrium generally defined in terms of a given Δ which even after changes in input prices, is assumed to be the same before and after technological change. Technological progress is said to be Harrod-neutral if, at any constant value of the capital-output ratio, the ratio of relative shares remain constant (it means the marginal product of capital remains unchanged). Harrod's classification compares points at which the capital output ratio is constant as opposed to the Hicks procedure, which compares points at which the K/L ratio is constant. Technical progress is said to be labour-saving (or capital-saving) in Harrod's sense if, at any constant value of the capital-output ratio, the ratio of relative shares is increasing (or decreasing). A third system of classification, usually attributed to Solow is identical to that of Harrod and Hicks except that it compares points on the new and old production functions at which the labour-output ratio is constant. It can be shown that Solow neutral technical progress corresponds to what we have called purely capital augmenting technological progress and that the only circumstances in which it is equivalent to Hicks and Harrod-neutrality is when the technology can be represented by a C-D aggregate production function. Analogously, capital augmenting neutral technical progress has been suggested by Solow. Here labour-output ratio remains constant at a constant wage rate.

Thus, Beckmann and Sato in their study on aggregate production functions have generated following classes of technological progress, these are defining as under:

1. Product augmenting

   1.1. Hicks Neutrality

   $$Y = A(t)\,F(K, L)$$

   (Where, Y = Output, K = capital, L = Labour and A (t) = technical progress).

Here product is increased in proportion to the output that would be obtained in the absence of technical change.

1.2. Labour additive

$$Y = A(t)L + F(K, L)$$

The increase in product is here proportional to the amount of labour used.

1.3. Capital additive

$$Y = A(t)K + F(K, L)$$

The increase in product is proportional to the amount of capital used.

2. Labour Augmenting

2.1. Harrod neutrality

$$Y = F(K, AL)$$

2.2. Labour combining

$$Y = F[K, A(t)K + L]$$

The augmentation of labour – as measured in efficiency units – is proportional to the amount of capital used.

3. Capital augmenting

3.1. Solow neutrality

$$Y = F(AK, L)$$

3.2. Capital combining

$$Y = F[K + A(t)L, L]$$

The augmentation of capital is proportional to the amount of labour used.

4. Input decreasing

4.1. Labour decreasing

The inverse production function is of the form

$$L = G(K,Y) + C(t)Y \quad dC/dt < 0$$

Where,

C (t) is decreasing with time. The reduction of the labour input is thus proportional to output Y.

4.2. Capital decreasing

This is the symmetric case when the inverse production function is of the form

$$K = H(K, Y) + C(t)Y \quad dC/dt < 0$$

Here the reduction of capital is proportional to output.

Thus, technical progress may operate on the input side or on the product side and its effect may be proportional to any of the basic magnitudes: output, labour, or capital input. Combinations of these types may also arise.

5. Factor augmenting technical progress

$$Y = F[A(t)K, B(t)L]$$

This combination of Harrod and Hicks, or Solow and Hicks neutrality is obtained when the capital output ratio is a separable function of labour's share and of time.

$$\frac{Y}{K} = A(t) \text{ o (Share)}$$

It indicates that, factor augmenting technical progress can be obtained in a more general way form the invariant relationship between the share and the elasticity of factor substitution.

Among the types of technological progress that we have obtained in this way are several, which are well known in the literature: Hicks, Harrod, Solow, and factor augmenting technological progress. They have defined these concepts of bias for disembodied technical change. Consider now that technical change is embodied. At the aggregate level, technological change may be embodied *(as viewed by* Kaldor, Arrow, and Solow *of 1960)* or disembodied *(as conceive by* Solow *of 1956)*. Generally, embodied and endogenous technological change goes together. The new machines produced are superior which incorporate innovations and various improvements and new internal (endogenous) to the economic system. Arrow's conception of learning by doing and Kaldor's technological progress functions are the classic examples of endogenous–embodied technological change. Embodied technical change refers to the introduction of new technology in the physical capital inputs. According to Burmeister and Dobell (1970) the new technology is built into or embodied in new capital equipment or newly trained or retrained labour. While disembodied technological change is defined as if, independent of any changes in the factor inputs the iso-quant contours of the production function shift towards the origin at time passes. In contrast to embodied technological change, most of the neo-classical regard technological change as outside the machines and external to the system. Disembodied progress is seen in the shape of improvements in organization, material as labour handling, new management techniques, *etc.* Thus, conclusion emerges that disembodied technological change is automatic but embodied is not.

2.2.6: Changes in Technology or Technological Advances

After great industrial revolution, Technology has becomes the single greatest factor that distinguishes modern economies from primitive ones. Because technology lowers the cost of production and provides new products, it increases both productive and allocative efficiency for all firms. However, technology is unpredictable. No one knows what will be discovered where or when. Indeed, it is very difficult even to measure the pace of new technology. In economics, the short run is considered a period time during which firms can change variable inputs but not fixed costs. The long-run is considered a long enough time period so that firms can change even fixed costs. Some economists try to take a very long run view in which technology changes, but it is very difficult to form such an analysis, even using statistics. Although it is called a very long run view, the time period could actually be short, depending on how fast technology is changing within an industry. Lipsey & Carlaw first consider the total factor productivity (TFP) approach to measuring changes in technology. They further argue that, contrary to widely held views, changes in total (or multi-) factor productivity do not measure technological change. Ideally, they measure only the super normal gains associated with such changes. The changes in technology itself are brought about by the combination of activities of research, invention (or knowledge creating activities), development and innovations (or the activities applying knowledge to production process), and the spread of technology depends upon the rate of its adoption and diffusion.

2.2.6 (A): Discovery-Invention-Innovation-Diffusion Mechanism

It has been seen that Research and development (*i.e.*, R&D) and technological innovations continuously create new knowledge and uncertainty. Schumpeter defined three phases in the process of technological change: invention, innovation and diffusion. But, considering pure and natural science together with social science, there are four processes to technological advances in an economy: discovery, invention, innovation, and diffusion.

First and foremost is known as Discovery which involves the elucidation of the fundamental processes of nature through observations of nature, reasoning, and experimentation. Science is the branch of knowledge that seeks to understand the fundamental nature and processes of the universe. Indeed, advances in science involve what is called the scientific method, first, a hypothesis is formed to explain some observation, and then an experiment is designed to test the hypothesis. The validity of the hypothesis is determined by the reproducibility of the experiment and how well it coheres with previous knowledge. When there is significant confidence in the correctness of a hypothesis, then it is often referred to as a Theory. Of course, economics is one of the sciences where testing hypotheses is not feasible in most cases, so most advances in economics are made by assessing how well hypotheses explain or predict the economy. Although discoveries are important for providing the fundamentals of knowledge, most businesses do not invest money to make discoveries because discoveries are often serendipitous and they rarely apply to a particular business, which is why a large part of science is financed by the government. Instead, businesses spend money on research and development, applying previous knowledge to either produce new products relevant to their business or to improve methods of doing their business, where results are more predictable and more pertinent. Moreover, most governments allow the business to patent the new product or process as an invention.

Invention has been considered as the first step to technological advantage by Schumpeter and he defined it as: the discovery of product or process of producing by using imagination, thinking and experimenting. Invention is a process and the result of it is also called invention. Invention is based in scientific knowledge and it is the result of work of individuals who work on their own or as members of Research and Development (R&D) departments in firms. Government encourages invention by providing patents, right to sell any innovative process of production, machines or products in a set time. Using Freeman's (1974) definitions, invention is can idea a sk*etc*h or a model for a new improved device, product process on system. Inventions may be defined as new ways of attaining given ends. They would include, therefore, the creation of things previously non-existent, using either new or existing knowledge: and also the creation of the things that have existed all the time, only these were to be discovered.

Innovation, on the other hand, in the economic sense, is accomplished only with the first commercial transactions involving the new product, process, system or device. Unfortunately, innovation is often used to describe either the whole process of technological change or an act that is original to the decision-maker under discussion but not to the economy as a whole. Thus, Innovation is directly related to invention. While invention is *"discovery and proof of workability"*, innovation is the successful introduction of new product (invention) in the market, the first use of a new method of producing, or the creation of new form of business firm. There are two types of innovation: product innovation, improving products and services, and process innovation, which is improved ways of production and spreading of these inventions in the market. In contrast to invention, innovations cannot be patented. Innovation needs not to weaken or destroy the existing firms. Because new products and processes threaten firms' survival, existing firms have a high incentive to engage into research and development (R&D) process continuously. These innovative products and processes enable firm to earn higher revenue or to maintain the present ones. Innovation can strengthen or weaken market power.

The imitators in the same field will eventually follow early innovations. This is the third and the most important step of technological change, *i.e.*, the diffusion stage, which occurs after the invention and innovation stages and refers to the process by which the innovation spreads across the market. In other words, Diffusion is the spread of innovation to other firms so that they can remain competitive. This diffusion occurs by either emulating or copying others' products or processes, which is achieved in several ways. First, a company can look at any patents, which are available for inspection by anyone, to see the essential design, and develop new ways of working around that, developing ideas or methods to achieve the same functionality but without infringing the patent. Since patents are narrow in scope, a patent may reveal ways around it, enabling a competitor to develop a process or product that is functionally equivalent to the patented product or process. In other cases, competing firms can reverse engineer the product to see how it works and to see how it could be improved. Thus, Diffusion is the process of spreading of inventions through imitating or copying. To take the advantage of new profits or to slow down disappearing of others, all firms try to implement the innovations. In most of the cases, innovation leads to widespread imitation (that's diffusion) of inventions. For example, soon after McDonald's introduced the fast-food hamburger, Burger Kings also started to produce it, since it offered high revenues for the firms that supplied this good.

Schumpeter made a sharp distinction between invention, innovation and imitation, with the greater emphasis being laid on the innovation and the charismatic figure of the entrepreneur who against all odds takes the bold step to innovate. To him innovation could take the form of a new product or a new process, opening up of new markets, the acquisition of new source of raw material, or a structural reorganization on an industry. Technological innovation thus, is a vital component in the performance of firms and of other organizations. That is not to say that the only effect of innovation is to make firms (or industry) perform better. The greater efficiency that firms achieve as a consequence of various types of innovation is only an intermediate step towards the achievement of economic growth and economic development. However, if one looks at technological change as more or less a continuous process, the distinction between these stages (*i.e.* invention, innovation and diffusion) becomes rather blurred.

The innovation process is sometime referred to as additions to the economy's book of blue prints; however to confuse matters further, this blueprint terminology is sometimes used to refer to invention, and then only certain blueprints are selected for development to the innovation stage. However, once we have innovation, these start to have an impact on the economy as they are used or produced. This is the process of diffusion. As diffusion proceeds we may think of the disembodied production function shifting. Alternatively, we may think of the embodied production function being different for different vintages because of innovation, and the process of diffusion concerns how machines of vintage V takes over from those of other vintages. The differences in the rate of technical change between different industries in an economy and between different economies depend upon the speed of initial adoption and the speed of diffusion.

Website development can be considered as a best example of diffusion. For instance, Groupon developed a new way for firms to market their products and services by selling what is known as a group coupon over the Internet, which gives the consumers a steep discount if enough of them subscribe to the product or service. Now, many other companies have also started offering major deals, many of them offering variations of the group coupon. Businesses engage in research and development to develop inventions, innovations, and to profit from diffusion. However, firms generally do not invest research efforts to find new discoveries, because they cannot be patented nor do they necessarily have an obvious business application, making it unlikely that a discovered process can be used to develop a new product or new process that would allow the firm to earn more profit. For this reason, governments generally subsidized basic research.

Thus, on the basis of above discussion, it can be summarized that Technological change occurs through a three chain relationships- invention, innovation and diffusion. Taking single line definition of all these three mechanisms of technological change, we can say that Invention is the creation of an item based on original ideas and knowledge more often described as "breakthrough" technology. Innovation is additional creativity that improves the features and usefulness of invented products. Diffusion refers to the spread of technological knowledge into various streams of economic activities that expands the space for further creativity to amplify the chain mechanism of invention, innovation and diffusion. Each aspect of this chain involves the appropriation of ideas through direct acquisition, "learning-by-doing" and R&D. Innovation is the pivot of this chain relationship in that innovation inspires invention and motivates diffusion, implying that both invention and diffusion posses some attributes of innovation.

2.2.6 (B): Research & Development

Research and development spending may be considered as the monetary equivalent of inputs to the process producing technological advances. R&D spending is an input to both invention and innovation. However, to the extent that R&D tends to be directed more towards development than research, we might argue that an analysis of R&D is concerned more with innovation than with invention. As already pointed out, research is the knowledge activity, whereas development is devoted to the capacity to produce. A new function, R&D becomes institutionalized in firms beginning at the end of the 19th century. The institutionalization of R&D has been one of the most central changes in the way in which firms compete and change their technologies (Freeman, 1982). R&D can be categorized into three branches:

(i) **The basic research:** that aims at the original investigations for advancement of scientific knowledge, without any commercial objectives;

(ii) **Applied research:** that incorporates investigations which are directed towards the discovery of new scientific knowledge having specific commercial objectives with respect to products and processes; and

(iii) **Development**: that is directed towards the technical activities of non-routine nature concerned with translating research findings or other scientific knowledge into products and processes.

The developed country especially in the field of applied research and development generally incurs heavy expenditure on R&D. The impact of R&D activity on growth is difficult to measure; many times falling to create anything new but when successful, spectacular results are achieved. The development of organised R&D in the modern firm has brought about a dramatic change in the capacity of industry to generate a stream of new and modified products, and to use variations in this capacity as a competitive weapon between firms and between national economies. The institutionalization of R&D has changed but it has not solved the problem of how to innovate. However, the institutionalization of R&D has had a number of important consequences for the way in which we think about technological change. First, we can no longer consider inventions and innovations as exogenous to the economic system. Resources have to be allocated to particular R&D projects and, therefore, the direction of inventive and innovative activity is decided form the very outset based on economic criteria. Second, since R&D generates new knowledge and new uncertainty, its existence is incompatible with the assumptions of perfect information and equal technology contained in the neo-classical theory of the firm. Third, the fact that firms choose between R&D projects implies that there is a spectrum of R&D strategies, which have become important components of overall firm strategy.

2.2.7: Technological Change Biased Effects on TFP

The economics of innovation and knowledge has basically ignored the legacies of the economics of technological change, as much as the economics of technological change has ignored the tools and the heuristics of the economics of innovation and knowledge. Little effort has been made to explore the relations between the debates upon the correct growth accounting methodology and the direction of technological change on the one hand and the characteristics of innovation processes at work within innovation systems on the other. As a matter of fact the integration of the microeconomics of innovation and knowledge with the analysis of the direction of technological change and its effects on total factor productivity is a fertile field of investigation that deserves to be better explored.

The historical evidence shows that technological change is strongly biased as it affects asymmetrically the output elasticity of either input such as capital, labour or specific intermediary inputs. Technological change can be capital-intensive, when it favours the usage of capital via the increase of its output elasticity, and hence labour-saving, or labour-intensive, or more specifically skill-intensive, when, on the opposite it favour the use of labour and more specifically skilled labor, when it increases more the output elasticity of labour than that of capital (Robinson, 1938).

Large Econometric History evidence suggests that the bias or direction of technological change differs across countries. As Habakkuk (1962) has shown, through the XIX century technological change has been mainly capital-saving in UK and labour- and raw material intensive in the US. Within the same country, different periods of economic growth can be identified by the changes in the direction of technological change (David, 2004). New and convincing evidence has been provided recently about the strong skill-bias of the gales of technological change based upon information and communication technologies introduced in the last decades of the XXI century (Goldin and Katz, 2008). This persistent and renewed evidence about the strong directionality of technological change contrasts the basic methodology elaborated so far to assess its effects.

The conventional methodology for the measurement of total factor productivity (TFP) assumes, in fact, the Hicks neutrality of technological change. Some methodological innovations are requested in order to measure properly the overall effects of technological change on productivity growth, when its directionality is acknowledged. This is all the more necessary when inputs are not equally abundant and hence the slope of the iso-cost differs from unity. When the introduction of biased technological change (BTC) enables the more intensive use of the more abundant production factors, in fact the consequent change in the slope of isoquants does affect output and hence productivity growth.

As a consequence, countries with different factors' endowments will take advantage of technological innovations that allow for a more intensive use of locally abundant production factors. It follows that countries better able to introduce technologies that are able to matching the local conditions of factor markets should show better productivity performances than countries that have put less effort in shaping technologies according to the relative scarcity of production factors. Standard TFP measurement à la Solow does not allow for fully grasping this phenomenon. Such issue becomes even more meaningful when one considers the distinction between innovation and creative adoption (Antonelli, 2006). Indeed, technologies originated in one country might well be poorly suited to exploit the local conditions of factor markets of other countries. Their productivity-enhancing effects would therefore be much reduced. The effective adoption of those technologies in other countries, featured by different conditions for factor markets, needs for creative efforts aiming at adapting them to the local conditions. A new methodology measuring the effects of BTC on productivity could therefore help investigating the appropriateness of innovation policies based on international technology transfer in follower countries.

The introduction of a bias in technological change can be considered as the result of an effective knowledge infrastructure that displays its effects in terms of technological command only in the long term. In other words, Biased TC can be considered as a meta-substitution process by means of which more expensive inputs are substituted by less expensive ones with positive effects on the efficiency of the production process. As much as the rate of technological change is the endogenous result of economic activity, the direction of the technological change should be treated as the product of both the intentional design of innovators and of the selection of market forces: in both cases the intensive use of locally abundant production factors act as a sorting device that activates learning procedures and stirs dedicated ingenuity and research efforts. The increase in TFP that stems from the bias of technological change towards the intense usage of locally abundant factors can take place as long as each country has an advanced knowledge infrastructure that makes it possible to command the direction of technological change (Antonelli *et al.*, 2009). The identification of biased effects of TC in terms of TFP enable to better grasp the typology of innovation process that are at the origin of its introduction in terms of (Antonelli, 2002, 2003, 2006, 2012; Antonelli & Quataro, 2010).

2.2.8: Technological Change and Economic Development

By and large, the relationship between technological change and economic development has been always intimate and complex. Many contemporary events have contributed to this complexity. In fact, economic development is a way to bridge the technological gap between industrialized and underdeveloped countries. Technological innovations are intimately connected with the economic transformation of any society, as the mainspring for economic progress has been technological growth.

Technological choice and adoption are important aspects of economic engineering. The ancient classification of land, labour capital and organization as factors of production has very much changed. Each factor has gained so much specialization that one unit of ant category of these factors is not easily substitutable for the other. Technology as a factor of production has made such an important contribution that the emerging industrial estate is radically different from the old one. Decision making apparatus, finance mobilization, skill requirements, production programming and market strategy are based now on completely new considerations.

Technological invention is different form technological innovations. The former is a scientific fact, whereas the latter is the result of entrepreneurship of entrepreneurial ingenuity. The former is abundantly available in industrialized countries; it incubates in research laboratories and installations of those countries.

Transfer of technology from research centers to commercial establishments depends upon entrepreneurial requirements. In developing countries, the task has become complicated due to three factors, *viz.*, the problems relating to transfer of technology, the mixed nature of the economy, and concentration of technological knowledge mostly in advanced countries.

The transfer of technology from industrialized countries to developing areas is beset with many technical, economic educational, sociological and political problems. Having imported technology, its duplication, extension and development pose another type of problem.

The establishment of specialized research centers in industrialized countries is no solution of these problems. Encouragement given by the planning authorities in a developing country could provide resources for the establishment of research installations, but the main source of inventions and innovations does not follow merely from such establishments. Innovations have obviously been the mainspring of industrial improvements. The fact remains, however, that the concentration of research efforts and the scientific infrastructure available in industrialized countries have created a gap between the advanced and the backward nations which cannot be easily filled. The basic urge for innovation comes from the desire to mobilize the forces of nature for the advantage of human society. Social motivation, the expectation of better standards of living by employing the available technological knowledge, the assurance of securing the fruits of one's efforts, the level of education and the general law and other situation all contribute to the urge to innovate. They are possible only when economic development has reached a certain level and people have become enlightened, educated and better off.

The history of the industrial revolution in the west shows that the process of rapid technological transformation has not been merely the result of new improved technical processes. It is true that the process is generally traced to the inventions preconditions of economic development if they are expected to create the desired consequences.

2.3: Production Functions

In general terms, the relationship between the physical inputs and the physical outputs of a firm is referred to as the production function. Basically a production function is a technological or engineering concept. The production function shows for a given state of technological knowledge and managerial ability, the maximum rates of output that can be obtained from different combinations of the productive factors during a given period of time. In brief, the production function is a catalog of different output possibilities. Hence, production function is the process by which inputs are converted into output. In other words, production function is defined as the technical or engineering relationship between factor inputs and output. The production function formalizes the relationship between the quantity of output yielded by a product process and the quantities of the various input used in that process. Inputs refer to the various factor prices, which are used in the production. Outputs on the contrary refers o the quantity of goods produced by the firm with the help of the various inputs.

In the form of mathematical equations, the production function can also be expressed in which output is the dependent variable and inputs are the independent variables. Therefore, relationship of these terms in general form can be stated as:

$$Q: f(a, b, c, \ldots\ldots n)$$

Where Q is the rate of output of a given commodity and a, b, c, ... n are the various factor of production used per unit of the time. If a small firm of matches produces 10000 match box per 8 hour shift, then its production consists the maximum quantities of raw wood, labour, power, fuel materials *etc.* A production is always specified for a period of time. It is flow of inputs resulting in a flow of outputs during a specified period of time.

Each firm has its own production function which is determined by the state of technical knowledge and managerial ability of that firm. If there is an improvement in that state of technical knowledge or managerial ability of the firm, the old production function is displaced and a new one takes its place. The new production function has a greater flow of outputs from the original inputs or it involves smaller quantities of inputs for the same original outputs. It is possible for the new production function to have a smaller flow of outputs for a given quantity of inputs, if in the meanwhile there has been deterioration in the state of physical assets of the firm.

Keeping view of studying theory of production, it is beneficial for us to classify the production function into two categories depending on the variability of factor inputs as short-run and long-run. A production function is short-run when at least one factor input cannot be changed during the period. In other terms, when not all factors of production is variable then a production function is short-run, and called short run production function. Hence, at least one factor inputs must be fixed in the short-run production function. On the other hand, all the factor inputs are variable in long-run production function. Long-run is defined as the time period in which all factor inputs can be changed. However, there is no specific law that stipulates the short- and long-run. It is all about the theory but, in practice, it depends on the nature of production activities that determines short- and long-run.

2.3.1: Main Kinds of Production Functions

On the basis of the study of production function, we are able to identify the main sources of the growth of industries. Economists use a variety of functional forms to describe production relationships. Thus, we discuss a few important production functions as under:

2.3.1 (A): Linear Homogeneous Production Function

When all the inputs are increased in the same proportion, the production function is said to be homogeneous. The degree of production function is equal to one. This is known as linear homogeneous production function. In order to estimate the production function, it is necessary to express the function in explicit functional form. Mathematically, this form of production function is expressed as:

$$nQ = f(nL, nK)$$

This production function also implies constant returns to scale. That is if L and K are increased by n-fold, the output Q also increases by n-fold. This form of production function is a well behaved production function. Which makes the task of the entrepreneur quite simple and convenient? He requires only Finding out just one optimum factor proportions. So long as relative factor prices remain constant, he has not to make any fresh decision regarding factor proportions to be used, as he expands his level of production. Moreover, this feature of the same optimum factor proportions is also very useful in input- output analysis, In India, farm management studies have emphasised the constant return to scale and homogeneous production function.

2.3.1 (B): Cobb-Douglas Production Function

In economics, the Cobb-Douglas functional form of production functions is widely used to represent the relationship of an output to inputs. It was proposed by Knut Wicksell (1851-1926), and tested against statistical evidence by Charles Cobb and Paul Douglas in 1928. In 1928 Charles Cobb and Paul Douglas published a study in which they modeled the growth of the American economy during the period 1899-1922. They considered a simplified view of the economy in which production output is determined by the amount of labor involved and the amount of capital invested. While there are many other factors affecting economic performance, their model proved to be remarkably accurate.

In its most standard form for production of a single good with two factors, the C-D function can be written as:

$$Y = AL^{\alpha}K^{\beta}$$

Where, Y = total production (the real value of all goods produced in a year)

L = labor input (the total number of person-hours worked in a year)

K = capital input (the real value of all machinery, equipment, and buildings)

A = total factor productivity

α and β are the output elasticities of capital and labor, respectively.

These values are constants determined by available technology.

Output elasticity measures the responsiveness of output to a change in levels of either labor or capital used in production, ceteris paribus. For example, if $\alpha = 0.55$, a 1% increase in capital usage would lead to approximately a 0.55% increase in output.

Further, if $\alpha+\beta = 1$, The production function has constant returns to scale, meaning that doubling the usage of capital K and labor L will also double output Y. If $\alpha+\beta < 1$, Returns to scale are decreasing, and if $\alpha+\beta > 1$, Returns to scale are increasing. Assuming perfect competition and $\alpha+\beta = 1$, α and β can be shown to be capital's and labor's shares of output.

Cobb and Douglas were influenced by statistical evidence that appeared to show that labor and capital shares of total output were constant over time in developed countries; they explained this by statistical fitting least-squares regression of their production function. There is now doubt over whether constancy over time exists.

Marginal product is the change in total production, when there is an infinitesimal change in the inputs. Marginal product is the first derivative of the production function with respect to an input.

Suppose, $\partial Q/\partial L$

Thus, in the case of the C-D production function:

$\partial Q/\partial L = A\,\alpha\,L^{(\alpha-1)} K^{\beta}$.

Then, if L or K increases, the total output will increase, that is, the marginal product is positive. Thus, the marginal products of labour and capital are the functions of the parameters A, α and β and the ratios of labour and capital inputs. These in functional form can also be written as:

$MPL = \partial Q/\partial L = \alpha A L\ \alpha\text{-}1 K\ \beta$

$MPK = \partial Q/\partial K = \beta A L\ \alpha K\ \beta\text{-}1$

Output elasticity is the perceptual change in output in respond to a change in levels of either L or K.

$(\partial Q/Q) / (\partial L/L) = (\partial Q/\partial L) / (Q/L)$

If output elasticity is greater than 1, the production function is elastic and vice versa. In the case of the Cobb-Douglas production function, output elasticity can be measured quite easily:

$$
\begin{aligned}
(\partial Q/Q) / (\partial L/L) &= (\partial Q/\partial L) / (Q/L) \\
&= [\,A\alpha\,L^{(\alpha-1)} K^{\beta}\,] / [\,A\,L^{\alpha} K^{\beta} / L\,] \\
&= [\,A\alpha\,L^{(\alpha-1)} K^{\beta}\,] / [\,A\,L^{(\alpha-1)} K^{\beta}\,] \\
&= \beta
\end{aligned}
$$

Output elasticity with respect to labor is constant and equal to β. If β is 0.2 and labor increases in 10%, output will increase 2%. α and β are output elasticities of capital and labor, and are constant.

Returns to scale measure how much additional output will be obtained when all factors change proportionally. If the output increases more than proportionally, we say we have increasing returns to scale. If the output increases less than proportionally, we say we have decreasing returns to scale. In the case of the Cobb-Douglas production function, to check how much will output increase when all factors increase proportionally, we multiply all inputs by a constant factor c. Y′ represents the new output level.

$$
\begin{aligned}
Y' &= A\,(cL)^{\alpha}\,(cK)^{\beta} \\
&= A\,c^{L^{\alpha}}\,c^{\alpha}\,L^{\beta} \\
&= c^{\alpha}\,c^{\beta}\,A\,L^{\alpha}\,K^{\beta} \\
&= c^{(\alpha+\beta)}Y
\end{aligned}
$$

Now as it is clear that, if all inputs change by a factor of c, output increases by $c^{(\alpha+\beta)}$, then:

If $\alpha+\beta=1$, the production function has constant returns to scale.

If $\alpha+\beta>1$, the production function has increasing returns to scale.

If $\alpha+\beta<1$, the production function has decreasing returns to scale.

Although the C-D production function is a multiplicative type and is non-linear in its general form, it can be transferred into linear function by taking it in its logarithmic form. That is why, this function is also known as log linear function, which is as under:

$$\text{Log } Q = \log A + a \log L + p \log K$$

It is easier to compute C-D function when expressed in log linear form.

Some Properties of C-D Function

The C-D production function has some properties, these are: (i) There are constant returns to scale; (ii) Elasticity of substitution is equal to one; (iii) A and p represent the labour and capital shares of output respectively; (iv) A and p are also elasticities of output with respect to labour and capital respectively; (v) If one of the inputs is zero, output will also be zero; (vi) The expansion path generated by C-D function is linear and it passes through the origin; (vii) The marginal product of labour is equal to the increase in output when the labour input is increased by one unit; (viii) The average product of labour is equal to the ratio between output and labour input; (ix) The ratio α/β measures factor intensity. The higher this ratio, the more labour intensive is the technique and the lower is this ratio and the more capital intensive is the technique of production.

Limitations of C-D Production Function

C-D function has some limitations too as well and these can be summarized as follows: (i) The function includes only two factors and neglects other inputs; (ii) The function assumes constant returns to scale; (iii) There is the problem of measurement of capital which takes only the quantity of capital available for production; (iv) The function assumes perfect competition in the factor market which is unrealistic; (v) It does not fit to all industries; (vi) It is based on the substitutability of factors and neglects complementarily of factors; and last (vii) The parameters cannot give proper and correct economic implication.

Advantage of C-D Function: Why Use Cobb-Douglas?

First I would like to present brief information about the C-D production function possesses the importance merits. These are: (i) It suits to the nature of all industries; (ii) It is convenient in international and inter-industry comparisons; (iii) It is the most commonly used function in the field of econometrics; (iv) It can be fitted to time series analysis and cross section analysis; (v) The function can be generalized in the case of 'n' factors of production; (vi) The unknown parameters α and β in the function can be easily computed; (vii) It becomes linear function in logarithm; (viii) It is more popular in empirical research.

On the basis of above discussions, it is clear that one of the big advantages of using C-D is undoubtedly its simplicity, in that it is easy to make sense out of the coefficients imposed. The C-D assumption greatly simplifies estimation of output elasticities, conditional on an assumption on returns to scale. With a high average degree of competition in the goods market, the output elasticities can be equated to their respective factor shares. Thus, there is only one parameter to estimate. While a large variety of views on alternative specifications to the C-D approach of constant factor shares are available, one needs to be aware of the implications associated with these alternatives. For example, if one chooses to adopt an elasticity of less than 1, one is left with the problem of explaining why wage shares have fallen recently. If one goes for the alternative assumption of using an elasticity of greater than 1, then the lack of econometric evidence to support using such a function needs to be taken into account.

Consequently, given the difficulties associated with the alternatives, the Cobb-Douglas assumption of unity appears to be a reasonable compromise. In addition, of course, if one were to use a CES function with an elasticity of 0.8 or 1.2 the results would not differ very strongly from Cobb-Douglas. Finally, the aggregation problem associated with having a mixture of low and high skilled workers in the workforce would also appear to lend support to the Cobb-Douglas view. In this regard, if you aggregate over both sets of workers, one would come close to Cobb-Douglas, with low skilled workers having a high elasticity of substitution (EoS) with capital (EoS > 1) balancing out the low EoS associated with high skilled workers (EoS < 1). High skilled workers have generally a low EoS since such workers are regarded as being more complementary to K. This view regarding the distinction between low and high skilled workers is supported in a paper by Krussell *et al.* published in Econometrica in September 2000.

The practical application of the production function method requires making certain assumptions, particularly on the functional form of the production technology, returns to scale, and characteristics of the technological progress, as well as of the functioning of markets. The neo-classical two-factor C-D production function with Hicks-neutral technology is frequently used, 2 including the assumptions of positive and diminishing marginal products with respect to inputs of labor and capital, constant returns to scale, no unobserved inputs and perfect competition. These assumptions restrict the elasticity of output with respect to labor and capital to values between zero and one and their sum to being equal to one. Given the assumptions, the theoretical technological coefficients are then in practice approximated with the help of the income share of labor in produced output. Since the technological coefficients are assumed to be stable, the share of factors in production should be stable in time. Furthermore, if there is no presumption that the aggregate technology would differ across countries, the labor shares should also be roughly similar across countries.

2.3.1 (C): C.E.S. Production Function

Substituting scarce factors of production by relatively more abundant ones is a key element of economic efficiency and a driving force of economic growth. A measure of that force is the elasticity of substitution between capital and labor which is the central parameter in production functions, and in particular CES (Constant Elasticity of Substitution) ones. Until recently, the application of production functions with non-unitary substitution elasticities (*i.e.*, non Cobb Douglas) was hampered by empirical and theoretical uncertainties. As has recently been revealed, "normalization" of production functions and production-technology systems holds out the promise of resolving many of those uncertainties and allowing considerations as the role of the substitution elasticity and biased technical change to play a deeper role in growth and business-cycle analysis. Normalization essentially implies representing the production function in consistent indexed number form. Without normalization, it can be shown that the production function parameters have no economic interpretation since they are dependent on the normalization point and the elasticity of substitution itself. This feature significantly undermines estimation and comparative-static exercises, among other things.

Due to the central role of the substitution elasticity in many areas of dynamic macroeconomics, the concept of CES production functions has recently experienced a major revival. The link between economic growth and the size of the substitution elasticity has long been known. As already demonstrated by Solow (1956) in the neoclassical growth model, assuming an aggregate CES production function with an elasticity of substitution above unity is the easiest way to generate perpetual growth. Since scarce labor can be completely substituted by capital, the marginal product of capital remains bounded away from zero in the long run.

The first rigid derivation of the CES production function appeared in the famous Arrow *et al.* (1961) paper (hereafter ACMS). However, there were important forerunners, in particular the explicit mentioning of a CES type production technology (with an elasticity of substitution equal to 2) in the Solow (1956) article (done, Solow wrote, to add a "bit of variety") on the neoclassical growth model.

The homogenous CES production function takes the form as:

$$Q=\gamma\{\alpha K^{-\varrho}+(1-\alpha)L^{-\varrho}\}^{-1/\varrho}\exp(U),\ \varrho\geq-1,\ 0\leq\alpha\leq1,\ \gamma>0, \quad (1)$$

Where K is capital, L is labor, Q is output, and U is an error term satisfying $E[U \mid K,L]=0$.

The CES production function was introduced by Arrow, K.J., H.B. Chenery, B.S. Minhas and R.M. Solow (1961). Formally, the elasticity of substitution measures the percentage change in factor proportions due to a percentage change in the marginal rate of technical substitution.

In particular, for a canonical production function $Q= f(K,L)$ with marginal products $K=\partial(K,L)/\partial K$ and $L=\partial(K,L)/\partial L$, the marginal rate of technical substitution is L/K, *i.e.*, minus the slope of the isoquant in point (L,K), hence the elasticity of substitution between capital and labor is given by: $\sigma = -d\ \ln(L/K)/d\ \ln(L/K)$.

In the case of the deterministic homogenous CES production function

$Q = \gamma\{\alpha K^{-\varrho}+(1-\alpha)L^{-\varrho}\}^{-1/\varrho}$. Now we have

$$fL = \gamma\{\alpha K^{-\varrho}+(1-\alpha)L^{-\varrho}\}^{-(1+\varrho)/\varrho}(1-\alpha)L^{-\varrho-1},$$

$$fK = \gamma\{\alpha K^{-\varrho}+(1-\alpha)L^{-\varrho}\}^{-(1+\varrho)/\varrho}\alpha K^{-\varrho-1},$$

Hence $dLn\ (fL/fK)=-(\varrho+1)dLn(L/K)$ and thus $\sigma =1/(\varrho+1)$.

In order to estimate the parameters of the CES production function (1) can be written as:

$$Ln(Q/L) = Ln(\gamma) - (1/\varrho)Ln\,\{\alpha[\exp(\varrho Ln(L/K)) - 1] + 1\} + U \qquad (2)$$

$$= Ln(\gamma) - \frac{Ln\{\alpha[\exp(\varrho Ln(L/K))-1]+1\}}{\alpha[\exp(\varrho Ln(L/K))-1]} \times \frac{\exp(\varrho Ln(L/K))-1}{\varrho Ln(L/K)} \times (\alpha Ln(L/K)) + U.$$

When $\varrho \to 0$, then equation (2) becomes

$$Ln(Q/L) = Ln(\gamma) - \alpha Ln(L/K) + U.$$

While for $\varrho=0$ the CES production function becomes a C-D function:

$$Ln\,(Q) = Ln(\gamma) + \alpha Ln(K) + (1-\alpha)Ln(L) + U. \qquad (3)$$

The CES production function (2) is now a nonlinear regression model,

$$Y = g(x,b) + U,$$

Where $Y= Ln(Q/L)$, $x = ln(L/K)$, $b = \{b(1), b(2), b(3)\}' = \{ln(\gamma), \varrho, \alpha)\}$, and

$$g(x,b) = b(1) - \frac{Ln\{b(3)\,[\exp(b2)x)-1]+1\}}{b(3)[\exp(b(2)x)-1]} \times \frac{\exp((b(2)x)-1}{b(2)x} \times (b(3)x). \qquad (4)$$

Some Properties of C.E.S. Function

Like C-D production function, CES production function has also some properties, these are:

(i) The CES function is standardized of degree one. Thus like C-D function, the CES function displays constant returns to scale.

(ii) The value of elasticity of substitution depends upon the value of substitution parameter.

(iii) In the CES production function, the average and marginal products in the variables K and L are standardized of degree zero like all linear standardized production functions.

(iv) From the above property the incline of an Iso-quant *i.e.* Marginal rate of Technical Substitution (MRTS) of K for L can be represented convex to the origin.

(v) As a consequence of the above, if L and K are substitutable (infinity) for each other an increase in K will require less of L for a given productivity. As a consequent, the MP of L will increase. Thus the MP of an input will increase when the other input is increased.

(vi) The estimation of the elasticity of substitution parameter requires the assumption of perfect competition.

Merits of C.E.S. Function

It has the following merits and they are listed below.

(i) The CES function is a standardized of grade 1, and it is more general, it covers all type of returns.

(ii) CES function takes account of a number of parameters.

(iii) This function takes account of raw materials among its inputs.

(iv) CES functions are very easy to approximate and are free from impractical postulations.

Limitations of C.E.S. Function

(i) It has some limitations which have been summarized as follows:

(ii) This production function regards only two inputs. It can be comprehensive to more than two units. However, it becomes very complex and intricate arithmetically to use it for more than two inputs.

(iii) The distribution parameter or capital intensity factor co-efficient α is not dimensionless.

(iv) If data are fitted to the CES function, the value of the competence a parameter cannot be made independent of the other units representing in this production function.

(v) If the CES function is used to explain the production of a firm, it cannot be used to explain the aggregate production function of all the firms in the industry. Thus it involves the problem of aggregation of production function of diverse firms in the industry.

(vi) It suffers from the drawback that elasticity of substitution amidst any part of inputs is the same which does not materialize to be realistic.

(vii) In approximation of the parameters of CES production function we may come across a large number of problems like choice of exogenous variables, judgment procedure and the problem of multi collinear.

(viii) There is little possibility of identifying the production function under technological change.

Despite these limits, CES functions is constructive in its application to prove Euler's theorem, to reveal constant returns to scale to show that average and marginal products are standardized degree zero and to determine the elasticity of substitution.

2.3.1 (C): V.E.S. Production Function

In production function approach, recently a new attempt has been made by Bruno, Knox Lovell and Revankar to get a new production function. This resulting production function is the generalization of ACMS function (*i.e.* CES) which possesses the desirable properties of variable elasticity substitution. Lu and Fletcher have filled a logarithmic relationship containing the wage rate (W) as well as the capital-labour ratio (K/L) to explain value added per unit of labour.

$$V/L = a + b \log W + c \log K/L$$

Where, V = Value added, W = Wage rate, K = Capital, L = Labour, and a, b and c are the parameters to be estimated.

The elasticity of substitution (σ) is

$$\sigma = b/1\text{-}c\ (1\text{+}WL/rk)$$

Where, WL and rk are the shares of labour and capital respectively.

While the standard notation to denote a general production technology as Y = F(K, L), where Y, K, and L stand for output, capital and labor, respectively. Following Revankar (1971), we consider the following specification (This is a very similar VES specification was developed by Sato and Hoffman (1968):

$$Y = AK^{av}[L+baK]^{(1-a)v}.$$

We mostly assume that the production function exhibits constant returns to scale, *i.e.*, v=1. This production function can be written in intensive form, y=(k) where y ≡ Y/L and k ≡ K/L, as

$$y = Ak^{a}\,[L+bak]^{1-a}.$$

The VES production function in an otherwise standard Solow-Swan model and derives necessary and sufficient conditions for unbounded endogenous growth. The properties of the VES are also shared by the CES. Exceptions include the elasticity of substitution which for the CES production function is constant along an isoquant, while for the VES considered here it is constant only along a ray through the origin. But some Properties of VES Production Function are: (i) VES satisfies the requirements of a neo-classical production function; (ii) VES function includes the fixed co-efficient models; and (iii) VES production function is more general.

2.3.1(D): Translog Production Function

The Translog production functions occurred in the context of researches related to the discovery and definition of new flexible forms of production functions and to the approximation of CES production function. In fact, the first form of a Translog production may be considered the proposal made in 1967 by J. Kmenta for the approximation of the CES production function with a second order Taylor series, when the elasticity of substitution is very close to the unitary value, which is the case of C-D production function. The form of the above-mentioned production function is:

$$LnY = LnA + \alpha LnK + \beta LnL + \gamma Ln^2 (K/L) \qquad (1)$$

Where: Ln= Natural logarithm; Y= Output (Gross domestic product): K= Fixed capital

L= Employed population; and A, α, β and γ are the parameters to be estimated.

In 1971, Grilichs and Ringstad proposed new forms of production function. The first one was obtained by imposing the condition that $\alpha+\beta=1$. This way, the production function became in fact a labour productivity function:

$$Ln(Y/L) = LnA + \alpha Ln(K/L) + \gamma Ln^2 (K/L) \qquad (2)$$

It is to be noticed that the above-mentioned function is one of a second order polynomial in the logarithms of the single input considered, capital-labour ratio, respectively. The second form of this production function was defined in conditions of relaxing the constraints imposed to the parameters in the Kmenta function in order to test the homotheticity assumptions, and was written as:

$$LnY = Ln.A_{KL} + \alpha_K.LnK + \alpha_L.LnL + \beta_K{}^2.LnK + \beta_L{}^2.LnL + \beta_{KL}.LnK.LnL \quad (3)$$

In fact, the same production function was used by Sargant also in 1971 and was called a log-quadratic one. It is important to mention that the term "translog production function", abridged from "transcendental logarithmic production function "was proposed by Christiansen, Jorgensn and Lau in two papers published in 1971 and 1973, which dealt with the problems of strong separability (additivity) and homogeneity of C-D and CES production functions and their implications for the production frontier.

The translog production function is a generalization of the C–D function (Berndt, Ernst R.; Christensen, Laurits R., 1973). The name translog stands for 'transcendental logarithmic'. This function has both linear and quadratic terms with the ability of using more than two factor inputs. The three factor input translog production function can be written in terms of logarithms as follows:

$$\ln(Y) = \ln(A) + a_L Ln(L) + a_K Ln(K) + a_M.Ln(M) + b_{LL} Ln(L).Ln(L) + b_{KK} Ln(K).Ln(K) + b_{MM} Ln(M).Ln(M) + b_{LK} Ln(L).Ln(K) + b_{LM} Ln(L).Ln(M) + b_{KM} Ln(K).Ln(M)$$

$$= f(L,K,M)$$

(Where, Y = output, A = total factor productivity, L = labor, K = capital, and M = materials and supplies).

The translog production functions represent in fact a class of flexible functional forms for the production functions (Ch. Allen, St. Hall, 1997). One of the main advantages of the respective production function is that, unlike in case of C-D production function, it does not assume rigid premises such as: perfect or "smooth" substitution between production factors or perfect competition on the production factors market (J. Klacek, *et al.*, 2007). The concept of the Translog production function also permits to pass from a linear relationship between the output and the production factors, which are taken into account, to a non-linear one. Due to its properties, the Translog production function can be used for the second order approximation of a linear-homogenous production, the Allen elasticities of substitution estimation, the estimation of the production frontier or the measurement of the TFP dynamics.

2.3.2: Some Other Issues Related to Production Functions

It is obvious phenomenon the production function represents a useful and powerful tool for the macroeconomic analysis and evaluation of the governmental structural policies. The supply side performance of an economy is often identified with the growth rate of potential output. The use of the production function method for the measurement of potential output growth takes into account different sources of an economy's productive capacity, namely the contributions of labour, capital and TFP, the latter containing information about technological and allocative efficiency and hence about the supply-side functioning. Using the production function, one can discuss changes in the supply-side performance on the basis of the observed simultaneous developments in the quantity of L, K and TFP. For instance, an increase in the rate of capital growth accompanied by a rise in trend TFP may signalize some improvement in the supply-side performance. Observing an increase in the rate of the capital growth while trend TFP stagnates, one can, in contrast, deduce that the supply side is functioning ineffectively.

In econometrics estimation of production functions, technical change can be estimated using the single time trend (TT) approach (*i.e.* inclusion of a deterministic time trend in the estimation of a production function) (Solow, 1957; Tinbergen, 1942; Christensen *et al.*, 1973) or alternatively by the general index (GI) approach (Baltagi & Griffin, 1988). {*An intermediate approach is multiple time trend approach where multiplicity of trends has been introduced to capture structural changes such as pre- and post-economic reform periods (See* Heshmati and Nafar, *1998)}*. With the TT approach, the trend may be linear or non-linear, and certain specifications such as flexible functional forms may allow interactions between time and other explanatory variables. This allows the rate of technical change to be non- constant and non-neutral (Gollop & Jorgenson, 1980, Jorgenson & Fraumenti, 1981, Gollop & Roberts, 1983). The derivative of the production function with respect to time provides measure of the rate of technical change.

A critical weakness of the TT approach is the smooth pattern of growth with indefinite progress or regress rates. In order to capture the year-to-year changes in technical change, the GI approach could be used instead. It uses a set of time-specific dummies and their interactions with other explanatory variables to estimate a general index of the technical change. With the estimates of input elasticities and rate of technical change, total factor productivity (TFP) can be calculated accordingly using both TT and GI approaches.

Production functions for the industrial sector as a whole as well as for seven important industries in India are worked out based on cross-section data relating to individual firms for the two years 1951 and 1952 (V. N. Murti & V. K. Sastry, 1957, De-Min Wu, 1975), derives the exact distribution of the indirect least squares estimator of the coefficients of the C-D function within the context of a stochastic production model of *Marschak-Andrews* type. The stochastic term in C-D type models is either specified to be additive or multiplicative (Stephen M. Goldfeld & Richard E. Quandt, 1976). They developed a model in which a C-D type function is coupled with simultaneous multiplicative and additive errors. This specification is a natural generalization of the "pure" models in which either additive or multiplicative stochastic terms are introduced. A C-D type function with both multiplicative and additive errors has been proposed by Goldfeld & Quandt (1976). They suggested a maximum likelyhood approach to the estimation of a C-D type model when the model includes both multiplicative and additive disturbance terms. As expected, an analytical expression for the solution to the maximization problem did not exist. Indeed, because of the complexity of the likelihood function, their maximization algorithm had to be used in conjunction with a numerical integration technique.

Kelejian (1972) generalize and simplify the work by Goldfeld & Quandt (1976). Specifically, an estimation technique is suggested which does not require the specification of the disturbance terms beyond their means and variances, which does not require the compounding of a maximization algorithm with a numerical integration technique, but yet leads to asymptotically efficient estimates of the parameters of the regression function. In addition, the procedure readily lends itself to interpretation. For instance, it will become evident that if the distribution of the multiplicative disturbance term is not known, the scale parameter of the model (unlike the other parameters) will not be identified. The origins of the Cobb-Douglas form date back to the seminal work of Cobb & Douglas (1928), who used data for the US manufacturing sector for 1899-1922 (although, as Brown 1966), Sandelin (1976), and Samuelson (1973), indicate, Wicksell should have taken the credit for its "discovery", for he had been working with this form in the 19th century). Cheema (1978), Kemal (1991), and Wizarat (1989), showed the performance of large-scale manufacturing sector of Pakistan.

Bhasin & Seth (1980), estimate production functions for Indian manufacturing industries and to find whether plausible and meaningful estimates can be obtained for returns to scale, substitution, distribution, and efficiency parameters. Some studies are based on data collected through surveys specially designed for estimate the levels of technical efficiency (TE). Many of the studies are concerned with estimating and explaining variations in TE only in Small-Scale Industrial Units by fitting either a deterministic or a stochastic production frontier (*e.g.*, Bhavani (1991), Goldar [20], Neogi and Ghosh (1994). All these studies, however, use data relating to years prior to the economic reforms. For instance, Bhavani (1991) uses data collected under the first Census of small scale industrial units in 1973 to estimate the TE of firms at the 4-digit level industries of metal product groups by fitting a deterministic Translog production frontier with three inputs-capital, labor and materials- and observes a very level of average efficiency across the four groups. Neogi and Ghosh (1994) examine the inter-temporal movement of TE using panel industry-level summary data for the years 1974-1975 to 1987-1988 and observe TEs to be falling over time. In recent years, there has been an increasing interest in the examination of productivity from different parts of the economy such as industry, agriculture, and services.

TFP growth in manufacturing has been examined by applied parametric and non-parametric approaches. In most of the studies have used non-parametric approach, wherein total factor productivity growth has decomposed into efficiency change and technological change. Efficiency change measures "catching-up" to the isoquant while technological change measures shifts in the isoquant. For example, see Weber & Domazlicky; Nemoto & Goto (1999); (Maniadakis & Thanassoulis (2004). The studies by Golder *et al.* and Mukherjee and Ray (2004), however, relate to the post-reform era. Using panel data for 63 firms in the engineering industry from 1990-1991 to 1999-2000 drawn from the Prowess database (version 2001) of the Centre for Monitoring Indian Economy, Goldar *et al.* (2003) fit a translog stochastic production frontier to estimate firm-level TE scores in each year. They find the mean TE of foreign firms to be higher than that of domestically owned firms but do not find any statistically significant variation in mean TE across public and private sector firms among the latter group. They can attempt to explain variation in TEs in terms of economic variables, including export and import intensity and the degree of vertical nitration. Md. Zakir Hossain, M. Ishaq Bhatti, Md. Zulficar Ali, (2004), reviews some models used in the literature and selects the most suitable one for measuring the production process of 21 major manufacturing industries in Bangladesh. They estimates and tests the coefficients of the production inputs for each of the selected manufacturing industries using Bangladesh Bureau of Statistics annual data over the period 1982-1983 through 1991-1992.

Cheng-Ping Lin (2002) analyzes the cost function of construction firms with due consideration of their available resources by using C-D Production and Cost Functions. Moosup Jung, *et al.* (2008) made a study on Total Factor Productivity of Korean Firms and Catching up with the Japanese Firms. They measured and compared the TFP of both Korean and Japanese listed firms of 1984 to 2004. They found that the average TFP of Korean firms grew about 44.1% between 1984 and 2005, with 2.1% annual growth rates. Industry was observed to be outstanding. M. Z. Hossain and K. S. Al-Amri, (2004) find that for most of the selected industries the C-D function fits the data very well in terms of labor and capital elasticity, return to scale measurements, standard errors, economy of the industries, high value of R^2 and reasonably good Durbin-Watson statistics. The estimated results suggest that the manufacturing industries of Oman generally seem to indicate the case of increasing return to scale. Of the nine industries, seven exhibit increasing return to scale and only the rest two show decreasing return to scale. They also find that no industry with constant return to scale.

References

Abramovitz, M (1956): "Resources and output trends in the U.S. since 1870", *American Economic Review*, 46: 5-23.

Abramovitz M. (1986): "Catching-up, foreign ahead and falling behind", Journal of Econ. History, Vol. 46.

A. Cheema (1978): "Productivity Trends in the Manufacturing Industries," *The Pakistan Development Review*, Vol. 17, No. 1, pp. 55-65.

Ackerberg, D. A.; Caves, K. & Frazer, G. (2005): "Structural Identification of Production Functions" (*mimeo: University of California at Los Angeles*).

Adam B. Jaffe et.al (2003): 'Technological Change and the Environment' in an edited Book entitled -*"Handbook of Environmental Economics"*, Volume 1, K.-G. Mäler and J.R. Vincent, 2003.

Adil Mohommad (2010): "Manufacturing Sector Productivity In India: All India Trends, Regional Patterns, And Network Externalities From Infrastructure On Regional Growth", Ph.D. Thesis, University of Maryland.

A. Hashim (2009): "Cost and Productivity in Indian Textiles- Indian Council for Research on International Economic Relations. Multan District," *Journal of Quality and Technology Management*, Vol. 5, No. 2, pp. 91-100.

Allen Ch., Hall St. (1997): 'Macroeconomic Modeling in a Changing World', John Willey & Sons, New York.

Andrews, F. (1979), "Scientific Productivity: The Effectiveness of Research Groups in Six Countries", Cambridge University Press.

Andrew Sharpe (1980): "A Disaggregated Analysis of Price Changes and Productivity in the Canadian Economy, 1961-1976: A Labour Value Approach", December 1980.

Antonelli, C. (2006): "Diffusion as a Process of Creative Adoption", *Journal of Technology Transfer*, 31, 211-226.

Arrow, K. (1962): "Economic Welfare and the Allocation of Resources for Inventions", in The Rate and Direction of Inventive Activity, Princeton University Press.

Arrow, K. (1962): "The economic implications of learning by doing", *Review of Economic Studies*, 29: 155-73.

Ahluwalia, I.J. (1991): Productivity and Growth in Indian Manufacturing, Oxford University Press, New Delhi.

Balakrishnan, P and Pushpangadan, K. (1994): "Total Factor Productivity Growth in Manufacturing Industry: A Fresh Look", *Economic and Political Weekly*, 29: 2028-35.

Balakrishnan, P. & Pushpangadan, K. (1998): "What do we know about productivity growth in Indian industry?", *Economic and Political Weekly*, 33: 2241–46.

Balakrishnan, P. (2004): "Measuring Productivity in Manufacturing Sector", *Economic and Political Weekly*, April 3-10.

Baltagi, B.H. & J.M. Griffin. (1988). "A Generalized Error Component Model with Heteroscedastic Disturbances." *International Economic Review* 29, 745-753.

Banerjee, A. V. & Duflo, E. (2008): "What is middle class about the middle classes around the world?", *Journal of Economic Perspectives*, 22, 3-28.

Barro, R. (1991): "Economic Growth in a Cross-Section of Countries", *Quarterly Journal of Economics*, 106: 407-444.

Barro, R., Mankiw, N. & X. Sala-i-Martin, (1995): "Capital Mobility in Neoclassical Models of Growth", *American Economic Review*, 85(1): 103-115.

Battese, G. E. & Coelli, T. J. (1995): "A Model for Technical Inefficiency Effects in A Stochastic Frontier Production Function for Panel Data", *Empirical Economics*, 20, 325-332.

B. Carlsson (1981): "The Content of Productivity Growth in Swedish Manufacturing," *Research Policy*, Vol. 10, No. 4, pp. 336-355.

B. Contini, R. Revelli & S. Cuneo (1992): "Productivity and Imperfect Competition: Econometric Estimation from Panel- Data of Italian Firms," *Journal of Economic Behavior & Organization*, Vol. 18, No. 2, pp. 229-248.

Berndt, Ernst R.; Christensen, Laurits R. (1973): "The Translog Function and the Substitution of Equipment, Structures, and Labor in U.S. manufacturing 1929–68". *Journal of Econometrics* 1 (1): 81–113.

Bhasin V. K. & Seth V. K. (1980): "Estimation of Production Functions for Indian Manufacturing Industries," *Indian Journal of Industrial Relations*, Vol. 15, No. 3, 1980, pp.395-409.

Bhavani T. (1991): "Technical Efficiency in Indian Modern Small Scale Sector: An Application of Frontier Production Function," *Indian Economic Review*, Vol. 26, No. 2, pp. 149-166.

Brahmananda, P.R. (1982): "Productivity in the Indian Economy: Rising Inputs for Falling Outputs", Himalaya Publishing House, Mumbai.

Brown M. (1966): "On the Theory and Measurement of Technological Change," Cambridge University Press, Cambridge.

Bruno, M. (1984): "Raw materials, Profits and the Productivity Slowdown", *Quarterly Journal of Economics*, 99: 1-29.

Brynjolfsson, E. & L. M. Hitt, (1995): "Information Technology as a Factor of Production: The Role of Differences among Firms", in *Economics of Innovation and New Technology*, 3: 183–99.

Cameron, G. & J. Muellbauer (1996): "Knowledge, Increasing Returns, and the UK Production Function", In: Mayes, D. (ed.), Sources of Productivity Growth in the 1980s, Cambridge: Cambridge University Press.

Chen, E.K.Y. (1979): "Hyper-Growth in Asian Economies: a comparative study of Hong Kong, Japan, Korea, Singapore, and Taiwan", New York: Holmes & Meier Publishers.

C. P. Lin (2002): "The Application of Cobb-Douglas Production Cost Functions to Construction Firms in Japan and Tai- wan," *Review of Pacific Basin Financial Markets and Policies (RPBFMP)*, Vol. 5, No. 1, pp. 111-128.

Cobb, C.W. & P.H. Douglas (1928), "A Theory of Production", in American Economic Review, 18: 139-165. Denison, E. (1985), "Trends in American Economic Growth", Washington: Brookings Institution.

Christiansen L.R., Jorgensen D.W, Lau L.J. (1971): "Conjugate duality and the transcendental logarithmic production function", in *Econometrica*, vol. 39

Christiansen L.R., Jorgensen D.W, Lau L.J. (1973): "Transcendental logarithmic production frontier", in *Review of Economics and Statistics*, vol. 55.

CSO (2007): National Account Statistics, Government of India, New Delhi.

Das, D. K. (2004): "Manufacturing productivity under varying trade regimes, 1980-2000", *Economic and Political Weekly*, 39, 423-433.

David, P. (2004), "The Tale of two Traverses. Innovation and Accumulation in the first two Centuries of U.S. Economic Growth", SIEPR Discussion Paper No 03-24, Stanford University.

Dholakia, R. H. & Dholakia, B. H. (1994): "Total Factor Productivity Growth in Indian Manufacturing", *Economic and Political Weekly*, 29, 342-344.

Diewert, W. E. (2000): "The Quadratic Approximation Lemma and Decompositions of Superlative Indexes", Discussion Paper 00-15, Department of Economics, University of British Columbia, Canada

D.M. Wu (1975): "Estimation of the Cobb-Douglas Production Function", *Econometrica*, Vol. 43, No. 4, pp.-739-744.

Fabricant, S. (1954): "Economic Progress and Economic Change", 34th Report of the National Bureau of Economic Research, New York: NBER.

Fagerberg, J., Verspagen, B. (1996): "Heading for divergence? Regional growth in Europe reconsidered", *Journal of Common Market Studies* 34: 431–448.

Fagerberg, J., Verspagen, B., Canièls, M. (1997): "Technology gaps, growth and unemployment across European regions", Regional Studies, 31.

Feldstein, Martin S. and Rothchild, Michael (1974): "Towards an Economic Theory of Replacement Investment", *Econornetrica*, XLII, May 1974, 393-423.

Ferguson C.E. (1979): "The Neo-Classical Theory of Production and Distribution", Cambridge, New York, Melbourne, Cambridge University Press.

Foster, R.N. (1985): "Improving the Return on R and D", in *Research Management*, No.1.

Freeman, C, Soete, L. (1997): "The Economics of Industrial Innovation", 3rd ed. Pinter, London.

Goldar B.N. (1985): "Unit Size and Economic Efficiency in Small Scale Washing Soap Industry in India," *Artha Vijnana*, Vol. 27, No. 1, 1985, pp. 21-40.

Goldar, B. N. (1986): "Productivity Growth in Indian Industry", Allied Publishers, New Delhi.

Goldar B. N. (1988): "Relative Efficiency of Modern Small Scale Industries in India," In: *K. B. Suri, Ed.*, 'Small Scale Enterprises in Industrial Development', Sage Publication, New Delhi, 1988.

Goldar, B. N. (2002): "TFP growth in Indian manufacturing in 1980s", EPW, 37, 4966–4968.

Goldar, B. (2004): "Productivity Trends in Indian Manufacturing in the Pre- and Post-Reform Periods", Working Paper No. 137, ICRIER, New Delhi, June.

Goldar, B. (2006): "Productivity growth in Indian manufacturing in the 1980s and 1990s". *In Tendulkar, S. D., Mitra, A., Narayanan, K. and Das, D. K. (eds.)* 'India: Industrialization in a Reforming Economy', Academic Publishers, New Delhi, 2006.

Goldar, B. & Kumari, A. (2003): "Import liberalization and productivity growth in Indian manufacturing industries in the 1990s", *The Developing Economies*, 41: 436-60.

Goldar B.N., Renganathan V. S. & R. Banga (2004): "Ownership and Efficiency in Engineering Firms: 1990-1991 to 1999-2000," *EPW*, Vol. 39, No. 5, 2004, pp. 441-447.

Goldfeld S. M. & Quandt R. E. (1970): "The Estimation of Cobb-Douglas Type Functions with Multiplicative and Additive," *International Economic Review*, Vol. 11, No. 2, pp. 251-257.

Goldfield S. M. & Quandt R. E. (1976): "Nonlinear Methods of Econometrics", North-Holland Publication Company, Amsterdam, New York.

Goldin, C. and Katz, L. (2008): "The race between Education and Technology", Belknap Press for Harvard University Press. Cambridge.

Griliches, Z. (1958): "Research Costs and Social Returns: Hybrid Corn and Related Innovation", in *The Journal of Political Economy*, 5: 419-431.

Griliches, Z. (1979): "Issues in Assessing the Contribution of Research and Development to Productivity Growth", in *Bell Journal of Economics*, 10: 92–116.

Griliches, Z. (1980): "Returns to R&D Expenditures in the Private Sector", *In: Kendrick, K.W. and Vaccara, B. (eds.)*, 'New Developments in Productivity Measurement', Chicago University Press.

Griliches, Z. (1980): "R&D and the Productivity Slowdown", *American Economic Review*, 70: 343-348.

Griliches, Z. (1988): "Productivity Puzzles and R&D: Another Non-explanation", *Journal of Economic Perspectives*, 2: 9-21.

Griliches, Z. and F., Lichtenberg (1984): "R&D and Productivity Growth at the Industry Level: Is There Still a Relationship?", In: *Griliches, Z. (ed.)* 'R&D, Patents and Productivity', Chicago University Press.

Griliches, Z. and J. Mairesse (1984): "Productivity and Research Development at the Firm Level", pp. 339-374, In: *Griliches (ed.)*, 'Research and Development, Patents and Productivity', Chicago University Press.

Griliches, Z. & J. Mairesse (1998): "Production functions: The search for identification", In: Econometrics and economic theory in the 20th century: the Ragnar Frisch Centennial Symposium, Cambridge University Press, pp.169-203.

Griliches, Z. & Ringstad, V. (1971): "Economies of Scale and the Form of the Production Function", Amsterdam: North Holland.

Grossman, T. & E. Helpman (1991): "Innovation and Growth in the Global Economy", MIT Press. Hall, R. (1990), "Invariance Properties of Solow's Productivity Residual", In: *Diamond, P., (ed.)*, 'Growth/Productivity/ Unemployment: Essays to celebrate Bob Solow's Birthday', Cambridge, MA: MIT Press, pp.71-112.

Griliches, Z. & Ringsted V. (1971): Economies of Scale and Form of the Production Function, North-Holland Publishing Company, Amsterdam.

Habakkuk, H.J. (1962): *American and British technology in the nineteenth century*, Cambridge University Press, Cambridge.

Harrod, R. F. (1961): "The Neutrality of Improvements", *Economic Journal*, LXXI, June 1961, 300-304.

Heshmati, A. & N. Nafar. (1998): "A Production Analysis of the Manufacturing Industries in Iran." *Technological Forecasting and Social Change* 59, 183-196.

Hulten, Charles R., & Nishimizu, M. (1979): "The Importance of Productivity Change in the Economic Growth of Nine Industrialized Countries", *eds. S. Maital and N. M. Meltz*, 'Lagging Productivity Growth', London: Macmillan, 1979.

Hart, A. (1966): "A Chart for Evaluating Product Research and Development", In: *Operational Research Quarterly*, 17

Hawke, G.R. (1970): "Railways and Economic Growth in England and Wales", Oxford Uni. Press.

Hess, S.W. (1962): "A Dynamic Programming Approach to R and D Budgeting and Project Selection", in IRE Transactions on Engineering Management, 9(4).

Hicks, J.R. (1932): "The Theory of Wages", McMillan and Company Ltd., London.

Hossain, M.A. & Karunaratne, N.D. (2004): "Trade Liberalization and Technical Efficiency: Evidence from Bangladesh Manufacturing Industries", *The Journal of Development Studies*, 40, 87-114.

Hulten, Charles K. (1975): "Technical Change and the Reproducibility of Capital", *American Economic Review*, XLV, December 1975, 956-965.

Hulten, Charles K. (1979): "On the 'Importance' of Productivity Change", *American Economic Review*, LXIX, March 1979, pp.126-136.

Industrial Credit & Investment Corporation of India Limited (1994): "Productivity is Indian Manufacturing: Private Corporate Sector 1972-73 to 1991-92", ICICI, Mumbai.

J. K. Mullen & M. Williams (1990): "Explaining Total Factor Productivity Differentials in Urban Manufacturing," *Journal of Urban Economics*, Vol. 28, No. 1, pp. 103- 123.

J. M. Page Jr. (1984): "Firm Size and Technical Efficiency: Application of Production Frontiers to Indian Survey Data", *Journal of Development Economics*, Vol. 16, No. 1-2, pp. 129-152.

J. Nemoto & M. Goto (2005): "Productivity, Efficiency, Scale Economies and Technical Change: A New Decomposition Analysis of TFP Applied to the Japanese Prefectures," *Journal of the Japanese and International Economies*, Vol. 19, No. 4, pp. 617-634.

Jorgenson, D.W. & Griliches, Z. (1967): "The explanation of productivity change", *Review of Economic Studies*, 34: 349–83.

Jorgenson, D. (1966): "The Embodiment Hypothesis", *Jour. of Pol. Economy, LXXIV*, Feb., 1966, pp.1-17.

Jorgenson, D. (1990): "Productivity and Economic Growth", in *Berndt, E. and Triplett, J.* 'Fifty Years of Economic Measurement', Cambridge, MA: NBER, pp. 19-118.

Kaldor, N. & J. Mirrlees (1962): "A New Model of Economic Growth", *Review of Economic Studies*, pp.174-245.

Kemal R. (1991): "Substitution Elasticities in the Large Scale Manufacturing Industries in Pakistan," *The Pakistan Development Review*, Vol. 20, No. 1, pp.11-36.

Kathuria, V., Raj, R.S.N & Sen, K (2010): "Organized versus Unorganized Manufacturing Performance in the Post-Reform Period", *Economic and Political Weekly*, 45: 55-64.

Kathuria, V., Raj, R.S.N & Sen, K (2011): "Productivity Measurement in Indian Manufacturing: A Comparison of Alternative Methods", Working Paper No. 31, Institute of Development Policy and Management, School of Environment and Development, University of Manchester, UK (http:// www.sed.manchester.ac.uk/idpm/research/publications/wp/depp/ depp_ wp31.htm).

Kelejian H. H. (1972): "The Estimation of Cobb-Douglas Type Functions with Multiplicative and Additive Errors: A Further Analysis," *International Economic Review*, Vol. XIII, pp. 179-182.

Kendrick, J. W. (1991): "Appraising the U. S. output and productivity estimates for government – where do we go from here?", *Review of Income and Wealth*, 37: 149-58.

Kiefer D.M. (1964): "Winds of Change in Industrial Chemical Research", in *Chemical and Engineering News*, 42: 88-109.

Klacek J., Vosvrda M., Schlosser S. (2007): "KLE Production Function and Total Factor Productivity", in Statistika, No. 4.

Kmenta J. (1967): "On Estimation of CES Production Function", in *International Economic Review.*

K. Mukherjee & S. C. Ray (2004): "Technical Efficiency and Its Dynamics in Indian Manufacturing: An Inter-State Analysis," Working Paper, No. 18, University of Connecticut, Storrs, 2004.

Krishna, P. & Mitra, D. (1998): "Trade liberalization, market discipline and productivity growth: new evidence from India", *Journal of Development Economics*, 56, 447-462.

Krugman, P. (1990): "The Age of Diminished Expectations", MIT Press, Massachusetts, Cambridge.

Kumar, S. A. (2004): "Decomposition of Total Factor Productivity Growth: A Regional Analysis of Indian Industrial Manufacturing Growth", Working Paper No. 22, NIPFP, New Delhi, India.

Kumbhakar, S.C. & Lovell, C.A.K. (2000): "Stochastic Frontier Analysis". Cambridge: Cambridge Uni. Press.

Kumbhakar S. (2000): "Estimation and Decomposition of Productivity Change when Production is not Efficient: A Panel Data Approach", *Econometric Reviews*, 19, 425- 460.

Kumbhakar, S.C., A. Heshmati & L. Hjalmarsson (1999): "Parametric Approaches to Productivity Measurement: A Comparison among Alternative Models." *Scandinavian Journal of Economics* 101, 405-424.

Levinsohn, J. & Petrin, A (2003): "Estimating Production Functions Using Inputs to Control for Unobservables", *Review of Economic Studies*, 70, 317–342.

Lichtenberg, F., R. (1995), "The Output Contributions of Computer Equipment and Personal: A Firm-Level Analysis.", in *Economics of Innovation and New Technology* No.3, pp. 201–217.

Lipsey, R. G. & K. Carlaw (2001): "What does Total Factor Productivity measure?", Study Paper version 2, Simon Fraser University, Vancouver BC (http://www.csls.ca/ipm/1/lipsey-e.pdf).

Lithwick, N. H., Post, G. & Rymes, T. K. (1967): "Postwar Production Relationships in Canada", *ed. M. Brown*, 'The Theory and Empirical Analysis of Production', New York: Columbia Uni. Press for the NBER.

Lovell, C. A. K. (1993): "Production frontiers and productive efficiency" in *H. O. Fried, C. A. K. Lovell and S.S. Schmidt (eds.)* 'The Measurement of Productivity Efficiency-Techniques and Applications', Oxford University Press, New York: pp. 3-67.

Lucas, R. (1988): "On the Mechanics of Economic Development", in *Journal of Monetary Economics*, 22(1): 3-42.

M. Daly, I. Gorman, G. Lenjosek, A. MacNevin & W. Phiriyapreunt (1993): "The Impact of Regional Investment Incentives on Employment and Productivity: Some Canadian Evidence," *Regional Science and Urban Economics*, Vol. 23, No. 4, pp. 559-575.

Mahadevan, R. (2003): "To measure or not to measure Total Factor Productivity Growth?" *Oxford Development Studies*, 31(3): 365-78.

Mahadevan, R. (2004): "The Economics of Productivity in Asia and Australia", Edward Elgar, Cheltenham.

Mairesse, J. (1978), "New estimates of Embodied and Disembodied Technical Progress", in Annales de l'INSEE, 30-31: 681-720.

Majumdar, R. (2004): "Productivity Growth in Small Enterprises: Role of Inputs, Technological Progress and learning by doing", *The Indian Journal of Labour Economics*, 47: 901-11.

Majumdar, S. K. (1996): "Fall and rise of productivity in Indian industry: has economic liberalization had any impact?", *Economic and Political Weekly*, 31, M-46-M-53.

Mansfield, E. (1971): "Social and Private Rates of Return from Industrial Innovations", in *Quarterly Journal of Economics*, Vol. 91.

Mansfield, E. (1980): "Basic Research and Productivity Increase in Manufacturing", *American Economic Review*, 70: 863-73.

Marjit, S and Kar, S. (2009): "A Contemporary Perspective on Informal Labour Market – Theory, Policy and the Indian Experience", *Economic and Political Weekly*, 44: 60-71.

M. D. Little, D. Mazumdar & J. M. Page Jr. (1987), "Small Manufacturing Enterprises: A Comparative Study of India and Other Economics," Oxford University Press, Oxford.

Md. Z. Hossain & K. S. Al-Amri (2010): "Use of Cobb-Douglas Production Model on Some Selected Manufacturing Industries in Oman," Education, Business and Society: Contemporary Middle Eastern Issues, Vol. 3, No. 2, pp. 78-85.

Md. Z. Hossain, M. I. Bhatti & Md. Z. Ali (2004): "An Econometric Analysis of Some Major Manufacturing Industries: A Case Study," *Managerial Auditing Journal*, Vol. 19, No. 6, 2004, pp. 790-795.

M. Jung, K. Lee & K. Fukao (2008): "Total Factor Productivity of Korean Firms and Catching up with the Japanese Firms," *Seoul Journal of Economics*, Vol. 20, No. 1, 2008, pp. 93-139.

Mohanty, D. (1992): "Growth and Productivity in The Indian Economy, Reserve Bank of India", Occasional Papers, 13: 55-80.

Muellbauer, J. (1986): "The Assessment: Productivity and Competitiveness in British Manufacturing", *Oxford Review of Economic Policy*, 2(3): 1-25

Mukherjee, K. & S. Ray (2004): "Technical efficiency and its dynamics in Indian manufacturing: an inter-states analysis", Department of Economics, Working Paper Series 2018, University of Connecticut.

Murti & V.K. Sastry (1957): "Production Functions for Indian Industry," *Econometrica*, Vol. 25, No. 2, p.205-221.

Nadiri, M. (1980): "Sectoral Productivity Slowdown", *American Economic Review*, 70: 349-355.

Nadiri, M. I. (1982): "Producers Theory", eds. *K.J. Arrow & M. D. Intriligator*, 'Handbook of Mathematical Economics', (Amsterdam: North Holland, 1982), 445.

Nair K.N.S. (1980): "Technological Changes in Agriculture: Impact on Productivity and Employment", Birla Institute of Scientific Research, Economic Research Division, Vision Books.

Nelson, R. (Ed.) (1993): "National Innovation Systems: A Comparative Study", Oxford University Press, New York.

Nelson, R.R. & Pack, H. (1999): "The Asian Miracle and Modern Economic Growth Theory", *Economic Journal* 109, 416-436.

Nelson, R. R. (1981): "Research on Productivity Growth and Productivity Differences: Dead Ends and New Departures", *Journal of Economic Literature*, XIX, September 1981, 1029-1064.

Neogi & B. Ghosh (1994): "Inter-temporal Efficiency Variations in Indian Manufacturing Industries," *Journal of Productivity Analysis*, Vol. 5, No. 3, pp. 301-324.

N. Maniadakis & E. Thanassoulis (2004): "A Cost Malmquist Productivity Index," *European Journal of Operational Research*, Vol. 154, No. 2, pp. 396-409.

Norsworthy, J. R. & Jang, S. L. Jang (1992): "Empirical measurement and analysis of productivity and technological change: applications in high technology and service industries", In *Jorgenson, D. W. & Laffont, J. J. (eds.)* 'Contributions to Economic Analysis Series', North-Holland, 1992.

Olley, G. S. & Pakes, A. (1996): "The Dynamics of Productivity in the Telecommunications Equipment Industry", *Econometrica,* 64 (6), 1263–1297.

Patel, P., Pavitt, K. (1994): "Technological Competences in the World's Largest Firms", Science Policy Research Unit, Brighton.

Paul Crabtree : "A Framework for Understanding Technology and Technological Change", *The Innovation Journal: The Public Sector Innovation Journal,* Vol. 11(1), pp.1-16.

Pradhan, G. & Barik, K. (1998): "Fluctuating total factor productivity in India: evidence from selected polluting industries", *Economic and Political Weekly,* 33: M25–M30.

Pasinetti, L. L., Structural Change and Economic Growth (Cambridge: Cambridge University Press, 1981).

Pasinetti, L. L. (1965): "A New Theoretical Approach to the Problems of Economic Growth", (Vatican: Pontificia Academia Scientiarum.

Peterson, William (1979): "Total Factor Productivity in the U.K.: A Disaggregated Analysis", *eds. K.D. Patterson and K. Schott,* 'The Measurement of Capital: Theory and Practice' (London: Macmillan, 1979).

Quah, D.T. (1996): "Regional convergence clusters across Europe". CEP-LSE Discussion Paper 274.

Raj, R.S.N. (2006): "Productivity and Technical Efficiency in the Indian Unorganized Manufacturing Sector: Temporal and Spatial Analysis", unpublished Ph.D. dissertation submitted to IIT Chennai.

Raj, R.S.N. (2011): "Structure, Employment and Productivity Growth in the Indian Unorganized Manufacturing Sector: An Industry Level Analysis", *Singapore Economic Review,* 56: 349-376.

Raj, R.S.N. & Babu, S.M (2011): "Productivity and Efficiency of Unorganised Manufacturing Sector", Project report submitted to ICSSR, CMDR, Dharwad.

Raj, R.S.N. and Sen, K. (2011): "Did International Trade Destroy or Create Jobs in Indian Manufacturing?", *European Journal of Development Research,* Forthcoming.

Revankar, Nagesh S. (1971): "A Class of Variable Elasticity of Substitution Production Functions", *Econometrica,* Vol. 39, No. 1, pp.61-71.

Rip A. & René Kemp (1997): "Technological Change", v2c6.fra, November 18, 1997.

Robinson, J. (1938): "The classification of inventions", *Review of Economic Studies* 5, 139-142.

Robinson, J. (1953): "The production function and the theory of capital", *Review of Economic Studies,* Vol. XXI, pp. 81-106.

Romer, P.M. (1986): "Increasing Returns and Long-run Growth", *Jour. of Pol. Economy*, 94(5): 1002-1037.

Romer, P.M. (1990): "Endogenous Technological Change", *Journal of Political Economy*, 98: S71-S102.

Romer, P.M. (1993): "Implementing A National Technology Strategy with Self-organizing Industry Investment board", Brookings Paper on Economic Activity: Microeconomics.

Rao, J.M. (1996a): "Manufacturing Productivity Growth: Method and Measurement", *Economic and Political Weekly*, 31: 2927-36.

Rao, J.M. (1996b): "Indices of Industrial Productivity Growth: Disaggregation and Interpretation", *Economic and Political Weekly*, 32: 3177-88.

Ray, S. C. (2002): "Did India's Economic Reforms Improve Efficiency and Productivity? A Non-parametric Analysis of the Initial Evidence from Manufacturing", *Indian Economic Review*, 37, 23-57.

RBI (2004): "Report on Currency and Finance, 2002-03", Reserve Bank of India, 2004.

Ruttan, V.W. (1960): "Research on the Economics of Technological Change in American Agriculture", *J. Farm Econ*, Vol. 42 (4): pp. 735-754.

Rymes T.K. (1972): "The Measurement of Capital and Total Factor Productivity in the Context of the Cambridge Theory of Capital", *Review of Income and Wealth*, XVIII, March 1972, 79-108.

Salim, R. A. & K.P. Kalirajan (1999): "Sources of output growth in Bangladesh food processing industries: a decomposition analysis", *The Developing Economies*, 37, 247-269.

Samuelson P. (1979): "Paul Douglas's Measurement of Production Functions and Marginal Productivities," *Journal of Political Economy*, Vol. 87, No. 5, pp. 923-939.

Sandelin B. (1976): "On the Origin of the Cobb-Douglas Production Function," *Economy and History*, Vol. 19, No. 2, pp. 117-125.

Sargent, T. C. & Rodriguez, E. R. (2000): "Labour or total factor productivity: do we need to choose?", International Productivity Monitor, 1: 41-44.

Scherer, F.M. (1971): "Industrial Market Structure and Economic Performance", Chicago: Rand McNally.

Scherer, F.M. (1982): "Inter-industry technology flows and productivity growth", *Review of Economics and Statistics*, 64: 627-34.

Schmidt, G. (1997): "Dynamic Analysis of a Solow-Romer Model of Endogenous Economic Growth", Working Paper no.68, COPS.

Schmookler (1966): "Invention and Economic Growth", Harvard University Press, Cambridge, U.K.

Schumpeter, J. A. (1939): "Business Cycles: A Theoretical, Historical and Statistical Analysis of Capitalist Processes", New York: Macmillan.

Stigler, G. (1947): "Trends in Output and Employment", New York: NBER. *Stiglitz, J.* (1992), "Endogenous Growth and Cycles", Stanford University Working Paper.

Shaw, S.L. (1979): "Technological Innovations and Growth in Indian Agriculture", Commerce, Vol.139, pp.3577 – 5963.

Solow, R. W. (1957): "Technical Change and Aggregate Production Function", *Rev. Econ. Stat.*, Vol. 39: pp.312 – 320.

Solow, R. M., Capital Theory and the Rate of Return (Amsterdam: North Holland, 1963), 10-11.

Srivastava, U.K., R.W. Crown & E.O. Heady (1975): "Green Revolution and Farm Income Distribution", *Development Digest*, 20: pp. 65 – 80.

Srivastava, V. (2000): "The Impact of India's Economic Reforms on Industrial Productivity, Efficiency and Competitiveness: A Panel Study of Indian Companies", Report of a project sponsored by the IDBI, NCAER, New Delhi.

Srivastava V. & A. Sengupta (2000): "Endogeneity of the Solow Residual: Some evidence from Indian Data", National Council of Applied Economic Research (NCAER), Discussion Paper 13, June.

Star, S. (1974): "Accounting for the growth of output", *American Economic Review*, 64: 123-35.

Stoneman, P. & N., Francis (1994): "Double Deflation and the Measurement of Output and Productivity in UK Manufacturing, 1979-1989", in *International Journal of the Economics of Business*, 1: 423-437.

S. V. Lall & G. C. Rodrigo (2001): "Perspective on the Sources of Heterogeneity in Indian Industry," World Development, Vol. 29, No. 12, pp.2127-2143.

Tata Services Limited (2003): "Reforms and Productivity Trends in Indian Manufacturing Sector", Department of Economics and Statistics, Tata Services Limited, Mumbai.

Terleckyj, N. (1974): "Effects of R&D on the Productivity Growth of Industries: An Exploratory Study", Washington DC: National Planning Association.

Tinbergen, J. (1942): "Zur Theorie des Langfristigen Wirtschaftsentwicklung", reprinted in Tinbergen, J. (1959) Selected Papers, Amsterdam. Zaman, G., Goschin, Z., Pârþachi, I and C. Herþeliu (2007), "The Contribution of Labour and Capital to Romania's and Moldova's Economic Growth", Journal of Applied Quantitative Methods, 2(1): 176-185.

Trivedi, P., A. Prakash & D. Sinate (2000): "Productivity in Major Manufacturing Industries in India: 1973-74 to 1997-98", Study No. 20, DRG, RBI, August.

Trivedi, Pushpa (2004): "An Inter-State Perspective on Manufacturing Productivity in India: 1980- 81 to 2000-01", *Indian Economic Review*, 39, 203-237.

Unel, Bulent (2003): "Productivity Trends in India's Manufacturing Sectors in the last Two Decades", IMF Working Paper No. WP/03/22.

Unni, J., Lalitha, N. & Rani, U. (2001): "Economic Reforms and Productivity Trends in Indian Manufacturing", *Economic and Political Weekly*, 36, 3915-22.

Usher, Dan (1980): "The Measurement of Economic Growth", New York: Columbia University Press.

Wagner, K. & Ark, B. (1996): "International Productivity Differences: Measurement and Explanations", North-Holland, Amsterdam.

Wizarat S. (1989): "Sources of Growth in Pakistan's Large Scale Manufacturing Sector: 1955-56 to 1980-81," *Pakistan Economic and Social Review*, Vol. 21, No. 2, pp. 139-159.

W.L. Weber & B.R. Domazlicky (1999): "Total Factor Productivity Growth in Manufacturing: A Regional Approach Using Linear Programming," *Regional Science and Urban Economics*, Vol. 29, No. 1, pp. 105-122.

Yeager, Leland B. (1979): "Capital Paradoxes and the Concept of Waiting", *ed. M. J. Rizzo*, 'Time, Uncertainty and Disequilibrium: Exploration of Austrian Themes', (Lexington, Mass.: D. C. Heath, 1979).

Y. Nikaido (2004): "Technical Efficiency of Small-Scale Industry: Application of Stochastic Production Frontier Model," *Economic and Political Weekly*, Vol. 39, No. 6, pp. 592-597.

Young, A. H. & Musgrave, J. C. (1980): "Estimation of Capital Stock in the United States", in *ed. Dan Usher*, 'The Measurement of Capital' (Chicago: University of Chicago Press for the NBER, 1980).

Zaman, G. & Z. Goschin (2007a): "Analysis of Macroeconomic Production Functions for Romania. Part one- the time-series approach", in *Economic Computation and Economic Cybernetics Studies and Research*, 41(1-2): 31-46.

Zaman, G. & Z. Goschin (2007b): "Analysis of Macroeconomic Production Functions for Romania. Part two - The cross-section approach", in *Economic Computation and Economic Cybernetics Studies and Research*, 42(3-4): 23-32.

Zarembka, P. (1976): "Real Capital and the Neoclassical Production Function", *eds. F. L. Altmann, 0. Kyn & H.-J. Wagener*, 'On the Measurement of Factor Productivities', Gijttingen: Vandenhoeck & Ruprecht.

Zellener, A. S., Kementa, J. & Dreze, J. (1966): "Specification and estimation of Cobb-Douglas production functions", *Econometrica*, 34: 784-95.

3

Review of Literature

The review of literature serves as a background for any research and scientific investigation which helps in understanding it in proper perspective. An exhaustive review of literature help any researcher to identify variables relevant for research, decide tools and techniques to be adopted, delineate a new area of study and relate the present study with the previous ones. Productivity and Technological change assumes a greater significance for developing nations like India, which are severely constrained by shortage of capital still in the era of 21st century. It therefore, become more imperative for a nation to make use of this scarce resource which a greater efficiency and in such a manner which could ensure maximum possible exploitation of existing factor endowments. A significant of literature is available, thus, an attempt has been made to review the literature available on the subject separately in this chapter. Keeping this in view, all the available literatures concerning the problem is being reviewed under the following heads/sub-heads:

3.1: General Theoretical Frame related to present study

3.2: Overview of India's Manufacturing Sector

3.3: Methodological Issues related to Productivity, Production Functions & Technological Change aspects

3.4: Wood & Forest based related aspects

3.1: General Theoretical Frame related to Present Study

As we know that Economists especially of developed nations and some developing countries have always been a search of the factors responsible for growth of a nation. The classical school laid on increases in quantity of conventional inputs (i.e. land, labour and capital) and treated them as factors of growth. The pioneering work of Schumpeter in early 20th century and findings of subsequent studies concerning developed nations have, however, unfolded the fact that there are some other factors also making significant contribution to growth of a nation; technological change happens to be the most crucial. Many researchers for U. S. economy have explored the field of technological change. Their contribution has enabled many more researchers to explore the alternative prospects further. In following paras the literature on theoretical frame of the area has been reviewed.

In 1957, Robert Solow published a paper wherein he found empirically that for the period 1909 to 1949, 87.5 percent of growth in the United States (US) gross output per man-hour was due to the technical change or productivity increase. He explained this by distinguishing two main sources of output growth: (i) growth due to the contribution of capital and labour inputs and (ii) due to the rate of productivity growth. In which, the former does not lead to any change in the production function while the latter lead to the shifting of the production function indicating technical efficiency which could be due to change in technology, learning by doing, capacity utilization, economies of scale etc (Ahluwalia, 1991).

The rate of technical change is, thus generally identified with the proportionate amount of shift over time in the aggregate production function. This could be measured empirically as a residual between the growth of output and the weighted sum of inputs (Haltmaier, 1984). This ice breaking revelation provided the theoretical foundation for almost all the subsequent work on productivity measurement (Hall, 1989). But, there is a great controversy regarding the methodology used for estimating the total factor productivity (Hulten, 2000; Trivedi *et al.*, 2000). Even, though, the rate of growth of productivity in the industrial sector has been put forward as the key phenomenon in determining the sectoral evolution (Pack, 1988). But to achieve the high productivity growth, the appropriate policy framework is required, regarding which, no consensus is found in the literature (ibid).

Their contributions have enabled many researchers to explore the potentiality further. Abramovitz (1956), Fabricant (1959), Kendrick (1961) and Solow (1957) established considerable research on the measurement and contributed significant in the literature on theoretical frame of technological change. Ferguson (1964-65) study related to empirical research on production functions and technological change in American manufacturing industry. Dhrymes (1965) has obtained elasticity of substitution in an interstate cross section study in which he used capital figures from the US census of manufacturers. Whereas, Jorgenson & Griliches (1967) has made an attempt to explain the ways the very existence of total factor productivity. They used the Divisia index to calculate the rate of growth of TFP as an index of the rate of growth in output. Jorgenson (1988) study related to technological innovation and productivity change in Japan and the US. More than a six decade ago Abramovitz (1956), Fabricant (1959), Kendrick (1961) and Solow (1957) established on the basis of measure such as equation given below as (1) and (2) that the conventionally measured inputs, capital and labour, leave a large portion of the growth of output unexplained.

$$\frac{dA}{A} = \frac{Q1/Qo}{(wL1 + rKl)/(wLo + rKo)} - 1 \qquad ....(1)$$

and

$$\frac{dA}{A} = \frac{dQ}{Q} - \propto \frac{dL}{L} + \beta \frac{dK}{K} \qquad ....(2)$$

$$\beta = 1 - \propto$$

Since then considerable research on the measurement, determinants and consequences of factor productivity has been undertaken. To give some idea about the empirical behavior of factor productivity and some related issues, the following facts based on Kendrick's work (1970) on the U.S. economy are worthwhile to mention. The rate of growth of total private output increased from 2.8% during 1919-48 to 4.0% during the years 1948-66, but with considerable variation over the period due to cyclical factors.

The rate of growth of labour input was fairly steady, though with some fluctuation. However, the growth rate of capital services varied considerably, mainly due to fluctuations in its rate of utilization. Though capital accumulation increased since 1910, the rate of return on capital remained constant and the share of labour rose considerably. Both of these were made possible by the advancement of technical change, permitting substituion of capital for labour. The rate of increase in labour productivity was generally low in the first phase and very high towards the end of contractions and the beginning of expansions. Short-run production functions generally indicate increasing returns to labour alone. TFP increased by about 2.1% per annum for the whole period, but with considerable yearly fluctuations. There was substantial variation among the rate of growth in various industries.

There was a mild acceleration in the rate of growth of factor productivity after World War II, perhaps due to the stability of the economy. In addition, the dispersion in the rate of growth among industries has declined. Technological change has been more widely diffused. The costs of both K and L have decreased and the capital-labour ratio has increased in most of the industries during the postwar period. Thus, the picture is one of diversity. The economy grew at different rates in different periods, and different industries have been responsible for larger or smaller contributions to the growth of the aggregate economy. There is also evidence that suggests interdependency between the overall growth rate and the industrial structure of the U.S. economy.

Massell (1961) has made an attempt to extend the analysis of the earlier paper by refining the catchall technological change category to distinguish among several component parts. He did this by disaggregating to the SIC two-digit industry level and considering the 19 industry groups within the manufacturing segment of the U.S. economy, for the period 1946-57. He has also given some notion of the importance of the aggregation problem. He pointed out that the disaggregation yields additional information regarding productivity change; for example, it reveals the dispersion around the average of industry rates of technical change, and a method of separating technical change into an inter-industry and intra-industry component. The discrepancy between the two measures, which he termed inter-industry technical change, amounts to nearly one-third of the change in aggregate technology. Most of this can be attributed to the shifting of capital to industries where its marginal productivity was higher.

Stigler (1963) estimated elasticities of substitution by comparing capital receipts ratios and wage rates for small and large firms. He obtained very high estimates of the elasticities averaging around 4.0. Further he also points out his assumption that capital costs are the same for a large and small firm gives his estimates an upwards bias.

Ferguson (1964-65) study related to empirical research on production functions and technological progress in American manufacturing industry. The SMAC function was fitted to time series data for the period 1949-61. The main purpose of the paper was to discover the extent of neutral technological progress and the existence of biased technological progress in two-digit American manufacturing industries. Further he concluded that the technological progress appeared, on balance, to be either neutral or capital using in 16 of the 19 industries. Only three industries, however, are large by any measure of size. Thus, he pointed out when the manufacturing sector is aggregated, capital-saving changes in these three industries may largely offset capital-using changes in many smaller industries.

Whereas Dhrymes (1965) has elasticities of substitution in an interstate cross-section study in which he used capital figures from the U.S. Census of manufacturers. He used a somewhat complicated model and runs regressions of labour requirements on the wage rate. The median of his estimates is 0.94.

Ramsuder (1965) study stated main conclusion that a 12% rate of growth per annum of industry is in balance with a rate of growth of agriculture in the range 5.7 to 5.9%. The term balance was used in this context to mean that there is balance between total supply (including imports) and total demand (including exports) at base year price. He has also worked out the changes in the overall direct tax-rate that will have to be made in order to make a 5% growth rate of 'agriculture' balance a 12% growth of 'industry'. In spite of this he pointed out that the direct tax rate as a proportion of Net National Product will have to be raised from 5% to a value in the range 7.4% to 7.9%.

The overall findings of Denison study (1967) on growth of total factor productivity in nine Western European countries for the period 1950-62 were: West Germany 7.31%, Italy 6.07%, the Netherlands 4.71%, Denmark 3.51%, Norway 3.51%, U. S. 3.31%, Belgium 3.21% and U. K. 2.31%. There were, however, differences in rankings of these countries by growth rates of national income and ranking by growth rate of factor input.

Where as, Jargenson & Griliches (1967) attempted to explain away the very existence of total factor productivity. They adopted the usual neoclassical assumption of competition, a constant returns to scale production function and they considered technical change as a shift in the production function which is used as an organizing scheme as in Dension's (1967) work. They have used the Divisia index to calculate the rate of growth of TFP ($\delta A/A$) as an index of the rate of growth in output. Their empirical calculation suggests that the rate of growth of TFP has been about 0.1% per annum for the period 1948-65.

A study has been conducted by Wolfe (1968) entitled –"Productivity and Growth in Manufacturing Industry: Some Reflections on Professor Kaldors Inaugural Lecture." According to him – "Kaldor asserts that there are increasing returns to size in manufacturing industry but not in non-manufacturing activities. He further asserts that the growth of the productivity of factor inputs in manufacturing will depend largely upon the rate of growth of manufacturing output. The productivity of factor inputs in non-manufacturing activities may also grow, but this growth will not depend on the growth of total output." In this paper Wolfe concluded Kaldor lecture's as the rate of growth of output as a whole may be increased by making labour more plentifully available to manufacturing industry.

Clague (1969) estimates elasticity of substitution using both engineering and accounting data from United Stated and Peru. The methodology is the same as the SMAC paper. Clague's studies reveals that the elasticities are less than unity except in the case of cotton spinning.

However, Pack (1974) estimated elasticity of substitution and capital labour ratios for many production operations and concludes that the production operation like grain milling, paints, tyres, cotton and woollen textiles, *etc.*, the elasticity is much above unity. The elasticity of substitution was calculated using the formula.

$$\frac{(K/L)i}{(K/L)j} = \frac{(W/R)i}{(W/r)j} \quad \text{(derived by Arrow \textit{et al.}, in 1961)}$$

Pack extends his study further to suggest that the industries in which the ratio is high, if these industries were to move into corner points of optional capital labour combination both employment and value added would be enhanced. He concludes that the preceding results suggest that considerable substitution possibilities exist in a number of manufacturing industries while there may be some branches in which substitution is not possible. The dictates of income maximization when capital is the relatively scarce factor requires consideration where technologies permit labour-intensive production.

Mansfield (1980) in his article of U.S. manufacturing companies have determine whether an industry's or firm's rate of productivity change in recent years had been related to the amount of basic research it performed, when other relevant variables (such as its rate of expenditure of applied R & D) were held constant. Besides investigating the relationship between basic research and productivity increase at the industry level, he also investigated this relationship at the firm level. His results indicate that there was a statistically significant and direct relationship between the amount of basic research out by and industry or firm and its rate of increase of total productivity, when its expenditures on applied R & D were held constant. It may also reflect a tendency for basic research for applied R & D to be more effective when carried out in conjunction with some basic research. His findings also indicate that the composition of many industries R & D expenditures had changed in the last decade. Practically all industries have cut the proportion of their R & D expenditure going for basic research.

Nishimizu & Robinson's (1984) study of the manufacturing sector of Korea, Turkey and Yugoslavia attempts to use a quantitative framework and concludes that a positive association does exist between outward orientation and productivity growth. The impact of trade regimes on sectoral TFP growth has been explored by Nishimizu and Robinson within a quantitative frameqork in a study which analyses inter-industry differences in TFPG in Korea, Turkey and Yugoslavia with Japan as the comparator. The authors that applying covariance analysis to our panel data indicates that there are significant difference in the estimated regressions among industries in each country and across countries in each industry. Their quantitative analysis at the two-digit level leads them to conclude that substantial portions of the variation in TFP growth rates are "explained" by output growth allocated to export expansion and import substitution in Korea, Turkey and Yugoslavia, but interestingly not in Japan. Further they conclude that import substitution regimes seem to be negatively correlated with TFP change, where as export expansion regimes are positively correlated with TFP change.

Where as, Jorgenson (1988) study related to technological innovation and productivity change in Japan and the United States. He presented the sources of economic growth for Japan and the United States over the period from 1960 to 1979 also. In spite of this he saw that the growth of output was 8.3% in Japan and only 3.5% in the United States. He pointed out that this growth in output in the two countries among its three sources, namely, the contribution of capital input, the contribution of labour input ant the rate of technical change. Further he pointed out that the most important contributor to economic growth was the growth of capital input. He also concluded that major contributor to economic growth in the two countries was labour input which was accounted 1.5% for the Japanese growth rate and 1.2% for U. S. growth rate. While growth capital input was accounted for both countries 5 and 1.5%, respectively. This amounts to 60% of Japanese growth and 40% of U. S. growth. The rate of technical change was an important contributor as well, at nearly 2% in Japan and 0.7% in the United States.

A review has been carried out by Chakraborty (1982) on "CES Production Function of Recent Literature". This survey gave a comprehensive view of the research carried out in one important area of the applied production analysis. The main objective of the Chakraborty paper was to survey the recent empirical studies on C.E.S. production function presenting a critical over-view of the estimates of the different parameters (*e.g.* elasticity of substitution, returns to scale, *etc.*) of the various studies across the different countries involving cross section, time-series and pooled time series cross-section data at different levels of aggregation.

The main differences between the specification of the estimation model for use with time series data and the model involving cross-section data lies I introducing technological change and other effects which may cause the production function to shift over time. In most of the studies where researchers are more interested in elasticity of substitution rather than the rate of technological change, the technological change is specified as being Hick s neutral and constant. Studies of the aggregate elasticity of substitution for manufacturing sector have been made by Asher and Krishna Kumar (1973), for USSR, Hungary, Yugoslavia, USA, Canada and Isreal; Desai (1976) for USSR; Behrman (1972) for Chile; Burley (1973) for Australia; Tsurumi (1970) for Canada; Kazi *et.al* (1976) for Pakistan; Berndt (1976) for USA. Table 3.1 summarizes time series estimates of the elasticity of substitution and technical change.

Table 3.1 yields the following interesting conclusions which can be summarized as: (i) there is a wide variety of difference in the size of the elasticities across the countries. (ii) Based on nonlinear techniques and almost for the same period with same technological assumption of Hicks-neutrality Asher & Krishna kumar (1973) and Desai (1976) show that for USSR the elasticity of substitution is less than one. (iii) Results for USA by Brandt (1976) differ from that of (i) (in 1973). (iv) The elasticity of substitution for Pakistan is negative, which may indicate that there is no substitution. (v) For Canada the elasticity of substitution obtained by Tsurumi (1970) differs from that obtained by Asher & Krishnakumar. This may be due to the fact that he has used longer period of data.

Table 3.1: "Time Series Estimates of Aggregate Elasticity of Substitution & Technical Change tor the Manufacturing Sector".

Reference	Country Period	Elasticity	Assumption as to Nature of Technical Change	Technical Progress
Desai (1976)	U. S. S. R. (1955-69)	0.2771	Hicks-neutral	0.0408
Behrman (1972)	Chile (1945-65)	0.76	Hicks-neutral	0.017
Burley (1973)	Australia (1949-50 to 67-68)	0.1783	–	–
Tsurumi (1970)	Canada (1926-39,1946-67)	0.8300 1.0000 1.0000	Hicks-neutral	0.0113 0.0209
Asher & Krishna Kumar	U.S. S. R. (1950-69)	0.1082- 0.4035	Hicks-neutral	–
(1973)	Hangari (1950-62)	3.2410 (108.69)	Hicks-neutral	–
	Yugoslavia	0.6322 (31.926)	Hicks-neutral	0.451 (0.0026)
	U. S. A. (1950-66)	4.0983 (26.188)	Hicks-neutral	0.0287 (0.0063)
	Canada (1926-55)	0.6307 (29.09)	Hicks-neutral	0.0142 (0.003)
	Isreal	0.1178		
Berndt (1976)	U. S. A. (1929-68)	1.148- 1.245	Hicks-neutral	0.0439 (0.0086)
Kazi *et al.* (1976)	Pakistan (1954-69-70)	0.22 0.13	– –	– –

The results on technological change {see Table 3.1(A)} reveal a marked diversity in economic performance of the various countries manufacturing sector. The growth rates of the (Hicks-neutral) technical change do not show a uniform pattern. Possibly one has to explain this diversity in terms of various factors economic as well as non-economic organization, changes in the qualities of inputs, changes in the institutional, social and cultural factors operating within each country.

Bairam (1988) have studied which study was more specific in the respect of Verdoorn's Law. In the paper his purpose was to show that Kaldor's specification and interpretation of Verdoorn's original model could be wrong. Wherever Kaldor suggested that, in his inaugural lecture (1966), the growth of industrial productivity might be explained by the principle of Vardoorn's Law (1949). And, he further argues that a sufficient condition for the presence of economics of scale is the existence of a significant relationship the rate of growth of labour productivity and the rate of growth of output, indicated by a regression coefficient, which is statistically significant less than unity. According to Bairam, it is clear that if Verdoorn's Law is accepted as the underlying structure of the law, certain specification and interpretation problems arise. The main problems which arises was that, unless one assume an unlimited supply of labour to manufacturing, it is not clear from the model whether verdoorn's coefficient is determined by labour supply conditions or by technological parameters, or by both. Furthermore, even if it can be assumed that surplus labour exists in the non/manufacturing sectors, so that the model can be specified as a single equation C–D production function, a Verdoorn's coefficient around 0.5 may not imply increasing returns to scale. This is because Verdoorn's Law, as specified by Kaldor, does not explicitly includes a variable designed to capture the contribution of the rate of growth of capital stock.

In their papers, Caves *et al.* (1982a,b) (hereafter, CCD) show that under certain circumstances, the Tornqvist index (which is the discrete counterpart of the Divisia index) is equivalent to the geometric mean of two Malmquist output productivity indexes. The conditions include technical efficiency, allocative efficiency, that the underlying technology must be Translog, and that all second-order terms must be identical over time. In contrast, the Malmquist index does not require any assumptions with respect to efficiency or functional form. Our specification of productivity change as the geometric mean of two Malmquist in- dexes stems from CCD. Moreover, they show that the Tornqvist index is "exact" for technology that is Translog (i.e., one can compute a nonparametric [in the sense that one need not estimate the parameters of technology] productivity index that is "exactly" consistent with the Translog form). Furthermore, since the Translog is flexible, the Tornqvist index is "superlative" in the terminology coined by W. Erwin Diewert (see *e.g.*, Diewert, 1976).

The Malmquist index was introduced by Caves *et al.* (1982a,b) who dubbed it the (output- based) Malmquist productivity index after Sten Malmquist, who earlier proposed con- structing quantity indexes as ratios of distance functions (see Malmquist, 1953). Distance functions are function representations of multiple-output, multiple-input technology which require data only on input and output quantities. Consequently, their Malmquist index was a "primal" index of productivity change that, in contrast to the Tornqvist index, does not require cost or revenue shares to aggregate inputs and outputs, yet is capable of measuring TFP growth in a multiple-output setting.

The main purpose of Sangha's (1992) study was to show that may or may not be a positive function of labour productivity, implying that under certain circumstances production and labour productivity may move in the opposite directions, thus rendering the Verdoorn Law partially inapplicable. The Verdoorn Law, which relates production to productivity, was valid only when the level of production was on the rise; it was not valid when output was on the decline. However, during economic slow-down or downturn author seems an inverse relationship between production and productivity.

Cheng-Ping Lin [2002] analyzes the cost function of construction firms with due consideration of their available resources by using Cobb-Douglas Production and Cost Functions. Moosup Jung, *et al.* [2008] made a study on Total Factor Productivity of Korean Firms and Catching up with the Japanese Firms. They measured and compared the TFP of both Korean and Japanese listed firms of 1984 to 2004. They found that the average TFP of Korean firms grew about 44.1% between 1984 and 2005, with 2.1% annual growth rates. Industry was observed to be outstanding.

Finn R. Forsund (2010) study reveals a philosophical problem for studies of inefficiency of firms is how to rationalize the inefficiency. Since economists do not have any theory for inefficiency, explaining the results of efficiency analyses are notoriously more difficult than carrying out the estimations. The literature points to measures of inputs and management as not including quality dimensions as a reason for measured efficiency differences, indicating that more work needs to be done on data collection. Strategic behavior in game situations between owners and management, and between management and labour may also show up as inefficiencies and another reason is technology differences. The frontier production function is the key to information on best-practice technology. Estimation of efficiency is usually done for units observed during the same time period, thus in this respect the measures are static. Interpretations of dynamic efficiency measurement are offered. The vintage model of substitutability between inputs including capital before investment, but no substitution possibilities after investment, and ex-post production possibilities characterized by fixed input coefficients, can rationalize inefficiency due to technology differences. Key elements in understanding structural change were the entering of capacity embodying new technology and exiting of capacity no longer able to yield positive quasi-rent. Three crucial production function concepts were identified: the ex-ante micro unit production function as relevant when investing in new capacity, the ex-post micro production function, and the short-run industry production function giving the production possibilities at the industry level. Productivity measurement, taking these types of production functions into consideration, leads to different interpretations of productivity change than traditional approaches not being clear about which production function concept is used.

Subal C. Kumbhakar, Hung-Jen Wang (2010) in their paper discusses the specification and estimation of technical efficiency in a variety of stochastic frontier production models. The focus is on cross-sectional models. They had started from the basic neoclassical production theory and introduce technical inefficiency in there. Various model specifications with several distributional assumptions on the inefficiency component were explored in detail. Theoretical and empirical issues were also illustrated with empirical examples using STATA.

Kirtti Ranjan Paltasingh. & R.K. Mishra (2013-14) in their study, an attempt has been made to through conceptual frame of a Heuristic Theory of J-Curve of productivity growth and to test that hypothesis at much disaggregate level of three Indian states. They discusses that the J-Curve pattern of productivity performance following liberalization has been popularized by Virmani (2005, 2006, and 2012) and Virmani & Hasim (2011), Hasim *et al.* (2009). The theory developed here, says that the productivity growth follows a J-curve pattern in different sub-periods. It means, the growth rate of productivity initially declines following liberalization policy and then starts rising in subsequent period which may be termed as maturity period of reform while the initial reform period is termed as drive to maturity. Findings of their study confirms that in case of little industrially developed states at the time of initiation of reforms like Maharashtra and Timil Nadu, productivity growth behaves in S-curve pattern in the sense that productivity increases in subsequent periods. But in case of Gujarat, it is like J-curve pattern in the sense the growth of productivity declines initially then increases in subsequent period. Thus, this study concludes that the impact of liberalization has not been similar on state manufacturing sectors.

3.2: Overview of India's Manufacturing Sector

Before taking past literature to review into consideration, firstly I would like to highlight the present scenario of Indian Manufacturing sector in brief. After rapid industrialization process which started since 1991, it has been observed that manufacturing production and exports have been driving the rapid growth of world biggest developing and dynamic emerging economies. However, it has not contributed perceptibly to India's growth story; nor has it been up to the urgent task of shifting surplus work force from the agriculture sector (Ministry of Commerce & Industry, Working Group Report, 2011). At present, India's manufacturing sector contributes about 16% to the GDP, and India's share in world manufacturing is only 1.8%. This is in stark contrast to China; where manufacturing contributes 34% to the GDP and is 13.7% of world manufacturing; up from 2.9% in 1991 (Planning Commission, the Manufacturing Plan 2013). In fact contribution of manufacturing to GDP for 2010 is higher for countries like Thailand (36%), Malaysia (25%) and Indonesia (25%) than India (15%) (World Data Bank, *http://databank.worldbank.org/ddp/home.do;*). The striking aspect of India's growth has been the dynamism of the service sector, while, in contrast, manufacturing has been less robust. In fact, Kochhar *et al.* (2006) point out that the change in the share of manufacturing in GDP in India between 1980 and 2000 has been 2.5 percentage points (pps) lower than the average country at the same stage of development, while the change in service share was 10 pps higher than average. If we consider contribution of Wood & Forest based industries to GDP, we have not found any suitable figure for the recent years, and hence, it can be said that WFBI is still a neglected area for the researchers. As per the report of FAO, 2002 (Food and Agriculture Organisation of the United Nations. 2002), in the year 2002, forestry industry contributed 1.7% to India's GDP while in 2010, the contribution to GDP dropped to 0.9%, largely because of rapid growth of the economy in other sectors and the government's decision to reform and reduce import tariffs to let imports satisfy the growing Indian demand for wood products.

Similarly like WFBI within manufacturing sector, manufacturing itself has not been the engine of growth for the Indian economy; it now needs to grow at a much faster rate to sustain in the external competitive environment. Further, the importance of manufacturing sector to the domestic and global economy is set to increase even further as a combination of supply-side advantages, policy initiatives, and private sector efforts set India on the path to a global manufacturing hub (IBEF 2012). Manufacturing is likely to contribute 25 percent to the GDP by 2025 as per the target set by the National Manufacturing Competitiveness Council (NMCC) report. However, in order to attain a ~25% share of the GDP by 2025, manufacturing would need to grow at a rate of ~2-4% higher than the GDP.

So far review of this section is concern, it has also been observed that a lot of studies have been done on performance of manufacturing sector of Indian economy and most of these studies focus on the productivity performance. Only few importance studies out of them have been taken up for review. Thus, in this section, we have tried to explore all the issues related with Indian manufacturing sector. The work of foreign-based scholars in the field of technological change and other related aspects enabled many more scholars to explore this field for Indian manufacturing. Some major works on Indian Manufacturing Sector by the Indian author and other scholars from the abroad has been considered to be review this section and their works are outlined as under.

In Dutta (1955) study an attempt has been made to analyse production function for 29 Indian manufacturing for the year 1946 and 1947. In this paper statistical data have been utilized for deriving the production function in terms of labour and capital and for studying the marginal productivities of labour and capital. The main conclusion have been arrived at out of Dutta study that the wage rate, net output per head and marginal productivity of capital have increased while average gross productivity of capital and marginal productivity of labour have decreased in 1947, in comparison with 1946. The increase in the value of the marginal productivity of capital indicates the anticipation of a further flow of capital into Indian industry.

Murty & Sastry (1957) have been studied Production Function for Industrial sector as a whole as well as for seven important industries in India for the period of 1951 & 1952 with the help of cross section data. An interesting application of production function of C-D type had also been used to estimate the total capital employed in 1952 in Indian industry as a whole and in the three major industries of Cotton, Jute and Tea, separately. They observed that the C-D type production function has provided a good fit to the data. The coefficients of multiple correlations was found highly significant and slow that during both the year 1951 and 1952 more than 90 percent of the variation in net value of output explained by the two factors of production, *viz.* labour and capital. The differences in the elasticities of output (with respect of L & K) for the both years cannot be regarded as significant.

Reddy and Rao (1962) cover the period 1946-57. During these twelve years the Solow index of TFP, computed by them, shows a fall by 9.5 cent. There was, however, no steady trend. They find empirical evidence in support of the hypothesis of Hick-neutrality. It has been observed in their study that the factor shares are approximately equal to their respective Cobb-Douglas parameters. This is treated as evidence in support of the applicability of the marginal productivity theory of distribution.

Sengupta (1963) using ordinary least squares fitted a first degree homogeneous CES production function to 7 industries for the period 1948 to 1958, to test for two hypotheses: (i) elasticity of substitution equals to zero and (ii) elasticity of substitution equal to one separately. He found neutral technical change not significant in the Cement Industry, significant in Iron and Steel, and Jute industries and quite significant in the Sugar industry.

Rudra (1964) made an attempt to examine the dependence of industry on agriculture in three of its aspects. In the paper, he concluded that the rate of growth of industries is directly dependent on the rate of domestic saving but has no relation with the rate of investment. And, in spite of certain rates of growth of industry which are in balance with specified rates of growth of agriculture, one of his important conclusions was that a 12 per cent annual rate of growth of industry over the decade 1960-61 to 1970-71 is in balance not with 5% growth of agriculture as is believed but with a 6.9 per cent growth.

Singh (1966) analyses productivity trends in Indian manufacturing for the period 1951-63. He observes a steady rise in labour productivity measured by value added (deflated) per worker and a fall in capital productivity measured by value added to productive capital ratio (both undeflated). It was also observed that the ratio of output to "input" (both in current prices) remains more or less constant. It was, accordingly inferred that there has been little improvement in overall productivity, so that most of the growth in labour productivity was attributable to capital deepening.

Where as, Yeong – Her Yeh (1966) in his paper, pointed out that economies of scale might exist for many industries in India; where he has uses two different techniques to test for economies of scale. According to him, since it is possible that the higher productivities of larger factories may be due to economies of scale, but to better quality of workers. The increasing productivities of both capital and labour for many industries lend support to the conclusion based on the Cobb–Douqlas function. Moreover, the results obtained from the factor shares method confirmed those yielded by the ordinary least squares method. His study covered the period 1953-58 and inferred increasing returns to scale.

Where as, Datta (1966) studied –"Productivity of Labour and capital in Indian Manufacturing during 1951-1961." He arrived at the constant returns to scale of total industry, on the basis of a time series study for the period 1951 to 1961. Gujarati (1967) in his paper have studied the relative importance of capital, labour, and technology in explaining the output growth in Indian manufacturing. The findings of Gujarati study was that in those industries for which the capital coefficient was significant, the elasticity of output with respect to capital was close to unity, whereas in those industries for which the labour coefficient was significant, the elasticity of output with respect to labour was substantially above unity.

Shivamaggi, Rajagopalan and Venkatachalam (1968) examined trends is wages in seven important industries during 1951 to 1961 and compared them with trends in labour productivity and costs of production during the same period. The seven industries were covered were cotton Textiles, Jute Textiles, Iron & Steel, Cement, Paper & Paper Boards, Chemicals & Chemicals Products and Sugar. The index of labour productivity was constructed from the figures obtained by dividing value added in constant prices by man-hours. He observed an increase in labour productivity and attributed the same to capital deepening.

Raj Krishna and Mehta (1968) in their study carefully take into account the incompatibilities between CMI and ASI data. They take aggregate of 27 CMI industries (out of 29) and comparable ASI industries. To form some judgement about overall productivity they consider cost-output ratios. These are computed on the basis of reported figures (in CMI and ASI) and no adjustment has been made for price change. Cost calculations have been done with and without capital cost. A significant rising trend is observed in both "not capital" costs per unit of output and total cost per unit of output. Raj Krishna & Mehta conclude according that overall productivity has been declining

In the paper Diwan & Gujarati (1968) has concerned with the relationship between output and employment in Indian industries. To study the problem they formulated a simple theory that employment is demand determined; demand for employment is derived from the demand for output. A number of conclusions emerge from the study. Employment in industry is demand determined. In some industries supply lags behind demand but the elasticity to adjustment is not very low. Elasticity of capital-labour substitution is quite low. Employment elasticity of output is also quite low. There are high economies of scale. Increase in wage-rate seems to be greater than the increase in marginal production, which this increase seems to have reduced imperfections in the labour market. Growth in output follows a rather unusual process, industry seems to plane for stabilizing existing output rather than expand it. If their assumption that output is frustrated for want of certain complementary factors is correct, then employment and output are complementary, in the sense that more output will lead to more employment without further investment in capital.

Sankar (1970) analyses elasticities of substitution and returns to scale for Indian manufacturing. He used CES production function with some modifications and obtained coefficients for 15 industries for 1946 through 1958. He found that elasticity of industries and close to zero for one. His main finding for Indian industries was that positive neutral technical progress in 6 industries and negative in two industries. He estimated aggregate CES production function for the periods 1953 to 1958 and 1959 to 1962 and found neutral shifts of the production function. He also found economies of scale for 15 industries.

Banerjee (1971) in his study has made an attempt to analyse the productivity trends in Indian manufacturing for the period 1946-64. The main conclusions, which may be drawn from the Banerjee study, are: The labour productivity showed a decline in the years immediately following 1946. Thereafter, it continued to rise, till the index rose to 133 in 1598. It showed and upward trend over the period. The capital productivity index showed steady decline over the period and its rates of decline was much more than the rate of increase of labour productivity. The increase in labour productivity in this sector was achieved mostly through capital deepening. The rate of return on capital declined steadily over the period, at a highly significant rate about 5 percent. Neither the productivity indices nor the production function analysis indicated the presence of "technical progress" in this sector. The hypothesis of constant returns to scale was not rejected. The elasticity of substitution between capital and labour was found, to be not significantly different from unity implying considerable substitution possibilities.

Hashim & Dadi (1973) stress the need for accurate measurement of capital and in their study make such an attempt. For the period 1946-64 they find a rise in the Solow index of TFP by 51%. Average annual growth rate in the Solow index turns out to be 2.8% and the contribution of TFP growth to output growth is found to be about 50%. They observe a rising trend in labour productivity and, in disagreement with almost all earlier studies, find a rising trend in capital productivity as well. The ratio of gross fixed assets to gross value added (with appropriated price adjustments) falls from 5.66 in 1946 to 4.16 in 1964. During the same period a fall in noted in the ratio of total productive capital to gross value added from 6.55 to 5.20.

In this paper Mehta (1974) have adjusted CMI-ASI data and made them consistent both respect to coverage and classification at the three-digit level of disaggregation. He has made a strictly comparable time series of CMI and ASI is possible only for the period 1953-65. In 18 Indian industries he observed that all the factors of production as well as output are increasing. In the rest of industries some factors of production are decreasing but still output is increasing in all cases. In case of 19 industries capital has faster rate of growth than labour, increases in output may be due to capital deepening. He also observed that when capital intensity is increasing in an industry, this is also reflected in increased capital-output ratio and declining output-capital ratio indicating falls in capital productivity. This reinforces the evidence of capital deepening. It is well know that when average is rising, marginal will be rising faster than average and when average is falling marginal will be falling faster than average. He find that in 19 industries average capital productivity is declining and therefore marginal productivity will also be declining. He further highlighted that the behavior of relative share of income between labour and capital varies from industry to industry and therefore Cobb-Douglas or any other single production function would not be appropriate for analysing different industries.

Narasimham and Fabrycy (1974) in their study of Indian industries for the period 1946 to 1958 measured the technological change in terms of simple time shifts of the production function. The time shift factor was introduced in both Cobb-Douglas and CES production function. According to them, the constant rate of technological progress was not significantly different from zero. In their studies elasticities of substitution (σ) obtained 0.782. They gave estimates of returns to scale for 28 Indian industries and showed constant returns to scale in all 28 Indian industries individually and together.

Banerji (1975) undertakes a comprehensive study on productivity and other related aspects for Indian manufacturing and five important selected industries covering the period 1946-64. This work distinguishes itself by making much more careful price adjustments in deriving the capital series. Three indexes of TFP are used, namely, the Solow index, the Kendrick index and the CES index. A significant falling trend in TFP was observed for the period 1946-58 (using CMI data) and also for the period 1959-64 (using ASI data). For the entire period 1946-64, disregarding data incompatibilities, Banerji finds a fall in the Solow index by 40% (at the average annual rate of 1.6%). His estimates, however, show a much steeper fall in capital productivity.

Wherever, Venkataswami (1975) have studied statistical estimation of production function and technical change in the 28 individual industries and the entire manufacturing sector of India as well. The empirical results reveal that although the elasticity of substitution varies from industry to industry, but it was not significantly different from unity in 21 out of 29 cases. In fact they has revalidated and confirmed the essential soundness of the C–D function. While SMAC admitted that the elasticity of substitution was not significantly different from one (in 10 out of 24 industry groups). The estimates of the majority of the labour and capital coefficients of C – D function were found to be statistically significant. In this study, one of the results was that the manufacturing sector as a whole had been operating under conditions of increasing returns to scale. Moreover the results of both the CES and the C–D production function show that there had been a notable of technical progress in most of the industries. Of course, the two functions have yielded different value of technical change and the degree of technical change is, in general, less under the C–D function than under the CES function.

In Dholakia (1977) study an attempt has been made to examine the impact of errors in the measurement of factor inputs, especially the capital input, on the estimated factor elasticity's of output and also the estimated rate of technical progress. The estimates of Cobb-Douglas production function have been derived from the time series data for the period 1946 to 1966 obtained from the various reports of CMI and ASI. The main conclusions, which may be drawn from, the Dholakia study is: The estimates of time series production function for Indian industries are highly sensitive to the measurement of capital input. The use of depreciated book value of capital or the net stock of capital at constant prices generally yields the set of estimates in which the regression coefficients of capital input or time or both turn out to be either negative or insignificant. The use of gross stock of capital at constant prices leads to a remarkable improvement in the precision of the estimated coefficients of capital as well as time, and also in the overall explanatory power of the estimated production function. The elimination of preliminary errors in the measurement of capital inputs alters significantly the estimates of the relative contributions made by various factors to the growth of output, in as much as it raises the relative importance of capital and technical progress and reduces the relative contribution of labour. The elimination of errors in the measurements of capital input reduces the intensity of the problem of multi-collinearity arising frequently in the estimation of time series production function.

Kazi (1978) studied the elasticity of substitution in 9 Indian industries for the year 1960, 1961 and 1962 with the help of cross-section data. His results on the estimates of elasticity of substitution by using CES and VES production function showed that there is an upward bias in the estimate of σ by the CES method. The distribution of elasticity of substitution by VES production function showed that 75 percent of them were below unity.

Whereas, in an another study Kazi (1980) estimated VES production function for Indian industries using cross section data for the years 1973, 1974 to 1975. The estimates of σ obtained from the VES and CES suggested that it varies between industries. In comparison with the higher estimates of elasticity of substitution obtained in the earlier works; his estimates suggested for low substitution possibilities in 8 out of 9 industries for the years 1974 and 1975.

Bhasin & Seth (1980) has studied –"Estimation of Production Functions for Indian Manufacturing Industries". They have estimated the Cobb-Douglas and the C.E.S. production functions also evolved and alternative method of estimation of C.E.S. production function. Some of the salient conclusions of the paper were as under: There are inter-industry differences in the rate of growth of technical change. The rate of growth of technical change was above 10 percent in very few industries. There is less scope for factor substitution and the elasticity of substitution is less than unity in most of the industries. Most of the industries have experienced decreasing returns to scale. It was also observed that 17 out of the 27 industries exhibit a pattern of distribution of income, which is favourable to capital. The C.E.S. production function seems to be an appropriate specification for most of the industries. In the recent years, many studies have taken post 1965 period (it was around this period when a deceleration in industrial growth had taken place) also into consideration. Although a conflicting trend can be observed, majority of studies find deceleration in overall efficiency and the absence of technological change.

Mehta (1980) have studies productivity, production function and technical change for Indian industries for the period of 1953-65 and 1965-70. In his study, finds evidence in favour of constant returns to scale, capital deepening and absence of any technical change. The detailed analysis brings out fairly satisfactory performance of industries with diversified product range, like bicycles, glass and glassware and electrical fans, but there seem to have been a trend of overall decline in traditional industries, like cotton textiles, jute textiles, matches and sugar. Assumption of unitary elasticity of substitution did not hold for majority of industries studied.

Results of similar type have been reported by Ahluwalia (1985), who empirically brought out the main causes testing with different measures of total factor productivity; it was found that the average annual growth of TFP ranged from –0.2% per annum to −1.3% annum for the manufacturing sector as a whole. Further, the average annual growth of –0.3% and –0.1% was noticed for the period of 1959-65 and 1965-78, respectively. As far as the industry-wise analysis is concerned, excepting footwear and furniture and fixtures, the decline in efficiency has been more or less across the board. A significant decline was found in rubber products, miscellaneous industries, food manufacturing, wood and cork, metal products, leather and fur products. Two industries, which is already mentioned above (namely footwear and furniture and fixtures) experienced significant growth in the TFP (Solow and Translog indices).

The earlier estimates, by and large, cover the period in the fifties, the sixties and the seventies; only one study covers the period up to the early eighties. Krishna (1987) provides a detailed review of the studies on total factor productivity growth for the Indian manufacturing sector and reports that with the exception of the study by Hashim & Dadi, all earlier studies for the fifties and early sixties report a decline in TFP over the period. He also makes a detailed scrutiny of the two more recent estimates. Ahluwalia's estimates of TFPG for the registered sector of Indian manufacturing (based on the Translog index) of –0.4% per annum over the period from 1959-60 to 1985-86 was very similar to the other available estimates of TFPG for the registered manufacturing sector for about the about the same period in the range of –0.2 to –0.6% per annum by Brahmananda (1982) and Ahluwalia (1985), Banerji's (1975) estimate of –1.6% per annum relates to a much earlier period. For the large scale registered manufacturing sector, Mehta's estimates for the period 1959-70 also shows a larger decline, *i.e.* –1.6% per annum and 2.5% per annum for the Solow index and the Kendrick index, respectively.

An attempt has been made by Kumar (1983) to analyse productivity trends in certain important industries in the country during the decade 1960 to 1970. Out of 39 industries, for which productivity indices were compiled by the Labour Bureau, 5 industries *viz.* – Cotton textiles, Jute textiles, Sugar, Iron & Steel (metal) and electric light & power have been selected for study of productivity trends, on the basis of their relative employment and valued added to the manufacturing sector. These five industries together accounted for 42.58 percent and 40.82 percent employment in the years 1960 and 1970 respectively and their share in value added was 40.63 percent and 34.50 percent in 1960 and1970 respectively. He has also highlighted that the trend in each variable in respect of each industry. From partial labour and total factor productivities point of view, years 1961 and 1966 were the best as labour productivity increased by 19.0% and 15.4%, respectively in two years and the total factor, productivity by 11.0% in both the years. Capital productivity increased by 5.7% in year 1966 only. Author further highlighted that, in Jute textiles from the productivity point of view, years 1961, 1963, 1964, and 1968 were the best periods when all the three productivity indicators increased over the respective previous years.

The main purpose of Verma (1985) was to examine the production structure of Jute industry of India. The main findings of production structure of jute industry are as: Annual growth rates of TFP vary according to Kendrick, Solow & Divisia methods. Divisia index shows higher growth rate than Solow index and Kendrick. But growth rate of TFP index of jute industry was little. The growth rate of SFP indices of labour and capital during the period under (i. e. 1950 to 1978) based on exponential function was 2.1 per cent and 0.6 per cent, receptively. Jute industry seems to be operating under constant returns to scale. Elasticity of substitution between labour and capital in jute industry was also found small.

Goldar's (1986) study covered the period 1951 -79 has been divided in to two sub-periods 1951-65 and 1959-79. Goldar computed both partial and total factor productivity indices for manufacturing sector as a whole. Goldar's study uses Kendrick index, Solow index and translog index of total factor productivity. During the period 1951-65 the labor productivity and capital intensity showed an unprecedentedly trend. The capital productivity recorded a decline of 14% per annum. The average annual rate of growth was 1.27% per annum during 1951-79. In the second sub period Goldar has observed similarity in the results of partial productivity and capital intensity as in the first period. Goldar's estimates of TFPG for a composite sector including a large scale registered manufacturing trend to be relatively higher than other estimates. The average annual rate of growth in the care of translog index was of the order of 1.31% per annum. Thus it was also higher as compared with Solow and Kendrick Indices, which was of the order of 1.29 and 1.06% per annum. Goldar's estimate for small scale registered manufacturing is very similar to that for large scale *i.e.* 12% per annum. Accordingly Goldar has concluded that technological progress has contributed to output growth though marginally and growth in factor productivity is sluggish.

Singh & Singhal (1986) have studied – "Economies of Scale and Technical change". The main purpose of the paper was to investigate the presence of economic of scale and technical change and also to identify the main sources of growth of the manufacturing sector of Punjab. The main findings of the study were as: The contribution of capital to output growth was quite low, and, the labour factor was mainly responsible for the growth of output in the industrial sector. The neutral technological progress does not seem to be an important contributing factor rather it was negative in most of the cases. The large manufacturing sector as a whole shows clear evidence of constant returns to scale. Among the all five individual industries, only basic metal and alloys industry was enjoying increasing returns to scale. While the cotton textile and electricity industry show constant returns to scale and other two decreasing returns to scale. Further, they have suggested that the output growth in this sector has been achieved through the increased factor input and not through the technological progress and economies of scale.

Gumaste Vasant, (1988), attempted to explain that the Indian industry has responded to the governments promotional measures to encourage in-house research and development units and the money they spend have grown considerably over the last four decades. But what are the concrete results? How strong is the technological capability of Indian industry today? How effective is it in enabling the country to be technologically self-reliant? This study is based on discussions with the principals, the men in the wings and those behind the scenes in the industry-the automobile and ancillary industry.

Agarwal, R.N. (1988), attempted to explain the objective of the study is to pinpoint the main causes of the sickness of the industry and then to suggest remedial measures. It appears to us prima-facie that the industry is caught in the vicious circle of small size of the market; for its products and near absence of innovations in technology over the three decades. The industry has not developed its own vehicles and the export demand for its vehicles is negligible.

Kesari Kumar Pradeep and Saggar Mridul, (1989), in their work attempted to analyze the determinants of export performance for fifty five units in the "Machinery and Transport equipment" industry of India. The methodology adopted applies pooling of cross section and time series data over the years 1980-81, 1982-83, and 1983-84. The study follows the neo-factor proportion and neo-technology approaches. These approaches came into vogue as the Heckscher (H-O) theorem, due to its restrictive assumptions, was found incapable of explaining real world phenomenon of monopolistic competition in the arena of International Trade and foreign investment. Under the assumptions of the H-O theorem, such as perfect competition and perfect foresight, constant returns to scale, absence of product differentiation, all firms in a n industry will have access to technology, factors and product markets. As a result they are expected to perform in similar fashion.

Sahana Ghosh (1990) has studies the pattern of growth of the Indian manufacturing sector during the period 1975-76 to 1985-86. He has chosen 49 major three-digit industry groups for the investigation and their growth rates over the ten years since 1975-76 are analysed Growth of these industries in the first half of the eighties as compared to the second half of the seventies is also discussed Performance of the manufacturing industries according to the use-based and the input-based classification are also looked into. On the other side, 'Total Factor Productivity Growth In Indian Manufacturing: A Fresh Look' have been studied by P. Balakrishnan & K. Pushpangadan (1994). They were the view that Productivity estimates are sensitive to the measure of real value-added adopted. One source of bias in estimation is that due to the assumption of constancy of the relative price of material inputs. This paper provides estimates of total factor productivity in aggregate manufacturing industry having adjusted for changes in this relative price. These results indicate that, contrary to what is believed, productivity growth in the 1980s may, actually, have been slower than in the earlier decade.

In his study Sengupta (1993) attempts a quick review of the gensis and process of growth in Indian manufacturing during the eighties. The main focus of this study was the factory manufacturing sector and its subsectors, Edible cover food, beverages and tobacco. Other agro-based industries comprise wood, paper and leather products. The main conclusions of the Sengupta's study were as: Output of the factor sector grew faster than domestic consumption, due largely to a gain from the cottage sector and to a smaller extent, from export expansion. Productivity of the factor sector increased in response to the growth in output. It was increased by 0.41% with every 1% growth in output. For the manufacturing sector as a whole, about 10% of the productivity gain was retained as increased return on capital employed and the balance passed on as a reduction (or lower increase) in ex factory prices. The source of productivity increase shows a fairly consistent pattern across subsectors.

According to Coondoo, Neogi & Ghosh (1993) the growth and compositions of industries have been fast changing in the LDCs mainly through foreign collaborations during the last few decades. This study aims at revealing some interesting phenomena regarding the performance of Indian manufacturing industries over the period 1974-75 to 1085-86. This study suggests that the use of modern technology developed in the industrially advanced nations in a country like India needs a minute assessment of the performance of individual industries where the role of learning effect is significant to improve the factor productivities. The diffusion of labour using technology, which is economically efficient, is the option to attain employment targets. On the other hand, capital-using methods may also be economically efficient if the appropriate technology could be found out through R and D, and used for selective industries. Therefore, they have suggested that a strong R&D wing for the industries should be encouraged through effective policy frame by the government to evaluate the adoptability of modern technology and indigenous technology as well.

Mukherjee Avinandam and Trilochan Satry, (1996), in their work attempted to explain the automobile industry in South Korea, Brazil, China, and India is currently going through impressive growth. Governments have played a key role in the evolution of the industry in all these countries. South Korea, a relatively amount to the automobile industry, has made the most significant progress, and is now exporting cars to developed countries. It is the only country that invested in research & development for product development, retained management control in ventures with multi-national companies, and had ambitious export targets. The industry in Brazil is much bigger than that in South Korea, but indigenous product development capabilities are lacking and manufacturing competitiveness is limited even though the industry is entirely controlled by NINCS. The Indian industry is experiencing with rapid growth and the entry of the largest number of MNCs.

Rao (1996) used (double-deflation) method for measuring TFP. The study suggests a rapidly declining TFP growth for manufacturing industries after 1981. Dholakia and Dholakia (1994) study TFPG for Indian manufacturing from 1970-71-1988-89. The study reports that the annual growth of real value added in Indian registered manufacturing sector when measured through single deflation method shows remarkable acceleration during the 1980s as comparison to 1970 (from 3% to 8%). On the other hand, when the same is measured through double deflation compared to 1970s (3.55% to 11.2%) when the weights for the 19 input groups based on WPI (1970-71) were used. The study reports negligible growth during the 1980s as compared to 1970s (7.5% to 8.1%) when weights for whole manufacturing sector as considered by Balakrishnan and Pushpangandan (1994) are used.

In Baghel & Pendse (1996) study entitled –"*An econometric analysis of productivity growth and technological change in total manufacturing sector of India*", an attempt has been made to analyse productivity growth and statistically estimation of production functions and technical change in the total manufacturing sector of India. They observed that labour is major source of output growth, and the time trend coefficient was found insignificant in most of the cases of production functions and suggests inferences in favors of technological retrogression.

Fikker and Hasan (1998) analyzed the returns to scale for a panel of selected Indian manufacturing industries for the pre liberalization period from 1976 to 1985 using a restricted cost function. Although these resources found large number of firms operating with increasing rate to return to scale, the results suggested that most of them were operating close to constant returns to scale. The study suggests that there are not significant gains in scale efficiency from the tentative steps in economic liberalization in the 1980's. Krishna and Mika (1998) show that there are increasing returns to scale in electronics, transport equipment and non-electrical industries and that there was an increase in exploitation of the scale economics after the economic liberalization.

Study by Kusum Das (1998) found that productivity response to the trade policy response to the trade policy reform is mixed. This study correlated the productivity growth with different measures of trade liberalization. However the results of this exercise show that in majority of the cases the trade liberalization variable has a statistically insignificant positive relationship with productivity growth. A Study on productivity trends in Indian manufacturing undertaken by Unel (2003) has concluded that total factor productivity (TFP) growth in aggregate manufacturing and many sub-sectors accelerated after the 1991 reforms.

Puran Mongia and Jayant Sathaye (1998) had made an attempt to survey the available literature on productivity growth and technical change in six energy intensive industries in India covering the period 1947-1998. The motivation for the survey was to assess the magnitude of the autonomous energy efficiency improvement (AEEI) parameter for India's industrial sector. The industries covered in their study are aluminum (non-ferrous metals), cement, fertilizers, glass and glass products, iron and steel, and paper and paper products. They have also survey studies relating to productivity growth in aggregate manufacturing in order to gain a broader perspective. The survey reveals productivity growth to be an active area of research in India, both at the aggregate level as well as at the industry level. Over time increasingly complex functional forms and sophisticated econometric techniques have been employed for analysis of productivity growth. Also, efforts had made by them to arrive at more accurate measurement of factor inputs, particularly for capital. The survey also indicates wide inter- and intra-industry variation in estimates of partial and total factor productivity due to differences in methodology, levels of aggregation, sources of data, time periods of analysis, and reporting procedures. In view of these differences, it is difficult to make a definitive judgement about the nature and magnitude of productivity growth in energy intensive industries in India. The overall impression is that of positive though imperceptible growth in productivity over time. The policy implication is that little reliance can be placed on the AEEI factor as the moderating influence on growth of energy demand. This also points to the need for estimating productivity growth using uniform methodology, common data source and the same time period for all industries.

Sanja Samirana Pattnayak & Shandre M. Thangavelu (2001) have studied the impact of Economic Liberalization of the Indian government in the 1990s on the manufacturing productivity. In particular, they examines in detail the variations of TFP across manufacturing sector at a more disaggregated level from 1980-97. Further, manufacture productivity growth is decomposed by the framework introduced by Arnold C. Harberger (1991 and 1998). The TFP sunrise/sunset analysis is used to study the structural changes that are occurring in the manufacturing sector due to the economic liberalization policies introduced by the Indian government since 1991. The empirical results of their study show that productivity gains are not evenly distributed across sectors. Some sectors experience very high rates of productivity as they are in the process of adopting new technology and new methods of production particularly after the liberalization.

Sanja S. Pattnayak & Thangavelu, S. M. (2003) have studies the effects of the key economic reforms of 1991 on the Indian manufacturing industries using a panel of manufacturing industries. A Translog cost function is used to analyze the production structure in terms of biased technical change and economies of scale. A panel consisting of 121 Indian manufacturing industries from 1982 to 1998 was used in our estimation. The results of their study support the evidence that there are economies of scale (only moderate) in the Indian manufacturing industries and it has been exploited after the key economic reforms in 1991. Most of the industries in our study revealed bias technology change and majority of the industries have experienced capital-using technical change. This suggests that the key economic reforms of liberalizing the capacity licensing regime that allows greater investment in capital goods will have a positive impact on productive performance of the industries if the price of capital does not substantially increase after the economic reforms. They observe TFP improvements for most of the industries after the 1991 reform initiatives, which support the evidence of improvements in economic efficiency after the key reform initiatives of the 1991.

Pushpa Trivedi (2004) have pointed out that Industrial performance of various states needs to be viewed in totality, *i.e.* with respect to growth of output, employment and productivity. Moreover, productivity levels are as important as productivity growth trends, as both are pertinent in the convergence process. This study is an attempt to interpret inter-state differences in productivity movements in organized manufacturing sector, in a larger perspective of employment and output trends. The time-span of the study is 1980-81 to 2000-01 and it encompasses 10 major states of India. The study empirically confirms the existence of inter-state differences in productivity levels and growth rates. It points out that states, such as, Bihar and West Bengal are diverging away from rather than converging to the growth rates of output of organized manufacturing sector at the national level. Though productivity growth in Bihar appears to be high, it has been mainly achieved by joblessness. M.P. and Rajasthan, which have been considered as BIMARU states, seem to be good performers from a wider perspective and show the promise to get them rid of their economically backward status.

Kalirajan *et al.* (2004) has explained that the two reasons for the rapid growth of unorganized manufacturing sectors are: urbanization and rural to urban migration. It was observed that in the post 1997-98 period, output in the organized sector has grown at a slower rate than in organized manufacturing. The reasons are the emergence of flexible production systems and substantial increase in outsourcing by the organized sector. They suggested that there is a need to study the size, structure and performance of unorganized manufacturing sector in India.

Rajesh Raj S. N. paper analyses the size, growth and productivity performance of the unorganized manufacturing sector in India during 1978-79 to 2000-01. The study shows evidence of increase in size with a slowdown in the reforms period. Evidence indicates that the rate of growth varies widely across the two-digit industries but the variation in growth rate is smaller in the 90s. Textiles and machinery goods were the fastest growing segments of Indian unorganized manufacturing sector in the reforms period. Both the partial factor productivity approach and total factor productivity approach reflect that productivity of the sector has improved during the period under study. The decomposition of productivity growth into technical change and efficiency change reveals that the latter has been the major contributor to TFPG during the period under study. It is also found that capital intensity and wage rate are essential factors for augmenting labour productivity levels in the sector.

The main focus of the B.N. Goldar (2004) was on Total Factor Productivity in (registered) manufacturing. He shows that the conclusions of a recent much quoted study by the IMF and another widely reported one by an industry institute are not correct. One of the important results in this paper is to show that the declining share of wages in registered manufacturing is due to a decline (rise) in the relative marginal productivity of labour (capital). Whether this is because the newly available and more productive technologies are capital biased or is driven by labour policy distortions needs to be explored in future research. The conundrum of lower average TFPG in registered manufacturing during the nineties relative to that in the eighties remains. His earlier hypothesis that it is due either to lower capacity utilization since the 1994-95 to 1996-97 investment boom (till 2003-04) or the change in relative prices within manufactured goods as a result of reduced variance in (QR/Tariff) protection also remains to be tested. It should be noted however that ICRIER Working Paper No. 131 has estimated that TFPG in the manufacturing sector as a whole doubled from 1.3% per annum during 1980-81 to 1991-92 to 2.8% per annum during 1992-93 to 2003-04. This implies either a sharp acceleration of TFPG in the unregistered manufacturing sector or outsourcing of inefficient stages of production (or shifts of products) from registered to unregistered manufacturing. Thus, to supplement the analysis of productivity trends in the pre- and post-reform periods, Goldar paper takes a close look at growth in employment and output in India's organized manufacturing sector in the period since the mid-1990s. The analysis reveals that the trend rate of growth in employment in the period 1997-98 to 2001-02 was significantly negative, at about –3.3 per cent per annum. The trend growth rate in real value added in the period 1996-97 to 2001-02 was very low at about 0.5 per cent per annum. This was much lower than the trend growth rates in real value of output and the Index Number of Industrial Production (manufacturing) in this period, both exceeding 5 per cent per annum.

A vast study of Pendse & Baghel (2008) has studied entitled-"Technological Change and Productivity Growth in Manufacturing Sector of India". In this book, we observed that the relationship between technological change and economic development is intimate and complex. Many contemporary events have contributed to this complexity. Economic development necessarily depends upon productivity of the factors of production. Economic development is essentially a social process. All factors must contribute at their best. When tested empirically with the tool of production, found technological change to be the most important source of output growth. Technological choice and adoption are important aspects of economic engineering. The history of the industrial revolution in the west shows that the process of rapid technological transformation has not been merely the result of new improved technical processes. The socio-economic development of an area is influenced by a large number of factors operating in an economy. The ASI definition of variable is used in the present study. The present study has covered trends in partial factor productivity ratios and total factor productivity indices, and estimate parameters of Cobb-Douglas and Constant Elasticity of Substitution production. Thus, the Kendrick, Solow and Divisia index of total factor productivity has been used in the present book. The Cobb-Douglas, Constant Elasticity of Substitution, Variable Elasticity of Substitution and Translog production function form of production function are used. The present book covers 161 industries group for the purpose of analysis. The data used for the analytical purpose are gathered through secondary sources for the period 1973-74 to 1997-98. Thus, the validity and reliability of the results will depend upon the validity and reliability of the data. Moreover, the implications of the present study confined to coverage of the industries.

Abhay Gupta (2008) paper performs detailed and exhaustive set of accounting exercises for the period 1970-2003 using production function, index number and envelopment analysis methods. TFP growth rate average is 1.1% for both gross output based and net value added based measures. In gross output production, share of materials is 0.6, much larger than the capital and labor shares. Share of capital is constantly increasing. For the period just after the reforms (1991-1997), input growth jumps but TFP growth is negative. But after 1998, the trend reverses and output grows slowly despite negative input growth due to large TFP growth. Aggregated TFP growth rates (Domar-weighted and Fisher index) also follow the same pattern; showing upward trends after mid- 1990s. There are no significant differences in TFP growth rates among different-sized firms. After the reforms, TFP growth increases substantially in the public corporations. Productivity transition seems to be random across different (3-digit NIC code) industries. Industries with focus towards services experienced higher productivity growth than others. These results show that the lack of productivity growth was the reason for unimpressive performance of Indian manufacturing earlier.

K. Pushpangadan, N. Shanta (2008) have examines empirical evidence on the dynamic view of competition in Indian manufacturing industries using Mueller's auto-profit equations. The competitive profit rate for all industries is estimated to be 5 % and the average firm-specific rent 4.2 %. The observed average profit rate, 7.5 %, is thus 50% higher than the competitive rate. The estimate of the average strength of competition in the manufacturing sector shows that it takes 0.9 years for the deviation of profit rates from the norm to reduce to its half level. When compared with the findings of the studies for the pre-liberalization period, the competitive environment in the Indian manufacturing sector is found to have deteriorated in the post-liberalization period.

Shivi Agarwal *et al.* (2009) has examines the total factor productivity (TFP) growth and its components of 34 State Road Transport Undertakings (STUs) of India for the period 1989-1990 to 2000-2001 using DEA-based Malmquist Productivity Index (MPI) approach. The study finds that on average, the TFP change is -1.9% per annum over the sample time period. In order to identify the sources of TFP change, technical efficiency change (catch-up effect) and technical change (frontier shift) for each STU (over the 12-year period) are also measured. The results show that there is no significant change in technical efficiency but remarkable change in technology. Thus, the decomposed index of TFP growth shows that much of the observed regress in total productivity is explained by the negative change in technology of the STUs, over the sample period. In order to determine the effect of several background and uncontrollable variables, which cannot be included in the productivity assessment of the STUs, multiple regression analysis using fixed effect model of the panel data is conducted. The results reveal that change in Employee-Bus ratio, change in Depreciation Cost, change in Total Earnings and change in Total Cost, turn out to be the most significant variables in having impact on the change in productivity.

Abhay Gupta (2010) article addresses the question of why productivity growth in Indian manufacturing was slow in the pre-reform period and analyzes how economic reforms in the 1990s accelerated productivity growth. The answer lies in two subtle but important distortion-inefficiency mechanisms, which affected productivity growth by distorting intermediate input allocation. The interaction of quantitative restriction policies and inflexible labour laws resulted in lower than optimal materials per worker usage. The combination of high inflation and unavailability of credit exacerbated this factor distortion and lowered productivity growth further. Using a panel dataset on Indian industries, this article finds widespread underutilization of materials compared to labour until recently, and this sub-optimal materials per worker usage lowered productivity growth.

An attempt has been made by Deb Kusum Das *et al.* (2010) to understand the sources of Indian growth experience for the period 1980-2004 using a newly developed INDIA KLEMS project database for 31 sectors. In particular, it examines the relative contributions of factor accumulation and productivity growth in the different sectors of the Indian economy. A sector perspective gains significance in the context of major reforms undertaken in several sectors in the past two decades. In addition, there has been significant structural transformation in the economy during the past decade suggesting a high and increasing share of service sector GDP. They have uses a growth accounting framework to document and analyze the sources of India's economic growth by industry. Following the KLEMS methodology due to Jorgenson *et al.* (1987), productivity performance of each of the industrial sectors is assessed for the period 1980-2004 and four sub periods. The TFP growth measure incorporates contributions of labor-quantity and quality and capital-ICT and non ICT assets in its measurements. The paper documents the evidence of service sector led productivity growth in the Indian economy. They also find evidence of factor accumulation in accounting for the sources of growth for the Indian economy and its various sectors as well as industries.

A study of Deb Kusum Das and Gunajit Kalita (2011) reveals that Productivity growth in Indian manufacturing is an important driver of overall growth, yet the issues related to its measurement have still not been resolved. The issue of how to compute an aggregate productivity measure holds significance for two reasons: one, the productivity of a firm should reflect the productivity of the lower levels, which comprise the aggregate; and two, aggregate productivity should also emphasize the importance of inter-industry transactions in an analysis of productivity growth. We have made an attempt to compute the aggregate productivity growth using the Domar aggregation technique. Comparing the estimates based on the Domar aggregation technique with those based on the traditional aggregate value added approach, we observe that the preferred estimates are about half of those obtained by the traditional aggregate value added method.

Danish A. Hasim *et al.* (2011) have measured productivity and output growth of Indian manufacturing for post reforms period. They argued that the majority of the studies estimating productivity in the Indian organized manufacturing sector have reported a slowdown in productivity growth in the 1990s, baffling the analysts as this came in backdrop of major economic reforms that were initiated in 1991. Some felt that productivity could respond to reforms with a time lag but could not explain theoretically as to why it plummeted. Tracing the phenomena to the J-curve hypothesis, they attempts to provide a theoretical explanation for the dip in productivity and output growth in early phases of reforms and estimate if the reforms displayed a positive impact in later period. The underlying reasons for J-curve hypothesis are provided mainly in terms of technological gap (compared to the global benchmark) and relatively weak mobility of factors of production (like labour, capital, industrial/urban land etc) including those that arise from the rigidity in labour laws and difficulty associated with acquisition of land for industrial development. They analyses both aggregate and two-digit ASI industries over a period 1992/93-2007/08 and observes the trends in the following three sub-periods: (i) 1992/93-1997/98, (ii) 1998/99 to 2001/02, and (iii) 2002/ 03-2007/08. Empirical results obtained in the study, in general, support the hypothesis of J-curve pattern of productivity and output growth in the Indian organized manufacturing sector. The J-curve effect is found to be relatively weak in globally competitive sectors (like textiles) and strong in the sectors with relatively large technological deficit (like Automobiles). In view of continuing technological gap in many sectors and weak mobility in factors of production, the paper concludes that productivity and output growth could respond to earlier economic reforms even better, if these issues were addressed adequately.

An attempt has been made by R. Mriappan (2011) to estimate the economic returns to scale, marginal productivities of labour and capital inputs for 2-digit level industries in India's Unorganised manufacturing sector. The results of this study show that the elasticity of output with respect to labour and capital has increased and significantly contributed to the outpur growth during the post-reform period. The sum of elasticites of labour and capital was found greater than unity in all types of industries in 2005-06 at the aggregate level. The estimated coefficient of the year dummy variable fro 1989-90 was observed negative in the C-D, Translog and CES production functions.

Anoopa S Nair (2012) study reveals that the recently legislated Competition Act of India (2002) marks a clear distinction from the erstwhile MRTP Act by shifting the focus from dominant firms to firms who are abusing their dominant positions. This shift is based on the understanding that only when large firms adopt anti-competitive practices, would it hinder competition and the largeness of the firms alone is no hassle in the operation of competitive forces in an economy. In fact, punishing a firm just for the fact that it is large would amount to punishing efficiency in some instances. An empirical attempt has been made to identify the presence of dominant enterprises operating in Indian manufacturing sector. It also attempted to distinguish dominant firms which are efficient in their operations from the inefficient ones. The results of Nair study reveal that, among the selected industries, there are only a handful of dominant firms in Indian manufacturing sector and majority of these dominant firms are in fact cost efficient in their operations.

Mulimani, A. A. *et al.*, (2012) have made an attempt to study the SSI units and their problems and prospects. These units are playing a key role in the gross root economy in the study region. The secondary data have been compiled from the different sources and analytical methods have been employed for analysis. The resource base and locational advantages and disadvantages are also considered to identify the problems and also paid an attention for prosperity of SSI units. The planning policies are also revealed and also suggested the strategy for the prospering of such units with effective suggestions. Hope so, the proper plan may be help to bring the changes towards the path of prospering in the study area.

Sandeep Kumar & Kavita (2012) have examined the Southern region state-wise trend of total factor productivity growth in Indian manufacturing sector for the periods 1984-85 to 2004-05. Kendrick index is used to compute TFP index. The resulting information is used to examine whether the post-reform period shows any improvement in productivity in comparison to the pre- reform one. Findings of the present study indicate that TFP growth of Indian manufacturing sector for all states combined and selected South region states have declined during the post-reforms period as compared to the pre- reforms period. Further, growth of GVA in India and most of the states in the study reveal that there has been decline in growth of GVA during the latter period as compared to the earlier one. It's implying that industrial sector failed to sustain the growth momentum in output during the period after 1991. It is also found in their study that there is a tendency of convergence in terms of TFP growth rate among Indian states during the post-reform years and only the states that were technically efficient at the beginning of the reform remain innovative.

Fulwinder Pal Singh (2012) paper endeavors to analyze the TFP growth trends in Indian manufacturing sector at both aggregated and disaggregated inter-state levels. Using the Malmquist productivity index for panel dataset of 16 major industrial state over a period of 29 years spanning over 1979-80 to 2007-08, the study observed manufacturing sector of India is growing with 9.1 percent per annum growth of Total Factor Productivity (TFP) during the entire study period.

Out of 16 Industrial states there are five states namely Uttar Pradesh, Madhya Pradesh, Gujarat, Orissa and Rajasthan where double digit TFP growth has been noticed. The manufacturing sector of Uttar Pradesh is growing with highest TFP growth at the rate of 12.8 percent per annum followed by Madhya Pradesh with TFP growth of 11.8 percent per annum. The analysis of the sources of the TFP growth in Indian manufacturing sector reveals that both technical progress and technical change are equally contributing TFP growth in sector under evaluation. It has also been observed that at all India level efficiency change is greater than technical progress.

Vinish Kathuria *et al.* (2013) paper was the view that very few other issues in Indian economic development has generated so much debate than the measurement of total factor productivity (TFP) growth in Indian manufacturing therefore, they contributes to the recent literature on TFPG estimation in India in four important ways. Firstly, it uses three different techniques – growth accounting (GA) (non-parametric), production function with correction for endogeneity–Levinsohn-Petrin (LP) (semi-parametric) and stochastic production frontier analysis (SFA) (parametric), from 1994-95 to 2005- 06 to compute TFPG in Indian manufacturing to see how sensitive are the results to different estimation methods. Secondly, it uses firm/plant level-data in the estimation of TFPG, and therefore, estimates TFPG at the most disaggregated level. Thirdly, it pays careful attention to data issues and to methods of variable construction which have been the source of some of the debate on TFPG estimates in India. Finally, it brings together TFPG estimates of both the formal and informal manufacturing sectors, the latter being an important part of the overall manufacturing sector in terms of employment. The results indicate that the TFP growth of formal and informal sectors has differed greatly over this period and the estimates are sensitive to the technique used. While the GA and SFA methods show a decline in TFPG in the formal sector in 1994-2001, the LP method shows an increase. In 2001-2005, the GA and LP methods show a decline in TFPG, while the SFA method shows an increase for the formal sector. In the case of the informal sector, all three methods show a decline in TFPG in 1994-2001. However, for 2001-2005, the GA and LP methods show a decline in 2001-2005 for the informal sector, while the SFA method shows an increase. Thus, there is a lack of convergence of the different methods on TFPG estimates for the formal and informal manufacturing sectors in India.

The main purpose of Himani Aggarwal *et al.* (2013) study was to find out that how many OAMEs, NDMEs; DMEs are there in unorganized manufacturing industries. They also aimed to find the use of machines in these units and the labor-capital ratio. Attempt was made to show the number of female and male workers in these units. Sample size was hundred unorganized manufacturing units and target area was Ghaziabad and Noida which is the industrial hub of Uttar Pradesh. This study was different from majority of work, researches done in unorganized manufacturing industries because it was based on primary data. Majority of the studies are done on secondary data collected from NSSO but in this study there have been direct interaction with the respondents.

Manju Bai *et al.* (2013) in their study pointed out that India manufacturing segment is a crucial cog in the wheel of economic progress; the sector contribution to the gross domestic product (GDP) being 16%. With the passage of time post 1990-economic liberalisation era, India has well realised the importance of manufacturing for the overall industrial development. In this wake, the Government has also been very pro-active, especially during the last decade. This paper has described many variables that determine the relation of organised manufacturing output and employment growth rate like labour productivity, emoluments of employees, employees- worker, non-worker and wage rates for time period 2000-01 to 2009-10. This is shown by simple percentage, coefficient of variation, compound growth rate and simple regression analysis. Through these statistical tools the output growth rates vary between a minimum of 4.02% per annum in Kerala and a maximum of about 40% per annum in Jammu & Kashmir. The overall growth rate of manufacturing sector output is about 12.92% per annum. The growth rates of employees in the manufacturing differ from a minimum of about -12.26% per annum in Andaman & N. Island to a maximum of 24.36% per annum in Uttaranchal. The growth rates of workers and non-workers follow the same pattern across the states or not, we have computed the growth rates of workers and non-workers in the manufacturing sector of different states. It has been seen that the growth rates of workers are higher as compared to that of total employees in all the states expect in Chandigarh (2.87%), Jharkhand (-1.31%), Madhya Pradesh (2.4%), Andhra Pradesh (2.26%), Karnataka (6.98%) and Tamil Nadu (6.73%) where the growth rates are less than the growth rates of total employees and that of non-workers. This difference between output and employment shares indicates a continuity of lower labour intensity over time in the manufacturing sector. Similarly is the case in Gujarat where the shares of output (12.89 %) and employment (9.42%) increased to (13.86%) and (9.83%) in 2000-01 and 2009-10 respectively. The study shows that inter-state variations in the shares of emoluments to employees range from a minimum of 0.13% in Tripura followed by Himachal Pradesh with 0.14% to a maximum of 0.82% in Andaman & N Island followed by West Bengal with 0.54% in manufacturing gross value added. In case of total employees, the growth rate of productivity is negatively and significantly related to the levels of productivity. The coefficient of productivity level is negative and significantly different from zero at 5.5% level of significance using one tailed test.

N. Srividya *et al.* (2014) paper highlights various issues that obstruct the growth of manufacturing sector and offers several workable approaches to address the issue. Healthy and robust manufacturing sector is inevitable for the employment rate to increase, exports to increase and the livelihood to increase. In the light of the researches done by many people in this area, the paper studies the various factors that led to this situation. The major one that was found mere responsible for this is the rigidity of the labour laws because of which the investors are not willing to employ more people despite of demand for their products. Instead they prefer automation or mechanization. This automation or mechanization is affecting the creation and growth of employment in the manufacturing sector.

Sonali Roy Choudhury & Sadhan Kumar Ghosh (2014) paper presents an empirical study of the regional specialization and geographical concentration of some selected manufacturing industries across the 3 administrative divisions of West Bengal for the period of 2004-05 to 2009-10. Their analysis points out to the divergence in the level of specialization and concentration among the divisions and the industries. It brings out the high inequality among the divisions in terms of development of the top industries in West Bengal. While M. Manonmani (2014) pointed out that four southern Indian states viz Andhra Pradesh, Karnataka, Tamil Nadu and Kerala contribute 24.3% of the GDP of the country. They are the emerging as the major destination for industrialization. In this paper, he analyses the growth and determinants of SFP and TFP in the aggregate manufacturing sector of these states and TFP was found to be increasing from the beginning of the post-liberalization period in all the southern states. The indices of Capital Productivity had shown increasing trend in all 3 states except Karnataka.

3.3: Methodological Issues: Productivity, Production Functions & Technological Change related aspects

The existing literature on the issues raised here and technology sources itself at the Indian state level is very poor and it does not provide sufficient understanding of the sources of technology changes. Few studies attempt to investigate technology and all other issues in general or specific to sectors of the economy. But there is a huge literature available in developed countries and developing countries like China for the role and the implications of innovation activities on economic growth and development level. Within this framework, measuring innovation activities and technical change is rather important in order to measure the corresponding effects on economic growth.

It is well-known that for a neoclassical growth model to exhibit steady-state growth, either the production function must be Cobb-Douglas or technological change must increase the relative productivity of labor vis-`a-vis other factors of production. Many models of endogenous growth (Romer 1986, Romer 1990, Lucas 1988) also assume labor-augmenting technological change, sometimes in the more specific form of human capital accumulation. A number of recent papers provide micro foundations for this extensive literature by establishing theoretically that profit-maximizing incentive can ensure that technological change is; at least in the long run, purely labor augmenting (Acemoglu 2003, Jones 2005). Whether this is indeed the case is an empirical question that remains to be answered. Recent research also points to biased technological change as a key driver of the diverging experiences of the continental European and U.S. and U.K. economies during the 1980s and 1990s (Blanchard 1997, Caballero & Hammour 1998, Bentolila & Saint-Paul 2004, McAdam & William 2013). The unemployment rate rose steadily in a number of continental European economies, most egregiously in Spain (Bentolila & Jimeno 2006), while the share of labor in income has fallen sharply. In contrast, the labor share has been stable in the U.S. and U.K. While these medium-run dynamics are certainly consistent with biased technological change in the continental European economies, there is little direct evidence for it. Despite the importance of assessing the bias of technological change, the empirical evidence is relatively scarce, perhaps owing to a lack of suitable data. Following early work by Brown & de Cani (1963) and David & van de Klundert (1965), economists have estimated aggregate production or cost functions that proxy for labor- and capital-augmenting technological change with time trends.

{see, *e.g.*, Lucas 1969, Kalt 1978, Antr`as 2004, Binswanger 1974a, Jin & Jorgenson 2008).

(**Note:** A much larger literature has estimated the elasticity of substitution in aggregate production functions whilst maintaining the assumption of Hicks-neutral technological progress, see, *e.g.*, Arrow, Chenery, Minhas & Solow (1961), McKinnon (1962), Kendrick & Sato (1963), and Berndt (1976) for early work and Hammermesh (1993) for a survey}.

While this line of research has produced some evidence of labor-augmenting technological progress, aggregation issues loom large in light of the staggering amount of heterogeneity across firms (see, *e.g.*, Dunne, Roberts & Samuelson 1988, Davis & Haltiwanger 1992), as do the intricacies of constructing data series from NIPA accounts (see, *e.g.*, Gordon 1990, Krueger 1999). Ulrich Doraszelski & Jordi Jaumandreu (2012) in his paper entitled –"Measuring the Bias of Technological Change" has combine available firm-level panel data with advances in econometric techniques to directly assess the bias of technological change by measuring, at the level of the individual firm, how much of technological change is Hicks neutral and how much of it is labor augmenting. Whereas, John Laitner & Dmitriy Stolyarov (2003) develop a new general equilibrium growth accounting framework that features increasing returns to scale, imperfect competition and incorporates technological revolutions into the description of technical progress.

Production functions for the industrial sector as a whole as well as for seven important industries in India are worked out based on cross-section data relating to individual firms for the two years 1951 and 1952 (V. N. Murti and V. K. Sastry, 1957). The stochastic term in Cobb- Douglas type models is either specified to be additive or multiplicative (Stephen M. Goldfeld and Richard E. Quandt, 1970). They developed a model in which a Cobb-Douglas type function is coupled with simultaneous multiplicative and additive errors. This specification is a natural generalization of the "pure" models in which either additive or multiplicative stochastic terms are introduced. A Cobb-Douglas type function with both multiplicative and additive errors has been proposed by Goldfeld and Quandt (1976). They suggested a maximum likelihood approach to the estimation of a Cobb-Douglas type model when the model includes both multiplicative and additive disturbance terms. As expected, an analytical expression for the solution to the maximization problem did not exist. Indeed, because of the complexity of the likelihood function, their maximization algorithm had to be used in conjunction with a numerical integration technique.

The application of econometric techniques to the study of Indian industries can be traced to the 1950s when Bhatia (1954), Dutt (1955), and Murty and Sastry (1957) estimated production functions for Indian manufacturing. Since then there have been a large number of econometric studies on Indian industries. While the production function has been the focus of most econometric studies, there have been studies on various other economic aspects of Indian industries, including the cost, labour demand, and investment functions. The CD function has been estimated for Indian industries in a large number of earlier studies. But in details, we take up for discussion the estimates obtained in two studies, namely, Ahluwalia

(1991) and Goldar (1986). Goldar has estimated the CD function using aggregate time-series data and Ahluwalia both from aggregate time-series data and panel data. Her estimates based on panel data are taken up for discussion first.

Ahluwalia has used cross-section, time-series panel data for 84 industries for the period 1959-60 to 1985-86. The ratio form of the CD function has been used for estimation purposes, which has the advantage that it provides a direct test of the hypothesis that the degree of homogeneity of the function (returns to scale) is one. In ratio form, the CD function may be written as:

$$\text{In}(Y/L) = a + \beta\,\text{In}(K/L) + (\alpha + \beta - 1)\,\text{In}\,L + \lambda t + u$$

Since data for different industries are pooled to estimate the CD function, Ahluwalia introduces industry specific intercept dummies, allowing the intercept to differ among industries. Further, Ahluwalia defines a dummy variable *D* as taking value zero up to 1982-3 and one thereafter. The product of this dummy variable and the time variable is included in the regression equation to capture the change in the rate of technological progress after 1982-3. Ahluwalia's estimates of the CD function based on time-series data and such estimates made by Goldar (1986) are also shown in Table 3.2. In both studies, the ratio form of the CD function has been used. It is seen from the table that the estimates of the returns to scale parameter are not significantly different from one, so that the hypothesis of constant returns to scale is not rejected. Therefore, we may compare only the equations that have been estimated assuming the returns to scale to be constant. The estimates of β obtained in Ahluwalia's, study are quite close to that in Goldar's (in the range of 0.3 to 0.4). There are, however, marked differences in the estimates of λ, the rate of technological progress. In Goldar's study, it is found to be positive and statistically significant, while in Ahluwalia it is very small and statistically insignificant.

Table 3.2: OLS Estimates of the Cobb-Douglas Production Function for Organized Indian Industry based on Time-series Data

		Estimates of			
Author	**Period**	β	$\alpha+\beta-1$	λ	R^{-2}
Ahluwalia	1960-82	0.321	–0.315	0.013	0.90
(1991)		(1.6)	(–0.6)	(0.5)	
	1960-82	0.387	–	–0.001	0.91
		(2.6)	–	(–0.1)	
	1960-79	0.281	–0.332	0.016	0.90
		(1.4)	(–0.6)	(0.7)	
	1960-79	0.357	–	0.002	0.91
		(2.4)	–	(0.3)	
Goldar	1959-79	0.281	0.313	0.007	0.95
(1986)		(2.4)	(1.0)	(0.7)	
	1959-79	0.314	–	0.017	0.95
		(2.8)	–	(3.9)	

Note: *t*-values in parentheses.

On the CD function estimates, some methodological comments presented in the two studies would be relevant. Ahluwalia's estimated of the CD function based on panel data involves a highly restrictive assumption that all the industries have an identical production function. She permits some flexibility in model specification by allowing the intercept to vary across industries; but the coefficients of labour and capital and the rate of technological progress are assumed to be the same for all industries. It needs to be noted, however, as Ahluwalia points out that this assumption is no different from what lies behind the estimation of the production function from aggregated time-series data. Another point to be noted in this context is that the application of the OLS technique for estimating the CD function from panel data involves the assumption of homoscedasticity. This assumption may be questioned because the variance of the error term need not be the same for different industries as the number of firms varies from industry to industry. There may be other reasons for the error term to be heteroscedastic. This could have been tested, and if the assumption of homoscedasticity was not found to be justified, the Generalized Least Squares (GLS) method could have been used.

For the CD function estimates based on time-series data, it is important to check the residuals for any possible serial correlation. This has been done in both the studies, and the tests carried out do not indicate any significant serial correlation. Another problem that may affect the CD function estimates (especially those based on time-series data) is of multicollinearity. The results of the two studies seem to have been affected by this problem, especially when the unrestricted CD functional form has been used. Consider Ahluwalia's estimates of the unrestricted CD function, it is seen that the overall explanatory power of the model is high, but none of the coefficients of the explanatory variables is statistically significant. There is similar problem arises with Goldar's estimates. It seems multicollinearity has caused the estimated elasticity of output with regard to labour to be above unity, which violates the conditions of well-behaved production functions.

Direct estimation of the CES function for Indian industries has been undertaken in only a few studies. Sankar (1970), Narasimham and Fabrycy (1974), and Barua and Leech (1987) have used non-linear estimation techniques to estimate the CES function. Narasimham and Fabrycy (1974), Bhasin and Seth (1977), and Barua and Leech (1987) have used the Kmenta approximation to the CES function. Much more common has been the use of the SMAC function. Besides being easier to estimate, this has the advantage that in its estimation data on capital input are not required. Some general comments may be made on the CD function estimates obtained in various studies for Indian industries and the way in which the estimated coefficients have been interpreted. It has been very common among the studies to interpret the sum of labour and capital coefficients as a measure of returns to scale. This interpretation can, however, be seriously questioned for studies based on aggregated data, especially time-series data. It may be argued that the concept of returns to scale can be given an unambiguous meaning only at the micro-level. The relationship holds at a point of time and applies to a situation in which the character of inputs does not change. This is very different from the conditions that prevail when the analysis is carried out at the aggregate level using time-series data. Aggregate data tend to combine economies of plant size with economies of the size of the market. It is evident that for a proper measurement of returns to scale, plant level data should be used for the estimation of the production function. The estimates of the production function based on industry-level data or data for the aggregate manufacturing sector may not correctly show the extent of economies of scale or diseconomies associated with plant size, and it is necessary to be very cautious in drawing inferences about returns to scale based on such estimates of the production function.

Thus, it can be mentioned here that in many studies based on time-series data, estimates of the CD function have been found to be poor, being affected by the problem of multicollinearity, especially those studies that have used a time trend variable to capture technological progress. Due to multicollinearity, the coefficient of capital has proved to be low and statistically insignificant, and in some cases even negative. The coefficient of labour, on the other hand, had proved to be greater than one in some studies, again as a result of multicollinearity. Estimates of CD function parameters have been affected also by errors in the measurement of capital. This is a bigger problem for models in which a time variable is included, because this eliminates the trend components from various series and as a result the estimation biases caused by measurement errors of explanatory variables are accentuated. This has probably led to an underestimation of the capital coefficient and an overestimation of the coefficient of time. Some other studies from Indian and abroad taking the issues into consideration are presented here in following paras.

The study conducted by Sastry (1966) was oriented towards measuring the 'productivity change'. Whereas, Mehta (1974) attempted to find out the sources of growth of output, returns to scale, the elasticity of substitution and rate of neutral technical change in Indian Sugar industry for the period 1953 to 1965 with the help of CMI & ASI data. Gupta and Patel (1976) has made an attempt to examine the degree of factor substitutability, estimates returns to scale and marginal factor productivities for the sugar industry in India with the help of annual time-series data for the period 1946 to 1966. While in another study of Mehta (1980) have studied productivity, production function and technical change for Indian Industries for the period of 1953-65 and 1965-70. In his study, finds evidence in favour of constant returns to scale, capital deepening and absence of any technical change.

Stier (1980) employed a Translog cost function with three inputs, capital, labour and sawlog, to provide an econometric investigation of the role of sawlogs in the U.S. lumber industry during the period 1950-1974. The conditions for a well-behaved cost function were tested. The concavity condition was violated at two observations. The Allen elasticities of substitution, price elasticities of demand and the technological change biases were calculated. The results indicated that the occurrence of capital-using technological change, which enabled the labor force to be reduced by almost fifty percent while output, remained virtually constant. However, the production technology had not proved sufficiently flexible to mitigate completely the effects of resource scarcity. The real price of lumber had continued to raise, stimulation further substitution in the product market.

As compared with Hicksian, Harrodian measures of the concept of total factor productivity which rigorously take into account the reproducibility of commodity capital inputs and the technological interdependence of modern production economies are advocated. A number of recent measures of total factor productivity are shown to be variants of the Harrodian approach, and certain problems of aggregation associated with the Hicksian measures are shown to be resolved by the Harrodian measures. An examination of the concepts of technical progress and vertically integrated sectors advanced by Prof. Luigi L. Pasinetti and their relation to the Harrodian measures of total factor productivity has been made in Rymes T.K. (1983) paper entitled- 'More on the Measurement of Total Factor Productivity'. In his earlier study (Rymes, 1972), he argued that "when the fact that capital inputs are produced means of production is rigorously and logically incorporated in the measurement of capital and technical change, support is provided for Harrod-Robinson concepts of technical change. The Hicks-Meade-Solow concepts of technical change are shown to be theoretically faulty". In this paper, Rymes reaffirm the theoretical superiority of the Harrod-Robinson measures, briefly review some of the estimates being made and showed how the developing data system at Statistics Canada should permit, as his colleague Prof. L.M. Read long ago suggested, the estimation of Harrod- Robinson-Read (hereinafter HRR) measures of technical change. In addition, he shows that the HRR measures resolve certain aggregation problems encumbering the standard measure. Finally he attempts to show the relationships between the HRR measures and the concepts of technical change and vertical integration advanced by Prof. L.L. Pasinetti.

A review has been carried out by Chakraborty (1982) on 'CES production function of recent literature'. He points out that in most of the studies where researchers are more interested in elasticity of substitution rather than the rate of technological change, the technological change is specified as being Hick's neutral and constant. Studies of the aggregate elasticity of substitution for manufacturing sector have been made by Asher & Krishna Kumar (1973), for USSR, Hungary, Yugoslavia, USA, Canada and Israel; Desai (1976) for Australia; Tsurumi (1970) for Canada, Kazi, *et al.* (1976) for Pakistan, and Berndt (1976) for USA.

Martinello (1985) studied three Canadian industries, pulp and paper, sawmills and shingle mills and logging, using annual data between 1963 and 1982. To represent the production structures of these industries, Marinello specified a Translog long run cost function with four inputs, labour, capital, energy and wood, for each industry. Four restrictions were imposed to ensure the Translog function was homogeneous of degree one in prices and the Hessian of the function was symmetric. The unrestricted model and five restricted models were tested by the log likelihood ratio test for each industry. The five restricts included hick's neutral technical change, no technical change, homotheticity, constant returns to scale and shares independent of factor prices. Finally, the unrestricted model was selected for all the industries. Then, the author calculated factor substitution, technical change and returns to scale in each industry. He took technical change as a measure of total factor productivity improvement. The results indicated that input-demand functions for these industries all sloped down and were inelastic. Factor substitution was not rejected in any of the industries but it was not large. Sawmills and shingle mills presented moderate increasing returns to scale, while they were large in logging and pulp and paper industries. Technical change was non-neutral, capital using, and labour saving in all industries. Negative technical change was detected for sawmills and shingle mills and pulp and paper so that all of the productivity gains during the research period were associated with changes in scale rather than the passage of time.

Merrifield and Haynes (1985) used the Translog cost models (four inputs: labour, stumpage, structures, equipment) to represent production technologies for the lumber and plywood industries in two Pacific Northwest sub-regions between 1950 and 1979. Two hypotheses were evaluated. The fist was that each industry production function was homogeneous and the other was that the industry production process could be represented by unitary elasticities of substitution between factors. The results demonstrated that both the westside lumber and the eastside plywood were constant over observed output levels, while eastside lumber industry and westside plywood industry had experienced varying degrees of scale economies over past production levels. However, none industry sector could accept the hypothesis of unitary elasticities of substitution between inputs. The conditions for a well-behaved cost function were checked. Monotonicity was satisfied in the westside and the eastside lumber industries, and the eastside plywood industry, but was violated at some observations in the eastside plywood industry. The concavity conditions were not met at all the observations in any industry. Factor substitution, derived demand and technological change were measured for each industry sector. The results indicated the presence of output economies in specific industries, a potential impact of reduced factor prices, and the lack of a consistent trend in technological changed across products and regional industries.

Singh and Nautiyal (1986) studied long-run productivity and demand for inputs in the Canadian lumber industry from 1955 to 1982, using a Translog cost function with four inputs (Labour, Capital, material and energy). Four restrictions, the homotheticity, homogeneity, unitary elasticity, and homogeneity combined with unitary elasticity were tested to select the best model. In addition, the hypothesis that factor demands in the industry were actually interrelated was also checked. The results of the tests demonstrated that all the restrictions imposed were rejected. Then the regulatory conditions which are necessary and sufficient for a well-behaved cost function were found to be met at every point of the observations. The AES (no MES), long-run elasticity of demand, elasticity of scale, technical progress and two kinds of factor productivity, which are the long-term path of factor productivity and the actual factor productivity, were measured. The results indicated that factor demands in the Canadian lumber industry were actually interrelated; there were economies of scale in production of lumber in Canada, but technological progress was not observed; simulation of the actual and the least-cost paths of factor utilizations revealed substantial mis allocation of each input during the most years; the observed labour productivity increased at the rate of 2.9% per annum while, net of short-run conditions, the rate was 3.7%; productivities of other three inputs declined both on the observed and the long-run productivity paths, but much slower on the latter.

Abt (1987) measured factor demands in the U.S. Lumber Industry from 1963 to 1978, in the Pacific Northwest, the Southeast and the Appalachian hardwood regions. The author used a restricted cost function with labor and roundwood as variable inputs, and capital as a quasi-fixed input. In order to determine if the three regional models were significantly different, the author estimated the three models as a simultaneous six-equation system to test regional homogeneity without any economic constraints. After confirming that all regions were significantly different, individual regional models were estimated respectively and the economic properties of the cost function were tested in each region. As a result, the southern model estimates were inconsistent with the restrictions. But after imposing homogeneity in the southern model, each regional model satisfied the monotonicity and price concavity constraints at each observation. Then, tests were conducted for constant returns to scale (CRS). Since CRS could not be rejected in the western and southern region, it was imposed on the models for the two regions. Finally, the input demand elasticities were calculated and then used to decompose actual changes in factor into actual growth, output effect, own price, cross price, capital effect, technical effect and residual. The results showed that in the Appalachian region, consumption of sawlogs increased slightly (0.4%) while labor consumption decreased significantly (2.2%). In the southern region, the absence of a price effect could be attributed to similar increases in labor and sawlog prices (9% per year). In the West region, average factor demand changes were near zero.

Martinello (1987) presented the estimates of factor substitution, technical change and returns to scale in the four wood products industries in the B.C. interior and the B.C. coast, including sawmills and planning mills, shingle mills and plywood and veneer mills, between 1963 and 1979. The author used a Translog cost function with four inputs (labour, capital, timber and energy) for all industries, and five restrictions were imposed. Two contributions were made to the estimation of capital stock in this study. First, estimates of aggregate capital stocks are available, as are estimates of each industry's energy consumption. Second, the study showed how to combine the data to produce estimates of each industry's consumption of capital services. All of the estimated cost functions satisfied the monotonicity properties and curvature property at the means level. The only hypothesis that could be tested for each industry separately, no bias in technical change, was rejected in each industry by likelihood ratio test. Then, the price elasticities, returns to scale, and technical change were estimated for each industry. The results revealed that all the inputs were substitutes; the input demand functions sloped downward and were inelastic; the technical change was capital using, labour saving, and materials saving or neutral in all industries; the technical change had not been sufficient to offset the effects of the declining size and quality of timber, and thus average costs increased over the sample.

Meil and Nautiyal (1988) worked on production structures and factor demands in the four Canadian softwood lumber producing regions, the B.C. Coast, the B.C. Interior, Ontario and Quebec, from 1968 to 1984. A Translog variable cost function which has four inputs (labor, capital, material and energy), and three main outputs (lumber, ties and wood chips), was used to estimate the regional production structure indicators of the industry. The Translog cost function was estimated together with cost share functions and five restrictions. Specially, the study employed a cross-sectional time-series data set, categorizing mills into four size classes. Therefore, the results were categorized not only by regions but also by mill sizes. Each pair of regions was estimated together to test regional coefficient homogeneity, showing that each pair was significantly different at the 99% confidence level, so aggregate studies were not regionally applicable. Similarly, a test for within-region coefficient homogeneity was conducted for each region and the results revealed that only the B.C. Interior was left at an inconclusive point without the availability of more data. Therefore, the separate intraregional models were pursued. After the estimation of each intraregional model, the conditions for a well-behaved cost function were confirmed to be satisfied. Last, Allen partial elasticities of substitution (AES), factor demand decomposition, total factor productivity and biases of technology change were obtained.

Compared to the recent studies, Meil and Nautiyal (1988) did not estimated MES. The results indicated that demand for production inputs was governed by offsetting dynamic effects. With few exceptions, all mills across regions exemplified material- and energy-using and labor-saving biases in technical change. Large mills consistently registered the greatest labor-saving technical change, which countered their lack of attaining significantly large cost-reducing scale economies. Mid-sized mills consistently exhibited the largest returns to scale.

Meil, Singh and Nautiyal (1988) applied a Tranlog variable cost function with three outputs (lumber, ties and the byproduct pulp chips) and four inputs (materials, labour, energy and capital) to conduct the dynamic empirical cost structure analysis of the B.C. interior softwood lumber industry between 1948 and 1983. They construct a stock adjustment model containing the dynamic cost function to distinguish between the actual and the fully adjusted demand for factors. All parameter estimates satisfied the conditions for a well-behaved cost function at each annual point of observation. The calculated log likelihood values for the full-adjustment homogeneous model, the restricted test model for interrelated factor demand, and the comparative static homogeneous model were obtained from the estimation results. These decreasing values indicated that the hypothesis regarding the adequacy of instantaneous adjustment of variable inputs was rejected and suggested that ignoring the time taken to adjust to the prevailing economic production environment could seriously affect our understanding of the long-term production structure. Substitution elasticities of inputs, cross substitution elasticities of inputs over time and production, the long-run elasticities of demand, technical progress, and the results of simulating the least-cost paths of the inputs, total net variable cost and the productivity of inputs were reported. The results indicated that demand for variable inputs in the regional industry were interrelated; the assumption of full and instantaneous adjustment in variable factor demands proved to be implausible; small, but significant economies of scale and technology progress were present in the industry; the technology was largely materials using and labour saving; all inputs were substitutes except for material and energy; labour productivity was positive on both the observed and least-cost productivity expansion paths; wood and energy productivities were declining on both paths.

Constantino and Haley (1988) used a translog restricted profit function to represent the structures of the sawmilling industry in the B.C. coast and the United States Pacific Northwest, west side, from1957 to 1982. The authors developed the model in a novel way. First, wood quality was added to the model as an independent variable. Second, lumber and pulp chips were treated as two separated outputs. The specified restricted profit function had two outputs (lumber and pulp) and three inputs (labor, capital and wood). The capital stock was treated as fixed in the short run. Besides, two sets of restrictions were added, which were implied by linear homogeneity in prices and symmetry of the Hessian matrix. The profit function was estimated by means of a translog functional form together with three share equations. Two hypotheses were tested to determine whether wood quality had a significant role in the model of the lumber industry. The result confirmed that wood quality was an important variable in the model. Then own- and cross-price elasticities of demand and supply, and the elasticities of variable inputs and outputs with respect to capital and wood quality were calculated. A 1% increase in lumber price increased lumber supply by 1.11% in B.C., and by 1.44% in the U.S. Pacific Northwest. In both regions an increase in wood quality led to an increase in lumber supply and sawlog and labor demands, but a decline in pulp chip supply. Elasticities of output and input ratios were also obtained. It was found that as the price of an input increased, there was an incentive to use it more efficiently, while with market upturns, less efficient marginal producers started operating. This study mainly focused on if wood quality was an important variable in lumber industry modeling. Therefore, there was no measure on technical change or productivity growth.

Bernstein (1994) employed a developed dynamic model of multiple output production and investment to test the existence of price-cost margins and decompose rates of total factor productivity growth in the Canadian softwood lumber industry, over the years 1963-1987. In this model the softwood lumber which was sold domestically was not assumed to be identical or perfect substitutes with the one sold on foreign markets. A translog profit function was specified to estimate the model, including three outputs (domestic, exported softwood lumber and other lumber products) and two inputs (labor and capital). Then the three models with three different hypotheses were estimated. The first is that non-competitive behavior existed in both the domestic and export markets for softwood lumber. The second is that it existed in either market. The third is that competitive behavior existed in both markets. The estimation results demonstrated that the softwood lumber industry priced competitively in both markets and this industry was in short-run equilibrium. Finally, total factor productivity growth and decomposition were obtained. The average rate of TFP growth was 3.2% per year over the last three decades and the major contributing element to TFP growth arose from the rate of technological change. Capital adjustment costs led to around a 0.7% per annum declined in the rate of TFP growth.

Richard G. Lipsey & Kenneth I. Carlaw (2004) have argued that TFP is interpreted in the literature in different, mutually contradictory ways. Changes in TFP are shown to measure not technological change, only the super-normal returns to investing in such change – returns that exceed the full opportunity cost of the activity. Thus, in the limit, technological change can proceed with unchanged TFP. Measuring the effects of technological change instead requires counterfactual estimates. Reasons why changes in TFP are imperfect measures of super normal returns are studied – reasons connected with the timing of output responses, the treatment of R&D in the national accounts, the omission of resource inputs, and two types of aggregation.

Aikaterini Kokkinou (2005) paper attempts to define the main statistical indicators of innovation and technical change, as well as to determine the factors of measuring innovation activities and limitations regarding the statistical estimation of innovation activities and the role of technical change on growth and development.

The non-parametric approach (the term non-parametric was first used by Wolfowitz, 1942) is used when there is no information about the parameters of the variable of interest in the population, because it does not require a functional form to represent the relationship between outputs and inputs. One method that takes a non-parametric approach to efficiency measurement is index numbers, which are defined as real numbers that measure changes in a set of related variables (Coelli, 2005). The index numbers, including the Laspeyres and Paasche Index, the Fisher Index, and the Törnqvist Index, are used to measure price and quantity changes over time, as well as to measure differences in the levels across firms, industries, regions, or countries. The other method is Data Envelopment Analysis (DEA), first developed by Charnes, Cooper, and Rhodes (CCR) in 1978, which built on the ideas of Farrell (1957). They utilized the linear programming method to construct a production frontier, by introducing the DEA technique to assess the relative efficiency of a set of comparable units, called decision-making units (DMU) (Salehirad and Sowlati, 2006). The CCR model in which the frontier is constant return to scale (CRS), estimates aggregate efficiency, including technical and scale efficiency. Another model, the BBC model introduced by Banker, Charnes, and Cooper (BBC) in 1984, has a variable return to scale (VRS) frontier, and measures the pure technical efficiency of DMU with reference to the efficient frontier. Today, the index numbers, DEA, and other non-parametric approaches have been applied to numerous efficiency measurement problems in the forest products sectors.

M. Parameswaran (2004) paper examines two important components of the total factor productivity growth, namely technical change and technical efficiency change of firms belonging to the capital goods producing industries. Empirical analysis of these two components is motivated by the policy changes that capital goods producing industries were subjected to during the 1990s as well as by the existing evidences on the total factor productivity growth in Indian manufacturing industry. Technical change and technical efficiency change has been estimated in a single step, using a stochastic frontier production function. The results of the study show that all the industries studied experienced a significant improvement in the rate of technological progress during the post reform period. However, the evidence on the technical efficiency shows that not only the level of technical efficiency is lower during the post reform period, but also the rate of decline in the technical efficiency is higher during this period in all industries except in one. Thus the paper provides further insight into the productivity performance of these industries during the post reform period.

The knowledge production function is central to R&D-based growth models. Expressing this view in their IMF Working paper, Yasser Abdih and Frederick Joutz (2005) has empirically investigated the knowledge production function and inter-temporal spillover effects using co-integration techniques. Time-series evidence suggests there are two long-run co-integrating relationships. The first captures a long-run knowledge production function; the second captures a long-run positive relationship between TFP and the knowledge stock. The results of this working paper indicate the presence of strong inter-temporal knowledge spillovers and that the long-run impact of the knowledge stock on TFP is small. This evidence is interpreted in light of existing theoretical and empirical evidence on endogenous growth.

The objective of Ali M. Khalil (2005) paper was to estimate the Transcendental Logarithmic Production Function of manufacturing industry in Jordanian economy. Manufacturing industry can be considered as the fourth large one in Jordanian economy. In last decade, the growth rate in this sector is the second one after transport and communications. The gross fixed capital formation in manufacturing is the third one. In this paper he used a cross sectional data for cost shares, factor inputs, factor prices, and output. Main source of data for manufacturing industry in Jordan is the Industry Survey - Department Statistics - the Hashemite Kingdom of Jordan. The data for factor inputs and output were transferred to logarithm. A point around which production function should be estimated was chosen. Cost shares of inputs have been calculated by dividing compensations to employees, operating surplus, and value of materials by value manufacturing output. The theoretical framework in this research is the flexible production function. So, the three-input transcendental logarithmic (Translog) production functions could be used. This function is approximated by second order Taylor series. The log of likelihood ratio test has been used to choose among different hypotheses. The symmetric Translog hypothesis has been found as an appropriate one among other hypotheses. It is well behaved, where positivity and concavity of production function have been satisfied. After testing for nested hypotheses, the symmetry restrictions were imposed. So, restricted and unrestricted functional forms are estimated. However, Iterative Zellner-Efficient Estimate has been used to get estimates equivalent to the maximum likelihood estimates. Allen partial elasticities of substitution, AES, and own and cross price elasticities for factor inputs have been calculated. AES measure the substitutability or complement among factors of production. On the other hand, AES can measure the curvature of isoquant. AES shows that capital-labor, capital–materials, and labor-materials are all substitutive. But, AES for labor-materials is more than a half of that for capital-labor. AES for capital-labor is one and three fourth of that for labor-materials. The Price elasticities of factor inputs show that capital and labor demand are more elastic than demand for materials.

Eric Miller (2008) in his study entitled –"*An Assessment of CES and Cobb-Douglas Production Functions*" surveyed the empirical and theoretical literature on macroeconomic production functions and assesses whether the constant elasticity of substitution (CES) or the Cobb-Douglas specification is more appropriate for use in the CBO's macroeconomic forecasts. The Cobb-Douglas's major strengths are its ease of use and its seemingly good empirical fit across many data sets. Unfortunately, the Cobb-Douglas still fits the data well in cases where some of its fundamental assumptions are violated. This suggests that many empirical tests of the Cobb-Douglas are picking up a statistical artifact rather than an underlying production function. The CES has less restrictive assumptions about the interaction of capital and labor in production. However, econometric estimates of its elasticity parameter have produced inconsistent results. For the purpose of forecasting under current policies, there may not be a strong reason to prefer one form over the other; but for analysis of policies affecting factor returns, such as taxes on capital and labor income, the Cobb-Douglas specification may be too restrictive.

Bart van Ark, Abdul Azeez Erumban, Vivian Chen & Utsav Kumar (2009), paper focuses on comparisons of productivity, (unit) labor cost and industry-level competitiveness for the manufacturing sector of China and India. They provide a comparison between India and China using a broad international perspective. They find that China has increased its labor productivity to a level above that of India, but due to a somewhat higher compensation level, China is still somewhat at a disadvantage in terms of unit labor cost in manufacturing relative to India. In the second half of the paper, they an analysis of industry level differences in productivity, labor compensation and unit labor costs at state and province level in the two countries from the mid 1990s to the early 2000s. They also find rapid declines in unit labor cost across industries and provinces in China, but increases in many instances in India. This suggests that productivity and compensation growth have become much more aligned across regions in China whereas this is not the case in India. They have related these results to differences in the implementation of market reforms between the two countries and removal of barriers to resource mobility eradicating inefficient manufacturing activity.

Almas Heshmati & Subal C. Kumbhakar (2010) have reviewed all about the sources of growth of TFP, a number of productivity studies on China's economy examined productivity differences by types of ownership (*e.g.* Jefferson, 1990; Dollar, 1992; Jefferson and Xu, 1994; Chen *et al.*, 1998; Xu and Wang, 1999; Hu, 2001; and Zheng *et al.*, 2003). A number of other categories of productivity research include the examination of sectoral productivity growth differences (*e.g.* Lin, 1992; Jefferson, Rawski and Zheng, 1992, 1996; Wu, 1995, 2000; Xu, 1999; and Zheng and

Zheng, 2001) and the investigation of productivity difference among regions in China (*e.g.* Lee, 2000; Song *et al.*, 2000; Cai *et al.*, 2001; Demurger, 2001; Bao *et al.*, 2002; and Demurger *et al.*, 2002). A few of the datasets used in the above studies are at the firm level, while majority of them are at the aggregate national, sectoral, regional or provincial levels. Maddison and Wu (2008) suggested that the official Chinese National Bureau of Statistics exaggerated GDP growth and adjustment to conform to international norms. They present and discuss the necessary adjustments by contributions along new volume indices for the industrial sector and for services. These authors use a measure of purchasing power parity instead of the exchange rate. The methodology used in the analysis of productivity in China is diverse. These include both non parametric and parametric approaches. The non-parametric can be divided into growth accounting and Malmquist productivity indices. A handful of studies have applied the growth accounting approach (*e.g.* Chow, 1993; Borensztein and Ostry, 1996; World Bank, 1996; Hu and Khan, 1997; Maddison, 1998; Woo, 1998; Ezaki and Sun, 1999; Demureger, 2000; Wang and Yao, 2003; and, Arayama and Miyoshi, 2004). The growth accounting approach involves the subtracting of the growth of factor accumulation at a constant rate from the output growth to obtain the TFP measure. In this case with constant returns to scale, TFP is equivalent to technical change. Some of the studies above used Cobb-Douglas average production function (such as Chow, 1993; Ezaki & Sun, 1999; and Wang and Yao, 2003), while others (like Hu and Khan, 1997; and Arayama and Miyoshi, 2004) applied the Translog production function. These studies focus on the estimation of factor input shares to be used in the computation of the aggregate productivity growth over time. All of these studies have found positive TFP growth in post-reform China.

Gheorghe Zaman & Zizi Goschin (2010) has emphasized that as technical change is nowadays largely accepted to be an engine of economic growth, researchers have tried to include it explicitly in the economic growth models, either as an exogenous or endogenous factor of influence. Using the framework of the aggregate Cobb-Douglas production function in its classical form, as well as in several refined variants, we estimated the elasticities of production factors for Romania over the 1990-2007 periods, finding that technical progress has had a small contribution to the economic growth.

In macroeconomics, technological progress is usually identified with growth of residual productivity, an umbrella term containing everything that could not be traced back to the accumulation of factors of production, included in the aggregate production function. It is however uncertain – and competing methodologies provide conflicting clues on that – what exactly this production function should be. Thus, the objective of the Jakub Growiec (2010) was to investigate this matter more closely. Rather than trying to provide direct empirical evidence on the actual shape of the production function, he was remain agnostic about it and instead carry out an indirect study focusing on the properties of 14 alternative specifications of technological progress (i.e., growth of residual productivity). The contribution of this article to the literature is to provide a synthetic, numerical assessment of the relative advantages and disadvantages of a number of approaches to the measurement of technological progress across countries.

Kazumi Asako & Miho Takizawa (2010) in their study entitled -"*Marginal Productivity Principle and Measurement Biases in TFP: Evidence from International Productivity Database*" pointed out that trends and changes in total factor productivity (TFP) have been the subject of a considerable amount of research from both theoretical and empirical approaches, at both the micro-economic and macroeconomic levels, as a driver of economic growth and fluctuations. A traditional growth accounting method employed in such research assumes the marginal productivity principle (MPP) for factors of production. In this paper, they conducted international comparison of TFP growth rates between those measured when assuming the MPP and those measured when not assuming it, to determine the degree of deviation in TFP growth rates under different assumptions and examine its causes from the perspective of business cycles, using data on the five developed nations (Japan, the United States, the United Kingdom, Germany, and France). When not assuming the MPP, TFP measurement requires identification and estimation of the production function. After comprehensive judgment of the results of various estimates conducted under differing assumptions, we decided to employ a Cobb-Douglas production function. Based on this, we compared traditional TFP growth rates assuming the MPP and technological or true TFP growth rates not assuming the MPP, to measure and assess the bias by which the marginal productivity of labor and capital deviate from the rewards of each. For Japan, we conducted similar analysis using panel data at an industry level. The results of this measurement show that, while conditions vary by country and over time, the size of the bias is of a level that cannot be ignored in comparison with the level of the traditional TFP growth rate, and consequently that the degree to which the traditional TFP growth rate deviates from the true technological TFP growth rate is, at some times and in some cases, clearly not inconsequential. Putting another way, the likelihood is high that the marginal productivity bias and deviation between the two TFP growth rate indicators is cyclical and characterizes the historical business cycle in each nation.

Md. Moyazzem Hossain *et al.* (2012) in their study has suggested the most suitable functional form of production process for the major manufacturing industries in Bangladesh. They consider C-D production function with additive error and multiplicative error term. The main purpose of their paper is to select the appropriate C-D production model for measuring the production process of some selected manufacturing industries in Bangladesh. They have uses different model selection criteria to compare the C-D production function with additive error term to Cobb-Douglas production function with multiplicative error term and finally, they have estimated the parameters of the production function.

3.4: Wood & Forest based related aspects

Forestry in India is a significant rural industry and a major environmental resource. India is one of the ten most forest-rich countries of the world along with the Russian Federation, Brazil, Canada, USA, China, Democratic Republic of the Congo, Australia, Indonesia and Sudan. Wood and forest based industries linkages are the best instrument to achieve balanced development of the country/ states. As well as to minimize regional inequalities in industrial disbursement they can play a positive role. The importance of forests in national economy requires no explanation. Thus, keeping these views in mind, an attempt has been made in this section to review the past literature related to section into the consideration and Small Scale Rural industrialization too as well somehow aspects the topic of the project.

H. F. Kaiser (1971) considered the evolution of labour productivity (LP) in all forest products sectors in the United States. His study provided estimates of the rate of change in output per unit of labour for the period of 1947 to 1967. He found that labour productivity rose in all forest products sectors at an average annual rate of 3.4%. Saw milling, lumber and pulp and paper had the highest LP growth. The author explained that the rise in LP in the saw milling and lumber sectors (particularly in plywood mills) was due to the automation of the production process. At the same time, the total number of producers in the American saw milling and lumber sectors dropped by half while output rose by 50 per cent. Similarly, Kaiser explained that the trend in LP in pulp and paper industry was due to increased research and development which led to better automated processing. In most forest products sectors, the average wage rates went up as the number of skilled workers employed grew. The study by Gregory Horvath (1980) was similar to the one by Kaiser except the period covered was longer, from 1947 to 1977. All U.S. forest product sectors experienced LP growth, but some more than others (2.6% on average annually in the saw mill industry and 4.6% on average annually in the plywood industry). The author also explained LP growth by a combination of automation and economies of scale.

John Duke and Clyde Huffstutler (1977) of the Bureau of Labor Statistics calculated labour productivity (LP) estimates for years 1958 to 1975 in the saw milling sector. The average annual rate of growth in LP was equal to that of manufacturing in general in the U.S., at 2.7 per cent. LP growth was higher than average before 1965 and then lower until 1970. LP tended to fluctuate with output, which in turn fluctuated with the demand for lumber for housing. As noted in other studies, new technology was responsible for the growth. The number of unskilled jobs dropped at the same time, especially in lumber handling.

The study of the veneer and plywood industry between 1958 and 1976 by Mary Farris (1978) of the Bureau of Labor Statistics comes to the exact same conclusion as Kaiser (1971). LP grew at an average annual rate of 4.5% over the time period. This above average growth rate was partly the consequence of an expansion of demand and therefore output arising from housing needs. Technical innovations permitting the use of different tree species of various sizes and reducing the amount of unskilled labour needed in the production process by automation also contributed to LP growth. Plywood plants tended to be larger on average in 1976 than in 1958.

The study by Greber (1982) considered the rise of the capital-labour ratio in the American lumber and wood products industries between 1951 and 1973. This was a result of factor substitution from technical change. During the period under study, the number of firms in the wood products industries decreased (and the average size increased). At the same time, heavy machinery and automation became the norm. Technical change was biased towards capital, which triggered the rise in labour productivity observed in that period.

Jack Veigle and Horst Brand (1982) from the Bureau of Labor Statistics conducted a study of the American millwork industry (window frames, doors, mouldings, etc.). They were interested in the progression of labour productivity between 1958 and 1980 in the U.S. For the time interval of 1958 to 1972, the growth was positive with an average of 2.6 per cent annually and then negative at -1.4 per cent annually from 1972 to 1980. Labour productivity varied greatly from year to year depending on the respective growth of output and labour. The period of 1972 to 1980 was characterized by lower demand from residential housing. The authors explained the low labour productivity growth by a lack of investment in new technology. The relatively small size of the typical millwork plant compared to other manufacturing sectors could have explained the lack of investment in automation of the production processes, because of too small a volume of production.

Desai (1983) stated that rapid industrialization in India depends on the growth of small scale industries. Most of the small scale industries are operating under certain handicaps like shortage of raw materials, low levels of technical knowledge and counseling, poor infrastructure, inadequate capital and credit facilities, improper distribution system, lack of facilities for market analysis, research and development.

Bruno De Borger & Joseph Buongiorno (1985) estimated the total factor productivity growth rate for the American paper and paperboard industries for the years between 1957 and 1981. To do so they used a variable cost function from which they derived two different versions of total factor productivity growth estimates for each industry. The total factor productivity annual growth rate was positive for all years. There was an upward trend from 1957 to 1973 and a downward trend subsequently. These results were observed in both industries. Depending on which version of TFP growth was used, the average annual growth rate was 2.89% or 4.54% in the paper sector while the average annual growth rate was smaller at around 1.0% for both versions of TFP in the paperboard sector. The authors suggested two reasons for this rather large difference in TFP growth rates. The first one was that the paperboard industry was more energy intensive and that the rise in energy prices after 1973 affected it more than the paper industry. Yet, the TFP growth rate for the paper industry has been systematically two percentage points higher than the paperboard industry TFP growth rate since 1957. The second reason was the faster labour productivity total growth in the paper industry compared to the paperboard industry (150% and 130% respectively).

Using two all-India sample surveys, conducted by the Reserve Bank of India and the National Small Industries Corporation, respectively, an attempt has been made in R Nagaraj (1985) paper to understand the rate, pattern and characteristics of the growth of small scale industries in India. Part I of the study, based on the Reserve Bank of India's sample survey of small scale industrial units conducted during 1976-77, discusses the broad trends in the growth in number of small enterprises and some characteristics of the structure of this sector. In Part II, using the National Small Industries Corporation sample survey data for 1979, some other aspects of the small scale sector like marketing of output, availability of inputs, nature of competition, pricing, etc, are discussed.

B.K. Singh & J. C. Nautiyal (1986) used a translog cost model to study factor demands and productivity trends in the Canadian lumber industry from 1955 to 1982. They found that the lumber industry experienced economies of scale over the time period but their model could not capture any significant technological progress. Nautiyal and Singh (1986) studied the evolution of the single factor productivities in the Canadian pulp and paper industry over the period of 1956-1982. They used a translog cost function to determine the least cost long run combinations of inputs from which they derive the hypothetical long run single factor productivities. They observed that labour and capital productivity average annual growth rates were positive (1.8% and 0.8% respectively) while the ones for material and energy were negative (-1.4% and –0.6% respectively). These growth rates were smaller than the long run rates estimated by the authors because the adjustments to new levels of output and factor prices took time and caused temporary misallocation of inputs. In fact, the difference between long run and observed productivities was higher in recessions. At the end of the paper, they explained the slowdown in the various productivity growth rates in the 1970s and early 1980s by cyclical factors.

Arnold, Chipeta & Fisseha, (1987) in their study reveals that the development of the forest-based sector in developing countries will depend on the way in which forest-based industrialization progresses. The importance of small-scale enterprises (SSEs) was discussed in an earlier article. Modern, large-scale enterprises (MLEs) are discussed in the present article, which revisits much of the ground covered by Westoby's seminal article in Unasylva almost 30 years ago (Westoby, 1962). While many of the conclusions still hold, there has also been a substantial evolution of thinking regarding the impacts of forestry-based development in such areas as income distribution, environment and trade. A number of groups are concerned about the economic, social and environmental impacts of large forestry and forest industry schemes. Some point to the fact that, although it is theoretically possible to guide their development in the direction of national goals, there is an increasing realization that past government policies have often introduced unacceptable distortions in the allocation of economic resources, resulting in serious environmental damage. Policy errors are more visible and often more costly if they involve a few large-scale, concentrated operations rather than a myriad of scattered SSEs. Yet, the economics of many production activities in forestry (*e.g.* economies of scale and lumpiness of capital) as well as market requirements (such as standardization and demand for large, secure volumes of product) often favour the establishment of MLEs in the forest-based sector. They had also explores the nature of the various impacts of large-scale forest-based industrial development and described policies that can be implemented to induce the materialization of positive impacts while minimizing negative ones.

Jamie Meil, B. K. Singh & J. C. Nautiyal (1988) used a Translog variable cost function to estimate SFPs (for labour, material and energy) in the absence of an adjustment period following fluctuations in input price or output demand. They studied the British Columbia interior softwood lumber industry for the 1950-1983 time periods. SFP average annual growth rates had the same sign as the least-cost annual growth rates but they were systematically lower: labour (3.8 % versus 4.7 %), material (-0.45% versus –0.19%) and energy (-0.86 % versus –0.54 %). The deviations from the least cost estimates were a sign of difficulty in adapting rapidly to changing market conditions. It is noteworthy that the B.C. interior softwood labour productivity growth was higher than that for the Canadian saw milling industry. The drop in energy productivity was also weaker than in the Canadian saw milling industry. The authors also found that the B.C. interior softwood lumber sector experienced technological change over the years. Although technological change was labour saving, it also was wood-using which could become a problem in a context of limited availability of wood resources.

Luis Constantino from the University of Alberta and David Haley from the University of British Columbia (1989) compared total factor productivity in the saw milling industries of British Columbia in Canada and the Pacific Northwest region of the U.S. over the period from 1957 through 1982. The authors found that when wood quality was ignored, the saw milling industry in the Pacific Northwest was, on average, 10% more productive than its B.C. competitor. But when the poorer quality of B.C. wood was taken into account, the B.C. industry was in fact more productive on average. Although the B.C. industry had a higher level of total factor productivity, its TFP growth was lower than in the Pacific Northwest. That region's saw milling industry therefore caught up with B.C.'s industry total factor productivity in the early 1980s.

Constantino and Haley (1989) employed an index number approach to compare the total factor productivity of sawmills in the B.C. coast from 1957 to 1982 with that in the U.S. Pacific Northwest. They found that the productivity level in the U.S. Pacific Northwest industry was higher than that in the B.C. coast industry, and was facilitated by the higher wood quality.

David Frank, Asghedom Ghebremichael, Tae Oum and Michael Tretheway (1990), all from the University of British Columbia, studied productivity trends in the Canadian pulp and paper industry for the 1963-1984 period. They estimated total and single factor productivities using a non-parametric method. Labour, energy and material average annual productivity growth rates were positive over the 1963-1984 periods (2.5%, 2.2% and 1.1% respectively). Since the average annual capital productivity growth rate was negative (-0.6 per cent) for the same period, this was a sign that the variable inputs productivities probably grew because of investment in capital. Furthermore, there were signs that the capital input growth was not optimal, implying excess capacity in the industry. The authors estimated the average annual total factor productivity growth rate to be 1.2%. They decomposed their TFP measure using Translog total and variable cost functions. Most of the TFP growth resulted from economies of scale (0.88 of the 1.2 percentage points) while technical change accounted for 0.32 of the 1.2 percentage points.

Oum *et al.* (1991) compared single factor and total factor productivities in the pulp and paper industries of Canada, the U.S. and Sweden over the 1970s. The authors used an index number approach in order to decompose total factor productivity growth into capacity utilization, output scale and technical efficiency change effects, the last effect calculated residually as the TFP growth rate minus the first two effects. In the evolution of the average capital productivity growth over the 1970-1980 period, Sweden and the U.S. had a negative average annual growth rate (-2% and -0.54% respectively) while Canada had a positive one (1.05%). Conversely, the average labour productivity growth rate was higher in Sweden and in the U.S. than in Canada at 3.51%, 2.97% and 2.58% respectively. These different rates were partly the result of different investment trends in each country. The annual capital input increase in Sweden was higher than in the U.S. and Canada (3.7 %, 2.8 % and 2.1 % respectively). The average total factor productivity growth rate was highest in Canada, followed by the U.S. and Sweden at 1.88 %, 1.7 % and 0.67 % respectively.

The comparative study on the American and Canadian pulp and paper industries by Oum and Tretheway (1992) builds on Oum *et al.* (1991) and on Frank *et al.* (1990). The authors estimated the average annual single factor productivity growth rates for the 1963-1984 time period using a Tornqvist methodology, a non-parametric approach. The growth rates of labour, material and capital productivity were higher in the U.S. than in Canada (3.5% against 2.6% for labour, 2.8% against 1.1% for material, and 1.3% against 0.6% for capital respectively) while the energy productivity growth rate was higher in Canada (2.6% against 0.6%). The authors calculated that capital inputs in the Canadian industry rose by 130 per cent during the period while they only grew 49% in the American industry. Despite the far greater volume of investments in the Canadian industry, the U.S. industry led the way in all single factor productivity growth rates with the exception of energy productivity. The investment by Canadian firms in energy-saving installations could not improve total factor productivity in a significant way. The authors suggested that the U.S. investment programs were more effective than the Canadian ones. The total growth of total factor productivity was 44% over the period in the U.S. and 29% in Canada. Put another way, the average annual TFP growth rate was 1.75% in the U.S. and 1.2% in Canada. Therefore, the Canadian industry became relatively less cost effective than its American competitors.

Jamie Brunet (1993) from Forestry Canada studied productivity levels and trends in the Canadian pulp and paper industry over the period of 1964-1988. According to Brunet, the regional differences in TFP growth rates could have been the result of different capital productivity trends. The annual average capital productivity growth rates for Quebec, Ontario and R.O.C. were negative over the 1964-1988 periods (-2.4%, -2.3% and -0.4% respectively) while it was positive for B.C. (2.2%). When Brunet used a variable factor productivity measure (from which capital inputs are excluded), the interregional differences in productivity growth are smaller, which suggested that capital productivity growth was in part responsible for TFP growth rate differences across regions. The author found that periods of productivity decline would be followed by periods of productivity advance and so on. He explained that this finding was due to the existence of excess productive capacity in the industry.

A study has been conducted by Seshaiah, *et al.* (1993) to examine the productivity trends in some industries, *viz.*, Cotton Textiles, Tobacco & Beverages, Food Products and Paper & Paper products of the Andhra Pradesh manufacturing sector for the period 1976-86. They have measured total factor productivity by using the Divisia Index and, they point out that the Divisia Index is the best among all other indices in terms of quality and efficiency to measure the TFP. Divisia Index is found to be very useful for measuring TFP in the case of multiple inputs and multiple outputs. The rate of growth of TFP of Tobacco and Beverages industry's declining growth rate (–7.44 per cent), indicates that there is technological retrogression in the industry during the period of study. As Tobacco & Beverages, same result came for the Paper and Paper Products, which shows there is technological retrogression in the industry.

Abt *et al.* (1994) applied the Törnqvist -Theil Index to measure productivity growths in the sawmilling industries from 1965 to 1988 in six U.S. and Canadian geographic regions (the B.C. Coast, the B.C. Interior, Ontario, Quebec, U.S. South and U.S. West). The results indicated that over the long-term, labor was the input that had experienced the highest growth in productivity. From 1980 to 1988, there were significant differences in the annualized growth rates in total factor productivity across regions. However, the growth over the 24-year research period was relatively uniform across most regions.

Jeffrey Bernstein from Carleton University (1994) used a dynamic model including price cost margins to evaluate and decompose the Canadian softwood lumber industry's total factor productivity growth for the 1963-1987 period and three sub-periods. Since certain studies have shown that price-cost margins affect TFP growth measures, he included those margins in his model. In doing so, he found an average total factor productivity growth rate over the 1963-1987 periods of 3.24% annually. The total factor productivity growth was also decomposed to show the effects of scale, technological change and capital adjustment. Technological change alone would induce a 2.35% total factor productivity growth (over 70% of the TFP) while returns to scale account for about 1.5% of growth. Capital adjustments reduced the total factor productivity growth by almost 0.7% a year on average.

Rolf Fare *et al.* (1994) analyzes productivity growth in 17 OECD countries over the period 1979-1988. A nonparametric programming method (activity analysis) has been used to compute Malmquist productivity indexes. These are decomposed into two component measures, namely, technical change and efficiency change. They find that U.S. productivity growth is slightly higher than average, all of which is due to technical change. Japan's productivity growth is the highest in the sample, with almost half due to efficiency change.

Jiing-Shyang Hseu (1994) studied the evolution of total factor productivity in Canada and the U.S. from 1961-1984 using three different non-parametric methods: the Tornqvist-Theil index and two linear programming techniques (translating hypothesis and distance function). The total factor productivity trends estimated using the Tornqvist and distance function approach gave similar results. The TFP fluctuations from year to year were much more accentuated with the translating hypothesis. The authors found that the average annual total factor productivity growth rates for Canada and the U.S. were almost the same, at 0.5%, with a small advantage for the U.S. The authors did not try to interpret their results or suggest the causes of the trends they observed.

Belton Fleisher, Keyong Dong & Yunhua Liu (1996) studied the effect of education and enterprise organization on labour productivity in the Chinese paper industry. They estimated that the average annual labour productivity growth rate was 6.4% between 1985 and 1990. When comparing different profit distribution methods, they found that none of them contributed more to labour productivity growth. However, the authors did find that the level of ownership had an impact on labour productivity. They suggested that ownership at a lower level of government (municipal versus provincial) induced more effort by workers since surpluses go to regional authorities.

Kristen Hoff *et al.* (1997) have made an attempt to review existing information on the performance of the U.S. secondary wood products industry, and wood furniture in particular; discuss divergent definitions and measures of competitiveness; summarize previously identified sources of competitiveness; and identify knowledge gaps that need to be addressed in order to better understand the factors affecting global competitiveness in this industry. They pointed out that more than 1 million U.S. workers in some 45,000 firms are employed in the lumber, wood products, furniture, and fixture industries. Wood household and office furniture are the largest manufacturing segments, adding $13.851 billion per year to raw product value. During the 1980s, U.S. furniture manufacturers lost sizeable market share to Pacific Rim countries. To improve their performance in increasingly global markets, U.S. manufacturers must have a clear understanding of how to assess their competitive position and how to affect its strategic determinants. This paper reviews existing information on the performance of the U.S. secondary wood products industry and summarizes current models regarding competitiveness and its sources. A review of the literature suggests that both internal firm processes and external market and government policy factors affect firm and industry competitiveness. However, these are rarely linked in a comprehensive analysis. This paper argues that in order to better understand the factors affecting global competitiveness in this industry, research is needed that combines engineering and economic analyses of competitiveness.

The study by Shashi Kant and Jagdish Nautiyal (1997), both from the University of Toronto, was the only one concerned with the Canadian logging sector exclusively. The authors were interested, among other things, in the total factor productivity growth of this sector from 1964 to 1992. They were able to decompose total factor productivity growth into two components, the scale effect and the technical change effect using a Translog cost function. The authors found that technical change led to labour and capital saving while increasing material and energy usage over the period studied. They also found a continuously negative technical change effect that pulled down total factor productivity growth, which was positive only nine of the 28 years. But this does not mean that technology is deteriorating.

Gupta, Pant & Baghel (1998) have studied- "*Growth and productivity trends in bidi industry of Madhya Pradesh*". In this study the SGR, CGR, SFPs and TFPG as well as variability pattern of bidi industry have been analysed. They concluded that share of capital is not performing well to the extent of technical and developmental progress in bidi industry as compared to share of labour. Whereas, Baghel & Gupta (2002) in their another study entitled -"*Econometric analyses of production functions and technical change in bidi industry of Madhya Pradesh*", have analyzed the extent of technical change and growth pattern of productivity in the bidi industry of the state Madhya Pradesh for the period of 1973-74 to 1992-93. They have also observed that labour is major source of output growth as compared to capital. In case of production functions, time trend coefficients either found negative or insignificant or suggests inferences in favor of technological retrogression.

Paulo Barreto, Paulo Amaral, Edson Vidal & Christopher Uhl (1998) studied the impact of forest management on productivity and the cost of logging in the Amazonian region. The objective was to show the economic benefits of planned logging activities compared to unplanned ones. The authors estimated three measures of productivity: labour productivity (volume of wood extracted per person hour), productivity of road making machines (machine hour per volume of wood extracted) and the productivity of skidders (volume of wood per machine hour). They found that the labour productivity level was 18% lower in a planned logging operation than in an unplanned one when the crew size was the same (two workers per crew). Labour productivity was lower in the planned operation because the crew tried to minimize the damage to other trees. Labour productivity rose when a worker was added to the crew, because of division of labour rather than planning.

The Indian pulp and paper industry was studied by Katja Schumacher & Jayant Sathaye, (1999). Single as well as total factor productivities were estimated. The authors used three types of estimates (Translog, Solow and Kendrick index) to evaluate total factor productivity growth rates from 1973 to 1993. Overall, the single input productivity average annual growth rates were negative except for labour (capital at -2.31%, energy at -2.68%, material at -0.82% and labour at 3.14%). But these averages did not give a good perspective on growth trends. The 1970s saw negative growth rates for all inputs while the 1980s saw strong positive growth for both labour and capital. These trends were the result of a particular national context. India suffered a paper shortage in the early 1970s that led the government to create a large number of small paper mills equipped with outdated capital from developed countries to end the crisis. Machinery decay over the years led to the negative single factor productivity growth rates. But in the 1980s, the equipment was modernized and larger paper mills benefiting from economies of scale were created, explaining the positive growth rates for capital and labour. Over the time period studied by the authors, the average annual TFP growth rate was negative (Translog at -2.2%, Solow at -3.6%, Kendrick at -3.4%). All three TFP estimates followed a similar time trend. The 1990s were a difficult time for the Indian paper industry since the TFP growth rate was between -13.3% and -13.4%. The unstable economic context, the scarcity of wood resources and environmental regulations could have explained the size of the decrease in productivity growth.

Priyanka Dembla (2000) highlighted that in the literature two approaches have been followed in the estimation of production relations - estimation of the conventional production function and estimation of the so-called augmented production function. This paper estimates both kinds of production relations using the more flexible Translog functional form for the registered manufacturing sector as a whole as well as for eachuse-based sector - *i.e.* consumer goods, intermediate goods, capital goods and basic goods using panel data for the period 1973-74 to1995-96 and attempts to reconcile and compare the two approaches. A distinguishing feature of this study is that it attempts to test for the presence of unit root in panel data of the relevant variables. The estimated results reveal the importance of the additional explanatory variables used in the augmented model besides labour and capital stock in explaining production behaviour in the sector. Another important observation relates to the suitability of the translog specification. This implies that all the variables used in the study exhibit significant nonlinearities and that there is significant interaction among them.

Andrew C. Worthington (2000) have studied the nature and extent of efficiency and productivity growth in deposit-taking institutions is investigated using nonparametric frontier techniques. Employing Malmquist indices, productivity growth is decomposed into technical efficiency change and technological change for a sample of Australian building societies. The results indicate that most building societies experienced productivity gain in the past several years, and this was largely the result of technological progress rather than efficiency improvements. That productivity growth which did occur due to an increase in efficiency over the period tended to be the result of improvements in scale efficiency, whilst efficiency gain was most pronounced in building societies with a high ratio of net interest income and non- interest income to total assets, low operating expense ratios, and relatively high expenditures on marketing and promotion.

Atakelty Hailu and Terrence Veeman (2000a, 2000b, 2001), both from the University of Alberta, produced a series of studies on productivity trends in the Canadian pulp and paper industry. The first one (2000a) used a parametric input distance function as well as a Tornqvist index method to estimate total factor productivity growth over the 1959-1994 period. On average, using the input distance function, the annual productivity growth rate was 0.19% but with some annual variations over the time interval. The authors found that there was negative productivity growth during the 1960s and 1970s (-1.55% and -0.74% annually respectively). Productivity growth was positive on average since the 1980-82 recession (0.99%) and especially high since 1990 (3.95%).The authors had difficulty explaining the negative growth of the 1960s but they attributed the negative growth of the 1970s to the two oil crises and their effect on output and to the environmental constraints introduced in that decade. The especially high productivity growth rate of the 1990s was due to the addition of several new mills equipped with modern production installations. At the same time, some of the older mills with outdated technology exited the market. A similar TFP growth trend was found using the Malmquist index measure, derived from the Tornqvist index method, but with an average annual growth rate of 0.41%. When the authors controlled for the effect of output expansion, they found a negative growth rate (-0.15%). Therefore, the TFP growth was mostly attributable to scale effect rather than to technological progress.

Raphael Kaplinsky *et al.* (2003) in their paper pointed out the fact that because of its resource and labour intensity, the wood furniture sector presents an opportunity for developing countries and their firms to participate effectively in the global economy. This paper begins with a brief description of the global wood furniture industry and highlights the importance of exports wood furniture products for developing countries and emerging and transitional economies. The paper then maps the wood furniture value chain and opens-up the nature of the buying function, since this function represents the key form of control over global production networks in this sector (that is, the wood furniture chain is what is increasingly referred to as a "buyer-driven chain"). The paper then asks what producers need to do in order to upgrade their activities, particularly in developing countries. In order to address these issues the authors describe the evolution of an initiative designed to promote the upgrading of one segment of the wood furniture industry in a middle- income country, South Africa. This experience is then used to generate a series of generic policy challenges, which might be transferred to other countries and to other sectors.

Jatinder S. Bedi (2003) have studies the structural changes in the composition of spun yarn, age of installed spindles and its impact on the efficiency and productivity of spindles. The methodology used here is different than the traditional production function approach. The percentage of excess installed spindles compared to the minimum required at the latest technology available during 1996 could be explained due to the under-utilization of spindles and the technological gap between the installed spindles and spindles of modern technology. The inverse of excess spindles used over time due to the technological gap shows the changes in productivity of working spindles. The productivity analysis along with the state of spinning industry and future requirement of spun yarn is used in forecasting the future requirement of spindles.

Productivity growth in the Paper industry has been studied relatively more extensively. Estimates by various authors are brought together in Table 5.4(A). Banerji reported negative but near zero growth of total factor productivity which as he said could be the result of mixed movements over time. His estimates of production function led Banerji to believe that capital deepening in the paper industry had been accompanied by some sort of technical progress. His results on technical progress go contrary to a contemporary study by Berthwal (1971) who found absence of technical progress in paper industry on the basis of time series estimation for the period 1949-64. Banerji attributed the variance in results to difference in the period of study as well as the data used. Sinha and Sawhney's reported estimate of TFPG at 0.9% per annum for 1950-63. This was a sub-period within Banerji's period of study. Mark the difference in the growth of K/L ratio. In spite of this the difference in productivity growth estimates can still be considered within the acceptable range and is possible additionally because of difference in length of the period and in choice of index. Mehta's estimates differ widely from those of others. This along with the results if his production function estimation which showed technical progress at the rate of 15.9% per annum, point to the possibility of unstated differences in the definition of variables. Mark again the large growth rate of capital-labour ratio.

Goldar's estimate for a slightly later though partially overlapping period pointed in the opposite direction from that of Mehta's. It is within the reasonable range when compared with that of CSO for the period 1960-71. If one compares Goldar's figures for growth of capital labour ratios with that of Mehta and of CSO, a possible reason for difference in estimates appears to be differences in measurement of capital. CSO's estimates for the longer period (1960-77) and a later sub-period (1969-77) indicate slackening of growth rates though the basic growth rates remain comparable with that of Goldar. Ahluwalia's estimate of low negative growth for a much longer period may be the result of slackening which had set in the later part of sixties and which is also reflected in the estimates of Dabir-Allai and also of Arora. It needs to be pointed here that Dabir-Allai's estimates is at the 2-digit level and hence implies a much larger variable aggregate being studied. It compares well with estimate of 0.5% per annum in Ahluwalia (1985) which is comparable in terms of level of aggregation and also of reporting procedure. Ahluwalia (1985) reports two additional estimates like in the case of aggregate manufacturing. See previous footnote. These are (i) Translog a = 0.5% per annum, and (ii) Translog b = 1.2% per annum. These show like in the previous case wide variation which differences in the measurement of capital can make. Arora's estimate would be less extreme if a more appropriate reporting procedure was used. Parhi's positive growth rates for a period when they appeared to be turning negative otherwise need to be discounted because of extreme values of growth of capital-labour ratios and of capital productivity. Pradhan's estimates though done in a different methodological framework do not contradict the broad tendencies indicated by the previous three authors.

The objective of the CSLS Research Report (2003) to review the literature and was to gather and synthesize all the studies on productivity levels and trends in the forest products sector from around the world. The result is that most but not all of the studies listed and summarized in this review are on the Canadian and American forest products sectors, since those two countries are the major forest products producing English speaking countries. The majority of the reviewed studies are published either in books, industrial publications or scientific journals. All the wood products industry sectors are covered, some more than others. The studies have been regrouped in sections: (1) general, (2) logging, (3) saw milling, lumber and wood products, and (4) pulp and paper studies. A Review of literature, for the categories listed as above, has been summarized by CSLS Report and presented in Table 3.3 (B1), 3.3 (B2), 3.3 (B3), 3.3 (B4) and 3.3 (B5).

The first section of this report contains studies concerned with the forest sector as a whole while the other three are about specific sub sectors. Each section is then subdivided by country and the studies are ordered by publication date. Each study has been summarized in order to present the main research results. When it has been possible, the authors' interpretation or explanations are also included in the summaries. The summaries are then followed by a synthesis of the major findings. There are two basic approaches to the empirical study of productivity. The index number or non-parametric approach develops indexes of inputs and output from data on employment, hours worked, capital stock and real output to directly calculate partial and total factor productivity estimates. The parametric approach, on the other hand, uses econometric techniques to develop productivity estimates based on indexes of inputs and output. The two approaches are complementary, with both having advantages and disadvantages (Hulten, 2001). About one half of the studies surveyed use the index number approach and one half the econometric approaches.

Salehirad and Sowlati (2005) measured the efficiency of 82 primary wood producers in B.C. in 2002 using DEA approach, and non-parametric statistical tests were applied for the post-hoc efficiency analysis. The results indicated that the aggregate efficiency of B.C. sawmills remained lower because of technical performance deficiencies, revealing that the emphasis on technical efficiency improvement policies could greatly reward the performance of the industry. The northern and central interior mills had the highest efficiency, followed by the southern interior mills, and the coastal mills were the least efficient. In a study on the efficiency changes in some wood products industry, Vahid & Sowlati (2006) applied DEA to examine the efficiency changes of six wood product manufacturing subsectors in Canada, from 1993 to 2003. The analysis showed that all the six subsectors had improved their technical efficiencies but the technical efficiency was much lower than their scale efficiency, which led to low aggregate efficiency. Hence, improving the technical and managerial knowledge would have greater effect on efficiencies, compared to increase the consumption of resources.

Using Törnqvist-Theil index, Zhang and Nagubadi (2006) analyzed the total factor productivity growth in the U.S. and Canadian sawmill and wood preservation industries between 1985 and 2003. The results revealed that the gap in productivity growth had widened further between the United States and Canada. Contrary to the theoretical expectations, the US-Canada Free trade Agreement had not worked in closing the productivity gap in this industry between the two countries.

Alice Shiu & Almas Heshmati (2006) have presented in their paper the panel econometrics estimation approach of measuring the technical change and total factor productivity (TFP) growth of 30 Chinese provinces during the period of 1993 to 2003. The random effects model with heteroscedastic variances has been used for the estimation of the Translog production functions. Two alternative formulations of technical change measured by the single time trend and the general index approach are used. Based on the measures of technical change, estimates of TFP growth could be obtained and its determinants were examined using regression analysis. The parametric TFP growth measure is compared with the non-parametric Solow residual. TFP has recorded positive growth for all provinces during the sample period. Regional breakdown shows that the eastern and central regions have higher average TFP growth when compared with the western region. Foreign direct investment (FDI) and information and communication technology (ICT) investment are found to be significant factors contributing to the TFP difference. While these two factors are found to have significant influence on TFP, their influence on production is relatively small compared to traditional inputs of production.

By using the Tornqvist-Theil index approach, Daowei Zhang & Rao V. Nagubadi (2006) analyze trends in total factor productivity (TFP) growth in the United States and Canadian sawmill and wood preservation industries (NAICS 3211) between 1958 and 2003. The results indicate that the TFP grew at an average annual compound rate of 1.11% due to higher growth in aggregate outputs by 1.42% and a smaller growth in aggregate inputs by 0.31% in the United States. In Canada, TFP grew at a smaller average annual compound rate of 0.61% as a result of growth in aggregate outputs and inputs by 3.57% and 2.94%, respectively. Although productivity growth for production worker input was similar in both countries, large gaps in productivity growth existed for nonproduction workers, material, capital, and energy inputs. The gap in TFP growth between the two countries appears to have widened since 1987, primarily due to the differences in capital stock growth. Volatility in lumber prices after 1991 might have also played a role, as the Canadian industry could not adequately respond to changing market situations in the face of uncertainty and trade barriers on softwood lumber exports to the United States.

Nanang and Ghebremichael (2006) studied and compared the production technologies in the timber harvesting industries in B.C., Ontario and Quebec, using annual data from 1961 to1999. The authors decided to separate the analysis for three reasons. First, separate analysis can examine whether the same technology described the production across the country and also examine policy implications more accurately by province. Second, the differences in technology resulted from the fact that different provinces have different laws or regulations regarding inputs into production. The authors specified the total translog cost function with four inputs (labor. Capital, material and energy) and one output, which was aggregated from three outputs (sawlogs, pulpwood and fuelwood and others). The restrictions for the linear homogeneity of the cost function in input prices and symmetry in cross-price effects were imposed on the translog cost function. In addition, four cost share equations were obtained by taking the first derivative of the function with respect to the natural logarithm of each input price. Six models with different restrictions were tested by the log likelihood ratio approach to determine which model best represented the industry.

The five restrictions were Hick's neutral technological change, existence of technological change, homotheticity and homogeneity, and unitary elasticity of substitution. In B.C. and Quebec industries, the restrictions on homotheticity and linear homogeneity could not be rejected. In Ontario industry, all restrictions were rejected. After estimation of the models, the standard Chow test was used, confirming that the differences between the coefficients across provinces were statistically significant. Finally, own- and cross-price elasticities, input substitution elasticities, scale effect, technological change and bias, and total factor productivity were measured. As a result, all three provincial industries exhibited biased technological changes, saving on capital and labour, and intensively using energy and materials. Scale effects ranging from 2.17 in B.C. to 1.44 in Quebec implying that the timber harvesting industries appeared to have had the greatest potential for cost reduction through output expansion. Policies that reduced the rental cost of capital, such as increase in investment tax credits and cuts in the corporate income tax rates had the potential to increase productivity.

Nagubadi and Zhang (2006) analyzed the production structure and input substitutions in the Canadian sawmill and wood-preservation industry from 1958 to 2003, using a Translog cost function. To correspond to a well-behaved production function, the restrictions for homogeneous of degree one in the input prices were imposed. The models with different restrictions were tested, including unrestricted, nonhomothetic Hicks-neutral technical change, homothetic non-neutral technical change, homothetic Hicks neutral technical change, and homogeneous and unitary elasticity. Only homotheticity and Hicks-neutral technical change hypothesis was accepted. After the estimation of the Translog cost function with cost share equations, Allen own- and cross- partial elasticities of factor substitution, the MES, the rate of technical change, the TFP growth and the scale effect were measured. According to the MES results, substitution of labor by other inputs was easier than the substitution of materials by other inputs. In contrast to the previous findings of zero or negative rate of technical change, this study found technical progress at the rate of 0.57% per annum and a TFP of 0.54% per annum in the Canadian sawmill and wood preservation industry.

This study was different than the previous studies in four aspects. First, previous studies covered only lumber and wood chips but this study treated the output as the sum of softwood lumber, hardwood lumber, wood chips, wood preservation products, wood ties-shingles-shakes, and other products. Second, previous studies used aggregated inputs but this study used seven more detailed inputs, including production labor, non production labor, machinery and equipment capital, plants and structures capital, fuels energy, electric power, and materials. Third, a few studies obtained the cost share of capital in valuing capital serviced, using value added minus labor cost. This study used opportunity cost principle to estimate the service prices of capital. Last, previous studies just calculated the AES, but this study also used the MES which is more accurate for the analysis.

The vast report of Gerald T. Fox *et al.* (2007) calculates the economic impact of the furniture industry cluster in the Triad Region and the High Point Area of North

Carolina. The Triad Region is based on the combined geographic area of Davidson, Forsyth, Guilford and Randolph counties. The High Point Area within the Triad Region refers to the cities of High Point, Archdale, Trinity and Thomasville. The furniture cluster includes furniture manufacturing, major supporting industries, furniture wholesale and retail, and the International Home Furnishings Market. This report, consisting of two parts, estimates the magnitude and economic influence associated with the overall home furnishings industry throughout the Triad Region of North Carolina, with emphasis upon the High Point Area. This analysis is the first study of its kind to measure the scale and impact of furniture-related industries throughout the Triad Region, and provides a benchmark for further research into the subject.

Part One of this report estimates the economic impact of the furniture industry cluster throughout the Triad and High Point areas of North Carolina for the year 2006. The furniture cluster consists of four expenditure categories: furniture manufacturing, major supporting industries, furniture wholesale and retail, and the International Home Furnishings Market. The Triad Region refers to Davidson, Forsyth, Guilford and Randolph counties, while the High Point Area within the Triad Region refers to the cities of High Point, Archdale, Trinity and Thomasville. The economic impact of the furniture cluster upon the Triad Region and the state is measured in terms of industry output and employment. Part Two of this report contains a description of survey results of some of the industries in the Triad Region that support the furniture industry. Industries that have an impact on the furniture industry are easy to find. "Wood veneers" and "foam production" have a direct relation to the furniture industry, while some of the other industry groups are supportive of the industry and are less obvious. Author was the intention to identify some of these pertinent industry groups, create a list of the companies in these areas, contact the companies, and then survey them about their relationship and support of the broader furniture industry. Some of the main results of this analysis are as follows:

- The furniture industry cluster contributes 20 percent or more of the total industry output produced in the High Point Area.
- The economic impact of the furniture cluster in the High Point Area upon the Triad Region is $3.93 billion in industry output with almost 31,000 jobs.
- The core furniture-related industries—consisting of furniture manufacturing, wholesale and retail—create an economic impact upon the Triad Region of $5.75 billion in industry output and 43,000 jobs.
- The economic impact of furniture manufacturing across the Triad Region is $4.8 billion in industry output with almost 34,000 jobs.
- The economic impact of furniture wholesale and retail is almost a billion dollars in industry output with over 9,300 jobs across the Triad Region.
- The economic impact of the furniture cluster upon the Triad Region is $8.25 billion in industry output and 65,362 jobs.

- The economic impact of the furniture cluster in the Triad Region upon the North Carolina economy is $8.94 billion in industry output with more than 69,000 jobs.
- The High Point Area contributes 11 percent of the total output in the Triad Region.
- The Triad Region contributes 16 percent of the total industry output in North Carolina.
- The Triad Region contributes 21.55 percent of the state's manufacturing output and 16.57 percent of the state's manufacturing jobs.
- About 44 percent of the industry output in the Triad Region comes from manufacturing.
- Furniture and related product manufacturing is the largest manufacturing subsector in the Triad Region with over 16,000 jobs; furniture manufacturing generates 17 percent of the total manufacturing jobs in the four-county Triad Region.

A. Sanatomba Singh (2007) has attempted to present the Growth and Potential of Forest-based Industries in Manipur. He observed that the State of Manipur is a repository of rich flora and fauna along with good potential of forest resources -Timber Forest Products (TFPs) and NTFPs, there exist only a few forest-based industries in the State. Also, the progress of development of forest-based industries in the State is very slow due to paucity of investment funds and inadequate supply of raw materials required by the various forest-based industries which can provide the basic requirement of the rural masses and also employment opportunities. He suggested that if the State Government takes care and gives financial and infrastructural facilities support, some of the forest-based industries may kindle hope for future development and industrialization of the various forest-based industries in the State.

Xianchun Liao; Yaoqi Zhang (2008) in their paper have examines determinants of softwood production in the U.S. South by comparing the forest industry (FI) and nonindustrial private forests (NIPF). An econometric model is derived tinder the framework of profit maximization. A two-stage least squares (2SLS) technique with time series data from 1953 to 2002 is employed in this study. The estimated results show that supply price elasticities of 0.70 for sawtimber and 0.90 for pulpwood for FI owners are larger than those of 0.29 for sawtimber and 0.32 for pulpwood for NIPF owners, which in general, are within the price elasticity range from previous studies.

Parag Dubey (2008) has pointed out in his study that India was a restricted market for wood products. The structural reforms initiated in the early 1990s with a cautious and calibrated approach to external sector liberalization has led to a step-up in economic growth and, a ban on most domesting logging created a unique market opportunity for global forest based industries. He concluded that most forest and wood lots of the country are producing far below their potential, the situation calls for directs measures for enhancing forest productivity.

Another study of Parag Dubey (2009) entitled –"*Role of Indian forest Products Industry in Climate Change Migration: A Managerial Perspectives*", focuses on outlining adaptive management strategies that enhance the ability of forests and their products to adapt to climate change and mitigate its effects through increased carbon sequestration and storage. Developing carbon credit markets that motivate true reductions in carbon emissions must address all carbon pools and their GHG emissions. To be effective in this area, the Indian management community must have a voice in defining the markets and policies, to the extent that it is of strategic interest to the future of the companies and the society. The Indian forest-based industry has a reasonable potential to sequester. However, domestic manufacturing is highly fragmented and unorganized, generally inefficiently managed, has low product quality, and lacks standardization. Nonetheless, there are various ways to positively influence the carbon balance; including sink enhancements and increasing the market share of the existing wood products. Globally, forest market is undergoing dramatic changes. The natural advantage in the forestry sector is gradually shifting away from countries with the highest levels of forest resources to countries where trees grow to commercial maturity at the fastest rate and where the cost of converting them into products is the lowest. It is thus obvious that many foreign companies view India as a country with a strong commercial appeal, both as an emerging market and as an economic partner in possible collaboration. These provide a unique link for dealing with climate change through the competitiveness of Indian forest industries and its livelihood impact.

Joshua Idassi *et al.* has conducted an impact analysis to examine the relative importance of the forestry related sectors to the overall Tennessee economy, utilizing the IMPLAN database and model. The 1994 data used for this study were the most recent available. The Tennessee input-output economic model results indicated that in 1994, the Tennessee forest products industry directly employed 66,947 people and paid about $2.28 billion in wages. Also, value-added generated directly by the forest products industry totaled over $3.9 billion. When the forestry sector of the Tennessee economy produces products or services to meet demand, the overall State economy is affected in three ways: directly, indirectly and with induced effects. The total effect on the state economy is the sum of these three separate effects. Therefore, in 1994, relative to other Tennessee industries, the total effect of the forest products industries was 158,833 jobs, over $3.5 billion in wages and salaries, $15.5 billion of industrial output, and over $7.5 billion of value-added.

Jingjing Li (2009) study uses a Translog cost function to specify the production structures of the softwood lumber industry in three U.S. regions (the West Coast, the Inland, and the South), and four Canadian regions (Ontario, the British Columbia Coast, the British Columbia Interior and Quebec), from 1988 to 2005. First, two separate production models are specified and analyzed, one is a "U.S. model" for the U.S. regions, and the other is a "Canada model" for the Canadian regions. Second, all seven regions are included in one production model, a "U.S.-Canada model". In the U.S.-Canada model, purchasing power parity over the Gross Domestic Product is used to convert cost and price data of Canada from Canadian into U.S. dollars. The Allen and Morishima elasticity of substitution, price elasticity of demand, rate of technical change, and total factor productivity growth are estimated in each model, and the results are presented and compared.

Mrosek *et al.* (2010) study reveals that Forests and the forest- and wood-based industry sector are very important from an ecological, economic and social perspective, especially when they form an industry cluster in a certain region or country. Considering increasing challenges for the forest sector worldwide, a cluster perspective can be very helpful for showing the complex structure and the various interactions between the different industry branches and along the chain of production and value-adding. In their research project, a framework for stakeholders' analysis of forest and wood-based industry clusters was developed and tested as a case study at the State of North Rhine-Westphalia (NRW) in Germany. The study is based on methods of cluster and stakeholder analysis as well as the analysis and assessment of stakeholder and corporate networks. The case study was carried out based on a questionnaire survey among 632 stakeholder organizations (response rate 35.9%). The framework consists of three modules: stakeholder identification, stakeholder analysis and assessment, and stakeholder networking and cluster management. The NRW forest sector is characterized by more than 630 stakeholder organizations, covering a wide range of industry branches and different spatial scales. The institutional characteristics vary highly. Stakeholder communication and cooperation mostly takes place between directly linked industry branches along the chain of production and value-adding. Recommendations for improved communication and cooperation include the establishment or further development of cooperation platforms or networks. The presented framework supports stakeholder analysis under the cluster perspective. If applied in a consistent and transparent way, the framework of stakeholder analysis can contribute to more holistic, standardized, replicable and comparable descriptions of forest and wood-based industry clusters in Europe and abroad.

Using the data of 86 sugar mills operating in the year 2003-04, Sunil Kumar & Nitin Arora (2011) study endeavours to gauge the extent of technical efficiency in the sugar industry of Uttar Pradesh (UP) by employing the technique of data envelopment analysis (DEA). The empirical results reveal that mean overall technical inefficiency (OTIE) is about 19 percent, and both managerial and scale inefficiencies contribute almost equal to observed OTIE. Also, a majority of firms need downsizing in the scale of their operations. The regression results indicate that public ownership adversely affects the operating efficiency of firms, and a weak learning-by-doing effect exists in the sugar industry of UP.

Nimai Das and Debnarayan Sarker (2011) have examines an empirical exercise on the impact of network externalities under social capital in a gender sensitive planning on joint forest management programme in West Bengal. One impact is that building up higher level of social capital has been more successful in programme participating villages as the network structure about the relational variables of social capital is comparatively dense in these villages. Albeit the overwhelming proportion of the households live below poverty line, food-livelihood insecurity cannot destroy the level of social capital of institutions as there already exists an underlying tendency for united actions derived especially from prior collective experience and reciprocal relation. Conversely, the positive complementary effect of network externalities is also higher for these villages. These two effects are more pronounced among poor category households in general and 'female forest protection committee' villages vis-àvis 'joint forest protection committee' villages as well as 'programme nonparticipating' villages in particular.

LMS Baghel (2012) attempted to present the overall scenario of the forest-based cottage and home industries and small scale industrial units located in and around the area of Sambalpur division, Orissa, and explore the potential of these income and employment generation industries, The overall analysis of his study reveals that Sambalpur division has a great potential for industrial development of SSI units and forest-based cottage and home industries. He points out that there is a need of concerted efforts should have been made in near future for the setting up of new FBCH industries and SSI units not only for strengthen the area's economy of Sambalpur division but also generate income and create employment opportunities for the rural population of forests areas.

Binu Ann Kuriachan (2013) have studied on Cement industry and observed that cement industry has played a significant role in the Indian economy, since independence, because of its role in terms of building up basic infrastructure for its citizens, production, trade and development. Indian cement industry has continuously improvised its production technology from inefficient wet process to the fuel efficient dry process over the years. This paper is an attempt to analyze the technological change of the Indian cement industry and its impact by estimating econometric production functions. The results of the study show that even though mean technical progress of the Indian cement industry is low it has improved over time. The estimated production functions (Cobb-Douglas and Constant Elasticity of Substitution) of the Indian cement industry reveals that they exhibits decreasing returns to scale and their technical change has been capital saving over the study period.

Hodges, D.G., *et al.* (2014) has studied "Recession Effects on the Forests and Forest Products Industries of the South". They pointed out that the economic recession affected southern forests and related industries substantially, particularly those sectors most closely related to home construction. Between 2005 and 2009, for example, the three primary forestry sectors – wood manufacturing, paper manufacturing, and forestry and logging – lost more than 110,000 jobs in the southern United States. This article assesses the effects of the recession on the southern U.S. by reviewing existing data related to economic and resource impacts, including employment, timber product output, production facilities, state economies, exports, and forest area and management activities. While all sectors were affected, wood products and furniture manufacturing experienced the greatest change. As a result of the downturn, the South's forest sector's direct contribution to the regional economies decreased by 24 percent between 2004 and 2009. Some developments such as rebounding paper consumption, expanding export markets, and bio-energy, however, offer potential growth opportunities for the future.

Thus, on the basis of overall discussion, it can be summarised that the existing literature on Wood & Forest based Industries sources at the India's provincial/regional/state level is very poor and it does not provide sufficient understanding of the subject into the consideration. Very few studies attempt to investigate productivity growth, technological change and production functions in general or specific to WFBI sectors of the Indian economy.

Table 3.3(A): Partial and Total Factor Productivity Growth: Paper

Author	Period	Capital Productivity	Labour Productivity	K/L Ratio	TFPG	Measure
Banerji	1946-64	-0.4	6.00	6.4 -0.	30	S*
Sinha	1950-63	1.61	3.90	2.29	0.9	K**
Mehta	1953-64	-6.9	0.9	7.8	-3.3	S*
Goldar	1960-70	2.61	6.16	3.55	3.76	K*
CSO	1960-77	3.11	3.78	0.67	3.41	K*
CSO	1960-71	3.71	6.11	2.40	4.58	K*
CSO	1969-77	2.21	0.26	-1.95	1.65	K*
Ahluwalia	1959-79	--	--	--	0.1	T***
Ahluwalia	1959-85	-2.0	1.5	3.6	-0.7	T**
Dabir-Allai	1973-78	-0.8	0.6	1.4	0.30	S***
Arora	1973-81	-3.98	-0.62	3.36	-3.32	T***
Parhi	1982-91	-18.88	3.6	22.48	1.61	T**
Pradhan	1963-92	--	--	--	-0.59	T*
Pradhan	1963-71	--	--	--	-0.2	T*
Pradhan	1972-81	--	--	--	-0.15	T*
Pradhan	1982-92	--	--	--	-1.67	T*

Notes: All numbers are growth rates per cent per annum.
K indicates Kendrick index. T indicates Translog index.
* indicates compound annual growth rates.
** indicates semi-log trend rate of growth.
*** indicates simple average. -- indicates not reported or not available.
# indicates total productivity measure

Source: Puran Mongia & Jayant Sathaye *(1998): 'Productivity Growth and Technical Change in India's Energy Intensive Industries: A Survey', Energy Analysis Program Environmental Energy Technologies Division Lawrence Berkeley National Laboratory Berkeley, CA 94720, Oct., 1998.*

Table 3.3 (B1): Synthesis Table on General Studies

Authors	Country	Period	Methodology	Major Findings
Sandoe Wayman (1977)	Canada	1965-1972	Output/input ratios	No trends in capital productivity but annual variations. As a result of economies of scale and automation, labour productivity is up in all forest products sectors.
Mohnen Jacques Gallant (1996)	Canada	1963-1988	Tornqvist index	The rate of return on R&D investment is low and does not contribute significantly to TFP growth. TFP growth is the result of economies of scale.
Kaiser (1971)	U.S.	1947-1967	Output/input ratios	Labour productivity is up in all forest products sectors, with a parallel rise in wages. Automation and economies of scale are accountable for both trends.
Horvath (1980)	U.S.	1947-1977	Output/input ratios	Same conclusion emerges as Kaiser study but for a wider time period.
Greber White (1982)	U.S.	1951-1973	Econometrics (following Sato-Batavia)	Labour-capital ratios were down in all forest products sectors. Technical change was biased towards capital and led to labour productivity rise.
Irland Maxcy (1991)	U.S. (Maine)	1982-1987	Value-added ratios	In forest product sectors, higher wages are associated with higher labour productivity. There is an inverse relationship between labour productivity and labour intensity.

Source: CSLS Research Report, *Centre for the Study of Living Standards for the Forest Products Association of Canada, Feb., 2003.*

Table 3.3 (B2): Synthesis Table on Logging Studies

Authors	Country	Period	Methodology	Major Findings
Kant Nautiyal (1997)	Canada	1964-1992	Translog cost function	Technical change was labour and capital saving but used energy and material. The technical change effect was systematically negative. The annual TFP growth rates were therefore rarely positive. This could be a consequence of changing harvesting site conditions.
Wear (1994)	U.S. (South)	1962-1988	Index number approach	The TFP average annual growth rate was much lower on industry owned land than on privately owned land. This could be the result of difficulty in measuring assets. One should not necessarily equate TFP and technological change in the timber industry because of the effect of climate.
Cubbage Carter (1994)	U.S. (South)	1979-1987	Survey	The industry was much more capital intensive in 1987 than in 1979. Average weekly production rose during the period. Predominant technology has changed from short to long wood systems.
Sedjo (1997)	U.S.	1970-1993		Technology has led to rising labour productivity. But that was not enough to compensate for the negative effect on productivity of the reduced accessibility of wood resources over the years.
Barreto Amaral Vidal Uhl (1998)	Brazil	N/A	Output/input ratios.	Labour productivity was lower in planned logging operations because loggers minimized damage to other trees. However, capital productivity was higher in planned logging operations because of reduced time loss and wood waste.
Greene Jackson Culpepper (2001)	U.S.	1987-1997	Survey	The period was characterized by heavy investments. Capital productivity was almost steady while labour productivity rose almost 80 per cent.

Source: CSLS Research Report, *the Centre for the Study of Living Standards for the Forest Products Association of Canada, Feb., 2003.*

Table 3.3 (B3): Synthesis Tables: Lumber and Saw Milling Studies

Authors	Country	Period	Methodology	Major Findings
Singh Nautiyal (1986)	Canada	1955-1982	Translog cost function	There were no signs of technological progress. Only the labour productivity growth rate was positive. There was input misallocation because of slow adaptation by firms.
Meil Singh Nautiyal (1988)	Canada (B.C. Interior)	1950-1983	Translog variable cost function	Labour productivity was up while material and energy productivity were down during the period. There were deviations from the least-cost paths, therefore showing input misallocations. B.C.'s labour productivity growth rate was higher than Canada's.
Constantino Haley (1989)	Canada & U.S.	1957-1982	Index number approach	When controlling for wood quality, B.C.'s industry was more productive. But the Pacific northwest industry narrowed the productivity gap over the years.
Ghebremichael Roberts Tretheway (1990)	Canada	1962-1985	Index number approach	Labour and capital productivities grew the fastest. The two B.C. regions had a productivity advantage but the gap has diminished since 1962. After 1980, Quebec and B.C. Interior had the greatest productivity gains.
Bernstein (1994)	Canada	1963-1987	Dynamic model of multiple output production and investment.	The model controls for price-cost margins. The average annual TFP growth rate was relatively high. Technological change accounts for almost 70 per cent of the TFP growth rate.
Abt Brunet Murray Roberts (1994)	Comparative study of Canada & U.S.	1965-1988	Non-parametric superlative index method.	Labour productivity was positive for most regions and especially high in the 1980s. Although there were high single factor productivity growth rates differences, the TFP measures were similar between regions, a sign of regional competitive markets.
Duke Huffstutler (1977)	U.S.	1958-1975	Output/input ratios	Labour productivity for the period was equal to the average of the manufacturing sector a whole. New technology explains the labour productivity growth.
Farris (1978)	U.S.	1958-1976	Output/input ratios	Labour productivity growth rate was relatively high compared to other sectors. Technological innovation and rapidly expanding demand were responsible for the higher labour productivity growth rate.

Source: CSLS Research Report, *the Centre for the Study of Living Standards for the Forest Products Association of Canada, Feb., 2003.*

Table 3.3(B4): Synthesis Table on Wood Products Studies

Authors	Country	Period	Methodology	Major Findings
Veigle Horst (1982)	U.S.	1958-1980	Output/input ratios	Labour productivity was lower than in other forest product sectors because of declining demand. There also have been fewer investments in automation.
York (1992)	U.S.	1977-1989	Output/input ratios	Labour productivity was up, faster before 1984. Investments in automation by larger producers are the main cause of the labour productivity rise.

Source: CSLS Research Report, *the Centre for the Study of Living Standards for the Forest Products Association of Canada, Feb., 2003.*

Table 3.3 (B5): Synthesis Table on Pulp and Paper Studies

Authors	Country	Period	Methodology	Major Findings
Nautiyal Singh (1986)	Canada	1956-1982	Translog cost function.	Labour and capital productivity were up but material and energy productivity were down. The observed productivities diverged from their least-cost paths, especially in recession periods, which suggest input misallocation.
Frank Ghebremichael Oum Tretheway (1990)	Canada	1963-1984	Non-parametric method.	Because of investments, all single input productivity growth rates were up except for capital. Most of the TFP growth was the result of economies of scale.
Brunet (1993)	Canada (Comparative study)	1964-1988	Superlative index number approach	B.C. had a higher TFP growth rate than competitors and gained a productivity advantage over its Canadian competitors, probably because of its positive capital productivity growth rate. Excess capacity could explain the volatile TFP growth.
Hailu Veeman (2000a)	Canada	1959-1994	Parametric input distance function. Tornqvist index.	There was a negative TFP growth rate in the 1960s and 1970s but a positive one in the 1980s and 1990s. Most of the TFP growth is a result of the output scale effect. The oil crises are responsible for negative growth in the 1970s and mill composition is responsible for positive growth in the 1990s.

Hailu Veeman (2000b)	Canada	1959-1994	Parametric input distance functions	Traditional measures of TFP growth underestimate productivity improvements because they ignore the reduction in undesirable outputs.
Hailu Veeman (2001)	Canada	1959-1994	Non parametric approach (Chavas-Cox)	Contrary to their 2000a results, TFP growth was not negative, but practically zero in the 1970s and positive in the 1960s. Again, traditional measures underestimate productivity growth.
DeBorger Buongiorno (1985)	U.S.	1957-1981	Variable cost function	The paper and paperboard industries enjoyed positive productivity growth, but the paper industry's growth rate was higher. A larger L.P growth rate and less energy intensity for the paper industry.
Oum Tretheway Zhang (1991)	Comparative study of Canada, the U.S. & Sweden	1970-1980	Index number approach	Sweden had the highest capital growth and labour productivity growth rates, but the U.S. had the highest TFP growth rate and Sweden the lowest. The U.S. also experienced more technological change while Canada benefited most from the scale effect.
Oum Tretheway (1992)	Comparative study of Canada and the U.S.	1963-1984	Non-parametric approach (Tornqvist approach)	The TFP growth rate was higher in the U.S. despite a much lower capital input growth rate. Most of the U.S. growth occurred after 1980 while most of the Canadian growth occurred before 1974. Canadian industry had more trouble adjusting input mix.
Hseu Buongiorno (1994)	Comparative study of Canada & the U.S.	1961-1984	Non parametric methods	The Canadian and American industries experienced very similar TFP growth rates with two of the models. The model with the translating hypothesis produced much more yearly fluctuations.
Fleisher Dong Liu (1996)	China	1985-1990	Translog production function.	Labour productivity was up for the period. Profit distribution does not affect labour productivity but the level of ownership does.
Schumacher Sathaye (1999)	India	1973-1993	Translog, Solow, Kendrick index.	TFP growth was negative in the 1970s, but positive in the 1980s with the creation of larger and more modern installations. TFP growth was negative in the 1990s.

Source: CSLS Research Report, *the Centre for the Study of Living Standards for the Forest Products Association of Canada, Feb., 2003.*

References

A.A. Mulimani *et al.* **(2012):** 'Problems & Prospects of Small Scale Industries of Goa: A Geographical Study', *Indian Streams Research Journal*, Vol.1, Issue.XII, Jan. 2012, pp.1-4.

Abhay Gupta (2008): 'Looking beyond the methods: Productivity Estimates and Growth Trends in Indian Manufacturing', University of British Columbia, Vancouver, Canada, web: http://www.abhayg.com

Abhay Gupta (2010): 'Indian Manufacturing Productivity: What Caused the Growth Stagnation before the 1990s?', International Productivity Monitor, Nov., 2010.

Abramovitz M. (1957): 'Resources and Output Trends in the United States', *Amer. Econ. Rev.;* Vol. 46, May 1957.

Abt, R.C., J. Brunet, B.C. Murray & D.G. Roberts (1994): 'Productivity Growth and Price Trends in the North American Saw Milling Industries: an Inter-regional Comparison', *Canadian Journal of Forest Research.* 24(1) 139-148.

Abt, R. C. (1987): 'An Analysis of Regional Factor Demand in the. U.S. Lumber Industry. *Forest Science*, 33(1): 164-173.

Agarwal, R.N. (1988): 'Indian Automobile Industry: Problems and Prospects', *The Indian Economic Journal*, Vol. 36, No. 2, Oct. –Dec., 1988.

Ahluwalia I.J. (1985): 'Industrial Growth in India: Stagnation since the Mid-Sixties'', Oxford University Press, New Delhi, 1985.

Aikaterini Kokkinou (2005): 'Reviewing The Statistical Measuring Of Innovation Activities And Technical Change', Ελληνικό Στατιστικό Ινστιτούτο Πρακτικά 18ου Πανελληνίου Συνεδρίου Στατιστικής (2005) σελ.477-484.

Alice Shiu & Almas Heshmati (2006): 'Technical Change and Total Factor Productivity Growth for Chinese Provinces: A Panel Data Analysis', Discussion Paper No. 2133 May 2006, IZA, Bonn Germany.

Ali M. Khalil (2005): 'A Cross Section Estimate Of Translog Production Function: Jordanian Manufacturing Industry'. Global Review of Business and Economic Research, New Delhi: Serials Publ., Vol. 1(2), p.121-130.

Andrew C. Worthington (2000): 'Technical Efficiency and Technological Change in Australian Building Societies', Abacus 36(2):pp. 180-197.

Almas Heshmati & Subal C. Kumbhakar (2010): 'Technical Change and Total Factor Productivity Growth: The Case of Chinese Provinces', IZA Discussion Paper No. 4784 February 2010.

Anoopa S Nair (2012): 'Monopoly Power and Firm Dominance in Indian Manufacturing Sector: An Empirical Analysis', *Indian Economic Review*, Volume: 47, Issue Number: 1, pp.109-141.

Antr`as, P. (2004): 'Is the U.S. Aggregate Production Function Cobb-Douglas? New Estimates of the Elasticity of Substitution', Contributions to Macroeconomics 4(1).

Arnold, J.E.M. Chipeta, M. & Fisseha, Y. (1987): 'The Importance of Small Forest-based Processing Enterprises in Developing Countries', *Unasylva,* 39: 157-158.

Arrow, K. J. and Starrett, D. A (1973): 'Cost- and Demand-Theoretical Approaches to the Theory of Price Determination', eds. J. R. Hicks and W. Weber, Carl Menger and the Austrian School of Economics, Oxford: Clarendon Press, 1973.

Arrow, K., Chenery, H., Minhas, B. & Solow, R. (1961): 'Capital-labor substitution and economic efficiency', *Review of Economics and Statistics* 43(3), 225–250.

A. Sanatomba Singh (2007): 'Growth and Potential of Forest-based Industries in Manipur', *Indian Forester,* Volume 133, Issue 3, March 2007.

Baghel L.M.S (1994): 'A study of Technological Change and productivity Growth in Agro-based Industries of Madhya Pradesh (Since 1973-74)', Ph.D. Thesis, Rani Durgavati University, Jabalpur, Oct., 1994.

Baghel L.M.S. & Pendse N.G (1996): "Econometric Analysis of Productivity Growth & Technological Change in Total Manufacturing Sector of India", *Ind. Econ. Jour.;* Vol, 44 (2); pp.39-58, Oct-Dec, 1996-97.

Baghel L.M.S. & Pendse N.G. (1996): 'Role of Agro-based Industries in Industrial Development in the State of Madhya Pradesh', *Khadi-Gramodyog;* Vol. 42 (10); pp.-474-482, July 1996.

Baghel L.M.S. & Gupta, A. (1998): 'Growth & Productivity Trends in Bidi Industry of Madhya Pradesh', Van-Vigan; Vol. 36(2-4), 1998.

Baghel L.M.S. (1999): 'Madhya Pradesh Ke Arthik Vikas Mein Udogon ki Bhumika', *Khadi-Gramodyog;* Vol. 6-7, pp. 273-279, Mar-Apr.1999.

Baghel L.M.S. & Gupta, A. (2002): 'Economic Analysis of Production Functions and Technical Change in Bidi Industries of Madhya Pradesh', *Indian Forester,* Dehra Dun, 2002.

Baghel L.M.S. (2012): 'Future Prospects & Potential of NTFPs, Forest Based Cottage & Small Scale Industries in the State of Orissa: Especially in Sambalpur Division', Paper Communicated for publication in *Journal of Forest Economics,* Sweden, 2012.

Bart van Ark, Abdul Azeez Erumban, Vivian Chen & Utsav Kumar (2009); 'The Cost Competitiveness of the Manufacturing Sector in China and India: An Industry and Regional Perspective', Economics Program Working Paper, The Conference Board and Growth and Development Center of the University of Groningen, January 2009.

Barreto, Paulo, Paulo Amaral, Edson Vidal & Christopher Uhl (1998): 'Costs and Benefit of Forest Management for Timber Production in Eastern Amazonia', *Forest Ecology and Management.* 108(1-2): 9-26.

Bentolila, S. & Jimeno, J. (2006): 'Spanish unemployment: The end of the wild ride?', in M. Werding, ed., *'Structural Unemployment in Western Europe: Reasons and Remedies',* MIT Press, Cambridge, pp. 317–344.

Bentolila, S. & Saint-Paul, G. (2004): 'Explaining movements in the labor share', Contributions to Macroeconomics 3(1).

Berndt, E. (1976): 'Reconciling alternative estimates of the elasticity of substitution', *Review of Economics and Statistics* 58(1), 59–68.

Bernstein, J.I. (1994): 'Exports, Margins and Productivity Growth: With an Application to the Canadian Softwood Lumber Industry', *Review of Economics and Statistics.* 76(2): 291-301.

B, Fikkert and Hassan (1998): 'Returns to Scale in a Highly Regulated Economy: Evidence from Indian Firms', *Journal of Development Economics,* Vol. 56, 51-79.

Bhat Sharipad & Prof Seth Araman, T.V. (1995): 'Technology transfer and export performance: A case study of Indian Automobile Industry', *The Indian Economic Journal,* Vol No 43, No 2, October-December 1995.

Bhasin V.K. & Seth V.K. (1980): 'Estimation of production functions for Indian Manufacturing Industries', *IJIR;* 15(3), Jan. 1980.

Binu Ann Kuriachan (2013): 'An Econometric Analysis of the Dynamics of Technical Change and Efficiency of Indian Cement Industry', *Econometrics,* Sciknow Publications Ltd., E 2013, 1(3): p.37-40.

Binswanger, H. (1974a): 'The Measurement of Technical Change Biases with many Factors of Production', *American Economic Review* 64(6), 964–976.

Blanchard, O. (1997): 'The medium run', Brookings Papers on Economic Activity: *Macro-Ecomomics* 1997(2), 89–158.

B.N. Goldar (2004): 'Productivity Trends In Indian Manufacturing In The Pre- And Post-Reform Periods', Working Paper No. 137, Indian Council For Research On International Economic Relations, New Delhi, June 2004.

Brahmananda P.R. (1982): 'Productivity in the Indian Economy: Rising Inputs for Falling Outputs', Himalaya Publishing House, Bombey, 1982.

Brunet, J. (1993): 'Productivity in Canada's Pulp and Paper Industry: an Inter-Regional Comparison', Working Paper. Policy and Economics Directorate, Forestry Canada, 4th Floor, Fuller Building, 75 Albert Street, Ottawa, ON K1A 1G5, Canada.

Caballero, R. & Hammour, M. (1998): 'Jobless growth: Appropriability, Factor Substitution, and Unemployment', Carnegie-Rochester Conference Series on Public Policy 48, 51–94.

Caves, Douglas W.; Christensen, Laurits R. & Diewert, W. Erwin. (1982a): 'Multilateral Comparisons of Output, Input, and Productivity Using Superlative Index Numbers', *Economic Journal,* March 1982a, 92(365), pp. 73-86. .

Caves, Douglas W.; Christensen, Laurits R. & Diewert, W. Erwin (1982b): 'The Economic Theory of Index Numbers and the Measurement of Input, Output, and Productivity', *Econometrica,* November 1982b, 50(6), pp. 1393-1414.

Chugan P. K. (1995): 'Foreign collaborations and Industrial R and D in developing countries: case of Indian Automobile Ancillary Industry', *Economic and Political Weekly*, Aug 26, 1995.

Constantino, L. F., & Haley, D. (1988): 'Wood Quality and the Input and Output Choices of Sawmilling Procedures for the British Columbia Coast and the United States Pacific Northwest, west side', *Canadian Journal of Forest Research*, 18(2): 202-208.

Contantino, L.F. & D. Haley (1989): 'A Comparative Analysis of Saw Milling Productivity on the British Columbia Coast and in the U.S. Douglas-fir Region: 1957 to 1982', *Forest Products Journal*. 39(4): 57-61.

C.-P. Lin (2002): 'The Application of Cobb-Douglas Production Cost Functions to Construction Firms in Japan and Taiwan', Review of Pac Policies (RPBFMP), Vol. 5, No. 1, 2002, pp. 111-128.

CSLS Research Report (2003): 'Productivity in the Forest Products Sector: A Review of the Literature', the Centre for the Study of Living Standards for the Forest Products Association of Canada, Feb., 2003.

Cubbage, F. & D. Carter (1994): 'Productivity and Cost Changes in the Southern Pulpwood Harvesting', *Southern Journal of Applied Forestry*, 18(2).

Danish A. Hashim, Ajay Kumar & Arvind Virmani (2011): 'J-Curve Hypothesis of Productivity and Output Growth: A Case Study of Indian Manufacturing in the Post Reforms Period', *Indian Economic Review*,**Vol.:** 46, **Issue No.:** 1, pp.1-21

Daowei Zhang & Rao V. Nagubadi (2006): 'Total Factor Productivity Growth in the Sawmill and Wood Preservation Industry in the United States and Canada: A Comparative Study', *Forest Science* 52(5) 2006, pp.-511-521.

Das Gupta Rajaram (1986): 'Liberalization of Automobile industry policy and demand for commercial vehicles', EPW, Vol. XXI, No. 08, February 22, 1986.

Das, O.K. (2003): 'Manufacturing Productivity under varying trade regimes: India in the *f*A, 1980s and 1990s', Indian Council for Research on International Economic relations, New Delhi, Working Paper No. 107.

Datta, M.M. (1955): 'The Production Function for Indian Manufacturing", Shankhya: 15(4), 1955.

David, P. & Van de Klundert, T. (1965): 'Biased Efficiency Growth and Capital-Labor substitution in the U.S., 1899-1960', *American Economic Review* 55(3), 357–394.

Davis, S. & Haltiwanger, J. (1992): 'Gross Job Creation, Gross Job Destruction, and Employment Reallocation', *Quarterly Journal of Economics* 107(3), 819–863.

Deb Kusum Das *et al.* (2010): 'Total Factor Productivity Growth in India in the Reform Period: A Disaggregated Sectoral Analysis', Paper prepared in 1st WORLD KLEMS Conference held at Harvard University on Aug 19-20, 2010.

Deb Kusum Das & Gunajit Kalita (2011): 'Aggregate Productivity Growth in Indian Manufacturing: An Application of Domar Aggregation', *Indian Economic Review*, Vol.: 46, Issue No.: 2, pp.275-302.

DeBorger, B. & J. Buongiorno (1985): 'Productivity Growth in the Paper and Paperboard Industries: a Variable Cost Function Approach', *Canadian Journal of Forest Research*. 15(6): 1013-1020.

Denison, E. F. (1962): 'The Sources of Economic Growth', New York: Committee for Economic Development, 1962.

Desai, Vasant (1983): 'Problems and Prospects of Small scale Industries in India', Himalaya Publishing House, Bombay.

Diwan & Gujarati (1968): 'Employment & Productivity in Industries', *Artha Vijnana*, 10.

Duke, J. & C. Huffstutler (1977): 'Productivity in Saw Mills Increases as Labor Input Declines Substantially', *Monthly Labor Review*. 10(4): 33-37.

Eric Miller (2008): 'An Assessment of CES and Cobb-Douglas Production Functions', Working Paper, Congressional Budget Office, June 2008.

Farris, M.R. (1978): 'The Veneer and Plywood industry: Above-Average Productivity Gains', *Monthly Labor Review*. 101(9): 26-30.

Finn R. Forsund (2010): 'Dynamic Efficiency Measurement', *Indian Economic Review*, Vol. 45, Issue 2, pp. 125-159.

Fleisher, B.M., Keyong Dong & Yunhua Liu (1996): 'Education, Enterprise Organization and Productivity in the Chinese Paper industry', *Economic Development and Cultural Change*. 44(3): 571-87.

Fulwinder Pal Singh (2012): 'Economic Reforms and Productivity Growth in Indian Manufacturing Sector – An Inter State Analysis', International *Journal of Marketing, Financial Services & Management Research*, Vol.1 Issue 12, Dec., 2012, pp.-1-22.

Frank, D.L., A. Ghebremichael, T. H. Oum & M.W. Tretheway (1990): 'Productivity Performance of the Canadian Pulp and Paper Industry', *Canadian Journal of Forest Research*. 20(6): 825-836.

Gerald T. Fox *et al.* (2007): 'The Economic Impact of the Home Furnishings Industry in the Triad Region of North Carolina', High Point University High Point, North Carolina, May, 2007.

Ghebremichael, A., D.G. Roberts & M.W. Tretheway (1990) Productivity in the Canadian Lumber Industry, an Inter-regional Comparison. *Forestry Canada Information Report* O-411, 73p.

Gheorghe Zaman & Zizi Goschin (2010): 'Technical Change As Exogenous Or Endogenous Factor In The Production Function Models. Empirical Evidence From Romania', *Romanian Journal of Economic Forecasting* –2, 2010.

Goldar, B.N, (1986): 'Impact Substitution, Industrial Concentration and Productivity Growth in Indian Manufacturing', *Oxford Bulletin of Economics and Statistic*, 48, No.2, May, PP, 143-164.

Gordon, R. (1990): 'The Measurement of Durable Goods Prices', National Bureau of Economic Research Monograph Series, University of Chicago Press, Chicago.

Greber, B. & D. White (1982): 'Technical Change and Productivity Growth in the Lumber and Wood Products Industry', *Forest Science,* 28(1): 135-147.

Greene, W.D., B.D. Jackson & J.D. Culpepper (2001): 'Georgia's Logging Businesses, 1987- 1997', *Forest Products Journal.* 51(1): 25-28.

Griliches, Z. and Jorgenson, D. W. (1967 & 1969): 'The Explanation of Productivity Change, Review of Economic Studies, XXXIV, July 1967, 249-285, reprinted in Survey of Current Business, XLIX, May 1969, 50-61.

Griliches, Z. (1979): 'Issues in assessing the contribution of R&D to productivity growth', *Bell Journal of Economics* 10(1), 92–116.

Griliches, Z. (1998): 'R&D and Productivity: The Econometric Evidence', University of Chicago Press, Chicago.

Griliches, Z. (2000): 'R&D, Education, and Productivity: A Retrospective', Harvard University Press, Cambridge.

Griliches, Z. & Mairesse, J. (1998): 'Production functions: The search for identification', in S. Strom, ed., *"Econometrics and Economic Theory in the 20th Century: The Ragnar Frisch Centennial Symposium"*, Cambridge University Press, Cambridge.

Gumaste Vasant (1988): 'Anatomy of In-House R and D: A Case Study of Indian Automobile Industry", *Economic and Political Weekly,* May 28, 1988.

Hailu, A. & T.S. Veeman (2000a): 'Output Scale, Technical Change and Productivity in the Canadian Pulp and Paper Industry', *Canadian Journal of Forest Research.* 30(7): 1041-1050.

Hailu, A. & T.S. Veeman (2000b): 'Environmentally Sensitive Productivity Analysis of the Canadian Pulp and Paper Industry, 1959-1994: An input Distance Function Approach', *Journal of Environmental Economics and Management.* 40(3): 251-74.

Hailu, A. & T.S. Veeman (2001): 'Non-parametric Productivity Analysis with Undesirable Outputs: An Application to the Canadian Pulp and Paper Industry', *American Journal of Agricultural Economics.* 83(3): 605-16.

Hammermesh, D. (1993): 'Labor Demand', Princeton University Press, Princeton.

Himani Aggarwal *et al.* (2013): 'Unorganized Manufacturing Industries in Uttar Pradesh: An Empirical Study', *International Journal of Innovative Research & Development,* Vol. 2 Issue 12 (Special Issue), Dec., 2013, pp.99-103.

Hasim D.A., Kumar, A. & Virmani, A. (2009): 'Impact of Major Liberalization on Productivity: The J-Curve Hypothesis', Working Paper No. 5/2009, DEA, Dept. of Economic Affairs, Min. of Finance, Govt. of India.

Hodges, D.G., *et al.* **(2014):** 'Recession Effects on the Forests and Forest Products Industries of the South', *Forest Products Journal,* Vol. 61, No. 8, pp.615-623

Horvath, G.M. (1980); 'Lumber, Pulp and Paper', In J. Ullman (ed) '*The Improvement of Productivity, Myths and Realities"*, pp.158-176 (NY, Praeger).

Hseu, J.S. & J. Buongiorno (1994): 'Productivity in the Pulp and Paper Industries of the U.S. and Canada: A Nonparametric Analysis', *Canadian Journal of Forest Research.* 24(12): 2353-2361.

IBEF, 'India Manufacturing Sector', Report (2012), http://www.ibef.org/artdisplay.aspx?cat_id=84&art_id=31588

Jakub Growiec (2010): 'On the Measurement of Technological Progress Across Countries', National Bank of Poland, Working Paper No. 73, Education and Publishing Department 00-919 Warszawa, 2010.

Jatinder S. Bedi (2003): 'Production, Productivity and Technological Changes in Indian Spinning Sector', *Indian Economic Review*, Vol. 38, Isuue-2, pp.205-233.

Jingjing Li (2009): 'Production Structure, Input Substitution, and Total Factor Productivity Growth in The Softwood Lumer Industries in U.S. and Canadian Regions', A Thesis for the Degree of Master of Science in Forestry, Faculty of Forestry, University of Toronto, 2009.

Jin, H. & Jorgenson, D. (2008): 'Econometric Modelling of Technical Change, Working paper, Harvard University, Cambridge.

Joshua Idassi, Jorge Huarachi, Paul Winistorfer & Burt English, 'Economic Impacts of the Forestry and Forest Products Industries on the Tennessee Economy'.

Kaiser, H.F. (1971): 'Productivity Gains in Forest Products Industries', *Forest Products Journal.* 21(5): 14-16.

Kalt, J. (1978): 'Technological change and factor substitution in the United States: 1929- 1967', *International Economic Review* 19(3), 761–775.

Kalirajan, K. & Bhide, S. (2004): 'The post-reform performance of the manufacturing sector in India', *Asian Economic Papers*, 3, 126-157, 2004.

Kant, S. & J.C. Nautiyal (1997): 'Production Structure, Factor Substitution, Technical Change, and Total Factor Productivity in the Canadian Logging Industry', *Canadian Journal of Forest Research.* 27(5): 701-710.

Kathuria Sanjay (1987): 'Commercial Vehicles industry in India: A Case History from 1928-1987', *Economic and Political Weekly*, October 17-24, 1987.

Kazumi Asako & Miho Takizawa (2010): 'Marginal Productivity Principle and Measurement Biases in TFP: Evidence from International Productivity Database', Policy Research Institute, Ministry of Finance, Japan, Public Policy Review, Vol.6, No.2, March 2010.

Kendrick, J. W. (1961): 'Productivity Trends in the United States', Princeton: Princeton University Press for the NBER, 1961.

Kendrick, J. & Sato, R. (1963): 'Factor prices, productivity, and economic growth', *American Economic Review*, 53(3), 974–1003.

Kesari Kumar Pradeep & Saggar Mridul (1989); 'A firm level study of the determinants of export performance in machinery and transport equipment industry in India', *The Indian Economic Journal*, Vol. 36, No. 3, January-March 1989.

Kirtti Ranjan Paltasingh. & R.K. Mishra (2013-14): 'Economic Reform and Industrial Productivity: Testing the J-Curve Hypothesis at State Level', *The Ind. Jour. of Econ.*, Vol. XIVC, No. 375, Part IV, 2013-14, PP.631-656.

K. Pushpangadan & N. Shanta (2008): 'Competition and Profitability in Indian Manufacturing Industries', *Indian Economic Review*, Vol. 43, Issue-1, pp.103-123.

Kristen Hoff *et al.* **(1997)**: 'Sources of Competitiveness for Secondary Wood Products Firms: A Review of Literature and Research Issues', *Forest Products Journal*, Vol. 47, No.2, pp.-31-37, Feb., 1997.

Krueger, A. (1999): 'Measuring labor's share', American Economic Review 90(2), 45–51.

Lucas, R. (1969): 'Labor-Capital substitution in U.S. manufacturing', in A. Harberger & M. Bailey, eds, *'The taxation of income from capital'*, Brookings Institution, Washington, pp. 223–274.

Lucas, R. (1988), 'On the mechanics of economic development', *Journal of Monetary Economics*, 22(1), 3–42.

Maddison Angus (1982): 'Phases of capitalist development', New York: Oxford University Press, 1982.

Maddison Angus (1989): 'The world economy in the 20th century', Paris: Organization for Economic Cooperation and Development, 1989.

Malmquist, Sten (1953): 'Index Numbers and Indif- ference Curves.' Trabajos de Estatistica, 1953, 4(1), pp. 209-42.

Manju Bai *et al.* (2013): 'Employment Growth Analysis in Indian Manufacturing Sector' *OIDA International Journal of Sustainable Development*, Vol. 06, No. 09, pp. 11-22, 2013.

Martinello, F. (1985): 'Factor substitution, techical change, and returns to scale in Canadian forest industries', *Canadian Journal of Forest Research*, 15: 1116–1124.

Martinello, F., (1987): 'Substitution, technical change and returns to scale in British Columbian wood products industries', *Applied Econometric*, 19: 483– 496.

McAdam, P. & William, A. (2013): 'Medium run redux: Technical change, factor shares and frictions in the euro area', *Macroeconomic Dynamics forthcoming*.

Md. Moyazzem Hossain, Ajit Kumar Majumder, Tapati Basak (2012): 'An Application of Non-Linear Cobb-Douglas Production Function to Selected Manufacturing Industries in Bangladesh', *Open Journal of Statistics*, 2012, 2, p.-460-468.

Meil, J.K., B.K. Singh & J.C. Nautiyal (1988) 'Short-run Actual and Least-Cost Productivities of Variable Inputs for the British Columbia Interior Softwood Lumber Industry', *Forest Science*, 34(1): 88-101.

Meil, J. K. & Nautiyal, J. C. (1988): 'An intraregional economic analysis of production structure and factor demand in major Canadian softwood lumber producing regions', *Canadian Journal of Forest Research*, 18: 1036-1048.

Merrifield, D. E. & Haynes, R. W. (1985): 'A cost analysis of the lumber and plywood industries in two Pacific Northwest sub-regions', *The Annals of Regional Science*, 19(3): 16-33.

Ministry of Commerce & Industry, Report of the Working Group on 'Boosting India's Manufacturing Exports', 2011.

M.Manonmani (2014): 'Impact of Economic Reforms on Productivity Performance of Manufacturing Sector in South India', *The Ind. Jour. of Industrial Relations*, Vol. 50, No. 2, pp.232-241, Oc., 2014.

Moosup Jung, *et al.* (2008): 'Total Factor Productivity of Korean Firms and Catching up with the Japanese Firms', *Seoul Journal of Economics*, Vol. 20, No. 1, 2008, pp. 93-139

M. Parameswaran (2004): 'Economic Reforms, Technical Change and Efficiency Change: Firm level Evidence from Capital Goods Industries in India', *Indian Economic Review*, Vol.-39, Issue-1, pp.239-260.

Mukherjee Avinandam & Trilochan Satry (1996): 'Automobile Industry in Emerging Economies: A Comparison of South Korea, Brazil, China and India", EPW, Nov. 30, 1996.

Nagubadi, R. V. & Zhang, D. (2006): 'Production Structure and Input Substitution in Canadian Sawmill and Wood Preservation Industry', *Canadian Journal of Forest Research*, 36: 3007-3014.

Nanang, D. M., & Ghebremichael, A. (2006): 'Inter-regional comparisons of production technology in Canada's timber harvesting industries', *Forest Policy & Economics*, 8(8): 797-810.

Nautiyal, J. C., and Singh, B. K., 1985. Production structure and derived demand for factors inputs in the Canadian lumber industry. *Forest Science*, 31: 871-881.

Nautiyal, J.C. & B.K. Singh (1986): 'Long-term Productivity and Factor Demand in the Canadian Pulp and Paper Industry', *Canadian Journal of Agricultural Economics*. 34(1): 21-44.

Nimai Das and Debnarayan Sarker (2011)**:** 'Network Externalities under Social Capital: Lessons from Gender Sensitive Forest Management Programme in West Bengal', *Indian Economic Review*, **Volume:** 46, **Issue Number:** 2, **pp.** 243-274.

N. Srividya *et al.* (2014): 'Indian Manufacturing Sector–An Overview', *Asia Pacific Journal of Marketing & Management Review*, Vol.3 (6), JUNE (2014), PP. 49-73.

Oum, T.H. & M.W. Tretheway (1992): 'A Comparison of the Productivity Performance of the U.S and Canadian Pulp and Paper Industries', in *Nemetz, P. (ed) Emerging Issues in Forest Policy*. Vancouver, UBC Press: 212-35.

Oum, T.H., M.W. Tretheway & Y. Zhang (1991) "Productivity Measurement, Decomposition and Efficiency Comparison of the Pulp and Paper Industry: Canada, the U.S and Sweden", UBC, *Forest Economics and Policy Analysis,* Working Paper 159: 27p.

P. Balakrishnan & K. Pushpangadan (1994): 'Total Factor Productivity Growth in Indian Manufacturing: A Fresh Look', Working Paper No. 259, Centre for Development Studies, Kerala, India, March 1994.

Pendse N.G., Baghel L.M.S. & Choubey S.K. (1996): 'Productivity Trends & Technological Change in Vanaspati Ghee Industry in Madhya Pradesh', *Ind. Jour. of Econ., LXXVI* (303): pp. 499-509, April 1996.

Planning Commission, The Manufacturing Plan (2013): "Strategies for Accelerating Growth of Manufacturing in India in the 12th Five Year Plan and Beyond'.

http://planningcommission.nic.in/aboutus/committee/ strgrp12/str_manu0304.pdf;

Priyanka Dembla (2000), 'Production in the Indian Manufacturing Sector: A Panel Data Analysis', *Indian Economic Review,* Vol. 35, Issue-1, pp. 41-65.

Puran Mongia & Jayant Sathaye (1998): 'Productivity Growth and Technical Change in India's Energy Intensive Industries: A Survey', Energy Analysis Program Environmental Energy Technologies Division Lawrence Berkeley National Laboratory Berkeley, CA 94720, Oct., 1998.

Pushpa Trivedi (2004): 'An Inter-State Perspective on Manufacturing Productivity in India: 1980-81 to 2000-01', *Indian Economic Review,* Vol.-39, Issue-1, pp.203-237.

Rajesh Raj S. N.: 'Structure, Employment and Productivity Growth in the Indian Unorganized Manufacturing Sector: An Industry Level Analysis', CMDR Monograph Series No.-50, Centre for Multi-Disciplinary Development Research, Dharwad Karnataka.

Raphael Kaplinsky *et al.* (2003): 'The Global Wood Furniture Value Chain: What Prospects for Upgrading by Developing Countries, The case of South Africa', Sectoral studies Series, United Nations Industrial Development Organization, Vienna, 2003.

Rao, J.M. (1996): 'Manufacturing Productivity Growth, Method and Measurement', *Economical and Political Weekly,* Vol 31, No.44, November 2, pp 2927-2936.

Richard G. Lipsey & Kenneth I. Carlaw (2004): 'Total factor productivity and the measurement of technological change', *Canadian Journal of Economics,* Vol. 37(4), Nov., 2004.

Rolf Fare, Shawna Grosskopf, Mary Norris, & Zhongyang Zhang (1994): 'Productivity Growth, Technical Progress, and Efficiency Change in Industrialized Countries', *The American Economic Review,* Vol. 84 No. 1, March 1994, pp.-66-83.

R. Mriappan (2011): 'Growth & Productivity of the Unorganised Manufacturing Sector in India', *The Ind. Jour. of Industrial Relations,* Vol. 47, No. 1, July 2011.

R Nagaraj (1985): 'Some Aspects of Small Scale Industries in India', *Economic and Political Weekly*, Vol. 20, No. 41 (Oct. 12, 1985), pp. 1741-1746.

Romer, P. (1986): 'Increasing returns and long run growth', *Journal of Political Economy* 94(5), 1002–1037.

Romer, P. (1990): 'Endogenous technological change', *Journal of Political Economy* 98(5), S71–S102.

Rymes T. K. (1972): 'The Measurement of Capital and Total Factor Productivity in the Context of the Cambridge Theory of Capital, *Review of Income and Wealth,* 18, March 1972.

Rymes T.K. (1983): 'More on the Measurement of Total Factor Productivity', *Review of Income and Wealth* Volume 29, Issue 3, pages 297–316, September 1983.

Sahana Ghosh (1990): 'Growth of Manufacturing Industries In India 1975-76 To 1985-86: A Disaggregated Study', NIPFP Working Paper Series, No.4, 1990.

Salehirad, N., and Sowlati, T. (2005): 'Performance analysis of primary wood producers in British Columbia using Data Envelopment Analysis', *Canadian Journal of Forest Research,* 35(2): 285-294.

Sandeep Kumar & Kavita (2012): 'Productivity And Growth In Indian Manufacturing Sector Since 1984-85 To 2004-05s: An Analysis Of Southern Region States', ZENITH *International Journal of Business Economics & Management Research* Vol.2 Issue 4, April 2012.

Sandoe, M. & M. Wayman (1977) 'Productivity of Capital and Labour in the Canadian Forest Products Industry, 1965 to 1972', *Canadian Journal of Forest Research.* 7(1): 85-93.

Sanja Samirana Pattnayak & Shandre M. Thangavelu (2001): 'Economic Liberalization and Productivity Growth: A Disaggregated Analysis of Indian Manufacturing Industries', *Australasian Meeting of the Econometric Society,* 2001.

Sanja S. Pattnayak & Thangavelu, S. M. (2003): 'Economic Reform and Productivity growth in Indian Manufacturing Industries: An interaction of technical change and scale Economies' Paper provided by National University of Singapore, Department of Economics in its series Departmental Working Papers Number wp0307.

Schumacher, K. & J. Sathaye (1999): "India's Pulp and Paper Industry: Productivity and Energy Efficiency", Ernest Orlando Lawrence Berkeley National Laboratory, paper # LBNL- 41843.

Seshaiah, S.V. & Reddy V.K. (1993): 'Productivity Trends in some Industries of Andhra Pradesh Manufacturing Sector', *IEJ*: 41(2): pp.111-118, 1993.

Shivi Agarwal, Shiv Prasad Yadav, S. P. Singh (2009): 'Total Factor Productivity Growth in the State Road Transport Undertakings of India: An Assessment through MPI Approach', *Indian Economic Review,* Vol. 44, Issue-2, pp.203-223.

S.M. Goldfeld & R. E. Quandt (1970): 'The Estimation of Cobb-Douglas Type Functions with Multiplicative and Additive', *International Economic Review*, Vol. 11(2), 1970, pp. 251-257.

S. M. Goldfield & R. E. Quandt (1976): 'Non-Linear Methods of Econometrics', North-Holland Publication Company, Amsterdam, New York, 1976.

Singh, B.K. & J.C. Nautiyal (1986): 'A Comparison of Observed and Long-run Productivity of and Demand for Inputs in the Canadian Lumber Industry', *Canadian Journal of Forest Research*. 16(3): 443-455.

Solow, R. (1957): 'Technical change and the aggregate production function', *Review of Economics and Statistics* 39(3), 312–320

Sonali Roy Choudhury & Sadhan Kumar Ghosh (2014): Geographic Concentration and Regional Specialization of Manufacturing Industries in west Bengal', *The Jour. of Industrial Statistics*, Vol. 3(1), pp.40-60.

Steir, J. C., (1980): 'Technological Adaptation to Resource Scarcity in the U.S. Lumber Industry', *Western Journal of Agricultural Economics*, 5(2): 165-175.

Subal C. Kumbhakar, Hung-Jen Wang (2010): 'Estimation of Technical Inefficiency in Production Frontier Models Using Cross-Sectional Data', *Indian Economic Review*, Vol. 45(2), pp.7-77.

Sunil Kumar and Nitin Arora (2011): 'Assessing Technical Efficiency of Sugar Industry in Uttar Pradesh: An Application of Data Envelopment Analysis', *Indian Economic Review*, Vol.: 46, No. 2, pp.323-353.

Thorsten Mrosek, Mechthild Amann, Uwe Kies, S. Denise Allen & Andreas Schulte (2010): 'A Framework for Stakeholder Analysis of Forest and Wood-Based Industry Clusters – Case Study at the State of North Rhine-Westphalia, Germany', *The Open Forest Science Journal*, Vol. 3, pp.-23-37.

Ulrich Doraszelski & Jordi Jaumandreu (2012): 'Measuring the Bias of Technological Change', Final Draft, Department of Economics, Harvard University, Littauer Center, 1875 Cambridge Street, Cambridge, USA.

Unel, B. (2003): "Productivity trends in India's manufacturing sector in the last two decades" IMF working paper WP/03/02.

Vahid, S., & Sowlati, T., (2006): 'Efficiency analysis of the Canadian wood product manufacturing sub-sectors: A DEA approach', *Forest Products Journal*, 57 (1/2): 71-77.

Varmani, A. (2005): 'Policy Regimes, Growth and poverty in India: Lessons of Government Failure and Entrepreneurial Success', Working paper No. 170, ICRIER, New Delhi.

Varmani, A. (2006): 'India's Economic Growth History: Fluctuations, Trends, Break Points and Phases', *Indian Economic Review*, Vol. XXXXI, No. 1, pp. 81-103.

Varmani, A. (2009): 'The Sudoku of India's Growth', BS Books, Business Standard Publications, New Delhi.

Varmani, A. & Hasim, D.A. (2011): 'J-Curve of Productivity and Growth: Indian Manufacturing Post-Liberalization', IMF Working Paper No. 11/163, International Monetary Fund, New York, USA.

Varmani, A. (2012): 'Accelerating and Sustaining Growth: Economic and Political Lessons', IMF Working Paper No. 12/185,

Vinish Kathuria *et al.* (2013): 'Productivity Measurement In Indian Manufacturing: A Comparison Of Alternative Methods', *Journal of Quantitative Economics,* Vol. 11 Nos.1&2 (Combined), January-July 2013, pp.-148-179.

V.N. Murti & V. K. Sastry (1957): 'Production Functions for Indian Industry,' *Econometrica,* Vol. 25, No. 2, pp.-221.

Xianchun Liao; Yaoqi Zhang (2008): 'An econometric analysis of softwood production in the U.S. South: a comparison of industrial and nonindustrial forest ownerships, *Forest Products Journal;* Vol. 58 Issue 11, p.69.

Yasser Abdih & Frederick Joutz (2005): 'Relating the Knowledge Production Function to Total Factor Productivity: An Endogenous Growth Puzzle', IMF Working Paper, WP/05/74, Middle East and Central Asia Department, International Monetary Fund, April, 2005.

Zhang, D., & Nagubadi, R. V. (2006): 'Total Factor Productivity Growth in the Sawmills and Wood Preservation Industry in the United States and Canada: A Comparative Study', *Forest Science,* 52(5): 511-521.

4

Research Methodology

I. Introduction

The purpose of this chapter is to present the research assumptions underpinning this project work, as well as to introduce the research strategy and the statistical and empirical techniques applied. The chapter defines the scope and limitations of the research design, and situates the research methodology amongst existing research in economics. The goal of methods adopted and procedures described in this work was to generate reproducible results that allowed for the formation of valid, generalizable conclusions in regards to development of Wood &Forest-based Industries of Indian and Indian States for long-term.

In virtually every country in the world with significant forest resources, forest industries (both wood and non-wood) have played and continue to play a key role in overall socio-economic development. Recognizing the significant influence of such industries on structural changes and growth in Indian industries, a study like this is highly considerable. In economic theory the production function is generally a concept which states quantitatively the technological relationship between the output and the various factors of production.

In the available literature, different functional forms like Leontief type fixed coefficient, C-D, CES, VES and Translog Production functions have been used in both theoretical and empirical aspect of production relationships. Measuring Technical progress and identifying the sources of output growth in terms of production function is an interesting and fruitful area of knowledge. The importance of an understanding of production function and productivity trends for purposes of planning needs no elaboration. The estimates of production function like Cobb-Douglas gives the estimates of the marginal product of labour and capital, sources of output growth and returns to scale. While through the CES production function the elasticity of substitution and scale parameters can be obtained. Despite its importance, the study of the economic relations of production in India at inter-state and inter-regional level has tended to be neglected field. Thus, it is essential to conduct and analyze productivity trends and production functions as well as technological change to understand the industrial situation in particular and the economic situation of the country in general. Partial productivity ratios show unit requirements on savings in the use of factors of production indicating shift in the production function.

A total factor productivity (TFP) index tells us about the overall efficiency. Total factor productivity growth encompasses the effect not only of technological progress but also of better utilization of capacities, learning–by-doing, improved skills of labour, *etc.* Thus, the present study have cover trends in partial (single) factor productivity ratios (SFP) and total factor productivity (TFP) indices, and estimate the parameters of the production functions. The reason for using all these methods of measuring technological changes is that technological progress cannot be measured precisely; only its broad trend can be traced.

The development of any area of economy has always been influenced by a large number of factors operating in that economy separately or altogether. The choices of these factors operating may be tangible, while some other might be intangible. But past studies observed that the choice of these factors has been subjected to data to data availability. Keeping these views in mind, and hence to understand and measure the contribution of individual inputs, and the factor other than the inputs like technical change or batter utilization of existing resources, it has becomes essential to procure data on output and inputs particularly those concerning capital and labour, the data on wages paid to the labour is also needed, besides appropriate price deflators to be used for making the time series data comparable. Therefore, the main purpose of this study is devoted to the productivity trends and statistical estimation of production functions and technical change in the wood and forest-based industries for Indian states and India as a whole.

II. Objectives of the Study

With the overall objective of the proposed study is to analyze the growth of technological change and productivity growth in wood and forest-based industries of Indian states as well as India as a whole the specific objectives are as under:

1. To examine the existing structure and pattern of development among the states.
2. To study the growth of Wood and Forest-based industries among the states of India.
3. To compare the extent of technological change in wood and forest-based industries among the states.
4. To identify the causes, responsible for the observed trend in the productivity of Wood and Forest-based industries.
5. To analyze inter-industry differences in wood and forest-based industries among the states and region.
6. To workout disparity in growth of productivity and technological change among the states.
7. To suggest measures for further growth and improvement.

III. Hypotheses of the Study

Keeping above theoretical framework and review of the past literature in mind following null hypotheses have been formulated:

1. Growth of productivity is not associated with growth in output and value added.
2. There is a No relationship between capital intensity and technological change.
3. There is No inequality in distribution of industrial set up among the states.
4. All endogenous and exogenous variables of technological change are neutral.

IV. Research Design & Methodology for the Present Study

Research involves formulating the problem to be investigated, selecting a suitable research design, choosing and applying appropriate procedures for data collection, and analyzing and communicating the process and findings through a written report. The research methodology refers to the research decisions taken within the framework of specific determinants unique to the research study (De Beer, 1999: 23). Although, quantitative data can be analyzed manually or by computer but in this study data analysis that has been undertaken have utilize the SPSS and MS Excel. Anyway, research procedures adopted in our study has been described here in detail with scientific and respective manner.

1. Data Source

The importance of the data base in an empirical study like the present one needs hardly be emphasized. The current study is based on statistical and econometric analysis mainly regression analysis for estimation of productivity growth and production function and thus requires time series data on output and the inputs mainly, capital, labor, raw material and fuel materials, depreciation, GVA, NVA, total emoluments, Gross fixed capital formation, Gross capital formation, price index of machine & equipments, and whole sale price index number for the industry groups covered in the proposed study.

The proposed study is based on secondary data, therefore no sampling is required, The Indian states are leading for decentralized planning *i.e.* District level, and Block level, Village level *etc.* therefore any study which takes cognition of technological change and productivity growth seems to be useful in policy formulation. The proposed study includes those wood and Forest-based industries of three digits (*i.e.* Minor Group) which have been classified by Central Statistical Organization (C.S.O.) and they are shown in Annexure I.

The data for arriving at growth rates (*i.e.* AGR & CAGR), partial (or single) factor productivity (SFP), total factor productivity (TFP) and various production functions has been gathered from summary results of annual Survey of Industries (ASI) for the factory sector. Data relating to WPI, production capacity, foreign trade, *etc.* has been collected from various published and unpublished sources of different ministries and other organizations, including industries association.

Thus, for the purpose of our UGC sponsored major research project, the section below discusses the measurement of the variables and the issues involved.

Output (Q or Y or V) Measurement: The Annual Survey of Industries provides data on both the output and the value added at the current prices. But our production, which is the basis of total factor productivity estimation, assumes output to be the function of labour and capital only. It is; therefore, appropriate to take value added as representative of output instead of the value of output. Out of the two measures of value added provided by the ASI namely, net value added and gross value added, we uses the latter one. In this study, the Wholesale Price Index (WPI) for manufactured products has been used to deflate the nominal values of gross value added. To arrive at value added measures at constant (*i.e.* base year) prices, the yearly valued at current prices has been deflated by using the Wholesale Prices Index gathered from the Office of Economic Advisor, govt. of India which is based on 1970-71 prices and are available from 1973-74.

Labour (L): The customary way of measuring labour input is either to use the number of hours worked or the number of workers employed. A large number of studies have used the former as it accounts more accurately for part-and full-time employees in terms of actual hours worked. Still the measure suffers from a limitation if a mix of skilled and unskilled workers is employed. This is because the contribution of skilled workers to production is much higher than that of unskilled workers (Vinish Kathuria *et.al.*, 2011). Thus, appropriate labour measure would require incorporating the quality of the labour inputs accounting for the sex, education, employment status of the worker *etc.* (Mahadevan, 2003).

So far as the measurement of labour input in our study is concerned, the ASI data on total employees (*i.e.* Total number of persons engaged) has been taken. Reason behind of such selection is, as both workers, working proprietors and supervisory/managerial staff can affect productivity, so number of persons engaged is preferred to number of workers.

Capital (K or C): In practice, the measurement of capital input is the most complex of all input measurements. There is no universally accepted method for its measurement and, as a result, several methods have been employed to estimate capital stock. In many studies, the capital unit is treated as a stock measured by the book value of fixed assets. Some studies have employed the perpetual inventory method to construct capital stock series from annual investment data.

Keeping all above views in mind, Gross fixed capital stock at constant price has been taken as a measure of capital input. ASI provides data on fixed capital stock at historical cost. It consists of land, buildings, plant and machinery, capital work in progress, furniture, fixtures and office equipments and others. For constructing the capital stock for the purpose of present study, CSO's data on fixed capital stock for 1973-74 have considered as the benchmark year of capital stock. Thus, capital stock series is constructed by using perpetual inventory accumulation method.

Measurement of Energy Input: The ASI publishes annual data on total value of fuels consumed at current prices. Fuels consumed represents total value of all items of fuels, lubricants, electricity, water *etc.* consumed by the factory during the accounting year except those that directly enter into products as materials consumed. To obtain the value of fuels consumed at constant prices the price index of energy *i.e.* the official Wholesale Price Index Number of Fuels, Lights and Lubricants from the RBI Bulletin has been used.

Measurement of Materials Input: Again the ASI is the source of published data on value of total materials consumed in current prices. Total materials consumed represents the total delivered value of all items of raw materials, chemicals, packing materials & stores which actually entered into the production process of the factory during the accounting year. To arrive at the value of intermediate goods consumed at constant prices the study uses a weighted constructed price index for materials which is derived for each industry separately. For each industry the ASI gives detailed quantity and expenditure data on a large number of items. The detailed data are available for broad groups of intermediate inputs consumed: Basic materials, chemicals & auxiliary materials, packing materials, consumable stores & materials consumed for repair and maintenance. After going through the list of basic materials consumed by each industry the study uses appropriate indices from the official series on Index Number of Wholesale Prices. For instance, for packing materials, a weighted average of Price Indices of Paper Products, Wood and Wood Products have been obtained. For materials consumed for repair and maintenance the price index of Machinery & Machine tools is used. For each industry, thus, weighted price indexes of intermediate goods consumed have constructed.

Emoluments: Total emoluments primarily constitute wages to workers, provident fund (PF) and other benefits and so on. To estimate real emoluments, the nominal value has been deflated by Consumer Price Index. In the present study, the share of emoluments in total value added considered as the share of labour. Assuming constant returns to scale, the share of capital is one minus the share of labour.

2. Period of the Study

The data for the purpose of the present study is related to period 1973-74 to 2012-13 (last year of study depends upon the availability of the data till aforesaid year). Per 1973-74 will not be considered because of the introduction of National Industrial Classification (NIC) in 1973-74 and comparability with the later period.

3. Specification Variables

For the purpose of the proposed study, the regions and states which have been taken are shown in Annexure II. While the Wood and Forest–based Industries which has been selected from the Annual Survey of Industries classification, C.S.O., New Delhi, are presented in Annexure I. Information variables proposed for the study are as under:

- Number of Factories
- Fixed Capital
- No. of Employees
- Mondays Employees

- No. of Workers
- Mondays Workers
- Total Inputs
- Total output
- Net Value Added (NVA)
- Depreciation
- Gross Value Added (GVA)
- Total Emoluments
- Fuel Consumed
- Material Consumed
- Gross Fixed Capital Formation
- Gross Capital Formation
- Price Index of Machine & Equipments
- Wholesale Price Index Number
- Time (as a proxy for technological change)

4. Definition of Variables

The ASI definition of variables which used in the present study is as under:

- **Fixed Capital**: Fixed capital includes factory land improvement in it and other construction, building, plant and machinery, miscellaneous assets like furniture and fixtures, fitting, transport equipment, tools and other fixed assets having a normal productive age of more them a year and assets under construction installation. These are net of depreciation.
- **Working Capital**: Working capital includes stock of materials, fuel, stores, *etc.* finished and semi finished goods, cash in hand and at bank, and net balance of amount receivable over the amount payable for the accounting year. The data on capital items are given in terms of book values of capital assets net of cumulative depreciation.
- **Worker**: A worker is defined as a person employed, directly or through an agency, whether for wages or not, in any manufacturing process or in cleaning any part of machinery or promises used for a manufacturing process, or in any other kind of work, incidental to, or connected with the manufacturing process, but exclude persons holding positions of supervision or management of employed in a confidential position.
- **Persons other than Workers**: Persons other than workers include persons holding supervisory and managerial positions, clerks in administrative office, stores and welfare section, match and ward staff.
- **Total Employees**: Apart from workers and persons other than workers total employees include unpaid working proprietors (in case of cooperative factory, working members), family workers, *etc.* (if paid these are included in the categories of workers or persons other than workers.

- **Total Emoluments**: Remuneration to employees given under two hands, wages and salaries and total emoluments. Data on both heads are provided for workers as well as total employees (except some years when data on total emoluments are given for total employees only). Besides wages and salaries and bonus, total emoluments include imputed value of benefits in king, old age benefits, social security and other benefits.
- **Total Inputs**: As per ASI Report, this comprises gross value of fuel materials *etc.* consumed (as defined above) and also other inputs *viz.* (a) cost of non-industrial services received from others (b) cost of materials consumed for repair and maintenance of factory's fixed assets including cost of work done by others to the factory's fixed assets (c) cost of contract and commission work done by others on materials supplied by the factory (d) cost of office supplies and products reported for sale during last year & used for further manufacture during the accounting year.
- **Fuel Consumed:** Fuel Consumed represent total purchase value of all items of fuels, lubricants, electricity, water (purchased to make steam) *etc.* consumed by the factory during the accounting year except those which directly enter into products as materials consumed. It excludes that part of fuels, which is produced and consumed by the factory in manufacture *i.e.*, all intermediate products and also fuels consumed by employees as part of amenities. It includes quantities acquired and consumed from allied concerns, their book value being taken as their purchase value and also the quantities consumed in production of machinery or other capital items for factory's own use.
- **Materials Consumed**: Materials consumed represent the total delivered value of all items of raw materials, components, chemicals, packing materials and stores which actually entered into the production process of the factory during the accounting year. It also includes the cost of all the materials used in the production of fixed assets, including construction work for factory's own use. Components and accessories fitted as purchased with the finished product during the accounting year are also to be included. It excludes intermediate products. Intermediate products in the above context mean all those products which are produced by the factory and consumed for further manufacturing process.
- **Depreciation**: Depreciation is consumption of fixed capital by the factory due to wear and tear and obsolescence during the accounting year and is taken as provided by the factory owner, or if not provided by the factory this is estimated on the basis of cost of installation and working life of the fixed assets. Depreciation is calculated at the rates allowed by income tax authorities for assessing taxable income. This rate varies according to the type of asset and industry.
- **Value Added**: Value added by manufacture if derived by deducting total inputs and depreciation from total output. Gross ex factory value of output comprises value of products and by-products, net value of semi finished goods, value of industrial or n-n-industrial services rendered to

others and net balance of goods sold in the same conditions as purchased, *etc.* This is net of excise duty paid or sales tax released (by the factory on behalf of government), transport changes from factory and selling agents commission.

Apart from the above variables, the data on Price indices for deflation purposes were also taken from various sources.

5. Measurement of Growth Rates

5 (A): Annual Growth Rate (AGR)

The annual average growth rate (AGR) shows the average percentage change from base year to current year. The formula is:

$$AGR \frac{P_1 - P_0}{P_0} \times \frac{n}{1} \times 100 \qquad \text{...... (5.1)}$$

To analyze the growth pattern over a period of time, semi log model have been used

$$Ln\,\gamma_t = Ln\,\gamma_o + t\,Ln\,(1+r) \qquad \text{...... (5.2)}$$

r = the compound rate of growth of γ

Now let, $\beta_1 = Ln\,\gamma_o$ $\beta_2 = Ln\,(1 + r)$ $r = \text{antilog}\,(\beta_2) - 1$ $Ln\,\gamma_t = \beta_1 + \beta_2 t$

Now if add the error term μ, to model we will get

$$Ln\,\gamma_t = \beta_1 + \beta_2\,t + \mu_t \qquad \text{...... (5.3)}$$

5 (B): Compound Annual Growth Rate (CAGR)

In the present study, we have estimated compound annual growth rate of variables taken here in this work and of Partial (Single) Factor Productivities (SFPs) and Total Factor Productivity (TFP) too as well. For this purposes, exponential function have been used to estimate compared annual growth rate (CAGR) symbolically it expressed as follows:

$$Y = ab^t \qquad \text{...... (5.4)}$$

Where, Y = is the dependent variable (*i.e.* total factor productivity, labor & capital productivity; t = is the independent variable (*i.e.* time period); a and b are the parameters. After getting b coefficient the following method has been used to estimate CAGR:

$$CAGR = [\text{Antilog } p - 1] \times 100. \qquad \text{...... (5.6)}$$

6. Measurement of Productivity

The need for studying measurement of productivity growth arises due to the intimate link between productivity growth and economic growth. In Economic literature, Productivity growth has long been recognized as an important driver of economic growth and a determinant of international competitiveness of a country relative to others. Productivity growth is the basis of efficient economic growth. According to Kuznets (1966), an essential element in the development and structural transformation of the developed economies was the fast growth in industrial productivity (Duraisamy, 2000).

Apart from the CAGR, to examine the growth of productivity, various methods have been employed such as partial factor productivity also known as single factor productivity *(i.e.* ratio of output to a single input), multi-factor productivity (ratio of output to more than one input) or total factor productivity (ratio of output to all the inputs used in the production process) methods. The single or partial factor productivity method considers only one factor of production at a time while assuming the contribution from other factors of production constant. Therefore, it fails to capture the contribution of all the factors as a whole in the total output. In order to circumvent this problem, TFP approach is suggested.

However, it is not rational to ignore partial factor productivity approach. Balakrishnan (2004) argues that labour productivity (LP) merit attention in its own right and serves a different purpose for which the TFP is not a substitute. It is argued that LP is a measure of potential consumption and a steady rise in the productivity of labour is necessary for a sustained increase in the standard of living of a population. An attempt to ignore changes in LP reflects an inadequate concern for potential increase in consumption. Thus, there is a strong case for measuring LP particularly in the Indian context. Taking cognizance of it, an attempt has been made in this study to capture the levels and trends in both partial and total factor productivity in Wood & Forest base Industries of India and Indian States separately as well.

6.1: Single Factor Productivity Ratio (S.F.P.)

This is a simplest measure of productivity, ratio of any factor of production to output is known as partial factor productivity of that factor; in other words, per unit output of any factor of production is known as partial factor productivity of that factor (Pendse & Baghel, 2008). Partial productivity ratios are the reflectors of value added by input factors separately, that is to say, of the extent of the effective utilization of materials, labour capital, machines, *etc.* and the cause for their low or high performance.

The more conventional measures of SFP are the LP and CP. Since trends in the SFPs are dominantly affected by the trends in factor intensity, *i.e.* the capital-labour (K/L) ratio. If the K/L ratio is increasing over time, the analysis of SFP changes would overstate the increase in LP and understate the increase in CP. "*The dominant feature that emerges from the analysis of the long-term trends in SFPs is that of a sharp increase in capital intensity accompanied by falling CP (i.e. rising K/O ratio) and moderately rising LP*".

Furthermore, the trends in the SFP indices with respect to individual factors include the effect of capital deepening in the context of capital accumulation in a growing economy. Thus, increases in LP may be due to improved efficiency and technological progress as well as more capital per unit of labour, and the two sets of influences does not operate independently of each other. More specifically, LP may increase because of factors such as learning-by-doing experience, improved skills, *etc.*, and because there are better and more machines to work with. It can be also said that when capital intensity is increasing, the measured increase in SFP will be larger than the increase in the pure productivity component since part of the measured increase is due to the rising K/L ratio.

Similarly, when analysing the change in CP, the observed increase in it will understate the increase in the pure productivity of capital. It is, therefore, reasonable to argue that in a situation where capital intensity is increasing over time, the analysis of partial productivity changes would overstate the increase in LP and understate the increase in CP. In order to measure of S.F.P. *viz.*, enterprise, labour, capital, raw-material productivity, capital intensity and emoluments per worker, gross value added (GVA) at constant prices per employees has been used as the measure of labour productivity (LP). Ratio of GVA to gross fixed capital (both at constant prices) has been used as the measure of capital productivity (CP). Gross fixed capital at constant prices (deflated by the machinery and machine tools index) per employees taken as the measure of capital intensity (IC). In addition to this, the efficiency of input lies in terms of ratio of GVA per unit of raw-material used has measured at constant prices by using appropriate deflator.

Thus, the study considered the following structural ratios (*i.e.* single factor productivity ratio) as follows:

1. **Enterprise Productivity (EP)**: Gross real value added / Number of enterprises

   $$EP = V/F \quad \text{... (6.1)}$$

   (It has been measured as the ratio of GVA to the number of factories).

2. **Labour Productivity(LP)**: Gross real value added / Number of workers

   $$LP = V/L \quad \text{... (6.2)}$$

   (It has been calculated as the ratio of GVA, at constant prices to the number of workers).

3. **Capital Productivity(CP)**: Gross real value added / Real fixed assets

   $$CP = V/K \quad \text{... (6.3)}$$

   (It has been measured as ratio of product to capital, both in constant price).

4. **Capital Intensity (CI or IC)**: Real fixed assets / Number of workers

   $$IC = K/L \quad \text{... (6.4)}$$

   (It has been measured as the ratio of gross fixed capital stock at constant prices to the total number of employees).

5. **Raw Material Productivity (RMP)**: Gross Value Added/Raw Material Consumed

   $$RMP = V/RMC \quad \text{... (6.5)}$$

   (It has been measured as the ratio of gross value of production to expenses for few materials, both in constant prices).

6. **Emoluments per Worker (EW)**: Real emoluments / Number of workers

   $$EW = RE/L \quad \text{... (6.6)}$$

   (It has been measured as the ratio of gross value of emoluments to the number of workers).

Total emoluments primarily constitute wages to workers, bonus paid and other benefits and so on. To obtain real emoluments, the nominal value has been deflated using Consumer Price Index (CPI) with base year. The definition and measurement of other variables namely, number of enterprises, employment, fixed capital stock and gross real value added are as discussed in separately earlier in this chapter. The variation across industries, an average worker in each industry and variation across the time periods has also been analyzed in terms of these structural ratios.

Factors such as economies of scale, increasing capital labour ratio and increase in wage rate have considerable influence on labour productivity growth (Salter, 1960; Hahn & Mathews, 1964). Some analysis has shown that relatively more capital-intensive industries are more productive and efficient than the less capital- intensive ones. Even though the emolument per employee has increased considerably, the workers in the sector did not receive emoluments commensurate with the contribution they made to value added. Therefore, it has become important to see how changes in these factors affect the productivity of labour (LP) in the Wood & Forest based Industries. For this purpose, a double logarithmic function is specified and then estimated by using multiple regression technique and the productivity growth function and is expressed as:

$$\text{Ln (APLG)} = \acute{a} + \text{Ln (CAPG)} + \text{Ln (EMOLG)} + \text{ln (GVAG)} + u \quad \ldots (6.7)$$

Where, APLG = growth of labour productivity, CAPG = growth of capital intensity

EMOLG = growth of emoluments per employee,

GVAG = growth of gross value added and u = error term

Labour productivity growth has been regressed on growth of value added, capital-labour ratio and emoluments per worker. A positive and significant relationship is expected between growth of LP and value added. Growth in LP might also be due to increase in IC through the substitution of capital for labour or the availability of more machines per worker. Increase in growth of emoluments per worker could positively influence the LP, particularly in sectors such as Wood & Forest based Industrial sector where emoluments paid is very low.

6.2: Total Factor Productivity Indices (T.F.P)

Very few other issues in explaining economic growth have generated so much debate than the measurement of total factor productivity (TFP) growth. The concept of TFP and its measurement and interpretation have proved a fertile ground for researchers after the initial work of Abramovitz (1956) and Solow (1957). Three different views exist on what TFP is (Lipsey & Carlaw 2001). The conventional view considers that TFP is the measure of the rate of technical change (*e.g.*, Law, 2000; Krugman, 1996; Young, 1992). The second view (Jorgensen & Griliches, 1967) regards that TFP measures only the free lunches of technical change, which are mainly associated with externalities and scale effects. The third view is highly skeptical whether TFP measures anything useful (Metcalfe, 1987; Griliches, 1995).

The differing views on TFP emerged to explain the long-run growth for the United States and other countries. Thereafter, this concept gained prominence after the realization that in the long run the input growth is subject to diminishing returns and will be insufficient to generate high output growth (Mahadevan, 2003). This also resulted in efforts to obtain more accurate estimates of TFP growth for different sectors as well as the economy as a whole.

Analysis of total factor productivity (TFP) measures the increase in total output which is not accounted for by increases in total inputs. The level of TFP can be measured by dividing total output by total inputs. The TFP index is computed as the ratio of an index of aggregate output to an index of aggregate inputs. Growth in TFP is therefore the growth rate in total output less the growth rate in total inputs. In other words, TFP growth refers to the amount of growth in real output that is not explained by the growth in inputs. As TFP levels are sensitive to the units of measurement of inputs and outputs, they are rarely estimated; instead TFP growth is preferred. Hence, it is common to use the notation "TFP" to refer to growth rather than levels, and thus, this is the convention adopted in our study.

Productivity is a relationship between production and the means of production, or, more formally a relation of proportionality between the output of a good or service and inputs which are used to generate that output. This relationship is articulated through the given technology of production. In economic literature, it has been seen that Productivity growth is crucially affected by technological change. Their relationship is so close that the two terms are often used interchangeably. Productivity is a wider concept. Even though a crucial one, technological change is only one of the many factors which affect productivity growth.

TFP extends the concept of single factor productivity such as output per unit labour or capital to more than one factor. This is the ratio of gross output to a weighted combination of inputs. For the case of production function like:

$$Q_t = f(L_t, K_t) \quad \text{....... (6.8)}$$

Standard forms of the TFP at time t would be given by as follows where the first form of equation is an *arithmetic index* and the second a *geometric index*:

$$\text{AGR} \ \frac{Qt}{\alpha\, Lt + \beta\, Kt} \ \text{OR At} \ \frac{Qt}{g(\alpha\, Lt, \beta\, Kt)} \quad \text{....... (6.9)}$$

Where,

At : Index of TFP at time t.

g : the aggregation procedure implicit in the specific production function adopted.

α, β are appropriate weights.

Although for the measurements of TFPG, there are three principal approaches in economics and these are: (i) The Index Number Approach; (ii) Parametric Approach, and (iii) Non-Parametric Approach.

In the present study, we focus primarily on studies which have estimated productivity growth using the first approach. Wherever appropriate, the results from the estimation of production functions have been mentioned in support of as alternative explanations to the results of the first approach. In this Index number approach, the observed growth in output is sought to be explained in terms of growth in factor inputs. The unexplained part or the residual is attributed to growth in productivity of factors. It consists in assuming a certain functional form for the producers' production function and then deriving an index number formula that is consistent (exact) with the assumed functional form. Preferred functional forms are the flexible ones. These indices differ from each other on the basis of underlying production function or the aggregation scheme assumed.

The reason behind the measure of TFP is that the SFP ratios may exhibit diverse movements over the time and it is often necessary to have a composite measure. Therefore TFP is such a measure which can be defined as output per unit of input. Thus, Solow (1957), Kendrick (1961), and Divisia indices have been considered for the measurement of TFPG and these are presented here in detailed as under:

6.2 (A): Solow Index

This well known index of total factor productivity growth was developed in 1957 by R.M. Solow which provides a geometric measure of rate of technological change. Solow assumes a more general neoclassical production function *i.e.* C-D production function with the assumption of constant elasticity of substitution, but technological change is of a Hick-neutral type. Assuming that the factors are paid the value of their marginal products, the growth of total factor productivity is measured as the difference between the rate of growth of value added and the rate of growth of total factor inputs. Under the assumptions of competitive equilibrium and constant returns to scale the following equation is obtained.

The production function used is

$$Q = A_{(t)} f(L, K) \quad \ldots\ldots. (6.10)$$

Where the K, L & Q represents capital, labour and output respectively and multiplicative factor $A_{(t)}$ measures the cumulated effects of shifts in the function. Taking the total differential of the equation (6.10) with respect to time and dividing by 'Q' yields

$$\frac{Q'}{Q} = \frac{A'}{A} + A\frac{fL'}{LQ}\ A\frac{fK'}{KQ} \quad \ldots\ldots. (6.11)$$

Where Q is physical units of the output and the primes (') indicate time derivatives, for example Q= ∂Q/∂t. If we state that ∂Q/∂t = q/p, i.e, the marginal products of capital is equal to the rental value of capital (q).

$$\partial Q/\partial L = w/p \quad \ldots\ldots. (6.12)$$

Where 'w' is the wage rate and 'p' is the price of output per unit. Therefore,

$$(\partial Q / \partial K) \times (K/Q) = Wc \quad \ldots\ldots. (6.13)$$

Here Wc denote relative share of the capital.

$$(\partial Q / \partial L) \times (L/Q) = Wn \quad \ldots\ldots. (6.14)$$

Where, Wn is the relative share of labour. Now substituting this in to the well known Cobb-Douglas production function: $V = AL^{\alpha}K^{\beta}$, we get

$$Q/L = q,\ K/L = k \text{ and } Wn = 1\text{-}Wc,\ \frac{Q'}{Q} = \frac{A'}{A} + Wc\frac{K'}{K}$$

$$\frac{Q'}{Q} = \frac{A'}{A} + Wc\frac{K'}{K} + Wn\frac{L'}{L} \qquad \text{....... (6.15)}$$

OR

$$\frac{\Delta Qt'}{Qt} = \frac{\Delta At'}{At} + Wc\frac{\Delta Kt'}{Kt} + Wn\frac{\Delta Lt'}{Lt} \qquad \text{....... (6.16)}$$

Where, Δs are the discrete approximation to time derivatives. Equation (6.15) is the basic equation with which it is possible to obtain time series concept of $\Delta A_t/A_t$, which is a measure of technological change. Now, for each year A_t can be calculated provided the other terms in equation (6.15) are evaluated. Solow simplifies his expression by dividing it by 'L'. If we take value added per man day (V/L) as an index of labour productivity instead of output per man day (Q/L) then the expression would be:

$$\frac{\Delta\left(\frac{V}{L}\right)t}{\left(\frac{V}{L}\right)t} = \frac{\Delta At'}{At} + \beta\frac{\Delta\left(\frac{K}{L}\right)t}{\left(\frac{K}{L}\right)t} \qquad \text{....... (6.17)}$$

Hence, Solow has assumed that there is a fixed share of capital (β) in the total value added per man day. Hence, β coefficient has been calculated by fitting the function.

$$Ln(V/L) = \alpha.\beta Ln(K/L) \qquad \text{....... (6.18)}$$

Thus, β will be the share of capital. Now rearranging terms, we get the expression as given below:

$$\frac{\Delta At'}{At} = \frac{\Delta\left(\frac{V}{L}\right)t}{\left(\frac{V}{L}\right)t} - \beta\frac{\Delta\left(\frac{K}{L}\right)t}{\left(\frac{K}{L}\right)t} \qquad \text{....... (6.19)}$$

OR

$$\frac{\Delta A}{A} = \frac{\Delta V}{V} - \left[(1-\beta)\frac{\Delta L}{L} + \beta\frac{\Delta K}{K}\right] \qquad \text{....... (6.20)}$$

Where, $\Delta A/A$, $\Delta Y/Y$, $\Delta L/L$ and $\Delta K/K$ are the relative share of TFP, and β is the income share of capital while (1-β) is the labour share, Once $\Delta A/A$ is computed using equation (6.20). The Solow index of TFP will be obtained using the following identity (taking A(o) as unity). Thus, from the estimated series of $\Delta A_t/A_t$, using the following procedure we can derive the total factor productivity growth trend *viz.*, A_t.

$$A_{(t+1)} = A_t \left[1 + \frac{\Delta At}{At}\right] \quad \text{....... (6.21)}$$

Though, Solow index is based on the restrictive assumption of unitary elasticity of substitution, it is not a serious drawback. Nelson (1965) has shown non-unitary elasticity of substitution is unlikely to make significant difference in the estimates of TFP. It is interesting to note that under the assumption of competitive equilibrium, the Solow and Kendrick index are equivalent for small changes in output and inputs (Levhari, E. 1966).

6.2 (B): Kendrick Index

This measure of TFPG was first employed by Kendrick in order to study the TFPG in American industries. This is an *arithmetic measure* of rate of technological change, which was developed in 1961 by *Kendrick* and it is based on linear production function. Kendrick index assumes infinite elasticity of substitution between the factors of production. The index is defined as the ratio of value added in production to a weighted average (arithmetic mean) of the two factors of production. In this index, two factors of production, labour and capital have used for the measurement purpose. Kendrick further assumed perfect competition in market and factors are paid according to their marginal product. Technological change is Hicks-neutral type and condition of constant returns to scale prevails. Thus, Kendrick's index of total factor productivity for the case of value added as output, and two inputs can be written as:

$$A = \frac{V}{\alpha L + (1 - \alpha)K} \quad \text{....... (6.22)}$$

Where, V is the index of output. L is the index of labour input and K is the index of capital input. While α and (1- α) are the income shares of labour and capital, respectively. This Kendrick index of TFP is based on the assumption of competitive equilibrium, constant returns to scale and Hicks- neutral technical change.

Suppose that there is one homogenous output denoted by Y and there are two factors of production, capital (K) and labour (L). Then production function may be written as:

$$Y = f(L, K) \quad \text{....... (6.23)}$$

Since the nature of production function is homogeneous (or linear), therefore,

$$Y = \alpha L + \beta K \quad \text{....... (6.24)}$$

Where α and β are constants.

Further, suppose w_0 and r_0 denote the factor rewards of labour and capital in the base year of the study, then the Kendrick index for year t may by written as:

$$A(t) = \frac{Y_t}{w_o L_t + r_o K_t} \quad \text{....... (6.25)}$$

The index for base year may be written as.

$$A(o) = \frac{Y_o}{w_o L_o + r_o K_o} \qquad \ldots\ldots\ (6.26)$$

By dividing equation (6.25) by (6.26), we get

$$\frac{A(t)}{A(o)} = \frac{Y_t / Y_o}{\dfrac{w_o L_t + r_o K_t}{w_o L_o + r_o K_o}}$$

$$A(t) = \frac{Y_t / Y_o}{\dfrac{w_o L_t}{(w_o + \dfrac{r_o K_o}{L_o})L_o} + \dfrac{r_o K_t}{(\dfrac{w_o L_o}{K_o} + r_o)K_o}}$$

Since, A (o) = 1

$$A(t) = \frac{Y_t / Y_o}{\dfrac{1}{\dfrac{w_o + r_o K_o}{w_o L_o} \cdot \dfrac{L_t}{L_o}} + \dfrac{1}{\dfrac{w_o + r_o K_o}{r_o K_o} \cdot \dfrac{K_t}{K_o}}}$$

$$A(t) = \frac{Y_t / Y_0}{S_o \cdot \dfrac{L_t}{L_o} + (1 - S_o) \cdot \dfrac{K_t}{K_o}}$$

Where, $$S_o = \frac{w_o L_o}{w_o L_o + r_o K_o}$$

$$A(t) = \frac{\text{Output Index}}{\text{weighted Sum of Input Index}} \qquad \ldots\ldots\ 6.27$$

Thus, the Kendrick index may be interpreted as the ratio of actual output to the output, which would have resulted from increased inputs in the absence of technology change. This index is easy to calculate and simple to understand and hence most widely used. For example, the raw material is important input with the L and K, the production function Y = αL + βK, and Kendrick index for t year will be as under:

$$A(t) = \frac{Y_t / Y_o}{w_o L_t + r_o K_t + h_o R_t} \qquad \square \cdots (6.28)$$

Therefore, Kendrick index is used generally for the analysis of TFP. The measure drawback of this index is assumption of linear production function. Thus, can't be used under the condition of diminishing marginal of factor. Moreover, another ground from criticism is assumption of Hicks-neutrality; whereas, Harrods neutrality is considered much better then Hicks-neutrality.

6.2 (C): Divisia Index

The need for using the Divisia index was spelt out by Solow (1957) and Jorgensen & Griliches (1967). Subsequently Christensen & Jorgenson (1969, 1970), Jorgenson

& Griliches (1972), Christensen, Cummings and Jorgenson (1981) and Gollop and Jorgenson (1980) have used approximations to the Divisia index (known as translog index) in their studies. It was on the strength that the rates of growth of the Divisia indexes of prices and quantities add up to the rate of growth of the value added (factor reversal test) and that such indexes are symmetric in different directions of time (time reversal test).

Divisia indexes also have the reproductive property that *"a divisia index of divisia index is a divisia index of the components"*. A discrete version of the continuous Divisia index is the translog index. Translog index numbers are symmetric in data of different time periods and also satisfy the factor reversal test approximately. But, they do not have the reproductive property (Diewert, 1976). The translog index of technological change is based on a translog production function characterized by constant returns to scale. It allows for variable elasticity of substitution and does not assume Hicks' neutrality.

Consider an aggregate production function with two factors of production

$$Y = f(K,L,T) \qquad \ldots\ldots (6.29)$$

Where, Y denotes aggregate output, K is aggregate capital, L denotes aggregate labour and T is time. It is assumed that f is continues, twice differentiable with a characteristic feature of constant returns to scale. These aggregates are taken as functions of their components.

$$Y = Y(Y_1, Y_2, Y_3, \ldots\ldots\ldots\ldots Ym) \qquad \ldots\ldots (6.30)$$

$$K = K(K_1, K_2, K_3, \ldots\ldots\ldots\ldots Kn) \qquad \ldots\ldots (6.31)$$

$$L = L(L_1, L_2, L_3, \ldots\ldots\ldots\ldots Lq) \qquad \ldots\ldots (6.32)$$

Corresponding to them, there are 'm' output prices, 'n' capital prices and 'q' labour prices denoted respectively.

$P_1, P_2, \ldots, P_m$; $r_1, r_2 \ldots, r_n$; $w_1, w_2, \ldots\ldots\ldots w_q$.

Corresponding aggregate prices are denoted by p, r, w. Under the assumption of perfect competition and profit maximization, the condition of producer's equilibrium requires the shares of the factors be equal to their elasticities, so that, we get:

$$V_K = r_k/P_y = \text{Log}Y/\text{Log}K \qquad \ldots\ldots (6.33)$$

$$V_L = w_l/P_y = \text{Log}Y/\text{Log}L \qquad \ldots\ldots (6.34)$$

Because of constant returns $V_K+V_L=1$, similarly, for individual components the conditions of producers' equilibrium require:

$$SY_i = P_iY_i / PY = \text{Log } Y/\text{Log}Y_i,\ i = 1\ldots\ldots\ldots m \qquad \ldots\ldots (6.35)$$

$$Sk_j = R_jKj / RK = \text{Log } K/\text{Log}K_j,\ j = 1\ldots\ldots\ldots n \qquad \ldots\ldots (6.36)$$

$$SL_u = W_uL_u/WL = \text{Log}L/\text{Log } L_u,\ u = 1\ldots\ldots q \qquad \ldots\ldots (6.37)$$

SY_i is the share of ith output components in aggregate output, similarly, SK_j and SL_u are the shares of the jth capital output and u[th] labour input respectively. Linear homogeneity requires:

SYi = SKj =SLu= 1

After Differentiation of equation (6.29) with respect to time and re arranging terms, we obtain,

$$\frac{d\ Log\,Y}{dT} = Vk\frac{d\ Log\,K}{dt} + VL\frac{d\ Log\,L}{dT} + VT \qquad \cdots\cdots(6.38)$$

The expression VT is called the Divisia quantity index of technological change. It should be noted that in the above expression.

$$\frac{d\log Y}{dT} = SYi\frac{d\log Yi}{dT} \qquad \cdots\cdots(6.39)$$

$$\frac{d\log K}{dT} = SKj\frac{d\log Kj}{dT} \qquad \cdots\cdots(6.40)$$

$$\frac{d\log L}{dt} = SLu\frac{d\log Lu}{dT} \qquad \cdots\cdots(6.41)$$

Thus, a weighted average growth rate of individual components gives the growth rate for the aggregate output. These are respectively called the Divisia quantity index of output, capital and labour. On the price side, the assumption of constant returns to scale, perfect competition and profit maximization require that the prices of output, capital and labour be consistent with the following equation.

PY= rK + wL .... (6.42)

Given this equation, we can express 'p' as a function of r, w and T

P= f(r, w, T) .... (6.43)

This is referred to as the price function. From the point of view of the price function, technological change is defined as:

$$VT = \frac{LogP}{T} \qquad \cdots\cdots(6.44)$$

Now, it can be written as:

$$\frac{d\ Log\,P}{dT} = VK\frac{d\ Log\,r}{dT} + VL\frac{d\ Log\,w}{dt} + VT \qquad \cdots\cdots(6.45)$$

This gives the Divisia price index of technological change. For application to data at discrete point of time, the Translog index is used. This Translog production function (6.39) is typically written as:

$$LogY = \alpha_o + \alpha_K LogK + \alpha_L logL + \alpha_T(T) + ½\beta_{kk}(logk)^2 + \beta_{kL}(logK)^2 + ½\beta_{LL}(logL)^2$$
$$+\beta_{kT}(logK)T + \beta_{LT}(logL)T + ½\beta_{TT}(T)^2 \qquad .... (6.46)$$

Constant returns requires

$$\alpha_K + \alpha_L = 1;\ \beta_{kk} + \beta_{kL} = 0;\ \beta_{Kl} + \beta_{LL} = 0 \qquad .... (6.47)$$

Corresponding to equation (6.32), we get here:

$$\Delta \log Y = V_Y (\Delta \text{Log } K) + V_L (\Delta \text{Log } L) + V_T \quad \ldots\ldots. (6.48)$$

Where,

$$\text{Log } Y = \text{Log } Y_{(T)} - \text{Log } Y_{(T-1)} \quad \ldots\ldots. (6.49)$$

$$\text{Log } K = \text{Log } K_{(T)} - \text{Log } K_{(T-1)} \quad \ldots\ldots. (6.50)$$

$$\text{Log } L = \text{Log } L_{(T)} - \text{Log } L_{(T-1)} \quad \ldots\ldots. (6.51)$$

and

$$V_K = ½ [V_{K(T)} + V_{K(T-1)}] \quad \ldots\ldots. (6.52)$$

$$V_L = ½ [V_{L(T)} + V_{L(T-1)}] \quad \ldots\ldots. (6.53)$$

This expression for V_T in equation (6.48) is termed as the average Translog quantity index of technological change.

7: Measurement of Production Functions

In general, the production function expresses the technological or engineering relationship between various inputs and output. A production function is a heuristic device that describes the maximum output that can be produced from different combinations of inputs using a given technology. This can be expressed mathematically as a mapping $Y = f(X)$, where X is a vector of factor inputs (X_1, X_2,···, Xn) and $f(X)$ is the maximum output that can be produced for a given set of inputs X_i. This formulation is quite general and can be applied at both microeconomic (*i.e.*, individual firm) and macroeconomic (*i.e.*, overall economy) levels (Eric Miller, 2008). By estimating the parameters of a production function, one can measure all the elements of technological change, such as, the efficiency of the technology, the capital intensity of technology, the elasticity of substitution and the economies of scale. However, concept of production function in economics is much wider and deals with entire production activity of a firm or economy.

The main purpose of applying production functions is to estimate marginal productivity and elasticity of inputs, returns to scale and elasticity of substitution between the inputs. Among the available literate of production functions, the choice in this study was in favour of Leontief type fixed coefficient, Cobb-Douglas, Constant Elasticity of Substitution, Variable Elasticity of Substitution and Translog production function because these are the most popular and widely used form of production functions. These production function involve treatment of elasticity of substitution constant and equal to unity in C–D specification and constant but not necessarily equal to unity in case of C.E.S. specification. While C.E.S. production function major limitation is, it assumes the elasticity of substitution to be constant for all input combinations. In the literature, several functional forms have been suggested and used in empirical studies in which elasticity of substitution is variable. These are called V.E.S. (variable elasticity of substitution) production function. Apart from these most commonly used specifications, Translog specification is also useful in working but the bias of technological change. A vast detailed discussion has been made in the following paras regarding the specification and from of above mentioned production function specification. A time trend variable will be included in production functions to represent technological progress.

However, Production functions for the Wood & Forest based Industries (WFBI) for India and Indian States has been estimated with a view to identify the contribution of factors other than capital and labour inputs to output growth. TFPG, which is an indicator of the change in efficiency in factor use, discussed earlier has also been measured for explaining the resource use efficiency. The conceptual issues associated with the problems of aggregation, valuation of capital, under utilization of capital and changing quality of labour in the specification of a production function have been extensively discussed while presenting conceptual framework and reviewing the available literature on the issues separately in Chapter II and Chapter III. Hence, it is reasonable to argue that estimation of production function provides a fairly adequate approximation to reality in specifying production environments with their degree of substitutability between factors, properties of returns to scale, kind of technical progress *etc.*

In case of growth accounting estimates, direct estimation of the production function has an advantage in that, it is not necessary to assume competitive equilibrium in order to derive an estimate of productivity growth. The efficiency parameter, the scale parameter and the extent of substitutability between factors can be obtained by directly estimating the parameters of a suitably production function through regression analysis. However, statistical problems associated with endogeneity of input variables, multi-collinearity among the explanatory variables and autocorrelation of the residuals tend to impinges upon the robustness and other desirable properties of the production function estimates. In the specification of a production function, intermediate inputs can be explicitly incorporated by relating changes in the gross value of output to changes in labour, capital and intermediate inputs. When intermediate inputs are explicitly recognized, it implies that the estimated TFP growth rates are lower than that base on the value added function. The analysis in the present study is based on the value added production function.

The general form of the Leontief type fixed coefficient can be written as follow:

$$K/L = A (V)^{\beta} \qquad \text{....... (7.1)}$$

If β is significantly different from zero, one may infer that there is no evidence for the Leontief type fixed coefficient production function for the industry under study. Leontief production function assumes zero elasticity of factor substitution among factors, but in reality factors of production in general are neither perfect substitutes nor zero. These rules out the Leontief from of production function in empirical research. In order to determine the relevant form of production function for the industry, it is wise to examine C-D, C.E.S., V.E.S. and Translog production functions.

As we begin with Leontief type of production function. So, for the estimation purposes, we have formulated two types of specification of Leontief type fixed coefficient production function, these are:

$$Ln (K/L) = Ln A + \beta Ln (V) \qquad \text{....... (7.2)}$$

$$Ln (K/L) = Ln A + \beta Ln (V) + \lambda t \qquad \text{....... (7.3)}$$

The Cobb-Douglas is a simple production function that is thought to provide a reasonable description of actual economies. It was created by labor economist Paul H. Douglas and mathematician Charles W. Cobb in an effort to fit Douglas's empirical results for production, employment, and capital stock in U.S. manufacturing into a simple function (Cobb and Douglas 1928). One of the most commonly estimated functional forms in the C-D production function written as:

$$V = A_0 L^{\alpha} K^{\beta} e^{\lambda t} \qquad \text{....... (7.4)}$$

Where V is output is in real terms, L is labour and K is capital input and **t** is time while α and β are constants to be empirically estimated. Technical change is assumed to take place at a constant rate (λ). As it is assumed disembodied and Hicks Neutral type of production function, therefore capital-labour ratio remains unchanged at constant relative prices following a shift in technology. The elasticity of substitution between L and K is implicity assumed to be one so that a 1% change in wage-rental ratio leads to the same percent change in the K-L ratio.

Differentiating V from equation (7.4) with respect to L and K respectively, we get:

$$\frac{\partial V}{\partial L} \times \frac{L}{V} = \alpha \qquad \& \qquad \frac{\partial V}{\partial K} \times \frac{K}{V} = \beta \qquad \cdots\cdots(7.5)$$

Thus, the elasticities of V (*i.e.* value added) with respect to L and K are equal to the parameters α & β, respectively, and are therefore constant in C-D function. If competitive equilibrium so that factors are paid the value of their marginal products and value added elasticities equal the respective factor share, it is clear from equation (7.5) above that in the case of C-D production function, this would imply that the factor shares are constant, i. e.:

$$\frac{sL}{pV} = \frac{\partial V}{\partial K} \times \frac{L}{V} = \alpha \qquad \& \qquad \frac{rK}{pV} = \frac{\partial V}{\partial K} \times \frac{K}{V} = \beta \qquad \cdots\cdots(7.6)$$

A natural logarithmic transformation of the C-D production function yields an equation that is linear in the natural logarithms of output, inputs and time and can be written as:

$$\text{Ln } V = a + \alpha \text{ Ln}(L) + \beta \text{ Ln}(K) + \lambda t \qquad \text{....... (7.7)}$$

By applying the Ordinary Least Square (OLS) estimation in the above equation, we can get estimates of α, β and λ, where λ provides a measure of the TFP growth, the sum of the estimates of α and β is a measure of the degree of homogeneity of the production function. Thus, returns to scale are decreasing, constant or increasing, depending on whether the degree of homogeneity is less than one, one, or greater than one.

The Cobb-Douglas production function with four inputs allowing for technological progress can be expressed as:

$$V = A_0 \, L^{\beta 1} K^{\beta 2} E^{\beta 3} M^{\beta 4} e^{\lambda t} \qquad \text{....... (7.8)}$$

V, L, K, E & M denotes output, capital, labor, energy and raw materials respectively. A_0 is the efficiency parameter, λ is the exponential rate of technological progress, t denotes time and the estimated β_i's (i=1,2,3,4) denotes the output elasticity of capital, labor, energy and raw material respectively. The logarithmic transformation of the above C-D function for the final estimation makes it linear in parameters. A natural logarithmic transformation of the C-D production function yields an equation that is linear in the natural logarithms of output and inputs and time and can be written as:

$$\text{Ln } V = a + \beta_1 \text{ Ln } (L) + \beta_2 \text{ Ln } (K) + \beta_3 \text{ Ln } (E) + \beta_4 \text{ Ln } (M) + \lambda t \qquad \text{....... (7.9)}$$

The OLS estimation to this equation yields estimates of β_1, β_2, β_4, β_4, and λ. While λ, provides a measure of the TFP growth, the sum of the estimates of α and β is a measure of the degree of homogeneity of production function.

If the function transformed so as to have Ln(V/L) as the dependent variable, the estimates of this function provide a direct test of whether the degree of homogeneity is significantly different from being constant. The ratio form of the production function can be written as:

$$\text{Ln } (V/L) = a + \beta \text{ Ln } (K/L) + (\alpha + \beta - 1)\text{Ln } L + \lambda t \qquad \text{....... (7.10)}$$

Testing whether the estimated $(\alpha+\beta-1)$ is significantly different from zero amounts to testing whether returns to scale are significantly different from being constant *i.e.*, $\alpha+\beta$ is significantly different from one. A restrictive form of the C-D function explicitly assumes constant returns to scale (*i.e.*, $\alpha+\beta=1$) and is written as:

$$\text{Ln } (V/L) = a + \beta \text{ Ln } (K/L) + \lambda t \qquad \text{....... (7.11)}$$

Without time trends this equation (7.11) can be written as:

$$\text{Ln } (V/L) = a + \beta \text{ Ln } (K/L) \qquad \text{....... (7.12)}$$

Thus, for the estimation purpose five specification of the C-d production function will be used as under:

$$\text{Ln } (V) = \text{Ln } A + \text{Ln } L + \text{Ln } K \qquad \text{....... (7.13)}$$

$$\text{Ln } (V) = \text{Ln } A + \text{Ln } L + \text{Ln } K + \lambda t \qquad \text{....... (7.14)}$$

$$\text{Ln } (V/L) = \text{Ln } A + \text{Ln } (K/L) + \lambda t \qquad \text{....... (7.15)}$$

$$\text{Ln } (V/L) = \text{Ln } A + \text{Ln } L + \text{Ln } (K/L) \qquad \text{....... (7.16)}$$

$$\text{Ln } (V/L) = \text{Ln } A + \text{Ln } L + \text{Ln } (K/L) + \lambda t \qquad \text{....... (7.17)}$$

During the estimation of C-D function, we observed that the C-D production function assumes elasticity of substitution between L and K is equal to unity. One way of testing for this assumption is to specify a production function with constant (but not necessarily unitary always) elasticity of substitution and test whether the elasticity is significantly different from one. Such type of production function is C.E.S. function and its standard form is:

$$V = \gamma_0 e^{\lambda t} [\, \delta L^{-\varrho} + (1-\delta)K^{-\varrho}]^{-v/\varrho} \qquad \text{......(7.18)}$$

Where, λ is efficiency parameter, δ is the distribution parameter,

ϱ is the substitution parameter and v is the scale parameter.

The elasticity of substitution (σ) is $1/(1+\varrho)$ and the permissible range of σ is between o and ∞. As in the C-D function, technological change is assumed to be

Hick-neutral and disembodied. The CES production function can't be transformed into a function that is linear in the parameters and thus requires a nonlinear estimation procedure.

Kmenta (1967) provides linearization methods of CES function by obtaining an approximation to the function. If the interest is centered on testing for the assumption of unitary elasticity of substitution, it is possible to derive from the CES function, assuming competitive equilibrium (so that the real wage rate *i.e.* 'w' equals the marginal product of labour *i.e.* MP_L), a linear relationship between LP, on the one hand, and the real wage rate and level of employment, on the other.

$$\text{Ln}\,(V/L) = a + b\,\text{Ln}\,(W) + c\,\text{Ln}\,(L) + dt \qquad \ldots\ldots (7.19)$$

Where, $b = v/v+\varrho$, $c = \varrho(v-1)/(v+\varrho)$, and $d = \lambda\,\varrho/(v+\varrho)$.

The SMAC form of CES production function can be written as:

$$\text{Ln}\,(V/L) = a + b\,\text{Ln}\,(W) \qquad \ldots\ldots (7.20)$$

If time variable in incorporated the above specification, the equation becomes:

$$\text{Ln}\,(V/L) = a + b\,\text{Ln}\,(W) + dt \qquad \ldots\ldots (7.21)$$

Felipe & McCombie (2001) examined ACMS's (or SMAC's) CES production function and found that regression results that are purportedly well explained by a CES production function can be equally well explained by an underlying accounting identity. The original motivation behind the CES was the observation that in a given industry, value added per unit of labor varied across countries with the wage rate (and hence the elasticity of substitution is not equal to unity). For the estimation purpose four types of specification of CES function has been formulated and these are:

$$\text{Ln}\,(V/L) = \text{Ln}\,A + \text{Ln}\,(W) \qquad \ldots\ldots (7.22)$$

$$\text{Ln}\,(V/L) = \text{Ln}\,A + \text{Ln}\,(W) + \lambda t \qquad \ldots\ldots (7.23)$$

$$\text{Ln}\,(V/L) = \text{Ln}\,A + \text{Ln}\,(W) + \text{Ln}\,L \qquad \ldots\ldots (7.24)$$

$$\text{Ln}\,(V/L) = \text{Ln}\,A + \text{Ln}\,(W) + \text{Ln}\,L + \lambda t \qquad \ldots\ldots (7.25)$$

Conceptually, C.E.S. type production function assumes a relationship between value added per labour and wage rate independent of capital/labour ratio *(i.e.* K/L). If, however, the K-L ratio varies due to changes in the factor price ratio, it is possible that the elasticity of substitution will vary as K-L ratio varies. On the other side, a major drawback of C.E.S. production function is that it assumes the elasticity of substitution to be constant for all input combinations. Several functional forms have been suggested in the literature and used in empirical studies in which elasticity of substitution is variable. These are called V.E.S. (variable elasticity of substitution) production function.

Accordingly Lu and Fletcher (1968) derived a more general form of CES production function which has the property of Variable Elasticity of Substitution (VES) and drops the assumption that for a given technology the elasticity of substitution (σ) as invariant to changes in the K-L ratio. The functional form of VES is as follows:

$$Y = A\,[\,(1\text{-}\delta)\,K^{n} + \delta K^{mn}\,L^{(1-m)\,n}]^{\,1/n} \qquad \ldots\ldots (7.26)$$

Where as usual Y is output (*i.e.* value added), K is capital and L is labour. It reduces to the CES form for m = 0. It has the properties of a neoclassical production function, *i.e.* positive marginal product and downward sloping marginal product curve, over the relevant range of inputs. This function is homogeneous of degree one (*i.e.* constant returns to scale) and has variable elasticity of substitution. VES production function derived by Lu & Fletcher can be written as:

$$Ln\,(V/L) = Ln\,A + b_2\,Ln(W) + b_3\,Ln\,(K/L) + e \qquad \ldots\ldots(7.27)$$

Where, A, b_2 and b_3 are the constants while **e** is a random error term. The VES production function with the inclusion of a time term (t) or trend can be written as under:

$$Ln\,(V/L) = Ln\,A + b_1 t + b_2\,Ln(W) + b_3\,Ln\,(K/L) + e \qquad \ldots\ldots(7.28)$$

Thus, with the help of above mentioned VES specification equations it becomes possible to find out whether the industry is having variable elasticity of substitution between its factors of production (*i.e.* K & L) over the period under study. Thus, for the estimation purposes two types of VES production function has been formulated and expressed here as follows:

$$Ln\,(V/L) = Ln\,A + Ln(W) + Ln\,(K/L) \qquad \ldots\ldots(7.29)$$

$$Ln\,(V/L) = Ln\,A + Ln(W) + Ln\,(K/L) + \lambda t \qquad \ldots\ldots(7.30)$$

At last, a more general specification of the production function is the Transcendental Logarithmic (Translog) production function developed in 1971, 1973 by Christensen, Jorgenson & Lau. It is a flexible functional form imposing relatively few a priory restrictions on the properties of the underlying technology. In particular, it does not assume a Hicks-neutral or a constant rate of technological change. The elasticity of substitution between inputs is also allowed to vary with the level of inputs. The functions written as:

$$Ln\,(V) = \alpha + \alpha_L\,(Ln\,L) + \alpha_K\,(Ln\,K) + \alpha_{tt} + \tfrac{1}{2}\beta_{LL}\,(Ln\,L)^2 + \tfrac{1}{2}\beta_{KK}\,(Ln\,K)^2 + \beta_{LK}\,(Ln\,L)\,(Ln\,K) + \beta_{Lt}\,(Ln\,L)t + \beta_{kt}\,(Ln\,K)t + \tfrac{1}{2}\beta_{tt}\,t^2 \qquad \ldots..(7.31)$$

Where, the α's and β's are the parameters of the production. As it is obvious from equation (7.31), the first four terms constitute nothing but the C-D production function, the last three terms relates to the assumptions about technical progress, including the nature of the bias in technical progress, and the other quadratic and interaction terms provide information on the curvature of the production function.

After Differentiation Ln(V) from equ.(7.31) with respect of Ln(L) and Ln(K), respectively, we get:

$$\frac{\partial\, LnV}{d\; LnL} = \alpha_L + \beta_{LL}(LnL) + \beta_{LK}(Lnk) + \beta_{Lt}\,t \qquad \square\ldots.\ (7.32)$$

$$\frac{\partial\, LnV}{d\; Lnk} = \alpha_K + \beta_{KK}(LnK) + \beta_{LK}(Lnk) + \beta_{Kt}\,t \qquad \ldots\ldots.\ (7.33)$$

The elasticity(s) of Output (*i.e.*, values added) with respect to L and K are not constant but depend on the input levels and time incorporated in the specification model. This is in contrast with the C-D case where these elsaticities are constant, equal to α_L and α_K, respectively. The β coefficients can be interpreted with the help of equations (7.32) and (7.33) and the equilibrium conditions.

The coefficients in equation (7.31) provide information on the possibilities of factor substitution within the framework of the Translog production function. When α and β coefficient is positive, under competitive equilibrium the factor share increases with the level of the input, assuming the level of the other inputs to be unchanged.

If Ln(V) is plotted against Ln(L) assuming Ln(K) to be constant, the C-D function will be a straight line, while the Translog will be a curve. The shape of this curve indicates that the Translog production function has a variable elasticity of value added with respect to an input (L), which is initially higher than that a C-D function but which declines as the input level increases. Assuming competitive equilibrium, the marginal productivity conditions imply that the elasticity of output with respect to each input is equal to the respective factor share in output. As elasticities of output as well as factor shares vary with input levels in the Translog production function, the σ is also a function of the input levels and is not a constant.

The rate of technical progress (or TFPG) in a Translog production function is given by

$$\frac{\partial LnV}{\partial t} = \alpha_t + \beta_{tt}t + \beta_{Lt}(LnL) + \beta_{Kt}(LnK) \qquad \ldots\ldots (7.35)$$

Where α t is the rate of autonomous TFPG, β_{tt} is the rate of change of TFPG, and β_{Lt} and β_{Kt} define the bias in TFPG. If both β_{Lt} and β_{Kt} are zero, then the TFP growth is of Hicks-neutral type. If β_{Lt} is positive, then the share of L increases with t then there is a labour using bias.

The main drawback of Translog specification is that it is not globally well behaved in the sense that the monotonicity, and quasi-concavity conditions may not be satisfied globally. As Berndt and Christensen (1973) have analyzed, when at least one Bij ≠ 0, where i and j refer to the inputs there exist configurations of inputs such that neither monotonicity nor concavity is satisfied. However, the well-behaved regions may be large enough so that the Translog function can provide a good representation of relevant production possibilities.

Monotonicity requires that the marginal products of all factors (∂V/ ∂L and ∂V/∂K) be positive. Since V, L and K are always positive; this amounts to requiring that the elasticities are positive. As for concavity (that the isoquants of the Translog function V are convex), the corresponding *Hessian matrix* of the second order partial derivatives must be negative semi-definite. The concavity can be assured if constant returns to scale are imposed and own share elasticities are non-positive. If the Translog production function is well behaved, it is possible to explore its characteristics with respect to homotheticity and constant returns to scale. Indeed, it is also possible to impose homotheticity and constant returns to scale by testable restrictions on the parameters.

A restricted form of Translog production function can be written as:

$$\text{Ln (V)} = \alpha + \alpha_L(\text{LnL}) + \alpha_K(\text{LnK}) + \alpha_{tt} + \tfrac{1}{2}\beta_{LL}(\text{LnL})^2 + \tfrac{1}{2}\beta_{KK}(\text{Ln K})^2 + \beta_{LK}(\text{LnL})(\text{LnK}) \quad \text{....... (7.36)}$$

As implication of a production function being homothetic is that marginal rates of substitution between K and L are constant at constant K-L ratios. In such situation, optimal factor proportions are independent of scale. The condition for homotheticity in a Translog production function is that:

$$\sum\beta ij=0 \quad \text{..... (7.37)}$$

Where, i = K, L and j = K, L

OR

$$\beta_{LL} + \beta_{LK} = 0, \text{ and } \beta_{LK} + \beta_{KK} = 0 \quad \text{..... (7.38)}$$

A constant return to scale in the Translog framework requires a restriction in addition to the restriction given above in equation (7.36). The additional restriction is:

$$\alpha_K + \alpha_L = 1 \quad \text{..... (7.39)}$$

If (7.38) and (7.39) are tested and not rejected, the Translog production function has constant returns to scale. Finally, in the next chapter(s), we present the estimates of production function for the Wood & Forest based Industries of India by using alternative specification outlined above in detail, (*e.g.* C-D, C.E.S., V.E.S. and Translog). The reason for using all these methods of measuring technological changes is that technical progress cannot be measured precisely; only its broad trends can be traced.

8: Measures used for Statistical Significance

The genesis of this concept of statistical significance was originated by eminent statistician Ronald A. Fisher when he developed statistical hypothesis testing, known as 'tests of significance', in his 1925 publication Statistical Methods for Research Workers. In their article, Fisher suggested a probability of one in twenty (0.05) as a convenient cutoff level to reject the null hypothesis. While in their 1933 paper, Neyman, J. & Pearson, E. recommended that the significance level (*e.g.* 0.05), which they called α or p, be set ahead of time, prior to any data collection. Statistical significance is determined by the size of the difference between the group averages, the sample size, and the standard deviations of the groups. For practical purposes, statistical significance suggests that the two larger populations from which we sample are *'actually'* different.

Thus, in statistical analysis, the level of statistical significance is often expressed as the previously mentioned so-called p-value. Depending on the statistical test what we have chosen and what we will calculate a probability (*i.e.*, the p-value) of observing sample results (or more extreme) given that the null hypothesis is true. Another way of phrasing this is to consider the probability that a difference in a mean score (or other statistic) could have arisen based on the assumption that there really is no difference. Every test of significance begins with a null hypothesis H_0. H_0 represents a theory that has been put forward, either because it is believed to be true or because it is to be used as a basis for argument, but has not been proved. Significance levels most commonly used in quantitative analysis of economic research are the .05 and .01 levels.

In statistics, the coefficient of determination denoted R^2 (or r^2) and pronounced R squared, is a number that indicates how well data fit a statistical model – sometimes simply a line or a curve. R^2 is a statistic that will give some information about the goodness of fit of a model. In regression analysis, the R^2 (coefficient of determination) is a statistical measure of how well the regression line approximates the real data points. An R^2 of 1 indicates that the regression line perfectly fits the data.

Adjusted R^2 (i.e. R^2) has been computed with the help of following equation:

$$(1\text{-}R^2) = [\{(n\text{-}1)/(n\text{-}k\text{-}1)\}x(1\text{-} R^2)]$$

The equations above show how the adjusted R^2 is computed. The sum-of-squares of the residuals from the regression line or curve have n-K degrees of freedom, where n is the number of data points (or years) and K is the number of parameters (i.e. variables) fit by the regression. The total sum-of-squares is the sum of the squares of the distances from a horizontal line through the mean of all Y values. Since it only has one parameter (the mean), the degrees of freedom equals n-1. The adjusted R^2 is larger than the ordinary R^2 whenever K is greater than 1. The adjusted R^2 will be always smaller than R^2 and would be best estimate of the degree of relationship in the underlying population.

An F statistic is a value we get when we run an ANOVA test or a regression analysis to find out if the means between two populations are significantly different. It's similar to a T-Statistic from a T-Test. A T-test will tell you if a single variable is statistically significant and an F-test will tell you if a group of variables are jointly significant. Thus, F-test in analysis of variance is used to test group variance against a null hypothesis, and is often used to determine whether any group of trials differs significantly from an expected value. For example, the null hypothesis could be set as the variance of two sample groups being equal, $s^1 = s^2$. To test whether $s^1 > s^2$ (sample 1 has significantly more variance than sample 2), take the ratio $s^1/ s^2$and compare it to an F-test value in a table of pre-computed critical values. To calculate the F-test value, find the degrees of freedom of each sample and the desired confidence interval. If the calculated ratio is less than the table value, accept the null hypothesis that the variance is not significantly different.

To investigate the model that is the model well fitted or not, we have considered the following null hypothesis:

$H_0: \theta = 0$, i.e., model is not well fitted against the alternative hypothesis,

$H_0: \theta \neq 0$, i.e., model is fitted well, where θ is vector of parameters, i.e.,

θ = (α's or β's, whatever chosen in the model).

Finally, F-Statistic has been computed with the help of following formula:

$$F = [\{R^2/k\text{-}1)/(1\text{-}R^2)/(n\text{-}k)\}]$$

Where k is the number of parameter and n is the number of observation.

We reject H_0 if F (Calculated.) > $F_{0.05'(k-1)(n-k)}$ (*i.e.*, Table value of F) which implies that model to be chosen for the study is fitted well.

Durbin-Watson (D-W) Statistic is a number which determines whether there is autocorrelation in the residuals of a time series regression. The D-W statistic ranges from 0 to 4 with 0 indicating positive autocorrelation and 4 indicating negative correlation. A value of 2 indicates no auto correlation in the sample. The Durbin-Watson (DW) statistic is used in a test for serial correlation of residuals (*i.e.*, error terms) in several types of regression models: (i) Simple regression models (ii) Multiple regression models, (iii) Time-series trend models. Thus, Durbin-Watson (D-W) Statistic has also been computed by using its standard form and formula is expressed as:

$$d = \sum_{t=2}^{T} \{(e_{t-1})\}^2 / \sum_{t=1}^{T} et^2$$

Where, **T** is number of data points (and, therefore, error terms).

'e' represents the OLS estimators. It is error term for data point t: if the independent variable is Yt, then $e_t = Y_t - \hat{Y}_t$

V. Significance & Relevance of the Present Study

India as a whole and all the States of India are basically agrarian economy. Health and status of the economy largely depends upon growth of agriculture and forests. Forests and industry are closely linked with each other. The linkages of above are important because any efforts made for tapping the potential of forest for industrial development is certainly a welcome step in this regard. India's industrial economy can be characterized as a large scale industrial economy along with some small scale set ups. During plan era, large number of Wood and Forest-based industries has been established in various parts of India. Wood and Forest based industrial establishment in the Indian states are very less to their potential. The total capital which is employed in these industries is quite very low while provides significant factory employment. The scope of utilize potential of forest for industrialization are always open and if tapped properly can boost the growth of Indian economy. Wood and forest based-industries are considered to be the most suitable agencies of technological break-though in forestry which is the key to increase productivity. However, this sector instead of becoming an effective instrument for generation of employment and development of India's economic growth has remained relatively stagnant. The experiences suggest that forest based industrial sector have laid more emphasis on employment even at cost of a growth and productivity. Dependence on traditional technology, disparity in inter-state forest based-industrial set up and growth of productivity of forest industries are another areas of serious concern.

On the basis of keeping above views in mind and in-depth elaboration of issues taken here in project which has already been discussed in Chapter-I, we once again certainly want to explore that Wood and forest based industries can play a vital role in the development process of a country like India because most of the rural population living in and around the forests. Thus, rural population and their dependency on agriculture and forest are still heavy. Economic prosperity of agrarian economy will not only depend upon the integration of agriculture with industry only but also upon the integration of forest with industry. Hence, for the development of rural areas, Wood and forest based industries have a special significance. In the midst of all these problems a special importance can be attached to the question of balance between forest and industry in development. The role of forest in its producer, producer-consumer and consumer aspect can be best utilized for framing policies for development. The analysis of technological change and productivity in wood and forest based industrial segment of economy is very much useful from development aspect of economic growth in a country like India. Technological improvements in wood and forest based industrial sector will enhance output of the economy on one hand and will uplift economic status of bulk of labour force, which is engaged in operation on the other. Balanced economic development can be ensured if wood and forest-based industrial units are established and spread over various part of this county. From this analysis it becomes clear that diagnosis of this industrial sector requires no explanation and its relevance to society and country is well established.

By 2030, India is likely to have GDP of USD 4 trillion dollars and a population of 1.5billions. To achieve sustainable growth of our economy, it is essential to develop forest sector and its related activities like forest based industrial development, to achieve double digit per annum growth of GDP. The study like proposed work has not been done earlier in India by any academicians, researchers, and also not by any institutions too. Therefore, there is an essential need to carry out such study for the economic growth and sustainable development of manufacturing sector like WFBI and country as well.

The literature reviewed separately in Chapter-III so for also reveals that although a number of studies are available to throw light on growth rates *(i.e.* AGR & CAGR), SFP, TFP and production function estimates, but still there exists a need of a comprehensive study which can take a stock of regional and interstate disparities in Wood and Forest based Industrial development. Very little efforts have been made by academicians and researchers in this direction. Thus, keeping above objectives in mind, any study which able to explore the problem of productivity growth and technological change in Wood and Forest based Industries for Indian states and India as a whole would be helpful in solving the problem of this sector. Hence the proposed study has been an effort to fill this gap.

VI. Conclusions

The ASI is the primary source of data for the present study. Despite several weaknesses, it remains an important data source for studies on the industrial sector. The present study has also paid particular attention to the measurement of the variables, especially to the measurement of capital stock and also to the construction of a deflator for intermediate materials input. Despite the attention paid to the measurement of the variables, it is important to keep in mind the limitations in the use of such a long time series for empirical purposes of obtaining trends in capacity utilization.

The proposed study has been conducted with the aims at finding out the extent of technological change in wood and Forest Industries and productivity growth. The partial factor productivity ratios have enabled us to quantify the unit factor requirements on saving in the use of factor of production. Thus, inter-state difference in SFP found useful in explaining the gap between developed and less developed states. The total factor productivity (TFP) have also throw light on overall efficiency, which would be helpful for the policy makers in framing policy for efficiency improvement in less developed states. The estimates of production function have given an idea of marginal product of factor, returns to scale elasticity of substitution among the states. The variation in all the above mentioned parameters among the states has given the idea of extent of disparity existing among the states.

At last, it can be said that this study would be helpful in framing the policies for complete overall of wood and forest -based industrial factor of India and balanced economic development of country. The observed trends in the various parameters would require identification of courses responsible for the phenomenon, and care of these causes responsible for in policy framing for administrators, scholars, and academicians. Finally, it can be said that this study will prove a mile stone in developmental literature of this country.

References

Abramovitz, M (1956): "Resources and output trends in the U.S. since 1870", *American Economic Review*, 46:5-23.

Arrow, K. J., H. B. Chenery, B. S. Minhas, & R. M. Solow (1961): "Capital-Labor Substitution and Economic Efficiency," Review of Economics and Statistics, XLIII (August, 1961), 225-50.

Balakrishnan, P. (2004) Measuring productivity in manufacturing sector, *Economic and Political Weekly*, 39, April 3-10, pp.1465- 71.

Christensen, L. & Jorgenson, D. (1969): "The measurement of US Real Capital Input, 1929-1967", *The Review of income & Wealth*, Series 15, No. 4, pp.19-50.

Christensen, L. & Jorgenson, D. (1969): "US Real Product and Real Factor Input, 1929-1967", *The Review of income & Wealth, Series* 16, No. 1, pp.293-320.

Christensen, L. Jorgenson, D. & Lau, L. (1973): "Transcendental Logarithmic Production Frontiers." *Review of Economics and Statistics*, Vol. 55, pp. 28-45.

Christensen, L., D. Cummings, & D. Jorgenson (1981): "Relative Productivity Levels, 1947-1973: An International Comparison," in *J. Kendrick and B. Vaccara, 'New Developments in Productivity Measurement and Analysis'*. Studies in Income and Wealth, 41, University of Chicago Press.

Cobb, C. & Douglas, P. (1928): "A Theory of Production." *American Economic Review*, Vol. 18, pp. 139-250.

De Beer, C.S. (1999): "Reading Texts and Understanding Meaning". In '*Reflective Public Administration: views from the South*', edited by JS Wessels & JC Pauw. Cape Town: Oxford University Press.

Diewert, W. Erwin (1976): "Exact and Superlative Index Numbers", *Journal of Econometrics* 4, 115-145.

Douglas, P. (1948): "Are There Laws of Production?" *American Economic Review*, Vol.38, p.1-41.

Duraisamy, M. (2000): "Growth and Productivity in the Unorganised Manufacturing Sector in India, 1984-1990", paper presented to a conference at New Delhi

Eric Miller, (2008): "An Assessment of CES and Cobb-Douglas Production Functions", Working Paper, Congressional Budget Office

Felipe, J. & McCombie, J. (2001): "The CES Production Function, the Accounting Identity and Occam's Razor." *Applied Economics*, Vol. 33, pp. 1221-1232.

Felipe, J. & Fisher, F. (2003): "Aggregation in Production Functions: What Applied Economists Should Know." *Metroeconomica*, Vol. 54, pp. 208-262.

Fisher, F. (1971): "Aggregate Production Functions and the Explanation of Wages: A Simulation Experiment." *Review of Economics and Statistics*, Vol. 53, pp. 305-325.

Fisher, F. Solow, R. & Kearl, J. (1977): "Aggregate Production Functions: Some CES Experiments." *Review of Economic Studies*, Vol. 44, pp. 305-320.

Goldar, B. (1986): "Productivity Growth in Indian Industry", Allied Publishers, New Delhi.

Gollop, Frank M. & Dale W. Jorgenson (1980): "U.S. Productivity Growth by Industry, 1947-73." in *J. Kendrick and B. Vaccara, 'New Developments in Productivity Measurement and Analysis'*. Studies in Income and Wealth, 41, University of Chicago Press. pp. 17-124.

Grifell-Tatje, E. & C.A.K. Lovell (1995): "A Note on the Malmquist Productivity Index, *Economic Letters*, 47, 169-175.

Griliches, S. (1995): "The Discovery of the Residual", NBER Working Paper #5348

Hahn, F. H. & Mathews, R.C.O (1964): "The Theory of Economic Growth: A Survey", *The Economic Journal*, 74, 779- 902.

Jorgenson, D.W. & Z. Griliches (1967): "The Explanation of Productivity Change", *Review of Economic Studies*, 34: 349–83.

Kendrick, J.W. (1957): "Productivity Trends, Capital and Labour". *Review of Economics and Statistics*, 39:3, pp. 248-257.

Kendrick, D. (1961): "Productivity Trends in the United States". Princeton University Press, Princeton, NJ.

Kmenta, J. (1967): "On Estimation Of The CES Production Function", *International Economic Review*, Vol. 8, No. 2, June, 1967.

Krugman P. (1996): "The Myth of Asia's Miracle" in Pop Internationalism, MIT Press, Cambridge.

Law M.T. (2000): "Productivity and Economic Performance: An Overview of the Issues", Public Policy Sources, No. 37, The Fraser Institute, Vancouver BC.

Leontief, W. (1941): 'The Structure of the American Economy, 1919-1939. New York: M.E. Sharpe Inc.

Lucas, R. (1969): "Labor-Capital Substitution in U.S. Manufacturing." In The Taxation of Income from Capital, ***Harberger, A. and Bailey, M.*** (ed.). Washington D.C.: The Brookings Institution, pp. 223-274.

Levhari, E. (1966): "The Relationship Between Two Measures of Total Factor Productivity", *The Review of Economics and Statistics.*

Lipsey, R. G. & K. Carlaw (2001): "What does Total Factor Productivity Measure?", Study Paper Version 2.

Lu, Y & L. Fletcher (1968): "A Generalisation of CES Production Function", *Review of Economic & Statistics.*

Mahadevan R. (2003): "To Measure or Not to Measure Total Factor Productivity Growth?", Oxford Development Studies, 31(3): 365-78.

Metcalfe, S. (1987): "Technical Change" in *Eatwell, J., M. Milgate and P. Newman (eds.) 'The New Palgrave, a Dictionary of Economics'*, Macmillan, London.

Mitra, A. (2001): "Employment in the Informal Sector", pp. 85-92. In ***Kundu, A. & Sharma, A. N. (eds.)*** *'Informal Sector in India: Perspectives and Policies'*, IHD, New Delhi, 2001.

Nelson, R. (1965): "The CES Production Function and Economic Growth Projections", *Review of Economics & Statistics*, Nov. 1965.

Nelson, R & Winter, S. (1982): "An Evolutionary Theory of Economic Change". Cambridge: Belknap Press.

Pendse, N.G., & Baghel, L.M.S., (2008): "Technological Change and Productivity Growth in Manufacturing Sector of India", Sarup & Sons Publishers, New Delhi.

Salter, W. (1960): "Productivity and Technical Change", Cambridge University Press, Cambridge.

Solow, R.M.(1957): "Technical Change and the Aggregate Production Function". *Review of Economics and Statistics*, 39:3, pp. 312-320.

Vinish Kathuria et.al. (2011): "Productivity Measurement In Indian Manufacturing: A Comparison of Alternative Methods", Inst. for Development Policy & Management (IDPM).

Unni J., Lalitha, N. & Rani U. (2001): "Economic Reforms and Productivity Trends in Indian Manufacturing, Working Paper No.119, Gujarat Institute of Development Research, Ahmadabad.

Young, A. (1992): "A Tale of Two Cities: Factor Accumulation and Technical Change in Hong Kong and Singapore" NBER Macroeconomic Annual, MIT Press, Cambridge.

Zellener, A. S., Kementa, J. & Dreze, J. (1966): "Specification and Estimation of Cobb-Douglas Production Functions, *Econometrica,* 34: 784-95.

www.businessdictionary.com/definition/Durbin-Watson-Statistic.html#ixzz3sQ1mVYYA.

https://en.wikipedia.org/wiki/F-test

https://en.wikipedia.org/wiki/Student%27s_t-test

Annexure- I

Wood and Forest Based Industries

(NIC Classification 1970: Industry From Minor Group)

Industry Code No.	Name of the Wood & Forest-based Industries
226	Manufacture of Bidi.
270	Manufacture of veneer, plywood and their products.
271	Sawing and Plaining of wood (other than plywood).
272	Manufacture of wooden and cane boxes, crates, drums, barrels and other wood containers, baskets and other rattan, bamboo, reed and willow wares made entirely or mainly of cane, rattan, reed , bamboo and willow,
273	Manufacture of structural wooden goods (including treated timber) such as beams, posts, doors and windows (excluding hewing and rough shaping of poles, bolts and other wood material which is classified under logging).
274	Manufacture of wooden industrial goods as bobbins, blocks, handles, sadding and similar equipment and fixtures.
275	Manufacture of cork and cork products.
276	Manufacture of wooden furniture and fixtures.
277	Manufacture of bamboo and cane furniture and fixtures.
279	Manufacture of wood, bamboo and cane products not elsewhere classified.
280	Manufacture of Pulp. Paper and paper board including newsprint.
281	Manufacture of container and boxes of paper and paper board.
282	Manufacture of paper products not elsewhere classified like dolls,
283	Manufacture of paper and paper board articles not elsewhere classified.
316	Manufacture of turpentine, synthetic resin, plastic materials and synthetic fibres like Nylon, terelene Except glass, *etc.*
317	Manufacture of matches.
378	Manufacture of bullock carts, push-carts, hand carts *etc.*
385	Manufacture of sports and athletic goods and play equipments.
386	Manufacture of Musical Instruments.
387	Manufacture of miscellaneous products not elsewhere classified like pencils.
226 to 387	Total Wood and Forest Based Industries.

Source: *NIC 1970, Annual survey of Industries, C.S.O., New Delhi.*

Annexure -II

Selected States & Regions of India

I. Eastern Region:

(i) Bihar

(ii) West Bengal

(iii) Orissa

(iv) Assam

II. Western Region

(i) Maharashtra

(ii) Gujarat

(iii) Madhya Pradesh

(iv) Rajasthan

III. Northern Region

(i) Haryana

(ii) Punjab

(iii) Uttar Pradesh

(iv) Delhi

IV. Southern Region

(i) Andhra Pradesh

(ii) Karnataka

(iii) Kerala

(iv) Tamil Nadu

V. Other States

VI. All India (I+II+III+IV+V).

5

Analysis of Productivity Growth: SFP & TFP

It is well acknowledged that economic growth of an economy mainly depends upon both i.e. first, on the proper and effective utilization of factors of production such as labour and capital; and secondly, depends on the efficiency in resource use, existing technology and technical progress. Such efficiency in use of resource is often referred to as productivity. Furthermore, it is also widely recognized in order to monitor the process of an industry or enterprises, it is essential to make a scientific appraisal of the trends in productivity efficiency with which resources are converted into goods and services (Pendse & Baghel, 2008). Some economists has also confirmed that growth in productivity is the only plausible route to increase the standard of living (Balakrishnan & Pushpangadan, 1998) and thus, it is also a measure of welfare too as well (Krugman, 1990). The relevance of economic growth of an economy is least in meaningful if it has not affected productivity growth and hence living standard of citizens of such country.

Productivity is defined as the contribution of all the inputs, they being considered as money, machine and human resources and is essential for achieving a higher level of productivity in industry. Several studies have been conducted for finding productivity in the past. In Economic literature, Productivity growth has long been recognized as an important driver of economic growth and a determinant of international competitiveness of a country relative to others. Productivity growth is the basis of efficient economic growth. According to Kuznets (1966), an essential element in the development and structural transformation of the developed economies was the fast growth in industrial productivity (Duraisamy, 2000).

Almost up to the end of 20th centuries, India has been relatively and industrially backward country. In late 70's gradual improvement in situation has been observed mainly because of promotional and productive measures, resulting in favorable environment for growth. Since 1991, India has undertaken major economic and industrial reforms. Being a low income poor nation characterized by shortage of capital, but rich nation characterized by the forest resources, it is imperative that the shortage of capital resources and still abundant forest resources can be utilized with optimum efficiency.

On the other side, Inter-regional disparity in levels of development and incomes has always been a major issue in the context of social, cultural, economic and political significance in India. There are wide disparities across the states is well known and is also recognized as a concern to be addressed separately through public policy. Well known economist Prof. T.S. Papola *et al.* (2011) in their study have argued that several mechanisms and instruments have been in use to reduce these disparities since independence. According to the dominant theory of modern economic development, industry is expected to play a major role in creating as well as mitigating disparities among different regions. Industry is seen as the main "engine of growth" (Kaldor, 1967) and industrial development subject to "cumulative causation" to a larger degree than development of other sectors (Myrdal, 1957). Industrial development, and consequently overall economic development of different regions, according to the typical conventional theory of regional development, is expected to take a path that finally leads to a "convergence" (See study of Barro and Salai-i-Martin, 1992 and 1995).

There have been very few studies examining the pattern of interstate disparities in industrial development in India, covering different periods of time. Some early studies, covering 1950's and 1960's, observed a decline in disparities (Dhar and studies, covering 1950's and 1960's, observed a decline in disparities (Dhar and Sastry, 1967; Sardamoni, 1969). Covering the period of 1950–51 to 1975–76, another study (Mathur, 1983) observed that while overall income disparity had narrowed down, the secondary sector has moved along an inverted U-shaped path (the primary sector having followed the reverse path). But not a single study has been examined till date on the inter-state and inter-region disparities in productivity growth and production functions for the Wood & Forest based Industries in India. Giving due weightage to above state of affair, the present study aims at measuring and analyzing the productivity growth in Wood and Forest based Industrial sector of India during the period 1973-74 to 2012-13.

This study attempts an assessment of the inter-state and inter-region comparison of productivity growth of wood and forest based industries of India and variations in the rates of productivity growth. In the process it also analyses the differences in the industrial structure and factors responsible for variations in the extent and structure of industrialization. Thus in the section first of this chapter Interstate variations in the levels of productivity and capital intensity have been studied. 'Industry' for the purpose of this study includes various 3 digit industries of minor group of WFBI. 40 years Period of study and various Indian states, regions and India as a whole have been covered in the study some time vary depending upon the availability of data. Major sources of data for study are: Annual Survey of Industries (ASI), CSO New Delhi. Thus, this chapter is divided into two sections. First section deals with the measurement and inter-state, inter-region analysis of single (i.e. partial) factor productivity growth in WFBI sector of India. In all only 16 states, 5 regions (i.e. Eastern, Western, Northern, Southern regions and Other States) and India as a whole has been considered for the purpose of investigation. In the next section, total factor productivity growth (*viz.* Kendrick, Solow and Divisia Index) has been analyzed and presented for the states and regions selected in the present study.

5.1: Growth of Partial Factor Productivity Ratios

AS it is previously discussed in the forth chapter that SFP is a simplest measure of productivity, ratio of any factor of production to output is known as partial factor productivity of that factor. In other words, per unit output of any factor of production is known as partial factor productivity of that factor ratios of Single (or partial) factor productivities such as labour, capital, raw-material, fuel and enterprise productivities as well as capital intensity and emoluments per workers for the wood and forest based industries at three-digit level for different years during the period 1973-74 to 2012-13 are presented in the end of the chapter as a Table form (See Table 5.1.1 to Table 5.1.22). Gross value added at constant prices per employee is taken as a measure of labour productivity (LP). Ratio of gross value added to gross fixed capital (both deflated for constant prices) is taken as a measure of capital productivity (CP). Gross fixed capital at constant prices per employee is taken as a measure of Capital intensity (CI). Gross fixed capital at constant prices to Number of Enterprise is taken as a measure of Enterprise Productivity (EP). Real Emoluments to number of workers has been taken as a measure of Emoluments per workers (EW) while gross value added to fuel consumed and Raw-material consumed has been considered as a measure of fuel productivity (FP) and raw-material productivity (RMP), respectively. For the purpose of estimates of growth of SFPs, Simple average annual growth rates have been computed with help of minimum least square technique for the period 1973-74 to 2012-13 for all the WFBI industries of India. The estimates of simple average annual growth rates are presented in Table 5.1.1 to Table 5.1.22.

To begin with it is appropriate to deal with trends in partial factor productivities i.e. enterprise, labour, capital, fuel and raw-material productivities, capital intensity and emoluments per workers for the industries at three-digit level for the Wood & Forest based industries under study. Increase in the partial factor productivity ratio signifies efficiency in the use of particular input over time. Labour and capital productivities indicate efficiency in the use of these inputs, the changes occurring in these are generated by various economic forces interacting simultaneously, whereas, the capital-labour ratio gives an idea of capital deepening over the period and its relationship with labour and capital productivities. While, emoluments per worker gives an idea about the economic status, living standard of workers and their economic viability over the period of time. The purpose of measuring these ratios at constant price was to study the real change.

Growth in the ratios of various partial factor productivities gives quite diversified picture. Table 5.1.1 presents estimates of growth in SFPs ratios for the state Andhra Pradesh. It is obvious from the table that labour productivity ranges between 0.979 to 1.84 per cent per annum, capital productivity ranges between 1.96 to more than 14 per cent per annum, and capital intensity ranges between 0.79 to almost 5 per cent per annum. For the purpose of industry-wise detailed analysis three-way classification of the growth rates were performed for labour and capital productivity, labour productivity and capital intensity, and capital productivity and capital intensity. A look at table gives us growth rate combinations of the various industries at three-digit level for labour and capital productivity. The growth rates of the LP ratios were observed as less than 1.0% per annum for industry code 316 and 317. The industries belong to this category are Manufacture of turpentine, synthetic resin *etc.* (316), Manufacture of matches (317). The highest growth in SFPs was observed for CP for Manufacture of paper products n.e.c. (283) 14.55 per cent per annum followed by manufacture of wooden furniture and fixtures 9.06 per cent per annum. While to measure productivity of fuel efficiency, two type productivity *viz.*, Raw-material productivity (RMP) and fuel productivity (FP) has been also measured for all the industries considered under WFBI. Table reveals that lowest growth in raw-material productivity (RMP) and fuel productivity (FP) were observed for Manufacture of paper products n.e.c. (283) while highest for the industry Manufacture of wood, bamboo and cane products n.e.c. (279). On the other side, highest growth in Enterprise productivity (EP) was measured for Manufacture of paper and paper board articles n.e.c. (283) and lowest for Manufacture of structural wooden goods (273). The growth in CP and IC of Wood & Forest based Industries as a whole for the southern state Andhra Pradesh was found 5.6 and 2.11 per cent per annum respectively. During the period under study it has been noticed that capital productivity was higher than all other SFPs ratios even increasing trend were found than LP. The trends in capital intensity found impressive in most of the WFBI and Total WFBI too as well. The above discussion reveals that high magnitude of CP measured at constant prices is partially associated with Capital Intensity (IC).

Table 5.1.2 presents estimates of growth in SFPs ratios for the Eastern state Assam. It is obvious from the table that labour productivity (LP) ranges between 0.528 to 1.89 per cent per annum, capital productivity ranges between 1.81 to 7.57 per cent per annum, and capital intensity (IC) ranges between 0.42 to 3.23 per cent per annum. The growth rates of the LP ratios were observed as less than 1.0% per annum for industry 4 industries of WFBI. The industries belong to this category are Manufacture of Wood, Bamboo and Cane products n.e.c (279); Manufacture of Wooden Furniture and Fixtures (276); Manufacture of Matches (317) and Manufactures of Pulp, Paper and paper board including newsprint (280). The highest growth in SFPs was observed for CP for Manufacture of Wooden Furniture and Fixtures (276) 7.57 per cent per annum; followed by Manufactures of Pulp, Paper and paper board including newsprint (280) almost 6 per cent per annum. Table further reveals that lowest growth in raw-material productivity (RMP) and fuel productivity (FP) were observed for Manufacture of structural wooden goods (273) while highest for the industry Manufacture of Container and Boxes of paper and paper board (281). On the other side, highest growth in Enterprise productivity (EP) was measured for Manufacture of Wooden and Cane boxes, crates, drums, barrels and other wood container baskets and ... (272) and lowest for Manufacture of Wood, Bamboo and Cane products n.e.c. (279). The growth in CP and IC of Wood & Forest based Industries as a whole for the Eastern state Assam was found 5.3 and 4.62 per cent per annum respectively. Similarly like state Andhra, in case of Assam too capital productivity was estimated higher than all other SFPs ratios even increasing trend were found than LP. Emoluments per workers were observed between 0.35 to almost 2 per cent per annum which shows not much impressive picture of the persons engaged in forest based manufacturing sector. The trends in capital intensity found impressive in most of the WFBI and Total WFBI too as well. The above discussion reveals that high magnitude of capital productivity measured at constant prices is associated with capital intensity.

Table 5.1.3 presents estimates of growth in SFPs ratios for the Eastern state Bihar. It is obvious from the table that labour productivity ranges between 0.75 to 2.05 per cent per annum, capital productivity ranges between 1.264 to 8.4 per cent per annum, and capital intensity ranges between 0.98 to 2.54 per cent per annum. The growth rates of the LP ratios were observed as less than 1.0% per annum only for a industry namely Manufactures of Pulp, Paper and paper board including newsprint (280). The highest growth in SFPs was observed for Capital Productivity for as it was found for Assam state *viz.*, Manufacture of Wooden Furniture and Fixtures (276) 8.41 per cent per annum; Manufacture of paper and paper board articles n.e.c. (283) almost 5 per cent per annum. Table further reveals that lowest growth in raw-material productivity (RMP) and fuel productivity (FP) were observed for Manufacture of turpepentine, synthetic resin *etc.* (316) while highest for the industry Manufacture of paper products n.e.c. (282) and Manufacture of paper and paper board articles n.e.c. (283), respectively for RMP and FP. On the other side, highest growth in Enterprise productivity (EP) was measured for Manufacture of paper products n.e.c. (282) and lowest for Sawing and Plaining of wood (271). The growth in Enterprise Productivity (EP) and IC of Wood & Forest based Industries as a whole for the highly populated Eastern state Bihar was found 3.5 and 2.5 per cent per annum respectively. Thus, in the state of Bihar, Enterprise productivity was estimated higher than all other SFPs ratios with significant increasing trend of IC were found than CP and LP. Emoluments per workers were observed between 1 to almost 2 per cent per annum which shows slighter impressive trend than Assam state. The trends in capital intensity found impressive in most of the WFBI and Total WFBI too as well.

Table 5.1.4 presents growth estimates of SFPs ratios for the state Delhi National Capital of India. The picture of growth trend of SFPs is quite different as it is discussed earlier for 3 States. To begin with, table reveals that labour productivity of WFBI of Delhi ranges between 0.53 to 2.7 per cent per annum, capital productivity ranges between 2.37 to 10.6 per cent per annum, and capital intensity ranges between 0.41 to 7.1 per cent per annum. The growth rates of the LP ratios were observed as less than 1.0% per annum only for one industry of WFBI i.e. manufacture of cork and cork products (275). The highest growth in SFPs was observed for CP for Manufacture of structural wooden goods (273) 11.06 per cent per annum; followed by Manufactures of wooden furniture and fixtures (276) above 10 per cent per annum. Table further reveals that lowest growth in raw-material productivity (RMP) and fuel productivity (FP) were observed for Manufacture of cork and cork products (275) while highest for the industry Manufacture of Vaneer, plywood and their products (270). On the other side, highest growth in Enterprise productivity (EP) was measured for Manufacture of structural wooden goods (273) and lowest for Manufacture of cork and cork products (275). The growth in CP and IC of Wood & Forest based Industries as a whole for the Northern state Delhi was found 3.92

and 3.23 per cent per annum respectively. Similarly like state Bihar, here in Delhi too Enterprise productivity was estimated higher than all other SFPs ratios even increasing trend were found than CP, IC and LP. Emoluments per workers were observed almost between 1 to 2.35 per cent per annum which shows impressive picture of the persons engaged in forest based manufacturing sector. The trends in capital intensity found impressive in most of the WFBI and Total WFBI too as well. The above discussion reveals that high magnitude of capital productivity measured at constant prices is associated with capital intensity.

In Table 5.1.5 estimates of growth in partial factor productivities ratios for the Gujarat state of Western region are presented. Table reveals that not a single industry has observed Labour Productivity below 1 per cent per annum. The highest growth in SFPs was observed for Capital Productivity for Manufacture of Musical Instruments (386) almost near to 10 per cent per annum; followed by Manufacture of wooden furniture & fixtures (276) 9.5 per cent per annum. Table further reveals that lowest growth in raw-material productivity (RMP) was observed for all categories of WFBI and it was below 1 per cent per annum while growth in fuel productivity estimated little bit impressive than RMP. On the other side, highest growth in Enterprise productivity (EP) was measured for Manufacture of cork and cork products (275) and lowest for Bidi industry (226); manufacture of wooden industrial goods as bobbins, blocks, handles sadding and similar equipment and fixtures (274) and manufacture of paper products n.e.c. (282). The growth in Enterprise Productivity (EP) and CP of Wood & Forest based Industries as a whole for the state Gujarat was found 3.74 and 3.28 per cent per annum respectively. Thus, in the state of Gujarat, Enterprise productivity was estimated higher than all other SFPs ratios with significant increasing trend of IC were found than LP. Emoluments per workers were observed between 1 to almost 2 per cent per annum which shows slighter impressive trend like Bihar and Assam state. The trends in capital intensity found impressive in most of the WFBI and Total WFBI too as well.

Table 5.1.6 presents estimates of growth in SFPs ratios for the Northern state Haryana. It is obvious from the table that labour productivity ranges between 1 to 2 per cent per annum, capital productivity ranges between 2.3 to 10.9 per cent per annum, and capital intensity ranges between 1.1 to 6.7 per cent per annum. The highest growth in SFPs was observed for Capital Productivity for as it was found for Assam and Bihar state *viz.,* Manufacture of Wooden Furniture and Fixtures (276) almost near to 11 per cent per annum; Manufacture of paper and paper board articles n.e.c. (283) above 8 per cent per annum. Similarly like the state Gujarat, lowest growth in raw-material productivity (RMP) was observed in Haryana too as well for all categories of WFBI and it was below 1 per cent per annum while growth in fuel productivity estimated little bit impressive than RMP. On the other side, highest growth in Enterprise productivity (EP) was measured for Manufacture of cork and cork products (275) and lowest for Sawing and Plaining of wood (271). The growth in Enterprise Productivity (EP) and Capital Intensity (IC) of Wood & Forest based Industries as a whole for the state Haryana was found 6.56 and 3.9 per cent per annum respectively. Thus, in the state of Haryana similarly like Bihar, Enterprise productivity was estimated higher than all other SFPs ratios with significant increasing trend of IC were found other than CP and LP. In Haryana state, Emoluments per workers were also observed between 1 to almost 2 per cent per annum which shows slighter impressive trend than Assam state. The trends in capital intensity found much impressive in most of the WFBI and Total WFBI too as well.

Growth estimates of SFPs for the Southern State Karnataka has been presented in Table 5.1.7. Table depicts that labour productivity (LP) ranges between more than 1.2 per cent and 2.41 per cent per annum, while capital productivity (CP) ranges between 2.35 to 9.4 per cent per annum, and capital intensity ranges between 1.66 to 9.12 per cent per annum. The highest growth in SFPs was observed for Enterprise Productivity (EP) for Manufacture of paper products n.e.c. (282) and it was almost near to 10 per cent per annum. Table further reveals that lowest growth in raw-material productivity (RMP) was found for all categories and it was below 1 per cent per annum except Manufacture of Bidi (226), while fuel productivity (FP) was observed impressive growth trend than RMP. The highest growth in Fuel productivity was found similarly as it was for RMP and it is Manufacture of Bidi (226). In the state Karnataka, the highest growth in SFPs for total Wood & Forest based Industries was found for Enterprise productivity (EP) and Capital Intensity (IC) and it was 4.64 and 4.216 per cent per annum, respectively. Thus, in Karnataka, capital factor have shown greater efficiency more compared to labour with significant increasing trend of CP than LP.

Table 5.1.8 presents estimates of growth in SFPs ratios for the Southern state Kerala. It is obvious from the table that labour productivity ranges between 0.72 to 1.76 per cent per annum, capital productivity ranges between 1.44 to 9.4 per cent per annum, and capital intensity ranges between almost 1 per cent and 8.2 per cent per annum. The growth rates of the LP ratios were observed as less than 1.0% per annum for industry 2 industries of WFBI. The industries belong to this category are Manufacture of turpentine, synthetic resin *etc.* (316), Manufacture of paper products n.e.c. (282). The highest growth in SFPs was observed for CP and it was for Manufacture of structural wooden goods (273) 9.41 per cent per annum; followed by Manufactures of Paper and paper board article n.e.c. (283) with 9 per cent per annum. Table further reveals that lowest growth SFPs were observed in raw-material productivity (RMP) and fuel productivity (FP). These two SFPs, measures the fuel efficiency of an industry and found very low in RMP for the industry code (316). On the other side, similarly like so many other states highest growth in SFPs for total WFBI was found for Enterprise productivity (EP). The growth in CP and IC of Wood & Forest based Industries as a whole for the southern state Kerala was found 3.17 and 3.87 per cent per annum respectively. Emoluments per workers were observed between 1.1 to almost 1.56 per cent per annum which shows not much impressive picture of the persons engaged in forest based manufacturing sector. The trends in capital intensity found impressive in most of the WFBI and Total WFBI too as well. The above discussion reveals that high magnitude of capital productivity measured at constant prices is associated with capital intensity.

Similarly like Bihar, Orissa, Assam *etc.*, Madhya Pradesh has also been relatively and industrially backward state of India. To know growth scenario of WFBI of the state M.P., ratios of single factor productivities (*viz.*, LP, CP, IC, RMP, FP, EP and EW) for the period of 1973-74 to 2012-13 are presented in Table 5.1.9. It is obvious from the table that labour productivity ranges between 1.09 to almost 2 per cent per annum, capital productivity ranges between 2.35 to 8.84 per cent per annum, and capital intensity ranges between 1.35 to 8.51 per cent per annum. The highest growth in SFPs was observed for Capital productivity (CP) in Manufacture of paper and paper board articles n.e.c. (283) almost near to 9 per cent per annum. Similarly like the other states, lowest growth in raw-material productivity (RMP) was observed in MP too as well for all categories of WFBI and it was below 1 per cent per annum while growth in fuel productivity estimated little bit impressive than RMP. On the other side, overall highest growth SFPs was found in Enterprise productivity (EP) and it was for Manufacture of turpentine...(316). The growth in Enterprise Productivity (EP) and Capital Intensity (IC) of Wood & Forest based Industries as a whole for the state Madhya Pradesh was found 4.61 and 3.99 per cent per annum respectively. Thus, in the state of Madhya Pradesh, Enterprise productivity was estimated higher than all other SFPs ratios with significant increasing trend of IC were found than CP and LP. In the state M.P., Emoluments per workers were also observed between 1.1 and 1.71 per cent per annum which shows not much impressive trend. The trends in capital intensity found much impressive in most of the WFBI and Total WFBI too as well.

In Table 5.1.10 estimates of growth in partial factor productivities ratios for the Maharashtra state of Western region are presented. Table reveals that not a single industry has observed Labour Productivity below 1 per cent per annum. Similarly like the state Gujarat, the highest growth in SFPs was observed for Enterprise Productivity (EP) for Manufacture of Cork and Cork products (275) 9.2 per cent per annum. Table further reveals that lowest growth in raw-material productivity (RMP) was observed for all categories of WFBI and it was ranges between 0.82 to 1.005 per cent per annum while growth in fuel productivity estimated little bit impressive than RMP and it was above 1 percent to below 2 percent per annum for all categories in FP. The growth in Enterprise Productivity (EP) and capital intensity (IC) for total Wood & Forest based Industries for the state Maharashtra was found 4.42 and 2.74 per cent per annum respectively. Thus, in the state of Maharashtra, Enterprise productivity was estimated higher than all other SFPs ratios with significant increasing trend of CP were found than LP. Emoluments per workers were observed between 1 to almost 2 per cent per annum which shows slighter impressive trend like Bihar than Assam state.

Industrially backward and Bimarau States is used for Bihar, Madhya Pradesh, Rajasthan, Orissa and Uttar Pradesh, which have lagged in terms of economic development despite their huge potential. In the era 21st century, to know their status and potential of industrial development especially considering WFBI, Table 5.1.11 presents estimates of growth in SFPs ratios for the Eastern state Orissa. It is obvious from the table that labour productivity (LP) ranges between almost near to 1 per cent to 2.45 per cent per annum, capital productivity ranges between

1.81 to 7.4 per cent per annum, and capital intensity ranges between 1.2 to 2.74 per cent per annum. The growth rates of the LP ratios were observed as less than 1.0% per annum only for one industry namely Manufactures of Matches (317). The highest growth in SFPs was observed for Enterprise Productivity for Manufacture of paper products n.e.c. (282) 9.2 per cent per annum; followed by Manufacture of container and boxes of paper and paper board (281) 8.1 per cent per annum. Table further reveals that lowest growth in raw-material productivity (RMP) and fuel productivity (FP) were observed. The highest growth in capital productivity (EP) was measured for Manufacture of wooden furniture and fixtures (276) and lowest for Manufacture of wood, bamboo and cane products n.e.c. (279). The growth in Enterprise Productivity (EP) and Capital Intensity (IC) for WFBI as a whole in the state Orissa was found 4.44 and 2.8 per cent per annum respectively. Thus, in the state of Orissa similarly like other eastern state Bihar, Enterprise productivity was estimated higher than all other SFPs ratios with significant increasing trend of IC were found than CP and LP. Emoluments per workers were observed between 1 to almost 2 per cent per annum which shows slighter improving trend so far as the state workers of WFBI are concerned. In Orissa too, the trends in capital intensity found impressive in most of the category of Wood & Forest based Industries and Total WFBI too as well.

In the state Punjab, diversification of industry started with the process of liberalization and economic reforms, while many of the established processing units, both in the small and medium and large sectors, came under pressure. Natural dynamism of Punjabi entrepreneurs helped it capture markets not only in the county but also abroad. The industry in the small-scale sector, by adopting a lower-level and labour-intensive technology provides employment to thousands of people in the state. To know growth scenario of WFBI of Punjab, category-wise details of ratios of SFPs for the period under study are given in Table 5.1.12. It is obvious from the table that labour productivity ranges between 0.31 to 1.87 per cent per annum, capital productivity ranges between 0.37 to more than 8.7 per cent per annum, and capital intensity ranges between 0.4 to almost 4.74 per cent per annum. The growth rates of the LP ratios were observed as less than 1.0% per annum for industry code 316. The industry belong to this category is Manufacture of turpepentine, synthetic resin *etc.* (316). For capital productivity (CP), highest growth was observed for Manufacture of wooden furniture and fixtures (276). Table further reveals that lowest growth in raw-material productivity (RMP) and fuel productivity (FP) were observed. On the other side, highest growth in Enterprise productivity (EP) was measured for Manufacture of paper products n.e.c. (282) and lowest for Manufacture of wood, bamboo and cane products n.e.c. (279). The growth in EP, IC and CP for total Wood & Forest based Industries Punjab state were observed 4.33, 3.55 and 3.14 per cent per annum respectively. During the period under study it has been noticed that Enterprise productivity was higher than all other SFPs ratios. The trend in capital intensity was also found impressive in most of the WFBI and Total WFBI too as well. The above discussion reveals that high magnitude of CP measured at constant prices is associated with IC.

Rajasthan started experiencing industrial development between 1950 and 1960. Large and small scale industries started springing up in the Kota, Jaipur,

Udaipur, Bhilwara and other Industrial Estates of Rajasthan. The main industries of Rajasthan include textile, rugs, woolen goods, vegetable oil and dyes. Heavy industries consist of copper and zinc smelting and the manufacture of railway rolling stock. The other industries related to Private Sector include steel, cement, ceramics and glass wares, electronic, leather and footwear, stone and other chemical industries. ***(www.mapsofindia.com/rajasthanindustry.htm).*** To know growth scenario of WFBI of Rajasthan, Table 5.1.13 presents estimates of SFPs ratios. Table reveals that labour productivity ranges between 1 to 2.3 per cent per annum, capital productivity ranges between 1.77 to 6.8 per cent per annum, and capital intensity ranges between almost 1.14 per cent and 4.37 per cent per annum. The lowest growth in LP ratios were observed for Manufacture of structural wooden goods (273), while highest growth in LP was found for Manufacture of wooden and cane boxes, crates, drums, barrels and other wood containers, baskets and other rattan, bamboo, reed and willow wares made entirely or mainly of cane, rattan, reed, bamboo and willow (272). The highest growth in SFPs was observed for IC and it was for Manufacture of cork and cork products (275) 9.13 per cent per annum. Table further reveals that lowest growth SFPs were observed in raw-material productivity (RMP) and fuel productivity (FP). These two SFPs, measures the fuel efficiency of an industry and found very low in RMP. On the other side, similarly like so many other states highest growth in SFPs for total WFBI was found for Enterprise productivity (EP). The growth in CP and IC in total Wood & Forest based Industries for Rajasthan was found 2.33 and 1.83 per cent per annum respectively. Emoluments per workers were observed between 0.88 to almost 2.02 per cent per annum which shows not much impressive picture of the persons engaged in forest based manufacturing sector. The trends in capital intensity also not found much impressive in most of the WFBI and Total WFBI too as well.

Growth estimates of SFPs for the Southern State Tamil Nadu has been presented in Table 5.1.14. Table depicts that labour productivity (LP) ranges between more than 1 per cent and 2.2 per cent per annum, while capital productivity (CP) ranges between 1.93 to 8.62 per cent per annum, and capital intensity ranges between 1.2 to 4.9 per cent per annum. The highest growth in SFPs was observed for Capital Productivity (CP) for Manufacture of paper and paper board articles n.e.c. (283) and it was 8.62per cent per annum. Similarly like other states, table reveals that lowest growth in raw-material productivity (RMP) was found for all categories and it was below 1 per cent per annum, while fuel productivity (FP) was observed impressive growth trend than RMP. The highest growth in Fuel productivity was found for industry code (274). On the other side, highest growth in Enterprise productivity (EP) was found for Manufacture of Sports & athletic goods and play equipments (385) and it was 8.3 per cent per annum and lowest for manufacture of wooden industrial goods as robins, blocks, handles, shading and similar equipments and fixtures (274). In the state Tamil Nadu, the highest growth was found in Capital Productivity (CP) for total Wood & Forest based Industries and it was 4.3 per cent per annum. Thus, in Tamil Nadu, capital factor have shown greater efficiency more compared to labour with significant increasing trend of IC than LP.

India's highly populated state Uttar Pradesh has also witnessed rapid industrialization in the recent past, particularly after the launch of policies of economic liberalization in the country. Uttar Pradesh is the third largest economy in India after Maharashtra and Tamil Nadu. The state has an abundance of natural resources. In 2011 the recorded forest area in the state was 16,583 km2 (6,403 sq mi) which is about 6.88% of the state's geographical area. To know growth scenario of WFBI of state Uttar Pradesh, Table 5.1.15 presents growth estimates of SFPs ratios. Table reveals that labour productivity ranges between 1.112 to 2.34 per cent per annum, capital productivity ranges between 0.75 to 1.31 per cent per annum, and capital intensity ranges between almost 0.63 per cent and 1.12 per cent per annum. The lowest growth in LP ratios were observed for Manufacture of container & boxes of paper and paper board (281), while highest growth in LP was found for Manufacture of cork and cork products (275). The highest growth in SFPs was observed for Enterprise Capital (EP) and it was the same industry which was found for LP and i.e. industry code (275). Table further reveals that lowest growth SFPs were observed in capital intensity and emoluments per worker (EW). Below 1 per cent growth in IC is very serious matter for the biggest state. On the other side, similarly like so many other states highest growth in SFPs for total WFBI was found for Enterprise productivity (EP). The growth in CP and IC in total Wood & Forest based Industries for UP state was found 1.01 and 0.86 per cent per annum respectively. Emoluments per workers were observed between 1 and 1.9 per cent per annum which show not much impressive picture of the persons engaged in forest based manufacturing sector in the state UP. The trends in capital intensity also not found much impressive in most of the category of WFBI and Total WFBI too as well.

Table 5.1.16 presents estimates of growth in SFPs ratios for the Eastern state West Bengal. It is obvious from the table that labour productivity (LP) ranges between 1.15 to 1.71 per cent per annum, capital productivity ranges between 2.1 to 7.9 per cent per annum, and capital intensity ranges between 1.23 to 4.88 per cent per annum. The growth rates of the LP ratios were observed less than 2.0% per annum for almost all 3-digit minor group industries of WFBI. The highest growth in SFPs was observed for CP for Manufacture of Wooden Furniture and Fixtures (276) 7.9 per cent per annum; followed by Manufactures of Musical Instruments (386) almost 6 per cent per annum. Table further reveals that lowest growth in raw-material productivity (RMP) was observed for sawing and plaining of wood (271) while highest for the industry Manufacture of wooden furniture and fixtures (276). On the other side, highest growth in Enterprise productivity (EP) was found for Manufactures of Musical Instruments (386). The growth in CP and IC of Wood & Forest based Industries as a whole for the Eastern state West Bengal was found 2.52 and 1.978 per cent per annum respectively. Table also reveals that in the state of West Bengal, Enterprise productivity (EP) has been observed higher growth trend in SFPs than all other SFPs ratios. Emoluments per workers were observed between 1 to almost 1.61 per cent per annum which shows not much impressive picture of the persons engaged in forest based manufacturing sector. The trends in capital intensity were also not found much impressive in most of the category of WFBI and Total WFBI too as well. The overall picture of state W.B. reveals that magnitude of capital productivity measured at constant prices although associated with capital intensity but not much impressive as compared to so many other states such as Gujarat, Haryana, Delhi, Punjab, Maharashtra.

However, Inter-state comparison of SFPs can be summarized on the basis of overall discussion of Table 5.1.1 to 5.1.16, and thus vast interpretation of these tables in terms of high and low growth of SFPs among the states for overall wood & forest based industries can be outlined as follows:

- High Labour Productivity (LP) observed for the state Delhi with 1.66% per annum.
- Low Labour Productivity (LP) observed for the state Assam with 1.25% per annum.
- High Capital Productivity (CP) observed for the state Andhra Pradesh with 5.6% p. a.
- Low Capital Productivity (CP) observed for the state Uttar Pradesh with 1.01% per annum.
- High Capital Intensity (IC) observed for the state Assam with 4.6% per annum.
- Low Capital Intensity (IC) observed for the state Uttar Pradesh with 0.86% per annum.
- High Raw-Material Productivity (RMP) observed for the state Assam with 1.17% p. a..
- Low Raw-Material Productivity (RMP) observed for the state Bihar with 0.78% p. a..
- High Fuel Productivity (FP) observed for the state Tamil Nadu with 1.67% per annum.
- Low Fuel Productivity (FP) observed for the state Assam with 1.14 % per annum.
- High Enterprise Productivity (EP) observed for the state Haryana with 6.56% per annum.
- Low Enterprise Productivity (EP) observed for the state Assam with 2.72% per annum.
- High Emoluments per worker (EW) observed for the state Delhi with 1.6% per annum.
- Low Emoluments per worker observed for the state Assam with 1.078% per annum.

On the other side, Overall discussion of Table 5.1.1 to 5.1.16 can be summarized as high and low growth of SFPs estimation among the industries within the selected states for various 3-digit minor group industries of wood & forest based industries are as follows:

- High Labour Productivity (LP) has been observed for industry code (282) of state Delhi and it was 2.57% per annum.
- Low Labour Productivity (LP) has been observed for industry code (316) of state Delhi and it was only 0.31% per annum.

- High Capital Productivity (CP) has been observed for industry code (273) of state Delhi and it was quite high 11.06% per annum.
- Low Capital Productivity (CP) has been observed for industry code (271) of state UP and it was 0.75% per annum.
- High Capital Intensity (IC) has been observed for industry code (385) of state Karnataka and it was 9.12% per annum.
- Low Capital Intensity (IC) has been observed for industry code (316) of state Punjab and it was very low only 0.37% per annum.
- High Raw-Material Productivity (RMP) has been observed for industry code (271) of state Delhi and it was 2.26% per annum.
- Low Raw-Material (RMP) has been observed for industry code (316) of state Punjab and it was also quite low i.e. only 0.15% per annum.
- High Fuel Productivity (FP) has been observed for industry code (274) of state Tamil Nadu and it was 3.11% per annum.
- Low Fuel Productivity (FP) has been observed for industry code (316) of state Punjab and it was also quite low i.e. only 0.23% per annum.
- High Enterprise Productivity (EP) has been observed for industry code (316) of state Punjab and it was 11.91% per annum.
- Low Enterprise Productivity (EP) has been observed for industry code (279) of state Assam and it was only 0.57% per annum.
- High Emoluments per worker (EW) has been observed for industry code (282) of state Delhi and it was 2.35 per cent per annum.
- Low Emoluments per worker (EW) has been observed for industry code (281) of state Assam and it was only 0.35% per annum.

Now, an attempt has been made for the region-wise assessment of the growth pattern of SFPs estimates of industry to the interstate variations in the rates of growth. In the process, it also analyses the differences in the industrial structure and factors responsible for levels of productivity and capital intensity are also studied. Table 5.1.17 presents estimates of growth in SFPs ratios for the Eastern Region of India. It is obvious from the table that labour productivity ranges between 0.334 to 1.88 per cent per annum, capital productivity ranges between 0.59 to 7.8 per cent per annum, and capital intensity ranges between 0.46 to 3.89 per cent per annum. The growth rates of the LP ratios were observed as less than 1.0% per annum for one industry of WFBI. The industries belong to this category are Manufacture of miscellaneous products n.e.c. like pencils (387). The highest growth in SFPs was observed for CP for Manufacture of Wooden Furniture and Fixtures (276) 7.8 per cent per annum; followed by Manufactures of Musical Instruments (386) 6.1 per cent per annum. The maximum growth in capital intensity was found for Manufacture of Pulp, paper and Paper board including Newsprint (280) while lowest for the same as it was found in case of LP and CP i.e. Manufacture of miscellaneous products n.e.c. like pencils (387). Table further reveals that lowest growth for all the other ratios of SFPs were observed for

Manufacture of miscellaneous products n.e.c. like pencils (387) too as well, while highest SFPs growth trend was observed for RMP in Manufacture of Wooden Furniture and Fixtures (276) and FP in Manufacture of Cork and cork products (275). On the other side, highest growth in Enterprise productivity (EP) was measured for Manufacture of Cork and cork products (275). The growth in CP and IC of Wood & Forest based Industries as a whole for the Eastern Region of India was found 3.16 and 2.97 per cent per annum respectively. Similarly like their own states Andhra and Assam, Enterprise productivity (EP) and capital productivity (CP) were estimated higher than all other SFPs ratios even increasing trend were found than LP. Emoluments per workers were observed between 0.316 to almost 1.74 per cent per annum which shows not much impressive picture of the persons engaged in forest based manufacturing sector. The trends in capital intensity found impressive in most of the WFBI and Total WFBI too as well. Thus, it can concluded that here in Eastern Region too, the above discussion of table 5.1.17 reveals that high magnitude of capital productivity measured at constant prices is associated with capital intensity.

To know growth scenario of WFBI of India's Western Region, category-wise details of ratios of SFPs for the period under study are given in Table 5.1.18. It is obvious from the table that labour productivity was found above 1 per cent per annum for each and every 3-digit minor group industry of WFBI. Capital Productivity (CP) ranges between lowest 2.35 to highest 12.05 per cent per annum, while capital intensity (IC) ranges between 1.3 to 5.6 per cent per annum. The growth rates of the LP ratios were observed least for Manufacture of Bidi (226) and maximum in Manufacture of paper products n.e.c. (282). For capital productivity (CP), highest growth was observed for Manufacture of Musical Instruments (386) and least growth in Manufacture of miscellaneous products n.e.c. like Pencils (387). Similarly like all the states, eastern region has also experienced low growth in RMP. On the other side, highest growth in Enterprise productivity (EP) was measured for Manufacture of Cork and cork products (275) and lowest for Manufacture of Bidi (226). The growth in EP, CP and IC for total Wood & Forest based Industries of Eastern Region were observed and i.e. 4.31, 2.93 and 2.77 per cent per annum respectively. The trend in capital intensity was also found impressive in most of the WFBI and Total WFBI too as well. Thus, it can be said that even in the eastern region too as well, high magnitude of CP measured at constant prices is associated with capital Intensity (IC).

Table 5.1.19 presents estimates of growth in SFPs ratios for the Northern Region of India. The table reveals that labour productivity ranges between 1 to 5 per cent per annum, capital productivity ranges between 1.64 to 27.32 per cent per annum, and capital intensity ranges between 1.21 to 14.65 per cent per annum. An interesting observation of the table emerges that in each and every category of SFP ratios (except only RMP), the highest growth was observed in Manufacture of Sports and Athletic goods and play equipments (385). While similarly like their own state Haryana, lowest growth in raw-material productivity (RMP) was observed in northern region for all categories of WFBI and total WFBI as well. The lowest growth in Enterprise productivity (EP) was measured for Industry code (316) and same lowest for productivity measures of fuel and raw-material productivities too as well. Thus, like its own state Haryana and similarly like Bihar, Enterprise productivity in northern region has also been observed higher than all other SFPs ratios with significant increasing trend of IC were found other than CP and LP. In the northern region, Emoluments per workers were also observed between above 1 per cent to 4.61 per cent per annum which shows impressive trend. The trends in capital intensity found much impressive in most of the category of WFBI and Total WFBI.

Table 5.1.20 presents estimates of growth in SFPs ratios for the Southern Region of India. It is obvious from the table that labour productivity ranges between 1.15 to below 2 per cent per annum, capital productivity ranges between 1.12 to 8.7 per cent per annum, and capital intensity ranges between 1.51 to almost 7 per cent per annum. The growth rates of the LP ratios were observed less for Manufacture of Matches while the same highest for manufacture of Sports and Athletic goods and play Equipments (385). The highest growth in SFPs were observed for CP and IC as same as it is found for LP and similarly lowest for the same as it was found in case of LP. Table further reveals that lowest growth for all the other ratios of SFPs were observed for industry code (386), while highest SFPs growth trend was observed for RMP in Manufacture of Bidi (276). On the other side, highest growth in Enterprise productivity (EP) was measured for Manufacture of paper and paper board articles n.e.c. (283). The growth in CP and IC of Wood & Forest based Industries as a whole for the Southern Region of India was found 5.7 and 1.387 per cent per annum respectively. Capital productivity (CP) followed by Enterprise productivity (EP) were estimated higher growth in SFPs than all other SFPs ratios even increasing trend were also found than LP. Emoluments per workers were observed above 1 per cent per annum but below to 2 per cent per annum which is the matter of serious concern in respect of the persons engaged in forest based manufacturing sector. The growth trend of capital intensity was also not found impressive in most of the category of WFBI and Total WFBI too as well. Thus, it can concluded on the basis of above discussion that Southern Region, reveals high magnitude of capital productivity measured at constant prices is not significantly associated with capital intensity.

Growth Estimates in partial factor productivities for Other States are presented in Table 5.1.21. This table is also reveals almost similar trend in growth of SFPs as it were found in most of the Indian States. Table reveals that LP in most of the Industries were found less than one percent which is the matter of serious concern. Only 6 industries out of 19 WFBI, have observed growth in LP more than 1 percent and industry code of these are: 226, 272, 274, 275, 279 and 282. It is also obvious from the table that the growth in CP and IC of Wood & Forest based Industries as a whole for the Other States of India was found 2.48 and 1.98 per cent per annum respectively. Enterprise productivity (EP) followed by Capital productivity (CP) were estimated higher growth in SFPs than all other SFPs ratios. Emoluments per workers were observed below to 1 per cent per annum which is also a matter of serious concern in respect of the persons engaged in forest based manufacturing sector. The growth trend of capital intensity was also not found impressive in most of the category of WFBI and Total WFBI too as well. Thus, it can concluded that Other States not performing well in terms of growth of SFPs ratios and also reveals that least magnitude of capital productivity measured at constant prices is not significantly associated with capital intensity.

In the last of this section, after discussing thoroughly SFP ratios growth pattern of inter-state, inter-region, now time is to conclude overall scenario of Wood & Forest based Industries for All India and it is presented in Table 5.1.22. Similarly like other states, this table also reveals almost similar trend in growth of SFPs. Table reveals that LP in most of the Industries were found less than one percent which is the matter of serious concern. Only 4 industries out of 19 WFBI, have observed growth in LP more than 1 percent and industry code of these are: 274, 275, 279 and 282. It is also obvious from the table that the growth in CP and IC of Wood & Forest based Industries as a whole for the All India was found 3.63 (which are higher than Other States) and 1.6 (lower than Other States) per cent per annum respectively. Capital productivity (CP) was found higher growth in SFPs than all other SFPs ratios. Emoluments per workers were also observed below to 1 per cent per annum as it was observed for the Other States. The growth trend of capital intensity was also not found impressive in most of the category of WFBI and Total WFBI too as well. Thus, it can concluded on the basis of above discussion that WFBI not performing well in terms of growth of SFPs ratios and also reveals that least magnitude of capital productivity measured at constant prices is not significantly associated with capital intensity.

However, on the basis of overall discussion of Table 5.1.17 to 5.1.21, Inter-state comparison of SFPs can be summarized and thus vast interpretation of these tables in terms of high and low growth of SFPs among the states for overall wood & forest based industries are outlined as follows:

- High Labour Productivity (LP) observed for the Northern Region with 1.52% per annum.
- Low Labour Productivity (LP) observed for the Other States with 0.73% per annum.
- High Capital Productivity (CP) observed for the Southern region with 5.73% per annum.
- Low Capital Productivity (CP) observed for the Other States with 2.48 % per annum.
- High Capital Intensity (IC) observed for the Northern region with 3.17% per annum.
- Low Capital Intensity (IC) observed for the Southern Region with 1.387% per annum.
- High Raw-Material Productivity observed for the Other States with 1.46% per annum.
- Low Raw-Material Productivity observed for the Western Region with 0.88% per annum.

- High Fuel Productivity (FP) observed for the Other States with 3.52% per annum.
- Low Fuel Productivity (FP) observed for the Eastern Region with 1.35% per annum.
- High Enterprise Productivity observed for the Northern Region with 5.09% per annum.
- Low Enterprise Productivity (EP) observed for the Other States with 1.85% per annum.
- High Emoluments per worker observed for the Northern Region with 1.44% per annum.
- Low Emoluments per worker observed for the Southern Region with 1.113% per annum.

On the other side, Overall discussion of Table 5.1.1 to 5.1.16 can be summarized as high and low growth of SFPs estimation among the industries within the selected states for various 3-digit minor group industries of wood & forest based industries are as follows:

- High Labour Productivity (LP) has been observed for industry code (378) of Northern Region and it was 5.15% per annum.
- Low Labour Productivity (LP) has been observed for industry code (276) of Other States and it was only 0.13% per annum.
- High Capital Productivity (CP) has been observed for industry code (385) of Northern Region and it was quite high 27.31% per annum.
- Low Capital Productivity (CP) has been observed for industry code (387) of Other States and it was 0.59% per annum.
- High Capital Intensity (IC) has been observed for industry code (385) of Northern Region and it was 18.65% per annum.
- Low Capital Intensity (IC) has been observed for industry code (387) of Other States and it was very low only 0.18% per annum.
- High Raw-Material Productivity (RMP) has been observed for industry code (385) of Northern Region and it was 3.23 % per annum.
- Low Raw-Material (RMP) has been observed for industry code (387) of Eastern Region and it was also quite low i.e. only 0.225% per annum.
- High Fuel Productivity (FP) has been observed for industry code (385) of Northern Region and it was 5.2% per annum.
- Low Fuel Productivity (FP) has been observed for industry code (387) of Eastern Region and it was also quite low i.e. only 0.36% per annum.
- High Enterprise Productivity (EP) has been observed for industry code (385) of Northern Region and it was 24.51% per annum.
- Low Enterprise Productivity (EP) has been observed for industry code (270) of Other States and it was only 0.77% per annum.

- High Emoluments per worker (EW) has been observed for industry code (385) of Northern Region and it was 4.61% per annum.
- Low Emoluments per worker (EW) has been observed for industry code (387) of Eastern States and it was only 0.32% per annum.

Overall conclusion of all the above tables (i.e. Table 5.1.1 to 5.1.22) can be summarized that in all the cases labour productivity have increased at a rate of about 1.0% per annum. The capital intensity has shown similar trends ranging from 0 % to 2.0% per annum. However, Capital productivity have shown impressive trends and growth rates were increasing between 0.0% to more than 11 per cent per annum for most of the industries of WFBI. Thus, it looks that increase in capital is resulting in technological progress in most of the industries of WFBI. This is quite natural since most of the economic reform measure aims at promoting use of capital for most of the manufacturing industries of India. The growth observed in case of these productivity ratios can't be treated in isolation with their relationship with other ratios. As discussed earlier high growth rate of capital productivity in most of cases can't be attributed nearly as an improvement in quality and efficiency of capital factor but quality and improvement in human labour and the technological specification may also be responsible for this growth. Most of the industries are labour intensive and employ large number of work force, therefore, role of latent potential of labour can't be ruled out.

A study of Singh (1987) on Punjab manufacturing for the period of 1967-82, found the similar movement in the three ratios and identified increased wage rate (w) as one of the major factors of increased labour productivity. A study conducted by Dhillon (1983) on the manufacturing sector in Karnataka for the period of 1968-77, found a decline in the two productivity ratios accompanied by significant increases in capital intensity. On the other hand, Gupta (1985) in case of Maharashtra manufacturing found an increase in capital productivity accompanied by a decline in both labour productivity and capital intensity during 1969-76, another study by Mujoo, R. (1991) found that the trend growth rate of capital intensity is not much above the trend growth rate of labour productivity. This is accompanied by only slight (insignificant) fall in capital productivity. She also found a highly significant correlation of 0.94 between labour productivity and capital intensity along with a slightly weak but significant correlation of 0.53 between labour productivity and capital productivity pointing towards efficiency in input use or technical progress in the organized industrial sector. Her study further indicates that a significant trend growth of wage rate is accompanied by an insignificant decrease in rate of returns (which simply measured surplus obtained after deducting wage bill from value added, per unit of capital). High wage rate in the organized manufacturing sector might be one of the important reasons for increased capital intensity during the period of the study.

5.2 Estimates of Total Factor Productivity Growth (TFPG)

As it is previously discussed in the Chapter-II that Total factor productivity index or the index of technical progress as it often referred, gives us an idea of overall efficiency in the factor use. Simply, this type of multi-factor productivity index measures the output per unit of capital and labour combined. The level of TFP is thus, measurable by diving total output by total factor of production (i.e. total inputs). TFP growth rates refers to the amount of growth in real output that is not explained by the growth in their inputs. In other words, it can be said that TFP index can be computed as the ratio of an index of aggregate output to an aggregate inputs (Kathuria et.al, 2013). In this study, for empirical purposes, we estimated 3 types of TFPG indices (i.e. Kendrick, Solow and Divisia Index) for the period 1973-74 to 2012-13 to provide estimates as comparable and consistent across the States and Regions as possible.

The measurement of TFP is beset with source well-know conceptual and empirical problem and also involves several restrictive assumptions, such as static competitive equilibrium and constant returns to scale *etc.* Of late years, the methodology of TFP measurement has advanced remarkably by incorporating the recent advances in production and cost theory and in index number theory and practice. With the time, new developments a rigorous theoretical frame work has evolved and new indexes of total factor productivity have been devised which require less rigid assumptions about the production technology and the nature of technical progress. But, still there many important questions remain regarding the methodology of its measurement still to be resolved adequately. Notwithstanding its measurement problems, the concept of total factor productivity has proved extremely useful in empirical work and a consequence has found favour widely (Pendse & Baghel, 2008).

Studies on total factor productivity acquire great significance in the context of growth in developing economies like India. These economies are characterized by acute, scarcity of resources (particularly capital and must use the available resources as best as they can. Also, generation of surplus, which plays a pivotal role in their growth, depends crucially on the efficiency with which resources are used. Evidently, the various studies B.N. Goldar, P.R. Bramhananda, I.J. Ahluwalia, Rajlakshmi, Mehta, Subramaniyan, Hasim & Dadi, Balakrishnan & Pushpangadan, Dholakia & Dholakia, Raj R.S.N., Unel B., Krishna & Mitra, Kumar, Pendse & Baghel, Kathuria & Sen *etc.* on productivity trends in Indian manufacturing industry would be of much interest, as it brought out how effective the investments made during the plans (in machinery, human capital and infrastructure) were in raising in overall efficiency in resource use.

Total factor productivity index or the index of technical progress as it often referred, gives us an idea of overall efficiency in the factor use. This index measures the output per unit of capital and labour combined. For empirical purposes, we estimated Kendrick, Solow and Divisia indices of total factor productivity for manufacturing sector of India for the period under study.

Studies on total factor productivity acquire great significance in context of growth in developing economics. These economies are characterized by acute scarcity of resources (particularly capital) and must use the available resources as best as they can. Also, generation of surplus, which plays a pivotal role in their growth, depends crucially on the efficiency with which resources are used. Evidently, the study (Goldar, 1983) on productivity trends in Indian manufacturing industry would be of much interest, as it brought out how effective the investments made during the plans (in machinery, human capital and infrastructure) were in raising in overall efficiency in resource use. Estimates of all three TFPG indices Kendrick, Solow and Divisia index) has been used in this study and is computed with the help of procedures and formulae as already discussed in detailed in the Research Methodological Chapter separately (i.e. Chapter-IV).

The estimates of total factor productivity growth indices for all selected States are presented separately in Table 5.2.1 to 5.2.16. Table 5.2.1 presents estimates of growth in TFPG indices for the state Andhra Pradesh. It is obvious from the table that most of the industries in Andhra, gained considerably in TFP during the period under study. In manufacture of structural wooden goods (272) industry, TFP grew the fastest, at a rate of 5.4 per cent per annum. Two industries namely manufacture of paper and paper board articles n.e.c. (283) and manufacture of refined petroleum products n.e.c. (316) significantly reported negative growth in the period under study. While Manufacture of matches (317) and manufacture of stationary articles n.e.c. (387) have registered TFP decline in study period. The trend growth rates for Kendrick index are found positive and significant at 1% level of significance but below 1 per cent per annum for industries 280, 276, 270, 226, 279, 271 code and total wood &forest based industries of state as well. If we consider Solow index, TFP estimates present somehow quite similar trend. In Solow measure, 9 industries have experienced significant but negative trend in TFP growth and Industry code of these industries are: 270, 271, 275, 279, 280, 281, 282, 283 and 386. In all 3 types TFP indices, Manufacture of Structural Wooden goods (272), manufacture of Bidi (226) and manufacture of wooden and cane boxes, crates etc (273) have reported positive growth performance with 1 per cent level of significance. The Kendrick and Divisia Index found positive and significant at 1 per cent level of significance for total wood & forest based industries in the state Andhra Pradesh with an increase of 0.88 and 0.9 per cent per annum respectively while Solow index for the same was recorded insignificant negative trend. Thus, overall analysis of table reveals that there was significant gain in TFP so far as the 3 industries and overall wood and forest based industries of Andhra Pradesh are concerned.

Table 5.2.2 presents estimates of growth in TFPG indices for the state Assam. Table reveals that few industries in Assam, gained considerably in TFP during the period under study. The trend growth of Kendrick index, in Manufacture of wooden and cane boxes, crates, drums, barrels *etc.* (273) and manufacture of paper and paper board articles and pulp products n.e.c. (282) industries, TFP grew the fastest, at a rate of 1.6 and 1.5 per cent per annum respectively. Three industries namely manufacture of structural wooden goods (272), manufacture of products of wood, bamboo cane, reed and grass (279) and manufacture of container and boxes of paper and paper board (281) significantly reported negative growth in the period under study. While Manufacture of matches (317) and manufacture of refined petroleum products n.e.c. (316) have registered TFP decline in study period. If we consider Solow index, TFP estimates present somewhat different trend. In Solow Index, 7 industries have experienced significant but negative trend in TFP growth and Industry code of these industries are: 270, 271, 273, 279, 281, 316 and 317. In all 3 types TFP indices, Manufacture of paper and paper board articles (282), manufacture of wooden and cane boxes, crates etc (273) and manufacture of pulp, paper and paper board (280) have reported positive growth performance with 1 per cent level of significance. For total wood & forest based industries in the state Assam, the Kendrick and Divisia Index found positive and significant at 1 per cent level of significance with an increase of 0.215 and 0.687 per cent per annum respectively while Solow index for the same was recorded negative trend. Thus, overall analysis of table reveals that there was not significant gain in TFP so far as the most of industries and overall wood and forest based industries of Assam are concerned.

For the state Bihar, estimates of growth in TFPG indices are presented in Table 5.2.3. Table reveals that most of the WFBI industries in Bihar, gained considerably in all TFPG indices during the period under study. One industry (i.e. Industry code 273) in Kendrick index and 2 industries (industry code 316 & 274) in Solow index have significantly reported negative growth in the period under study. While Manufacture of matches (317) and manufacture of refined petroleum products n.e.c. (316) in Kendrick and Divisia indices and manufacture of matches in Divisia index have registered TFP decline in study period. For total wood & forest based industries in the state Bihar, the Kendrick, Solow and Divisia Index found positive and significant with an increase of 0.614, 0.755 and 0.975 per cent per annum respectively. Thus, it can be concluded that in the state Bihar, there was significant gain in TFP in most of industries and total wood and forest based industries.

Table 5.2.4 presents estimates of growth in TFPG indices for the state Delhi. It is obvious from the table that in majority of the industries of WFBI in all types of TFPG indices gained considerably and found positive and significant at 1 per cent level of significance. Even, total WFBI too as well have experienced highly significant at 1% level positive trend in TFPG growth with an increase more than one per cent per annum. In manufacture of cork and cork products (275) industry, TFP grew the fastest, at a rate of 4.1 per cent per annum in the study period. Overall scenario of TFPG growth trend of state Delhi is quite impressive so far as the industries of WFBI are concerned. On the other side, almost similar feature have been observed in the state of Gujarat as well and their estimates of growth trend of TFPG are presented in Table 5.2.5. It is obvious from the table that similarly like the state Delhi, most of industries in Gujarat state TFPG indices have reported positive and significant at 1 per cent level of significance. Industry experiences TFP growth more than one percent per annum in all types of indices are the manufacture of Paper and paper board articles n.e.c. (283); manufacture of paper and paper board articles and pulp products n.e.c. (282) and manufacture of containers and boxes of paper and paper board (281). Thus, similarly like the state Andhra Pradesh, and Bihar, the state Delhi has also experiences significant gain in TFP in most of industries and total wood and forest based industries.

Table 5.2.6 presents estimates of growth in TFPG indices for the Haryana state. It is obvious from the table that the trend growth of Kendrick, Solow and Divisia index, in 12 out of 19 industries of WFBI gained considerably and found positive and significant at 1 per cent level of significance. Even, total WFBI too as well have experienced highly significant at 1% level positive trend in TFPG growth with an increase of 1.39, 1.52 and 1.71 per cent per annum respectively for Kendrick, solow and Divisia index. The trend growth rates for Kendrick and Solow index are found positive and significant at 5% but below the one per cent per annum for manufacture of wooden furniture and fixtures (276) while for Divisia index the same level of significant positive growth for manufacture of cork and cork products (275). Similarly like Delhi, manufacture of cork and cork products (275) industry in Solow TFP index grew the fastest, at a rate of 4 per cent per annum in the study period. It can be seen from table that TFPG growth trend of state Haryana state also observed quite impressive and positive growth performance so far as the industries of WFBI are concerned.

Table 5.2.7 presents estimates of growth in TFPG indices for the southern state Karnataka. Table reveals that few industries in the state, gained considerably in TFP during the period under study. The trend growth of Kendrick index, in Manufacture of structural wooden goods such as beems, posts, doors & windows (272) and manufacture of containers and boxes of paper and paper board (281) industries, TFP grew the fastest, at a rate of 3.77 and 3 per cent per annum respectively. If we consider Solow index, TFP estimates present somewhat different trend. In Solow Index, 5 industries have experienced either negative or insignificant trend in TFP growth and Industry code of these industries are: 270, 273, 276, 316, 317 and 387. In all 3 types TFP indices, Manufacture of structural wooden goods such as beems, posts, doors & windows (272); manufacture of containers and boxes of paper and paper board (281); manufacture of paper and paper board articles (283); manufacture of paper and paper board articles and pulp products n.e.c. (282); and manufacture of vaneer sheets, plywood and their products (271) have reported more than one per cent per annum positive growth trend with 1 per cent level of significance. Not only this, even for total wood & forest based industries in the state Karnataka, the Kendrick, Solow and Divisia Index found positive and significant at 1 per cent level of significance with an increase of 1.176, 0.87 and 0.94 per cent per annum respectively. Thus, overall analysis of table reveals that there was significant gain in TFP so far as the most of industries and overall wood and forest based industries of Karnataka state are concerned.

For the state Kerala, estimates of growth in TFPG indices are presented in Table 5.2.8. Table reveals that most of the WFBI industries in the state, gained considerably in all TFPG indices during the period under study. One industry (i.e. Industry code 316) in Kendrick and Divisia index and 4 industries (industry code 271, 276, 316 & 387) in Solow index have reported negative growth in the period under study. In manufacture of wooden industrial goods n.e.c. (274) industry, TFP grew the fastest in all indices, at a rate of 2.735, 2 and 1.82 per cent per annum respectively in the study period while industry code 316 have experiences insignificant and declining trend for all indices. For total wood & forest based industries in the state Kerala, the Kendrick, Solow and Divisia Index found positive and significant with an increase of 1.13, 0.77 and 0.62 per cent per annum respectively. Thus, it can be concluded that in the state Kerala, there was significant gain in TFP in most of industries and total wood and forest based industries.

Table 5.2.9 presents estimates of growth in TFPG indices for the state Madhya Pradesh. It is obvious from the table that each and every industries of WFBI in all types of TFPG indices gained considerably and found positive and significant. Even, total WFBI too as well have experienced highly significant at 1% level positive trend in TFPG growth with an increase more than one per cent per annum. TFPG trend for industry code 316, have observed more than 2 per cent per annum growth. Not a single Wood & Forest based Industries of MP have observed negative or insignificant trend in the period of study. Overall scenario of TFPG growth trend of state reveals quite impressive and experiencing accelerated growth so far as the industries of WFBI are concerned. On the other side, almost similar feature have been observed in the state of Maharashtra as well and their estimates of growth trend of TFPG are presented in Table 5.2.10. It is obvious from the table that most of industries in the state TFPG indices have reported positive and significant at 1 per cent level of significance. 10 Industries experiences TFP growth more than one percent per annum in Kendrick and Solow index, and 11 in Divisia index. The trend growth rates for Kendrick index found positive but below the one per cent per annum for the industry code 387, 270, 272, 279, 276, 280, for Solow index, industries belongs to 270, 275, 273, 276, 385, 387 while for Divisia index the same level of significant positive growth for industry code 272, 273, 275, and 279. Manufacture of paper and paper board article n.e.c. (283), Manufacture of paper and paper board articles and pulp products (282) and manufacture of petroleum product n.e.c. (316) industry have observed more than 2 per cent per annum growth for the all types of TFPG indices while Solow TFP index grew the fastest, at a rate of 4.4 per cent per annum in the study period. Thus, similarly like the state Andhra Pradesh, Bihar and Delhi, Gujarat, Haryana and MP, state Maharashtra have also experiences significant gain in TFP in most of industries and total wood and forest based industries.

Growth trend of TFPG for the state Orissa are presented in Table 5.2.11. Here in this state too, TFPG indices in most of industries have reported positive and significant at 1 per cent level of significance. 7 Industries experiences TFP growth more than one percent per annum in Kendrick and Solow index, and 8 in Divisia index. The trend growth rates for Kendrick index found positive but below the one per cent per annum for the industry code 226, 274, 276, 280, 316 and for 317; for Solow index, industries belongs to 270, 274, and for 273; while for Divisia index the same level of significant positive growth for industry code 273, 274, 279 and 316. Negative growth trend was observed in Solow index for the industry code 276, 280, 316 and for 317; while the same in Divisia index was found for industry code 270 only. Thus, state Orissa have experiences significant gain in TFP in majority of industries and total wood and forest based industries.

Table 5.2.12 presents estimates of growth in TFPG indices for the **Punjab** state. It is obvious from the table that the trend growth of Kendrick, Solow and Divisia index, in 11 out of 19 industries of WFBI gained considerably and found positive and significant at 1 per cent level of significance. Even, total WFBI too as well have also experienced highly significant at 1% level positive trend in TFPG growth with an increase of 1.1, 1.08 and 0.99 per cent per annum respectively for Kendrick, Solow and Divisia index. It can be seen from table that TFPG growth trend of Punjab state also observed positive growth performance so far as the industries of WFBI are concerned.

For the state Rajasthan, estimates of growth in TFPG indices are presented in Table 5.2.13. Table reveals that majority of the WFBI industries in the state, gained positive trend in all TFPG indices during the period under study but below one per cent per annum growth were observed in most of the cases. Only one industry i.e. manufacture of structural wooden goods such as beams, posts, doors & windows (272) have experienced more than one percent TFPG growth in all types of Index. 9 industry in Kendrick and Divisia index and 4 industries in Solow index have reported significant positive growth in the period under study but below the one percent per annum. Even total Wood and forest based Industries of Rajasthan have also experiences below one per cent growth in TFPG. Thus, it can be concluded that in the state Rajasthan, although, there was significant gain in TFP in most of industries and total wood and forest based industries too as well but not with high magnitude and impressive in the perspective of technological change.

Table 5.2.14 presents estimates of growth in TFPG indices for the southern state Tamil Nadu Pradesh. It is obvious from the table that except 4 industries, every industries of WFBI in all types of TFPG indices gained considerably and found positive and significant. Insignificant trend in Kendrick Index were observed for the industry code 316, 317, 378 and total WFBI which confirm that these industries are not experiencing technological change even in the era of technology and industrial growth. In the state of Tamil Nadu, 9 industries are experiencing more than one per cent per annum growth and gained considerably in all TFPG indices during the period under study.

Table 5.2.15 presents estimates of growth in TFPG indices for the state Uttar Pradesh. It is obvious from the table that most of the industries of WFBI in all types of TFPG indices gained considerably and found positive and significant. Even, total WFBI too as well have also experienced significant at 1% level positive trend in TFPG growth but below one per cent per annum of increase. The trend growth of Kendrick and Divisia index, in Manufacture of Bidi (226) industry, TFP grew the fastest, at a rate of 3.47 and 2.24 per cent per annum respectively. 2 categories (i.e. Industry code 316 and 317) of Wood & Forest based Industries of state UP have observed insignificant trend in the period of study, while 5 industries (Industry code 273, 274, 282, 316, 385 and 387 have experiences significantly negative trend in growth of Solow Index. Overall scenario of TFPG growth trend reveals that state Uttar Pradesh although experiencing positive trend in TFPG growth in most of cases but not accelerated growth considerably.

The estimates of total factor productivity growth indices for West Bengal are presented in Table 5.2.16. The trend growth rates are found positive and significant at 1% level of significance for 12 industries, 9 industries and 10 industries respectively for Kendrick, Solow and Divisia index. Two industries (industry code 272 & 316) in Kendrick index; seven industries (industry code 272, 276, 281, 283, 316, 386 and 387) in Solow Index and only one industry (272) in Divisia index are experiencing either negative or insignificant growth for the study period. On the other side, not a single industry experiencing more than one per cent per annum growth in all 3 TFP but more than 1 per cent trend growth for Kendrick index are found for 7 industries (Industry code 271, 273, 274, 275, 281, 282 and 283), while 2 per cent per annum growth in Kendrick Index are found for 2 industries namely Manufacture of Wooden Furniture and Fixtures (276) and Manufacture of Musical Instruments, excl. Toys (386). For Solow Index, 2% p.a. growth was found for manufacture of cork and cork products (275). It can be seen from table that TFPG growth trend of West Bengal state observed positive growth performance so far as the industries of WFBI are concerned.

Overall observation of above discussion reveals somewhat zigzag growth pattern of TFPG so far as the 16 states are concerned. Although some states are experiencing technological change with positive growth in TFPG but can't be associated in the level of high magnitude. After analyzing TFP growth trend in various states, an attempt has also been made for an assessment of the contribution of various categories of WFBI to the interstate variations in the rates of TFP growth. In the process it also analyses the differences within the industry and regions too as well. Thus, inter-state variations and trends at regional levels in the total factor productivity growth have been studied as the earlier not a single studies have made such exercises in India and very few studies have studied for few industries comes under WFBI and even only for a short period.

To begin with it is appropriate to deal with trends in total factor productivities growth indices *viz.* Kendrick, Solow and Divisia index for the industries at three-digit level for the Wood & Forest based industries under study. Table 5.2.17 presents estimates of growth in TFPG indices for the Eastern Region of India. It is obvious from the table that some industries in eastern region, gained considerably in TFP during the period under study. While 4 industries (Industry code 270, 272, 316, and 317), have registered TFP decline in study period. The trend growth rates for Kendrick index are found positive and significant but below 1 per cent per annum for industries 226, 273, 276, 279, 280 and 386 industry code and total wood & forest based industries of state as well. If we consider Solow index, TFP estimates present somehow quite different trend. In Solow measure, 4 industries have experienced significant but negative trend in TFP growth and Industry code of these industries are: 276, 316,386 and 387. In all 3 types TFP indices, manufacture of paper and paper board articles and pulp products (282) manufacture of Bidi (226), manufacture of vaneer sheets, plywood and their products (271), manufacture of wooden and cane boxes, crates etc (273), manufacture of cork and cork products (275), manufacture of containers and boxes of paper and paper board (281), and manufacture of paper and paper board articles n.e.c. (283) have reported significant positive growth performance either at 1% or 5% level of significance. The Solow and Divisia Index were found positive and significant at 1 per cent level of significance for total wood & forest based industries in eastern region with an increase of 0.7 and 0.9 per cent per annum respectively while Kendrick index for the same was recorded significant at 5% level. Thus, overall analysis of table reveals that there was significant gain in TFP so far as the most of eastern region WFB industries and total WFBI are concerned.

Table 5.2.18 presents estimates of growth in TFPG indices for the Western Region. It is obvious from the table that the trend growth of Kendrick, Solow and Divisia index, in 15 out of 19 industries of WFBI gained considerably and found positive and significant either at 1% or 5% or 10% level of significance. Even, total WFBI too as well have experienced highly significant at 1% level positive trend in TFPG growth with an increase of 1.29, 0.74 and 1.2 per cent per annum respectively for Kendrick, solow and Divisia index. The trend growth rates 11 industries for Kendrick have recorded more than one percent positive trend significantly but below one per cent per annum TFP growth were observed for 5 manufacturing unit of WFBI. Manufacture of Musical Instruments excl. Toys (386) and manufacture of cork and cork products (275) industry in Kendrick TFP index grew the fastest, at a rate of almost near to 3 per cent per annum in the study period. For Solow Index, 4 Industries (Industry code 275, 276, 316 and 387) have experienced negative growth trend in TFP, while negative and insignificant trend was observed for industry code 387. It can be seen from table that TFPG growth trend of Western region found quite impressive and positive growth performance so far as the industries of WFBI are concerned.

Table 5.2.19 presents estimates of growth in TFPG indices for the Northern region. It is obvious from the table that most of the industries of WFBI in all types of TFPG indices gained considerably and found positive and significant. Even, total WFBI too as well have also experienced significant at 1% level positive trend in TFPG growth. The trend growth of Kendrick and Divisia index, in Manufacture of Bidi (226) industry, TFP grew the fastest, at a rate of 3.2, 2.14 and 2.54 per cent per annum respectively. The trend growth rates are found insignificant for manufacture of matches (317) for Kendrick and Divisia indices while 2 categories (i.e. Industry code 386 and 270) of Wood & Forest based Industries of northern region have observed insignificant trend for the Solow index. Table further reveals that total WFBI have experienced significant at 1% level positive trend in TFPG growth with an increase of 1.06, 1.1 and 0.98 per cent per annum respectively for Kendrick, Solow and Divisia index. Thus, overall scenario of TFPG growth trend reveals that Northern region although experiencing positive trend in TFPG growth in most of industries and total Wood & Forest based Industries as well.

Growth trend of TFP indices for Southern Region are presented in Table 5.2.20. Here in this region too, TFPG indices in most of industries have reported positive and significant at 1 per cent level of significance. 9 Industries experiences TFP growth more than one percent per annum in Kendrick, 4 in Solow index, and 8 in

Divisia index. The trend growth rates for Kendrick index found positive but below the one per cent per annum for the industry code 226, 270, 276, 279, 280, 283 and for 387; for Solow index, industries belongs to 226, 271, 273, 274, 280, 281, 282, 316 and for 385; while for Divisia index the same level of significant positive growth for industry code 226, 273, 275, 279 and 385. Negative or insignificant growth trend was observed in Kendrick index for industry code 378; in Solow index, for the industry code 270, 317, 378 and for 387; while the same in Divisia index was found for industry code 316, 317, 378 and for 387. Thus, it can be said that most of the industries are experiencing either positive trend in growth of TFP but below the one per cent per annum, or found insignificant trend growth. Although, TFP growth for total wood and forest based industries have observed positive and significant with one per cent per annum growth, but not with high magnitude.

For the Other States, estimates of growth in TFPG indices are presented in Table 5.2.21. Table reveals that majority of the WFBI industries in the Other states, gained positive trend in all TFPG indices during the period under study but below one per cent per annum growth were observed in most of the cases. Only Three industries i.e. sawing and plaining of Wood (270), Manufacture of wooden industrial goods n.e.c. (275) and manufacture of veneer sheets, plywood and their products (271) have experienced more than one percent TFPG growth in all types of Indices. Out of 19, 15 industries in Divisia index, 12 industries in Kendrick, and 4 industries in Solow index have reported significant positive growth in the period under study but below the one percent per annum. 3 industries (Industry Code 276, 386 and 387) observed negative trend in growth of TFP. Thus, it can be concluded that in the Other States, although, there was significant gain in TFP in most of industries and total wood and forest based industries too as well but not with high magnitude and impressive in the perspective of technological change.

At last, the estimates of total factor productivity growth indices for the country as a whole (i.e. for All India) are presented in Table 5.2.22. The trend growth rates are found positive and significant at 1% level of significance for 15 industries, 17 industries and 16 industries respectively for Kendrick, Solow and Divisia index. 4 industries (industry code 316,317,378 & 387) in Kendrick index; 2 industries (industry code 316 and 387) in Solow Index and 3 industries (Industry code 316, 378 and 387) in Divisia index have experiences insignificant growth for the study period. On the other side, only 3 industries are experiencing more than one per cent per annum growth in all 3 TFP. More than 1 per cent trend growth for Kendrick index are found for 7 industries (Industry code 273, 274, 275, 281, 282, 283, 385 and 386), while 2 per cent per annum growth in Kendrick Index are found for only one industry group e.g. manufacture of Musical Instruments excl. Toys (386). For Solow Index, more than 3.6% p.a. growth was found for manufacture of cork and cork products (275) while in case of Divisia index not a single industry have experiences more than 2 or 3 per cent per annum growth in TFP. For the Kendrick and Divisia Indices, 4 industries (Industry Code 316, 317, 378 and 386) and for Solow index, 3 industries (Industry code 316 & 387) have observed insignificant trend in growth of TFP. It can be seen from table that TFPG growth trend of All

India, observed positive growth performance so far as the industries of WFBI under study are concerned.

Comparing the results of this study, we have not found any study which has studied in the context of Wood and forest based Industries for Inter-state or Inter-Region comparison. But for the comparison, we find that some of the earlier studies in the context of Indian manufacturing found a sluggish growth rate in total factor productivity (Singh, 1968). Against these, a steady decline in total factor productivity is observed in studies of Raj Krishna and Mehta (1968) and Benerji (1975) for the time period of 1946-64 and also that of Reddy and Rao (1962) for the period 1946-57. In a study, Alhuwalia (1985) noticed a significant decline in both Solow and Translog indices of total factor productivity for Indian manufacturing during the period of 1959-80. Besides these, many studies (Singh, 1987), Dillon (1983), Gupta (1985) on the manufacturing sector of different states have also found the similar trend in total factor productivity.

As against these studies, Hasim and Dadi (1973) in their study observed an increase of 2.8 per cent in Solow index with technical progress contributing about 50 per cent to output growth during 1946-64. Similarly Goldar's (1986) study covering the period (1951-65) witnessed positive growth rates in Kendrick, Solow and Translog indices, although the contribution of total factor productivity to output growth (21-24 percent) was much less as compared to the previous study.

Some studies have found new industries to be more efficient, while some other have identified traditional industries qualifying for higher efficiency. These are some studies, which have attempted to measure total factor productivity of individual industries on all India level, and a few are found to have dealt with certain industries at the State-level. In case of the former Sinha and Sawhney (1970) observed increased efficiency during the 1950-63 in respect of five industries, namely cotton textiles, jute textiles, sugar, cement, and paper and paper board. Banerji (1975) in his study did not found any appreciable increase in the overall efficiency; although the performance of sugar and bicycles was better then that of cotton, jute and paper industries. So far as Solow index and partial factor productivities are concerned, the technical progress and economics of scale were identified as factors responsible for the better performance of the first two industries.

Further, Mehta (1980) studied twenty-seven comparable ASI and CMI industries for the period of 1953-65. Based on the analysis, he found that there was an overall decline in efficiency. However, in certain industries like biscuit making, vegetable oils, tanning, glass and glassware he noticed increase in the total factor productivity. A significant decline was observed in sugar, cement, paper and paperboard, matches, iron and steal, cotton textiles and ceramics. On the whole traditional industries have shown a decline in efficiency, most of the growth being attributed to capital deepening. An overall decline in efficiency was found in most of the industries studied by Ahluwalia (1985) for the time period of 1959-80. A significant decline was found in rubber products, miscellaneous industries, food manufacturing, wood and cork, metal products, leather and for products.

Two industries (i. e. footwear and furniture and fixtures) experienced significant growth in the total factor productivity (Solow and Translog indices). Goldar (1986) in his study for the period 1960-70 noticed a significant trend growth rate in Kendrick index in tobacco manufacture, paper and paper products, and leather and fur. On the other hand, a significant decline was experienced in respect of food products, rubber products, petroleum products and basis metals. Industries with productivity losses outweigh the industries with productivity grains.

Slow growth of productivity for any industry is basically, the result of complex factors operating on the industry, some of which may be industry-specific while others relates to the overall industrial environment or, indeed, the overall economic environment. It is not the intension of this study to present an exact, quantitative and exhaustive account of the factors that have influenced the growth of productivity in each of the individual industry groups of manufacturing.

Broad-ranging constraints on productivity growth are many. For example, constraints on productivity growth may arise from supply–side factors, e.g. infrastructure. No amount of micro-level planning on productivity can succeed if there are shortage or transport, power and telecommunications. Similarly, shortages of critical intermediate inputs may arise due to an overall foreign exchange constraint and may adversely affect productivity. On the demand side, slow growth of incomes or slow growth of investment may limit the size of the domestic market. At the same time, there are other important factors, which very across firms, and perhaps also industries, which we cannot measure, Management input, and the quality of industrial organization are the obvious examples of such factors. Data availability imposes serious constraints on our experimentation in this direction. It is not possible to quantify these factors within an econometric framework for the large cross- section of manufacturing industries needed for the present analysis. Study by Kumar (1983) observed a decline in both partial productivity and total factor productivity in cotton, jute, iron and steel excepting the increase in sugar industry. However, a significant capital deepening is observed in all the four industries.

Inter-States analysis of individual industries has been carried out by Subramaniyan (1979) and Rajlakshmi (1983). Subramaniyan analysed the trends of total factor productivity in sugar industry in selected states, namely, Uttar Pradesh, Bihar, Maharashtra, Tamil Nadu and Andra Pradesh for the period 1953-69. Labour productivity showed a significant increase in all the states, whereas, capital productivity experienced a decline. Much increase in labour productivity was attributed to significant capital deepening experienced by this industry in almost all the states analyzed. Excepting Tamil Nadu, the total factor productivity (Kendrick and CES index) showed a general decline. The study of Rajalakhmi found significant growth of total factor productivity (Divisia Index) in electrical machinery in Karnataka and Maharashtra, while the index experienced the decline in west Bengal.

Table (5.1.1): Estimates of Growth in Partial Factor Productivity Ratios

State: Andhra Pradesh

Industry Code	Growth in SFPs						
	LP	CP	IC	RMP	FP	EP	EW
226	1.101	5.365	1.063	1.102	1.881	1.856	1.063
270	1.333	6.54	2.756	0.981	1.359	4.672	1.23
271	1.411	5.848	2.199	0.916	1.442	4.248	1.317
272	1.108	4.281	1.559	0.758	1.415	3.157	1.297
273	1.03	5.671	0.792	1.444	0.889	0.787	0.722
274	1.257	4.359	1.226	0.834	1.321	4.277	1.476
275	DNF	DNF	DNF	DNF	DNF	DNF	DNF
276	1.294	9.061	1.966	0.918	1.747	2.653	1.218
279	1.844	3.882	2.738	1.844	1.714	1.432	1.324
280	1.159	5.622	3.362	1.151	1.023	0.952	1.114
281	1.586	6.659	2.992	0.89	1.399	4.89	1.333
282	1.305	4.262	0.665	1.361	0.838	1.043	0.771
283	1.061	14.55	4.904	0.617	0.96	5.229	1.298
316	0.979	1.965	0.956	0.886	0.698	0.789	1.283
317	0.983	3.612	2.644	0.74	1.314	1.667	1.363
378	DNF	DNF	DNF	DNF	DNF	DNF	DNF
385	DNF	DNF	DNF	DNF	DNF	DNF	DNF
386	DNF	DNF	DNF	DNF	DNF	DNF	DNF
387	1.377	2.346	1.82	0.909	1.976	3.547	1.225
Total WFBI	1.255	5.602	2.11	1.023	1.34	2.774	1.237

Note: **LP:** Labour Productivity, **CP:** Capital Productivity,
IC: Capital Intensity, **RMP:** Raw-Material Productivity,
FP: Fuel Productivity, **EP:** Enterprise Productivity,
EW: Emoluments per Worker.
DNF: means data not found for all the years.
WFBI: Wood & Forest based Industries

Table (5.1.2): Estimates of Growth in Partial Factor Productivity Ratios

State: Assam

Industry Code	Growth in SFPs						
	LP	CP	IC	RMP	FP	EP	EW
226	DNF	DNF	DNF	DNF	DNF	DNF	DNF
270	1.335	5.45	1.884	0.95	1.308	3.864	1.122
271	1.324	4.038	1.579	0.864	1.091	3.3	1.262
272	1.87	3.735	1.134	1.11	1.573	5.872	1.989
273	1.358	3.294	1.511	0.635	0.863	4.33	1.626
274	DNF	DNF	DNF	DNF	DNF	DNF	DNF
275	DNF	DNF	DNF	DNF	DNF	DNF	DNF
276	0.874	7.573	1.132	1.969	0.603	0.968	0.825
279	0.528	1.808	0.425	0.886	0.585	0.575	0.467
280	0.953	5.963	3.234	0.856	0.817	0.702	1.105
281	1.085	3.604	1.387	2.494	1.555	1.232	0.354

282	1.296	4.233	1.413	1.296	1.296	1.296	1.296
283	DNF	DNF	DNF	DNF	DNF	DNF	DNF
316	1.851	2.013	2.241	0.959	1.239	8.668	1.59
317	0.949	4.412	1.617	1.291	0.84	0.902	0.818
378	DNF	DNF	DNF	DNF	DNF	DNF	DNF
385	DNF	DNF	DNF	DNF	DNF	DNF	DNF
386	DNF	DNF	DNF	DNF	DNF	DNF	DNF
387	DNF	DNF	DNF	DNF	DNF	DNF	DNF
Total WFBI	1.14	5.309	4.622	1.173	1.136	2.724	1.078

Note: **LP:** Labour Productivity, **CP:** Capital Productivity,
IC: Capital Intensity, **RMP:** Raw-Material Productivity,
FP: Fuel Productivity, **EP:** Enterprise Productivity,
EW: Emoluments per Worker.
DNF: means data not found for all the years.
WFBI: Wood & Forest based Industries.

Table (5.1.3): Estimates of Growth in Partial Factor Productivity Ratios

State: Bihar

Industry Code	Growth in SFPs						
	LP	CP	IC	RMP	FP	EP	EW
226	1.248	4.71	1.377	0.887	1.451	1.788	1.073
270	1.361	4.129	1.921	0.887	1.264	3.113	1.068
271	1.277	3.864	1.522	0.861	1.184	1.818	1.266
272	1.376	3.021	1.561	0.692	1.257	2.84	1.587
273	1.575	5.165	1.752	0.662	1.149	2.332	1.933
274	1.376	4.425	1.18	0.774	1.314	3.821	1.239
275	DNF	DNF	DNF	DNF	DNF	DNF	DNF
276	1.289	8.408	1.67	0.838	1.994	3.222	1.341
279	1.219	2.236	0.982	0.704	1.281	4.197	1.448
280	0.75	2.563	2.543	0.592	0.678	2.562	1.145
281	1.657	4.407	2.118	0.874	1.444	5.303	1.482
282	2.051	3.526	1.726	0.988	1.29	6.905	1.622
283	1.702	5.895	2.085	0.813	2.704	5.133	1.276
316	1.116	1.264	1.351	0.524	0.836	4.187	1.479
317	1.223	2.612	2.083	0.851	1.5	4.291	1.186
378	DNF	DNF	DNF	DNF	DNF	DNF	DNF
385	DNF	DNF	DNF	DNF	DNF	DNF	DNF
386	DNF	DNF	DNF	DNF	DNF	DNF	DNF
387	DNF	DNF	DNF	DNF	DNF	DNF	DNF
Total WFBI	1.299	2.307	2.474	0.781	1.414	3.499	1.308

Note: **LP:** Labour Productivity, **CP:** Capital Productivity,
IC: Capital Intensity, **RMP:** Raw-Material Productivity,
FP: Fuel Productivity, **EP:** Enterprise Productivity,
EW: Emoluments per Worker.
DNF: means data not found for all the years.
WFBI: Wood & Forest based Industries.

Table (5.1.4): Estimates of Growth in Partial Factor Productivity Ratios

State: Delhi

Industry Code	Growth in SFPs						
	LP	CP	IC	RMP	FP	EP	EW
226	DNF	DNF	DNF	DNF	DNF	DNF	DNF
270	2.167	10.63	3.058	1.439	1.432	7.023	1.893
271	2.053	8.509	2.447	2.266	1.445	4.773	1.905
272	1.403	5.419	1.592	0.821	1.609	5.102	1.628
273	2.01	11.06	2.38	0.899	1.593	9.216	1.665
274	1.123	3.896	0.964	0.625	1.348	3.439	1.352
275	0.53	2.37	0.413	0.283	1.05	1.539	0.916
276	1.46	10.22	1.891	0.949	1.618	4.3	1.352
279	1.98	4.168	1.595	0.813	1.451	8.158	1.842
280	2.106	10.21	7.143	0.908	2.456	4.732	1.626
281	1.638	6.879	2.094	1.132	1.501	3.49	1.382
282	2.569	8.391	2.162	0.842	1.306	8.926	2.349
283	1.779	9.459	2.179	0.756	1.698	4.607	1.478
316	1.569	3.151	1.899	0.676	1.312	4.621	1.387
317	DNF	DNF	DNF	DNF	DNF	DNF	DNF
378	DNF	DNF	DNF	DNF	DNF	DNF	DNF
385	DNF	DNF	DNF	DNF	DNF	DNF	DNF
386	1.319	9.651	5.478	1.021	1.795	5.817	1.658
387	1.238	2.346	1.82	0.805	1.407	3.419	1.241
Total WFBI	1.663	3.918	3.231	0.949	1.535	5.278	1.578

Note: **LP:** Labour Productivity, **CP:** Capital Productivity,
IC: Capital Intensity, **RMP:** Raw-Material Productivity,
FP: Fuel Productivity, **EP:** Enterprise Productivity,
EW: Emoluments per Worker.
DNF: means data not found for all the years.
WFBI: Wood & Forest based Industries.

Table (5.1.5): Estimates of Growth in Partial Factor Productivity Ratios

State: Gujarat

Industry Code	Growth in SFPs						
	LP	CP	IC	RMP	FP	EP	EW
226	1.248	4.741	1.289	0.887	1.451	2.329	1.092
270	1.35	6.622	1.904	0.74	1.383	3.272	1.184
271	1.465	6.071	1.746	0.826	1.394	3.837	1.298
272	1.348	5.206	1.529	0.792	1.441	3.338	1.306
273	1.553	8.545	1.727	0.858	1.68	3.68	1.404
274	1.34	4.647	1.149	0.904	1.449	2.329	1.206
275	1.568	7.015	1.222	0.8	1.097	7.687	1.38
276	1.369	9.581	1.773	0.897	1.619	3.355	1.202
279	1.535	3.231	1.236	0.844	1.693	3.73	1.321
280	1.351	6.553	4.583	0.946	1.075	3.392	1.18
281	1.473	6.186	1.883	0.865	1.545	4.096	1.302

282	1.34	4.377	1.128	0.904	1.449	2.329	1.206
283	1.274	6.77	1.56	0.872	1.313	3.563	1.218
316	1.433	2.878	1.735	0.945	1.13	5.611	1.284
317	DNF	DNF	DNF	DNF	DNF	DNF	DNF
378	DNF	DNF	DNF	DNF	DNF	DNF	DNF
385	DNF	DNF	DNF	DNF	DNF	DNF	DNF
386	1.358	9.937	2.34	0.644	1.19	4.649	1.528
387	1.275	2.346	1.82	0.844	1.401	2.726	1.155
Total WFBI	1.392	3.28	2.516	0.846	1.39	3.74	1.267

Note: **LP:** Labour Productivity, **CP:** Capital Productivity,
IC: Capital Intensity, **RMP:** Raw-Material Productivity,
FP: Fuel Productivity, **EP:** Enterprise Productivity,
EW: Emoluments per Worker.
DNF: means data not found for all the years.
WFBI: Wood & Forest based Industries.

Table (5.1.6): Estimates of Growth in Partial Factor Productivity Ratios

State: Haryana

Industry Code	Growth in SFPs						
	LP	CP	IC	RMP	FP	EP	EW
226	DNF	DNF	DNF	DNF	DNF	DNF	DNF
270	1.589	7.795	3.075	0.935	1.262	5.152	1.341
271	1.513	6.273	2.435	0.798	1.419	2.566	1.394
272	1.717	6.633	2.481	0.813	1.432	4.662	1.443
273	1.735	9.547	2.798	0.885	1.389	5.588	1.721
274	1.137	3.945	1.102	0.893	1.139	13.08	1.066
275	1.672	7.48	2.062	0.687	0.969	8.824	1.954
276	1.559	10.91	2.654	0.825	1.515	5.979	1.381
279	1.574	3.315	1.783	0.799	1.321	4.593	1.485
280	1.385	6.717	6.715	0.922	1.094	4.401	1.244
281	1.734	7.283	3.154	0.881	1.412	5.67	1.471
282	2.013	6.575	1.914	0.828	1.037	8.428	1.913
283	1.514	8.047	2.28	0.831	1.135	7.315	1.403
316	1.647	3.307	3.113	0.891	1.214	4.408	1.35
317	DNF	DNF	DNF	DNF	DNF	DNF	DNF
378	DNF	DNF	DNF	DNF	DNF	DNF	DNF
385	1.624	8.996	3.329	0.943	1.286	13.75	1.38
386	DNF	DNF	DNF	DNF	DNF	DNF	DNF
387	1.869	2.346	1.82	0.848	1.828	4.01	1.595
Total WFBI	1.619	3.814	3.916	0.852	1.296	6.562	1.476

Note: **LP:** Labour Productivity, **CP:** Capital Productivity,
IC: Capital Intensity, **RMP:** Raw-Material Productivity,
FP: Fuel Productivity, **EP:** Enterprise Productivity,
EW: Emoluments per Worker.
DNF: means data not found for all the years.
WFBI: Wood & Forest based Industries.

Table (5.1.7): Estimates of Growth in Partial Factor Productivity Ratios

State: Karnataka

Industry Code	Growth in SFPs						
	LP	CP	IC	RMP	FP	EP	EW
226	2.406	9.139	2.194	1.645	3.192	4.816	1.125
270	1.272	6.242	2.182	0.93	1.249	3.662	1.154
271	1.345	5.575	2.081	0.912	1.249	3.004	1.239
272	1.395	5.388	2.455	0.724	1.285	2.475	1.386
273	1.415	7.79	2.428	0.842	1.362	3.288	1.273
274	1.456	5.048	1.806	0.944	1.544	5.025	1.306
275	1.319	5.901	1.661	0.814	1.679	4.331	1.059
276	1.168	8.179	1.739	0.955	1.576	3.483	1.109
279	1.416	2.982	1.911	0.733	1.505	4.505	1.343
280	1.344	6.519	6.668	0.973	1.063	4.596	1.245
281	1.519	6.378	3.043	0.845	1.287	3.307	1.313
282	2.409	7.869	5.423	0.832	1.805	9.959	2.022
283	1.771	9.415	4.176	0.846	1.266	8.392	1.527
316	1.366	2.743	2.554	0.827	0.987	5.201	1.319
317	1.213	4.461	2.525	0.859	1.457	2.975	1.114
378	DNF	DNF	DNF	DNF	DNF	DNF	DNF
385	1.909	10.57	9.124	0.863	1.797	5.961	1.431
386	DNF	DNF	DNF	DNF	DNF	DNF	DNF
387	1.306	2.346	1.82	0.687	1.359	3.84	1.496
Total WFBI	1.531	3.608	4.216	0.896	1.51	4.636	1.321

Note: **LP:** Labour Productivity, **CP:** Capital Productivity,
IC: Capital Intensity, **RMP:** Raw-Material Productivity,
FP: Fuel Productivity, **EP:** Enterprise Productivity,
EW: Emoluments per Worker.
DNF: means data not found for all the years.
WFBI: Wood & Forest based Industries.

Table (5.1.8): Estimates of Growth in Partial Factor Productivity Ratios

State: Kerala

Industry Code	Growth in SFPs						
	LP	CP	IC	RMP	FP	EP	EW
226	1.088	4.133	0.941	0.901	1.825	2.211	1.112
270	1.32	6.476	2.451	0.942	1.225	3.04	1.237
271	1.239	5.135	1.714	0.915	1.37	2.007	1.181
272	1.736	6.705	2.921	1.021	2.001	3.52	1.284
273	1.71	9.411	3.141	0.863	2.62	5.144	1.558
274	1.437	4.984	2.234	0.87	1.774	3.051	1.315
275	DNF	DNF	DNF	DNF	DNF	DNF	DNF
276	1.327	9.289	2.28	1.327	1.327	1.327	1.327
279	1.029	2.167	2.027	0.454	0.67	2.762	1.492
280	1.494	7.245	8.182	0.991	1.071	7.525	1.34
281	1.641	6.891	3.614	0.851	1.47	4.28	1.431
282	0.915	2.99	2.32	0.469	0.583	2.641	1.274

283	1.759	9.353	3.646	0.981	1.442	5.71	1.534
316	0.717	1.439	1.436	0.42	0.52	3.276	1.41
317	1.144	4.205	1.509	0.81	1.946	2.939	1.068
378	DNF	DNF	DNF	DNF	DNF	DNF	DNF
385	DNF	DNF	DNF	DNF	DNF	DNF	DNF
386	DNF	DNF	DNF	DNF	DNF	DNF	DNF
387	1.641	2.346	1.82	0.934	1.363	5.938	1.486
Total WFBI	1.346	3.172	3.871	0.85	1.414	3.691	1.336

Note: **LP:** Labour Productivity, **CP:** Capital Productivity,
IC: Capital Intensity, **RMP:** Raw-Material Productivity,
FP: Fuel Productivity, **EP:** Enterprise Productivity,
EW: Emoluments per Worker.
DNF: means data not found for all the years.
WFBI: Wood & Forest based Industries.

Table (5.1.9): Estimates of Growth in Partial Factor Productivity Ratios

State: Madhya Pradesh

Industry Code	Growth in SFPs						
	LP	CP	IC	RMP	FP	EP	EW
226	1.228	4.664	1.353	0.907	1.408	2.114	1.11
270	1.345	6.601	2.541	0.954	1.279	6.659	1.231
271	1.151	4.774	1.584	0.689	1.099	2.163	1.158
272	1.634	6.312	2.922	0.91	1.392	4.868	1.499
273	1.088	5.988	1.792	0.901	1.825	2.211	1.389
274	1.689	5.858	1.894	0.905	1.95	4.802	1.479
275	1.297	5.802	1.363	0.955	1.611	3.509	1.034
276	1.18	8.259	1.949	0.872	1.574	3.688	1.223
279	1.802	3.793	3.192	0.738	1.744	5.002	1.656
280	1.531	7.427	8.511	0.919	1.08	4.67	1.384
281	1.779	7.472	4.042	0.877	1.471	4.801	1.478
282	1.992	6.507	3.326	0.841	1.367	7.049	1.711
283	1.662	8.836	2.879	0.917	1.786	4.734	1.314
316	1.558	3.129	3.076	0.888	1.113	9.03	1.407
317	DNF	DNF	DNF	DNF	DNF	DNF	DNF
378	DNF	DNF	DNF	DNF	DNF	DNF	DNF
385	DNF	DNF	DNF	DNF	DNF	DNF	DNF
386	DNF	DNF	DNF	DNF	DNF	DNF	DNF
387	1.506	2.346	1.82	0.925	1.879	3.85	1.314
Total WFBI	1.496	3.525	3.993	0.88	1.505	4.61	1.36

Note: **LP: Labour Productivity,** **CP: Capital Productivity,**
IC: Capital Intensity, **RMP:** Raw-Material Productivity,
FP: Fuel Productivity, **EP:** Enterprise Productivity,
EW: Emoluments per Worker, DNF means data not found for all the years.

* Significant at 1% level of significance
** Significant at 5% level of significance
*** Significant at 10% level of significance
@ Insignificant

Table (5.1.10): Estimates of Growth in Partial Factor Productivity Ratios

State: Maharashtra

Industry Code	Growth in SFPs						
	LP	CP	IC	RMP	FP	EP	EW
226	1.109	5.457	1.223	0.927	1.581	2.094	1.058
270	1.339	4.745	1.89	0.918	1.248	6.139	1.24
271	1.413	4.43	1.685	0.91	1.444	3.897	1.258
272	1.445	3.935	1.639	0.886	1.538	3.506	1.386
273	1.781	6.381	1.981	1.055	1.763	3.868	1.326
274	1.687	4.178	1.447	1.027	1.528	4.512	1.415
275	2.117	5.224	1.65	0.93	1.296	9.157	1.865
276	1.366	7.363	1.769	0.955	1.507	3.753	1.26
279	1.228	1.938	0.989	0.821	1.087	3.257	1.399
280	1.445	4.728	4.899	0.997	1.128	3.882	1.262
281	1.451	4.286	1.855	0.888	1.278	2.867	1.317
282	1.93	4.113	1.624	0.734	1.583	6.076	1.627
283	1.501	5.74	1.839	0.938	1.464	4.723	1.335
316	1.479	1.983	1.79	0.975	1.087	4.611	1.354
317	DNF	DNF	DNF	DNF	DNF	DNF	DNF
378	DNF	DNF	DNF	DNF	DNF	DNF	DNF
385	1.524	5.932	4.066	0.908	1.397	5.076	1.41
386	DNF	DNF	DNF	DNF	DNF	DNF	DNF
387	1.398	2.346	1.82	0.96	1.376	3.248	1.246
Total WFBI	1.513	2.573	2.734	0.926	1.394	4.417	1.36

Note: **LP:** Labour Productivity, **CP:** Capital Productivity,
IC: Capital Intensity, **RMP:** Raw-Material Productivity,
FP: Fuel Productivity, **EP:** Enterprise Productivity,
EW: Emoluments per Worker.
DNF: means data not found for all the years.
WFBI: Wood & Forest based Industries.

Table (5.1.11): Estimates of Growth in Partial Factor Productivity Ratios

State: Orissa

Industry Code	Growth in SFPs						
	LP	CP	IC	RMP	FP	EP	EW
226	1.217	4.824	1.342	0.745	1.354	3.201	1.128
270	1.123	3.281	1.584	0.83	1.029	4.996	1.255
271	1.109	5.955	1.322	0.927	1.581	2.094	1.058
272	2.049	4.033	2.324	0.928	1.681	4.051	1.726
273	1.987	6.705	2.211	1.328	2.175	3.439	1.957
274	1.496	3.377	1.283	0.917	1.562	4.614	1.322
275	DNF	DNF	DNF	DNF	DNF	DNF	DNF
276	1.219	7.385	1.578	1.008	1.729	2.734	1.194
279	1.624	1.811	1.308	0.847	1.02	5.143	1.621
280	1.445	5.31	4.9	1.063	1.203	8.068	1.223
281	2.141	4.227	2.736	0.932	1.526	6.442	1.727
282	2.442	4.126	2.055	1.01	1.933	9.173	1.709

283	1.168	2.99	1.431	0.602	1.431	4.297	1.404
316	1.109	2.885	1.342	0.927	1.581	2.094	1.058
317	0.981	3.721	1.192	0.891	1.349	1.856	0.915
378	DNF	DNF	DNF	DNF	DNF	DNF	DNF
385	DNF	DNF	DNF	DNF	DNF	DNF	DNF
386	DNF	DNF	DNF	DNF	DNF	DNF	DNF
387	DNF	DNF	DNF	DNF	DNF	DNF	DNF
Total WFBI	1.508	2.507	2.796	0.925	1.511	4.443	1.378

Note: LP: Labour Productivity, CP: Capital Productivity, IC: Capital Intensity, RMP: Raw-Material Productivity, FP: Fuel Productivity, EP: Enterprise Productivity, EW: Emoluments per Worker, DNF means data not found for all the years.

* Significant at 1% level of significance

** Significant at 5% level of significance

*** Significant at 10% level of significance

@ Insignificant

Table (5.1.12): Estimates of Growth in Partial Factor Productivity Ratios

State: Punjab

Industry Code	Growth in SFPs						
	LP	CP	IC	RMP	FP	EP	EW
226	DNF	DNF	DNF	DNF	DNF	DNF	DNF
270	1.751	4.881	2.471	0.935	1.291	3.786	1.614
271	1.36	3.981	1.621	0.822	1.072	3.246	1.452
272	1.604	4.045	1.819	0.787	1.47	2.779	1.519
273	1.868	7.189	2.078	1.022	2.328	4.635	1.647
274	1.109	4.982	1.776	0.927	1.581	2.094	1.478
275	1.378	5.036	1.074	0.946	1.679	9.108	1.125
276	1.227	8.72	1.588	0.833	1.927	4.739	1.152
279	1.109	2.334	0.99	1.109	1.109	1.109	1.109
280	1.304	4.153	4.423	0.892	1.007	3.089	1.234
281	1.109	6.032	3.017	0.927	1.581	2.094	1.427
282	1.779	2.609	1.497	0.716	1.156	9.301	1.803
283	1.774	4.631	2.173	0.762	0.962	11.91	1.459
316	0.308	0.369	0.373	0.153	0.228	1.144	1.325
317	DNF	DNF	DNF	DNF	DNF	DNF	DNF
378	DNF	DNF	DNF	DNF	DNF	DNF	DNF
385	1.109	7.955	4.739	0.927	1.581	2.094	1.257
386	DNF	DNF	DNF	DNF	DNF	DNF	DNF
387	1.193	2.346	1.82	0.849	1.602	3.832	1.066
Total WFBI	1.332	3.138	3.55	0.84	1.372	4.331	1.378

Note: **LP:** Labour Productivity, **CP:** Capital Productivity, **IC:** Capital Intensity, **RMP:** Raw-Material Productivity, **FP:** Fuel Productivity, **EP:** Enterprise Productivity, **EW:** Emoluments per Worker.
DNF: means data not found for all the years.
WFBI: Wood & Forest based Industries.

Table (5.1.13): Estimates of Growth in Partial Factor Productivity Ratios

State: Rajasthan

Industry Code	Growth in SFPs						
	LP	CP	IC	RMP	FP	EP	EW
226	1.148	5.096	1.266	0.814	1.494	2.537	1.165
270	1.78	5.028	2.512	0.797	1.396	3.539	1.563
271	1.64	4.187	1.955	0.895	1.347	3.095	1.403
272	2.285	3.664	2.592	0.879	1.232	6.687	2.017
273	1.026	4.727	1.142	0.849	0.895	1.682	0.88
274	1.975	3.517	1.694	0.862	1.735	4.966	1.612
275	1.634	4.677	1.274	1.186	1.853	9.134	1.053
276	1.367	6.773	1.77	0.797	1.4	3.191	1.442
279	1.888	1.767	1.521	0.857	1.42	4.616	1.647
280	1.288	4.154	4.37	0.867	1.051	3.425	1.269
281	1.919	4.149	2.452	0.824	1.329	5.46	1.669
282	1.942	2.926	1.635	0.837	1.266	5.919	1.698
283	1.539	5.245	1.885	0.86	1.371	5.167	1.467
316	1.604	2.103	1.942	0.787	1.47	2.779	1.519
317	DNF	DNF	DNF	DNF	DNF	DNF	DNF
378	DNF	DNF	DNF	DNF	DNF	DNF	DNF
385	DNF	DNF	DNF	DNF	DNF	DNF	DNF
386	DNF	DNF	DNF	DNF	DNF	DNF	DNF
387	1.338	2.346	1.82	0.891	2.34	4.736	1.125
Total WFBI	1.625	2.335	1.833	0.867	1.44	4.462	1.435

Note: **LP:** Labour Productivity, **CP:** Capital Productivity,
IC: Capital Intensity, **RMP:** Raw-Material Productivity,
FP: Fuel Productivity, **EP:** Enterprise Productivity,
EW: Emoluments per Worker.
DNF: means data not found for all the years.
WFBI: Wood & Forest based Industries.

Table (5.1.14): Estimates of Growth in Partial Factor Productivity Ratios

State: Tamil Nadu

Industry Code	Growth in SFPs						
	LP	CP	IC	RMP	FP	EP	EW
226	1.222	4.291	1.348	0.936	1.669	3.407	1.153
270	1.432	4.902	2.021	0.916	1.387	2.701	1.333
271	1.423	4.584	1.696	0.976	1.47	2.547	1.257
272	1.406	3.87	1.595	0.861	1.584	3.59	1.438
273	1.517	5.422	1.688	0.842	1.788	5.989	1.556
274	1.108	3.037	1.128	1.302	3.111	1.326	1.108
275	1.656	4.629	1.291	0.863	1.317	6.442	1.467
276	1.222	7.907	1.582	0.936	1.669	3.407	1.153
279	1.496	2.581	1.205	0.92	1.798	4.594	1.465
280	1.444	4.635	4.896	1.014	1.086	4.147	1.252
281	1.302	3.598	1.664	0.769	1.12	3.433	1.336

282	1.496	3.262	1.259	0.937	1.293	2.992	1.36
283	2.203	8.62	2.698	1.448	2.128	6.636	1.267
316	1.551	1.929	1.878	0.928	1.114	6.021	1.349
317	1.15	4.119	1.96	0.914	1.328	2.3	1.075
378	1.448	1.475	1.464	0.784	1.132	4.224	1.404
385	1.741	6.764	4.646	0.936	1.635	8.298	1.364
386	DNF	DNF	DNF	DNF	DNF	DNF	DNF
387	1.317	2.346	1.82	0.891	1.464	3.838	1.264
Total WFBI	1.222	4.291	1.348	0.936	1.669	3.407	1.153

Note: **LP:** Labour Productivity, **CP:** Capital Productivity,
IC: Capital Intensity, **RMP:** Raw-Material Productivity,
FP: Fuel Productivity, **EP:** Enterprise Productivity,
EW: Emoluments per Worker.
DNF: means data not found for all the years.
WFBI: Wood & Forest based Industries.

Table (5.1.15): Estimates of Growth in Partial Factor Productivity Ratios

State: U.P.

Industry Code	Growth in SFPs						
	LP	CP	IC	RMP	FP	EP	EW
226	1.173	1.307	0.924	0.897	1.509	1.843	1.1
270	1.604	1.047	0.787	1.102	1.47	2.779	1.153
271	1.285	0.746	0.775	1.723	1.086	3.042	1.391
272	2.138	1.369	1.117	1.562	1.88	5.027	1.48
273	1.604	1.047	0.787	1.597	1.47	2.779	1.525
274	1.594	1.03	0.884	1.547	1.422	4.918	1.502
275	2.338	0.892	0.868	2.622	1.366	8.329	1.903
276	1.282	0.929	0.947	1.381	1.603	3.607	1.194
279	1.604	1.047	0.787	1.622	1.47	2.779	1.457
280	1.463	0.955	0.718	1.253	1.341	2.534	1.108
281	1.112	0.787	0.63	1.49	1.06	2.688	1.319
282	1.811	0.953	0.862	1.901	1.214	4.092	1.572
283	1.501	0.88	0.779	1.706	1.001	4.15	1.345
316	1.358	0.931	0.883	1.459	1.086	6.559	1.3
317	1.495	1.023	0.945	1.46	1.307	7.005	1.355
378	DNF	DNF	DNF	DNF	DNF	DNF	DNF
385	1.408	1.181	0.937	1.192	1.697	6.385	1.278
386	DNF	DNF	DNF	DNF	DNF	DNF	DNF
387	1.483	1.002	0.918	1.48	1.47	3.125	1.237
Total WFBI	1.553	1.01	0.86	1.536	1.387	4.229	1.372

Note: **LP:** Labour Productivity, **CP:** Capital Productivity,
IC: Capital Intensity, **RMP:** Raw-Material Productivity,
FP: Fuel Productivity, **EP:** Enterprise Productivity,
EW: Emoluments per Worker.
DNF: means data not found for all the years.
WFBI: Wood & Forest based Industries.

Table (5.1.16): Estimates of Growth in Partial Factor Productivity Ratios

State: West Bengal

Industry Code	Growth in SFPs						
	LP	CP	IC	RMP	FP	EP	EW
226	1.343	5.381	1.481	0.845	1.52	5.118	1.222
270	1.207	4.541	1.703	0.871	1.213	3.228	1.146
271	1.153	3.88	1.375	0.742	1.038	1.942	1.27
272	1.438	4.94	1.631	0.874	1.645	3.226	1.42
273	1.516	5.563	1.687	0.856	1.182	4.648	1.428
274	1.585	3.946	1.359	0.922	1.287	3.579	1.468
275	1.665	5.613	1.298	1.02	1.678	7.86	1.481
276	1.265	7.899	1.639	0.93	1.685	3.839	1.242
279	1.709	2.361	1.376	0.924	1.494	7.363	1.585
280	1.604	5.078	4.882	0.787	1.47	2.779	1.278
281	1.36	4.549	1.738	0.821	1.271	3.345	1.323
282	1.741	3.393	1.465	0.814	1.062	4.104	1.608
283	1.312	4.38	1.607	0.771	1.161	2.946	1.313
316	1.604	2.103	1.839	0.787	1.47	2.779	1.399
317	DNF	DNF	DNF	DNF	DNF	DNF	DNF
378	DNF	DNF	DNF	DNF	DNF	DNF	DNF
385	DNF	DNF	DNF	DNF	DNF	DNF	DNF
386	1.227	6.01	2.115	0.773	1.035	8.454	1.457
387	1.322	2.346	1.82	0.891	1.412	3.111	1.25
Total WFBI	1.44	2.519	1.978	0.851	1.351	4.27	1.368

Note: **LP:** Labour Productivity, **CP:** Capital Productivity,
IC: Capital Intensity, **RMP:** Raw-Material Productivity,
FP: Fuel Productivity, **EP:** Enterprise Productivity,
EW: Emoluments per Worker.
DNF: means data not found for all years.
WFBI: Wood & Forest based Industries.

Table (5.1.17): Estimates of Growth in Partial Factor Productivity Ratios

Eastern Region

Industry Code	Growth in SFPs						
	LP	CP	IC	RMP	FP	EP	EW
226	1.269	4.972	1.4	0.826	1.442	3.369	1.141
270	1.256	4.35	1.773	0.885	1.204	3.8	1.148
271	1.216	4.434	1.449	0.848	1.224	2.289	1.214
272	1.683	3.932	1.663	0.901	1.539	3.997	1.68
273	1.609	5.182	1.79	0.87	1.342	3.687	1.736
274	1.485	3.916	1.274	0.871	1.388	4.005	1.343
275	1.791	6.039	1.396	1.097	1.806	8.457	1.594
276	1.162	7.816	1.505	1.186	1.503	2.691	1.151
279	1.27	2.054	1.023	0.84	1.095	4.32	1.28
280	1.188	4.728	3.89	0.825	1.042	3.528	1.188
281	1.561	4.197	1.995	1.28	1.449	4.08	1.222

282	1.882	3.819	1.665	1.027	1.395	5.369	1.559
283	1.435	4.442	1.758	0.707	2.067	4.715	1.34
316	1.42	2.066	1.693	0.799	1.282	4.432	1.381
317	1.051	3.582	1.631	1.011	1.23	2.35	0.973
378	DNF	DNF	DNF	DNF	DNF	DNF	DNF
385	DNF	DNF	DNF	DNF	DNF	DNF	DNF
386	1.243	6.089	2.142	0.783	1.049	8.563	1.476
387	0.334	0.593	0.461	0.225	0.357	0.787	0.316
Total WFBI	1.347	3.16	2.967	0.933	1.353	3.734	1.283

Note: **LP:** Labour Productivity, **CP:** Capital Productivity,
IC: Capital Intensity, **RMP:** Raw-Material Productivity,
FP: Fuel Productivity, **EP:** Enterprise Productivity,
EW: Emoluments per Worker.
DNF: means data not found for all years.
WFBI: Wood & Forest based Industries.

Table (5.1.18): Estimates of Growth in Partial Factor Productivity Ratios

Western Region

Industry Code	Growth in SFPs						
	LP	CP	IC	RMP	FP	EP	EW
226	1.208	4.833	1.303	0.869	1.451	2.327	1.122
270	1.453	5.749	2.212	0.852	1.326	4.902	1.305
271	1.417	4.866	1.743	0.83	1.321	3.248	1.279
272	1.678	4.779	2.17	0.867	1.401	4.6	1.552
273	1.362	6.41	1.661	0.916	1.541	2.86	1.25
274	1.673	4.55	1.546	0.925	1.665	4.152	1.428
275	1.654	5.679	1.377	0.967	1.464	7.372	1.333
276	1.32	7.994	1.815	0.88	1.525	3.497	1.282
279	1.613	2.682	1.734	0.815	1.486	4.151	1.506
280	1.404	5.715	5.591	0.932	1.083	3.842	1.274
281	1.656	5.523	2.558	0.863	1.406	4.306	1.441
282	1.801	4.481	1.928	0.829	1.416	5.343	1.56
283	1.494	6.648	2.041	0.897	1.484	4.547	1.334
316	1.518	2.523	2.136	0.899	1.2	5.508	1.391
317	DNF	DNF	DNF	DNF	DNF	DNF	DNF
378	DNF	DNF	DNF	DNF	DNF	DNF	DNF
385	1.851	7.208	4.941	1.103	1.697	6.168	1.713
386	1.647	12.05	2.839	0.781	1.443	5.64	1.854
387	1.379	2.346	1.82	0.905	1.749	3.64	1.21
Total WFBI	1.507	2.928	2.769	0.88	1.432	4.307	1.356

Note: **LP:** Labour Productivity, **CP:** Capital Productivity,
IC: Capital Intensity, **RMP:** Raw-Material Productivity,
FP: Fuel Productivity, **EP:** Enterprise Productivity,
EW: Emoluments per Worker.
DNF: means data not found for all years.
WFBI: Wood & Forest based Industries.

Table (5.1.19): Estimates of Growth in Partial Factor Productivity Ratios

Northern Region

Industry Code	Growth in SFPs						
	LP	CP	IC	RMP	FP	EP	EW
226	1.063	3.733	1.173	0.814	1.452	2.964	1.003
270	1.735	7.053	2.656	1.056	1.343	4.665	1.545
271	1.587	5.837	2.05	1.215	1.351	3.283	1.502
272	1.533	4.992	1.872	0.821	1.524	4.033	1.507
273	1.782	8.305	2.236	0.912	1.774	6.357	1.647
274	1.119	3.965	1.242	0.937	1.795	4.986	1.251
275	1.309	4.879	1.21	0.695	1.254	6.478	1.366
276	1.367	9.439	1.929	0.886	1.682	4.606	1.26
279	1.54	3.1	1.393	0.91	1.42	4.613	1.475
280	1.56	6.429	5.794	0.934	1.411	4.092	1.339
281	1.446	5.948	2.482	0.927	1.404	3.672	1.404
282	1.964	5.209	1.708	0.831	1.198	7.412	1.856
283	1.818	7.69	2.332	0.949	1.481	7.618	1.402
316	1.269	2.189	1.816	0.662	0.967	4.048	1.353
317	1.576	5.642	2.685	1.252	1.819	3.151	1.473
378	1.611	1.64	1.628	0.872	1.259	4.697	1.561
385	5.155	27.32	14.65	3.232	5.188	24.51	4.609
386	1.6	11.71	6.644	1.239	2.177	7.056	2.011
387	1.729	3.078	2.388	1.169	1.921	5.036	1.658
Total WFBI	1.516	3.349	3.17	0.899	1.441	5.097	1.436

Note: **LP:** Labour Productivity, **CP:** Capital Productivity,
IC: Capital Intensity, **RMP:** Raw-Material Productivity,
FP: Fuel Productivity, **EP:** Enterprise Productivity,
EW: Emoluments per Worker.
DNF: means data not found for all years.
WFBI: Wood & Forest based Industries.

Table (5.1.20): Estimates of Growth in Partial Factor Productivity Ratios

Southern Region

Industry Code	Growth in SFPs						
	LP	CP	IC	RMP	FP	EP	EW
226	1.454	5.732	1.387	1.146	2.142	3.072	1.113
270	1.339	6.04	2.352	0.943	1.305	3.519	1.239
271	1.354	5.286	1.923	0.93	1.383	2.952	1.249
272	1.411	5.061	2.133	0.841	1.571	3.185	1.351
273	1.418	7.073	2.012	0.998	1.665	3.802	1.277
274	1.314	4.357	1.599	0.987	1.937	3.42	1.301
275	1.488	5.265	1.476	0.838	1.498	5.386	1.263
276	1.253	8.609	1.892	1.034	1.58	2.718	1.202
279	1.446	2.903	1.97	0.988	1.454	3.426	1.536
280	1.36	6.005	5.777	1.032	1.061	4.305	1.238
281	1.512	5.881	2.828	0.839	1.319	3.978	1.353

282	1.531	4.596	2.417	0.9	1.13	4.159	1.357
283	1.699	10.49	3.856	0.973	1.449	6.492	1.406
316	1.153	2.019	1.706	0.766	0.83	3.822	1.34
317	1.123	4.099	2.159	0.831	1.511	2.47	1.155
378	1.495	1.522	1.511	0.809	1.168	4.359	1.449
385	1.825	8.669	6.885	0.899	1.716	7.13	1.398
386	DNF	DNF	DNF	DNF	DNF	DNF	DNF
387	1.41	2.346	1.82	0.855	1.54	4.291	1.368
Total WFBI	1.454	5.732	1.387	1.146	2.142	3.072	1.113

Note: **LP:** Labour Productivity, **CP:** Capital Productivity,
IC: Capital Intensity, **RMP:** Raw-Material Productivity,
FP: Fuel Productivity, **EP:** Enterprise Productivity,
EW: Emoluments per Worker.
DNF: means data not found for all years.
WFBI: Wood & Forest based Industries.

Table (5.1.21): Estimates of Growth in Partial Factor Productivity Ratios

Other States

Industry Code	Growth in SFPs						
	LP	CP	IC	RMP	FP	EP	EW
226	1.03	5.022	1.116	1.332	2.54	1.488	0.976
270	0.15	3.048	0.125	0.768	0.918	0.771	0.301
271	0.762	3.158	0.943	0.845	0.936	1.204	0.714
272	1.092	4.216	1.77	1.114	1.249	1.676	0.972
273	0.813	4.472	0.905	0.836	0.989	1.404	0.755
274	1.391	4.824	1.375	1.079	1.51	3.225	1.224
275	1.363	6.098	1.278	0.988	1.522	5.354	1.281
276	0.133	7.11	0.111	0.941	2.268	1.19	0.587
279	1.328	2.795	1.182	0.972	1.537	2.83	1.12
280	0.441	2.14	4.223	0.725	1.76	1.176	0.874
281	0.875	3.673	1.194	0.987	1.486	1.322	0.804
282	1.282	4.188	1.125	0.994	1.414	2.171	1.175
283	0.306	4.193	0.167	1.113	2.124	1.243	0.815
316	0.925	1.858	2.331	1.382	1.653	1.388	0.552
317	0.816	2.998	1.242	1.821	2.019	2.596	1.539
378	0.204	3.102	0.26	2.614	2.932	3.93	2.28
385	0.205	5.938	0.203	1.525	1.804	2.56	1.377
386	0.45	6.399	0.181	0.989	1.384	2.956	1.121
387	0.364	6.186	1.82	2.097	3.229	11.36	2.717
Total WFBI	0.733	4.285	1.134	1.217	1.751	2.623	1.115

Note: **LP:** Labour Productivity, **CP:** Capital Productivity,
IC: Capital Intensity, **RMP:** Raw-Material Productivity,
FP: Fuel Productivity, **EP:** Enterprise Productivity,
EW: Emoluments per Worker.
DNF: means data not found for all years.
WFBI: Wood & Forest based Industries.

Table (5.1.22): Estimates of Growth in Partial Factor Productivity Ratios

All India

Industry Code	Growth in SFPs						
	LP	CP	IC	RMP	FP	EP	EW
226	0.706	3.443	0.565	0.914	1.744	1.021	0.669
270	0.171	3.477	0.071	0.876	1.047	0.879	0.35
271	0.839	1.976	0.973	0.93	1.031	1.326	0.786
272	0.881	3.405	1.274	0.9	1.009	1.353	0.785
273	0.899	4.947	0.93	0.925	1.094	1.553	0.835
274	1.166	4.044	1.014	0.905	1.266	2.703	1.026
275	1.283	5.74	1.13	0.93	1.433	5.04	1.206
276	0.077	4.144	0.029	0.549	1.322	0.694	0.342
279	1.242	2.614	0.932	0.909	1.438	2.645	1.048
280	0.295	1.43	5.588	0.484	1.176	0.786	0.584
281	0.782	3.285	0.982	0.883	1.33	1.182	0.72
282	1.188	3.882	0.956	0.921	1.31	2.012	1.089
283	0.317	4.341	0.179	1.152	2.199	1.287	0.844
316	0.826	1.659	1.242	1.234	1.476	1.239	0.493
317	0.587	2.157	1.687	1.31	1.452	1.868	1.108
378	0.099	1.505	0.141	1.268	1.422	1.907	1.106
385	0.175	5.075	0.546	1.303	1.542	2.188	1.177
386	0.58	8.246	0.582	1.275	1.784	3.809	1.445
387	0.228	3.868	1.82	1.311	2.019	7.102	1.699
Total WFBI	0.65	3.644	1.086	0.999	1.426	2.137	0.911

Note: **LP:** Labour Productivity, **CP:** Capital Productivity, **IC:** Capital Intensity, **RMP:** Raw-Material Productivity, **FP:** Fuel Productivity, **EP:** Enterprise Productivity, **EW:** Emoluments per Worker.
DNF: means data not found for the years.
WFBI: Wood & Forest based Industries.

Table (5.2.1): Growth in Total Factor Productivity Indices

Industry Code	Description of 3 Digit Asi Industries (Based On Nic 1970 & 1987)	Andhra Pradesh		
		Kendrick Index	Solow Index	Divisia Index
226	Manufacture of Bidi	0.893*	1.355*	1.347*
270	Sawing and Planing of Wood (Other Than Plywood)	0.639*	-0.304***	0.987*
271	Manufacture of Veneer Sheets, Plywood and their Products	0.977*	-0.872*	1.185*
272	Manufacture of Structural Wooden Goods such as Beams, Posts, Doors & Windows (Excluding Hewing and Rough Shaping of Poles, Bolts & other Materials)	5.368*	1.89*	1.42*
273	Manufacture of Wooden and Cane Boxes, Crates, Drums, Barrels and other Containers, Baskets, and other Wares Made Entirely or Mainly of Cane, Rattan, Reed, Bamboo, Willow, Fibres, Leaves and Grass	1.592*	0.82*	0.97*
274	Manufacture of Wooden Industrial Goods n.e.c.	1.039*	-0.910	1.164*
275	Manufacture of Cork and Cork Products	DNF	DNF	DNF
276	Manufacture of Wooden Furniture And Fixtures	0.502**	0.364**	1.24*
279	Manufacture of Products Of Wood, Bamboo Cane, Reed and Grass (Incl. Articles Made from Shell etc) n.e.c.	0.963*	-0.396**	0.70*
280	Manufacture of Pulp, Paper and Paper Board incl. Manufacture of Newsprint	0.478**	-0.441**	0.785*
281	Manufacture of Containers and Boxes of Paper and Paper Board	1.464*	-0.94*	1.22*
282	Manufacture of Paper and Paper Board Articles and Pulp Products n.e.c.	1.445*	-0.556**	1.35*
283	Manufacture of Paper and Paper Board Articles not elsewhere classified	-1.737*	-0.491**	1.101*
316	Manufacture of Refined Petroleum Products n.e.c. (Obtained from Products or Residues from Petroleum Refining)	-0.57**	0.356***	0.035@
317	Manufacture of Matches	0.0028@	0.156***	0.015@
378	Manufacture of Bullock Carts, Push-Carts, Hand Carts *etc.*	DNF	DNF	DNF
385	Manufacture of Sports and Athletic Goods	DNF	DNF	DNF
386	Manufacture of Musical Instruments (Excl. Toys)	DNF	DNF	DNF
387	Manufacture of Stationery Articles n.e.c.	0.1086@	-0.22***	0.0156@
Total Wood & Forest Based Industries		0.877*	-0.013@	0.902*

(DNF means data not found for all the years).

Significant at 1% level of significance Significant at 5% level of significance*

**** Significant at 10% level of significance @ Insignificant*

Table (5.2.2): Growth in Total Factor Productivity Indices: Assam

Industry Code	Description Of 3 Digit ASI Industries (Based on NIC 1970 & 1987)	Kendrick Index	Solow Index	Divisia Index
226	Manufacture of Bidi	DNF	DNF	DNF
270	Sawing and Planing of Wood (Other Than Plywood)	0.6442*	-0.989*	0.994*
271	Manufacture of Veneer Sheets, Plywood and their Products	0.9844*	-0.889*	1.195*
272	Manufacture of Structural Wooden Goods such as Beams, Posts, Doors & Windows (Excluding Hewing and Rough Shaping of Poles, Bolts & other Materials)	-1.886*	-1.7314*	1.43*
273	Manufacture of Wooden and Cane Boxes, Crates, Drums, Barrels and other Containers, Baskets, and other Wares Made Entirely or Mainly of Cane, Rattan, Reed, Bamboo, Willow, Fibres, Leaves and Grass	1.604*	0.824*	0.982*
274	Manufacture of Wooden Industrial Goods n.e.c.	DNF	DNF	DNF
275	Manufacture of Cork and Cork Products	DNF	DNF	DNF
276	Manufacture of Wooden Furniture And Fixtures	0.506**	0.367**	1.25*
279	Manufacture of Products Of Wood, Bamboo Cane, Reed and Grass (Incl. Articles Made from Shell etc) n.e.c.	-0.969*	-2.086*	-0.707*
280	Manufacture of Pulp, Paper and Paper Board incl. Manufacture of Newsprint	0.482**	0.354**	0.791*
281	Manufacture of Containers and Boxes of Paper and Paper Board	-0.459**	-0.113***	0.231***
282	Manufacture of Paper and Paper Board Articles and Pulp Products n.e.c.	1.456*	1.59*	1.36*
283	Manufacture of Paper and Paper Board Articles not elsewhere classified	DNF	DNF	DNF
316	Manufacture of Refined Petroleum Products n.e.c. (Obtained from Products or Residues from Petroleum Refining)	0.0039@	-0.177***	0.0216@
317	Manufacture of Matches	0.0023@	-0.022@	0.0125@
378	Manufacture of Bullock Carts, Push-Carts, Hand Carts *etc.*	DNF	DNF	DNF
385	Manufacture of Sports and Athletic Goods	DNF	DNF	DNF
386	Manufacture of Musical Instruments (Excl. Toys)	DNF	DNF	DNF
387	Manufacture of Stationery Articles n.e.c.	DNF	DNF	DNF
Total Wood & Forest Based Industries		0.215***	-0.261***	0.687*

(DNF means data not found for all the years).

**Significant at 1% level of significance ** Significant at 5% level of significance*

**** Significant at 10% level of significance*

@ Insignificant

Table (5.2.3): Growth in Total Factor Productivity Indices: Bihar

Industry Code	Description of 3 Digit ASI Industries (Based on NIC 1970 & 1987)	Kendrick Index	Solow Index	Divisia Index
226	Manufacture of Bidi	0.884*	1.34*	1.334*
270	Sawing and Planing of Wood (Other Than Plywood)	0.634*	0.3014**	0.977*
271	Manufacture of Veneer Sheets, Plywood and their Products	0.967*	0.873*	1.173*
272	Manufacture of Structural Wooden Goods such as Beams, Posts, Doors & Windows (Excluding Hewing and Rough Shaping of Poles, Bolts & other Materials)	0.957*	1.701*	1.405*
273	Manufacture of Wooden and Cane Boxes, Crates, Drums, Barrels and other Containers, Baskets, and other Wares Made Entirely or Mainly of Cane, Rattan, Reed, Bamboo, Willow, Fibres, Leaves and Grass	-2.149*	0.027@	0.964*
274	Manufacture of Wooden Industrial Goods n.e.c.	1.029*	-0.424**	1.152*
275	Manufacture of Cork and Cork Products	DNF	DNF	DNF
276	Manufacture of Wooden Furniture And Fixtures	0.497**	-0.29***	1.225*
279	Manufacture of Products Of Wood, Bamboo Cane, Reed and Grass (Incl. Articles Made from Shell etc) n.e.c.	0.953*	2.05*	0.695*
280	Manufacture of Pulp, Paper and Paper Board incl. Manufacture of Newsprint	0.473*	0.702*	0.777*
281	Manufacture of Containers and Boxes of Paper and Paper Board	1.45*	1.22*	1.21*
282	Manufacture of Paper and Paper Board Articles and Pulp Products n.e.c.	1.430*	1.564*	1.3361*
283	Manufacture of Paper and Paper Board Articles not elsewhere classified	1.463*	1.56*	1.367*
316	Manufacture of Refined Petroleum Products n.e.c. (Obtained from Products or Residues from Petroleum Refining)	0.0038@	-0.212***	0.021@
317	Manufacture of Matches	0.0022@	0.123***	0.012@
378	Manufacture of Bullock Carts, Push-Carts, Hand Carts *etc.*	DNF	DNF	DNF
385	Manufacture of Sports and Athletic Goods	DNF	DNF	DNF
386	Manufacture of Musical Instruments (Excl. Toys)	DNF	DNF	DNF
387	Manufacture of Stationery Articles n.e.c.	DNF	DNF	DNF
Total Wood & Forest Based Industries		0.614**	0.755*	0.975*

(DNF means data not found for all the years).

**Significant at 1% level of significance*

*** Significant at 5% level of significance*

**** Significant at 10% level of significance*

@ Insignificant

Table (5.2.4): Growth in Total Factor Productivity Indices: Delhi

Industry Code	Description of 3 Digit ASI Industries (Based on NIC 1970 & 1987)	Kendrick Index	Solow Index	Divisia Index
226	Manufacture of Bidi	DNF	DNF	DNF
270	Sawing and Planing of Wood (Other Than Plywood)	0.8004*	0.3812*	1.235*
271	Manufacture of Veneer Sheets, Plywood and their Products	0.962*	0.869*	1.168*
272	Manufacture of Structural Wooden Goods such as Beams, Posts, Doors & Windows (Excluding Hewing and Rough Shaping of Poles, Bolts & other Materials)	0.951*	1.69*	1.397*
273	Manufacture of Wooden and Cane Boxes, Crates, Drums, Barrels and other Containers, Baskets, and other Wares Made Entirely or Mainly of Cane, Rattan, Reed, Bamboo, Willow, Fibres, Leaves and Grass	1.571*	0.812*	0.967*
274	Manufacture of Wooden Industrial Goods n.e.c.	1.028*	0.901*	1.152*
275	Manufacture of Cork and Cork Products	2.135*	4.080*	0.442**
276	Manufacture of Wooden Furniture And Fixtures	0.497**	0.588**	1.225*
279	Manufacture of Products of Wood, Bamboo Cane, Reed and Grass (Incl. Articles Made from Shell etc) n.e.c.	0.936*	2.0146*	0.683*
280	Manufacture of Pulp, Paper and Paper Board incl. Manufacture of Newsprint	0.471**	0.984*	0.773*
281	Manufacture of Containers and Boxes of Paper and Paper Board	1.455*	1.382*	1.215*
282	Manufacture of Paper and Paper Board Articles and Pulp Products n.e.c.	1.404*	1.535*	1.31*
283	Manufacture of Paper and Paper Board Articles not elsewhere classified	1.47*	1.607*	1.37*
316	Manufacture of Refined Petroleum Products n.e.c. (Obtained from Products or Residues from Petroleum Refining)	0.0037@	-0.2084*	0.021@
317	Manufacture of Matches	DNF	DNF	DNF
378	Manufacture of Bullock Carts, Push-Carts, Hand Carts *etc.*	DNF	DNF	DNF
385	Manufacture of Sports and Athletic Goods	DNF	DNF	DNF
386	Manufacture of Musical Instruments (Excl. Toys)	DNF	DNF	DNF
387	Manufacture of Stationery Articles n.e.c.	0.148***	0.0090*	0.021@
Total Wood & Forest Based Industries		1.19*	1.36*	1.43*

(DNF means data not found for all the years).

**Significant at 1% level of significance*

*** Significant at 5% level of significance*

**** Significant at 10% level of significance*

@ Insignificant

Table (5.2.5): Growth in Total Factor Productivity Indices: Gujarat

Industry Code	Description of 3 Digit ASI Industries (Based on NIC 1970 & 1987)	Kendrick Index	Solow Index	Divisia Index
226	Manufacture of Bidi	0.883*	1.341*	1.334*
270	Sawing and Planing of Wood (Other Than Plywood)	0.784*	-0.373**	1.209*
271	Manufacture of Veneer Sheets, Plywood and their Products	0.964*	0.87*	1.169*
272	Manufacture of Structural Wooden Goods such as Beams, Posts, Doors & Windows (Excluding Hewing and Rough Shaping of Poles, Bolts & other Materials)	0.953*	1.694*	1.4004*
273	Manufacture of Wooden and Cane Boxes, Crates, Drums, Barrels and other Containers, Baskets, and other Wares Made Entirely or Mainly of Cane, Rattan, Reed, Bamboo, Willow, Fibres, Leaves and Grass	1.592*	0.838*	0.974*
274	Manufacture of Wooden Industrial Goods n.e.c.	1.027*	0.8997*	1.149*
275	Manufacture of Cork and Cork Products	5.2813*	1.474*	0.622*
276	Manufacture of Wooden Furniture And Fixtures	0.499**	-0.572**	1.230*
279	Manufacture of Products of Wood, Bamboo Cane, Reed and Grass (Incl. Articles Made from Shell etc) n.e.c.	0.9428*	2.0281*	0.688*
280	Manufacture of Pulp, Paper and Paper Board incl. Manufacture of Newsprint	0.480**	-0.590**	0.788*
281	Manufacture of Containers and Boxes of Paper and Paper Board	1.443*	1.214*	1.205*
282	Manufacture of Paper and Paper Board Articles and Pulp Products n.e.c.	1.445*	1.58*	1.349*
283	Manufacture of Paper and Paper Board Articles not elsewhere classified	1.64*	1.793*	1.531*
316	Manufacture of Refined Petroleum Products n.e.c. (Obtained from Products or Residues from Petroleum Refining)	0.0061@	-0.287***	0.0336@
317	Manufacture of Matches	DNF	DNF	DNF
378	Manufacture of Bullock Carts, Push-Carts, Hand Carts *etc.*	DNF	DNF	DNF
385	Manufacture of Sports and Athletic Goods	DNF	DNF	DNF
386	Manufacture of Musical Instruments (Excl. Toys)	3.235*	0.351**	0.035@
387	Manufacture of Stationery Articles n.e.c.	0.152***	0.00927@	0.0218@
Total Wood & Forest Based Industries		1.33*	0.767*	0.921*

(DNF means data not found for all the years).

**Significant at 1% level of significance*

*** Significant at 5% level of significance*

**** Significant at 10% level of significance*

@ Insignificant

Table (5.2.6): Growth in Total Factor Productivity Indices: Haryana

Industry Code	Description of 3 Digit ASI Industries (Based on NIC 1970 & 1987)	Kendrick Index	Solow Index	Divisia Index
226	Manufacture of Bidi	DNF	DNF	DNF
270	Sawing and Planing of Wood (Other Than Plywood)	0.784*	-0.078@	1.209*
271	Manufacture of Veneer Sheets, Plywood and their Products	1.232*	1.113*	1.496*
272	Manufacture of Structural Wooden Goods such as Beams, Posts, Doors & Windows (Excluding Hewing and Rough Shaping of Poles, Bolts & other Materials)	1.211*	2.151*	1.778*
273	Manufacture of Wooden and Cane Boxes, Crates, Drums, Barrels and other Containers, Baskets, and other Wares Made Entirely or Mainly of Cane, Rattan, Reed, Bamboo, Willow, Fibres, Leaves and Grass	1.835*	0.943*	1.123*
274	Manufacture of Wooden Industrial Goods n.e.c.	1.212*	1.061*	1.356*
275	Manufacture of Cork and Cork Products	2.109*	4.029*	0.436**
276	Manufacture of Wooden Furniture And Fixtures	0.597**	0.433**	1.473*
279	Manufacture of Products Of Wood, Bamboo Cane, Reed and Grass (Incl. Articles Made from Shell etc) n.e.c.	2.452*	3.350*	2.926*
280	Manufacture of Pulp, Paper and Paper Board incl. Manufacture of Newsprint	0.781*	1.159*	1.282*
281	Manufacture of Containers and Boxes of Paper and Paper Board	1.798*	2.916*	1.50*
282	Manufacture of Paper and Paper Board Articles and Pulp Products n.e.c.	1.711*	1.871*	1.598*
283	Manufacture of Paper and Paper Board Articles not elsewhere classified	1.724*	1.886*	1.611*
316	Manufacture of Refined Petroleum Products n.e.c. (Obtained from Products or Residues from Petroleum Refining)	0.006@	0.332**	0.033@
317	Manufacture of Matches	DNF	DNF	DNF
378	Manufacture of Bullock Carts, Push-Carts, Hand Carts *etc.*	DNF	DNF	DNF
385	Manufacture of Sports and Athletic Goods	1.685*	0.965*	1.347*
386	Manufacture of Musical Instruments (Excl. Toys)	DNF	DNF	DNF
387	Manufacture of Stationery Articles n.e.c.	1.193*	1.042*	1.28*
Total Wood & Forest Based Industries		1.39*	1.52*	1.71*

(DNF means data not found for all the years).

**Significant at 1% level of significance*

*** Significant at 5% level of significance*

**** Significant at 10% level of significance*

@ Insignificant

Table (5.2.7): Growth in Total Factor Productivity Indices: Karnataka

Industry Code	Description Of 3 Digit ASI Industries (Based on NIC 1970 & 1987)	Kendrick Index	Solow Index	Divisia Index
226	Manufacture of Bidi	0.715*	1.085*	1.079*
270	Sawing and Planing of Wood (Other Than Plywood)	0.796*	-0.03@	1.23*
271	Manufacture of Veneer Sheets, Plywood and their Products	1.204*	1.088*	1.462*
272	Manufacture of Structural Wooden Goods such as Beams, Posts, Doors & Windows (Excluding Hewing and Rough Shaping of Poles, Bolts & other Materials)	3.77*	3.26*	1.71*
273	Manufacture of Wooden and Cane Boxes, Crates, Drums, Barrels and other Containers, Baskets, and other Wares Made Entirely or Mainly of Cane, Rattan, Reed, Bamboo, Willow, Fibres, Leaves and Grass	1.279*	-0.2***	0.783*
274	Manufacture of Wooden Industrial Goods n.e.c.	0.735*	0.644*	0.823*
275	Manufacture of Cork and Cork Products	1.729*	3.303*	0.358**
276	Manufacture of Wooden Furniture And Fixtures	1.806*	-2.2*	0.893*
279	Manufacture of Products of Wood, Bamboo Cane, Reed and Grass (Incl. Articles Made from Shell etc) n.e.c.	0.718*	1.544*	0.523**
280	Manufacture of Pulp, Paper and Paper Board incl. Manufacture of Newsprint	0.75*	1.113*	1.231*
281	Manufacture of Containers and Boxes of Paper and Paper Board	3.001*	2.36*	2.836*
282	Manufacture of Paper and Paper Board Articles and Pulp Products n.e.c.	1.131*	1.24*	1.057*
283	Manufacture of Paper and Paper Board Articles not elsewhere classified	1.159*	1.27*	1.083*
316	Manufacture of Refined Petroleum Products n.e.c. (Obtained from Products or Residues from Petroleum Refining)	0.006@	-0.12***	0.031@
317	Manufacture of Matches	0.003@	-0.18***	0.018@
378	Manufacture of Bullock Carts, Push-Carts, Hand Carts *etc.*	DNF	DNF	DNF
385	Manufacture of Sports and Athletic Goods	1.01*	0.579*	0.808*
386	Manufacture of Musical Instruments (Excl. Toys)	DNF	DNF	DNF
387	Manufacture of Stationery Articles n.e.c.	0.183***	0.011@	0.026@
Total Wood & Forest Based Industries		1.176*	0.868*	0.938*

(DNF means data not found for all the years).

**Significant at 1% level of significance*

*** Significant at 5% level of significance*

**** Significant at 10% level of significance*

@ Insignificant

Table (5.2.8): Growth in Total Factor Productivity Indices: Kerala

Industry Code	Description of 3 Digit Asi Industries (Based on NIC 1970 & 1987)	Kendrick Index	Solow Index	Divisia Index
226	Manufacture of Bidi	0.822*	1.085*	1.079*
270	Sawing and Planing of Wood (Other Than Plywood)	0.440**	0.379**	1.229*
271	Manufacture of Veneer Sheets, Plywood and their Products	0.6629*	-0.489**	1.462*
272	Manufacture of Structural Wooden Goods such as Beams, Posts, Doors & Windows (Excluding Hewing and Rough Shaping of Poles, Bolts & other Materials)	0.7076*	2.069*	1.71*
273	Manufacture of Wooden and Cane Boxes, Crates, Drums, Barrels and other Containers, Baskets, and other Wares Made Entirely or Mainly of Cane, Rattan, Reed, Bamboo, Willow, Fibres, Leaves and Grass	1.2177*	0.657*	0.78*
274	Manufacture of Wooden Industrial Goods n.e.c.	2.735*	1.996*	1.823*
275	Manufacture of Cork and Cork Products	DNF	DNF	DNF
276	Manufacture of Wooden Furniture And Fixtures	0.636*	-0.251***	0.893*
279	Manufacture of Products of Wood, Bamboo Cane, Reed and Grass (Incl. Articles Made from Shell etc) n.e.c.	1.074*	1.55*	0.523**
280	Manufacture of Pulp, Paper and Paper Board incl. Manufacture of Newsprint	0.75*	1.113*	1.23*
281	Manufacture of Containers and Boxes of Paper and Paper Board	1.001*	0.842*	0.836*
282	Manufacture of Paper and Paper Board Articles and Pulp Products n.e.c.	1.489*	1.24*	1.056*
	Manufacture of Paper and Paper Board Articles not elsewhere classified	1.741*	1.268*	1.083*
316	Manufacture of Refined Petroleum Products n.e.c. (Obtained from Products or Residues from Petroleum Refining)	-0.0046@	-0.312***	-0.031@
317	Manufacture of Matches	2.659*	0.239***	0.018***
	Manufacture of Bullock Carts, Push-Carts, Hand Carts *etc.*	DNF	DNF	DNF
385	Manufacture of Sports and Athletic Goods	1.113*	0.579**	0.81*
386	Manufacture of Musical Instruments (Excl. Toys)	DNF	DNF	DNF
387	Manufacture of Stationery Articles n.e.c.	0.329***	-0.011*	0.026***
Total Wood & Forest Based Industries		1.127*	0.769*	0.625*

(DNF means data not found for all the years).

**Significant at 1% level of significance*

*** Significant at 5% level of significance*

**** Significant at 10% level of significance*

@ Insignificant

Table (5.2.9): Growth in Total Factor Productivity Indices: M.P.

Industry Code	Description of 3 Digit Asi Industries (Based on NIC 1970 & 1987)	Kendrick Index	Solow Index	Divisia Index
226	Manufacture of Bidi	1.3099*	1.989*	1.978*
270	Sawing and Planing of Wood (Other Than Plywood)	0.7823*	0.373**	1.207*
271	Manufacture of Veneer Sheets, Plywood and their Products	1.196*	1.0798*	1.451*
272	Manufacture of Structural Wooden Goods such as Beams, Posts, Doors & Windows (Excluding Hewing and Rough Shaping of Poles, Bolts & other Materials)	1.206*	2.145*	1.77*
273	Manufacture of Wooden and Cane Boxes, Crates, Drums, Barrels and other Containers, Baskets, and other Wares Made Entirely or Mainly of Cane, Rattan, Reed, Bamboo, Willow, Fibres, Leaves and Grass	1.947*	1.001*	1.192*
274	Manufacture of Wooden Industrial Goods n.e.c.	1.3051*	1.143*	1.461*
275	Manufacture of Cork and Cork Products	1.60*	-0.661*	0.332**
276	Manufacture of Wooden Furniture And Fixtures	0.588**	0.427**	1.45*
279	Manufacture of Products of Wood, Bamboo Cane, Reed and Grass (Incl. Articles Made from Shell etc) n.e.c.	1.21*	2.608*	0.88*
280	Manufacture of Pulp, Paper and Paper Board incl. Manufacture of Newsprint	1.074*	1.594*	1.76*
281	Manufacture of Containers and Boxes of Paper and Paper Board	1.84*	1.548*	1.536*
282	Manufacture of Paper and Paper Board Articles and Pulp Products n.e.c.	1.76*	0.898*	1.645*
283	Manufacture of Paper and Paper Board Articles not elsewhere classified	1.797*	1.965*	1.679*
316	Manufacture of Refined Petroleum Products n.e.c. (Obtained from Products or Residues from Petroleum Refining)	2.042*	2.72*	2.772*
317	Manufacture of Matches	DNF	DNF	DNF
378	Manufacture of Bullock Carts, Push-Carts, Hand Carts *etc.*	DNF	DNF	DNF
385	Manufacture of Sports and Athletic Goods	DNF	DNF	DNF
386	Manufacture of Musical Instruments (Excl. Toys)	DNF	DNF	DNF
387	Manufacture of Stationery Articles n.e.c.	0.945*	0.657*	0.797*
Total Wood & Forest Based Industries		1.405*	1.299*	1.462*

(DNF means data not found for all the years).

**Significant at 1% level of significance*

*** Significant at 5% level of significance*

**** Significant at 10% level of significance*

@ Insignificant

Table (5.2.10): Growth in Total Factor Productivity Indices: Maharashtra

Industry Code	Description of 3 Digit Asi Industries (Based on NIC 1970 & 1987)	Kendrick Index	Solow Index	Divisia Index
226	Manufacture of Bidi	2.845*	1.351*	1.344*
270	Sawing and Planing of Wood (Other Than Plywood)	0.776*	0.389**	1.198*
271	Manufacture of Veneer Sheets, Plywood and their Products	1.224*	1.1052*	1.485*
272	Manufacture of Structural Wooden Goods such as Beams, Posts, Doors & Windows (Excluding Hewing and Rough Shaping of Poles, Bolts & other Materials)	0.652*	-1.777*	0.958*
273	Manufacture of Wooden and Cane Boxes, Crates, Drums, Barrels and other Containers, Baskets, and other Wares Made Entirely or Mainly of Cane, Rattan, Reed, Bamboo, Willow, Fibres, Leaves and Grass	1.183*	0.608**	0.724*
274	Manufacture of Wooden Industrial Goods n.e.c.	1.581*	-1.252*	1.77*
275	Manufacture of Cork and Cork Products	2.288*	0.696*	0.473**
276	Manufacture of Wooden Furniture And Fixtures	0.603*	-0.283***	1.485*
279	Manufacture of Products of Wood, Bamboo Cane, Reed and Grass (Incl. Articles Made from Shell etc) n.e.c.	0.491**	2.707*	0.918*
280	Manufacture of Pulp, Paper and Paper Board incl. Manufacture of Newsprint	0.770*	1.142*	1.264*
281	Manufacture of Containers and Boxes of Paper and Paper Board	1.980*	1.6656*	1.65*
282	Manufacture of Paper and Paper Board Articles and Pulp Products n.e.c.	2.1335*	3.422*	2.7134*
283	Manufacture of Paper and Paper Board Articles not elsewhere classified	2.256*	2.467*	2.1078*
316	Manufacture of Refined Petroleum Products n.e.c. (Obtained from Products or Residues from Petroleum Refining)	3.485*	4.379*	3.987*
317	Manufacture of Matches	DNF	DNF	DNF
378	Manufacture of Bullock Carts, Push-Carts, Hand Carts *etc.*	DNF	DNF	DNF
385	Manufacture of Sports and Athletic Goods	1.546*	0.886*	1.236*
386	Manufacture of Musical Instruments (Excl. Toys)	DNF	DNF	DNF
387	Manufacture of Stationery Articles n.e.c.	0.187***	-0.172***	0.027@
Total Wood & Forest Based Industries		1.50*	0.536**	1.459*

(DNF means data not found for all the years).

**Significant at 1% level of significance*

*** Significant at 5% level of significance*

**** Significant at 10% level of significance*

@ Insignificant

Table (5.2.11): Growth in Total Factor Productivity Indices: Orissa

Industry Code	Description of 3 Digit Asi Industries (Based on NIC 1970 & 1987)	Kendrick Index	Solow Index	Divisia Index
226	Manufacture of Bidi	0.822*	1.248*	1.241*
270	Sawing and Planing of Wood (Other Than Plywood)	-2.199*	0.928*	-1.122*
271	Manufacture of Veneer Sheets, Plywood and their Products	1.317*	1.189*	1.598*
272	Manufacture of Structural Wooden Goods such as Beams, Posts, Doors & Windows (Excluding Hewing and Rough Shaping of Poles, Bolts & other Materials)	1.206*	2.145*	1.77*
273	Manufacture of Wooden and Cane Boxes, Crates, Drums, Barrels and other Containers, Baskets, and other Wares Made Entirely or Mainly of Cane, Rattan, Reed, Bamboo, Willow, Fibres, Leaves and Grass	1.137*	0.585**	0.696*
274	Manufacture of Wooden Industrial Goods n.e.c.	0.786*	0.689*	0.880*
275	Manufacture of Cork and Cork Products	DNF	DNF	DNF
276	Manufacture of Wooden Furniture And Fixtures	0.502**	-0.805*	1.237*
279	Manufacture of Products of Wood, Bamboo Cane, Reed and Grass (Incl. Articles Made from Shell etc) n.e.c.	1 351*	2.905*	0.985*
280	Manufacture of Pulp, Paper and Paper Board incl. Manufacture of Newsprint	0 808*	-0. 5905*	1.326*
281	Manufacture of Containers and Boxes of Paper and Paper Board	1.854*	1.56*	1.548*
282	Manufacture of Paper and Paper Board Articles and Pulp Products n.e.c.	1.469*	1.61*	1.373*
283	Manufacture of Paper and Paper Board Articles not elsewhere classified	1.59*	1.744*	1.490*
316	Manufacture of Refined Petroleum Products n.e.c. (Obtained from Products or Residues from Petroleum Refining)	0.0069@	-0.208***	0.038***
317	Manufacture of Matches	0.0342**	-0.069@	0.018@
378	Manufacture of Bullock Carts, Push-Carts, Hand Carts *etc.*	DNF	DNF	DNF
385	Manufacture of Sports and Athletic Goods	DNF	DNF	DNF
386	Manufacture of Musical Instruments (Excl. Toys)	DNF	DNF	DNF
387	Manufacture of Stationery Articles n.e.c.	DNF	DNF	DNF
Total Wood & Forest Based Industries		0.764*	1.0397*	0.934*

(DNF means data not found for all the years).

**Significant at 1% level of significance*

*** Significant at 5% level of significance*

**** Significant at 10% level of significance*

@ Insignificant

Table (5.2.12): Growth in Total Factor Productivity Indices: Punjab

Industry Code	Description of 3 Digit Asi Industries (Based on NIC 1970 & 1987)	Kendrick Index	Solow Index	Divisia Index
226	Manufacture of Bidi	DNF	DNF	DNF
270	Sawing and Planing of Wood (Other Than Plywood)	0.7823*	-0.359**	1.207*
271	Manufacture of Veneer Sheets, Plywood and their Products	-2.0478*	-0.8554*	-1.45*
272	Manufacture of Structural Wooden Goods such as Beams, Posts, Doors & Windows (Excluding Hewing and Rough Shaping of Poles, Bolts & other Materials)	1.326*	2.358*	1.948*
273	Manufacture of Wooden and Cane Boxes, Crates, Drums, Barrels and other Containers, Baskets, and other Wares Made Entirely or Mainly of Cane, Rattan, Reed, Bamboo, Willow, Fibres, Leaves and Grass	2.604*	1.497*	1.181*
274	Manufacture of Wooden Industrial Goods n.e.c.	1.312*	1.149*	1.469*
275	Manufacture of Cork and Cork Products	0.024***	-0.168***	0.015***
276	Manufacture of Wooden Furniture And Fixtures	1.874*	1.577*	1.415*
279	Manufacture of Products of Wood, Bamboo Cane, Reed and Grass (Incl. Articles Made from Shell etc) n.e.c.	1.313*	2.824*	0.957*
280	Manufacture of Pulp, Paper and Paper Board incl. Manufacture of Newsprint	0.7305*	1.084*	1.199*
281	Manufacture of Containers and Boxes of Paper and Paper Board	2.022*	1.701*	1.688*
282	Manufacture of Paper and Paper Board Articles and Pulp Products n.e.c.	1.677*	1.833*	1.566*
283	Manufacture of Paper and Paper Board Articles not elsewhere classified	1.63*	1.783*	1.523*
316	Manufacture of Refined Petroleum Products n.e.c. (Obtained from Products or Residues from Petroleum Refining)	0.788*	0.634*	0.664*
317	Manufacture of Matches	DNF	DNF	DNF
378	Manufacture of Bullock Carts, Push-Carts, Hand Carts *etc.*	DNF	DNF	DNF
385	Manufacture of Sports and Athletic Goods	1.64*	0.940*	1.312*
386	Manufacture of Musical Instruments (Excl. Toys)	DNF	DNF	DNF
387	Manufacture of Stationery Articles n.e.c.	0.789*	0.1356***	0.113***
Total Wood & Forest Based Industries		1.098*	1.076*	0.987*

(DNF means data not found for all the years).

**Significant at 1% level of significance*

*** Significant at 5% level of significance*

**** Significant at 10% level of significance*

@ Insignificant

Table (5.2.13): Growth in Total Factor Productivity Indices: Rajasthan

Industry Code	Description of 3 Digit Asi Industries (Based on NIC 1970 & 1987)	Kendrick Index	Solow Index	Divisia Index
226	Manufacture of Bidi	0.822*	1.248*	1.241*
270	Sawing and Planing of Wood (Other Than Plywood)	0.446**	0.212***	0.689*
271	Manufacture of Veneer Sheets, Plywood and their Products	0.735*	-0.664*	0.892*
272	Manufacture of Structural Wooden Goods such as Beams, Posts, Doors & Windows (Excluding Hewing and Rough Shaping of Poles, Bolts & other Materials)	1.197*	2.1283*	1.759*
273	Manufacture of Wooden and Cane Boxes, Crates, Drums, Barrels and other Containers, Baskets, and other Wares Made Entirely or Mainly of Cane, Rattan, Reed, Bamboo, Willow, Fibres, Leaves and Grass	1.792*	-0.678*	1.097*
274	Manufacture of Wooden Industrial Goods n.e.c.	0.747*	-0.194***	0.836*
275	Manufacture of Cork and Cork Products	1.587*	-3.032*	0.328**
276	Manufacture of Wooden Furniture And Fixtures	0.593**	-0.431**	1.463*
279	Manufacture of Products of Wood, Bamboo Cane, Reed and Grass (Incl. Articles Made from Shell etc) n.e.c.	1.221*	2.63*	0.89*
280	Manufacture of Pulp, Paper and Paper Board incl. Manufacture of Newsprint	0.72*	-0.75*	1.198*
281	Manufacture of Containers and Boxes of Paper and Paper Board	0.987*	0.83*	0.82*
282	Manufacture of Paper and Paper Board Articles and Pulp Products n.e.c.	1.069*	1.169*	0.998*
283	Manufacture of Paper and Paper Board Articles not elsewhere classified	0.896*	0.98*	0.84*
316	Manufacture of Refined Petroleum Products n.e.c. (Obtained from Products or Residues from Petroleum Refining)	0.006@	0.336**	0.034***
317	Manufacture of Matches	DNF	DNF	DNF
378	Manufacture of Bullock Carts, Push-Carts, Hand Carts *etc.*	DNF	DNF	DNF
385	Manufacture of Sports and Athletic Goods	DNF	DNF	DNF
386	Manufacture of Musical Instruments (Excl. Toys)	DNF	DNF	DNF
387	Manufacture of Stationery Articles n.e.c.	0.190***	-0.504**	0.027@
Total Wood & Forest Based Industries		0.942*	0.348	0.945*

(DNF means data not found for all the years).

**Significant at 1% level of significance*

*** Significant at 5% level of significance*

**** Significant at 10% level of significance*

@ Insignificant

Table (5.2.14): Growth in Total Factor Productivity Indices: Tamil Nadu

Industry Code	Description of 3 Digit Asi Industries (Based on NIC 1970 & 1987)	Kendrick Index	Solow Index	Divisia Index
226	Manufacture of Bidi	0.566**	0.859*	0.855*
270	Sawing and Planing of Wood (Other Than Plywood)	0.784*	0.373**	1.21*
271	Manufacture of Veneer Sheets, Plywood and their Products	1.23*	1.112*	1.494*
272	Manufacture of Structural Wooden Goods such as Beams, Posts, Doors & Windows (Excluding Hewing and Rough Shaping of Poles, Bolts & other Materials)	0.67*	1.198*	0.99*
273	Manufacture of Wooden and Cane Boxes, Crates, Drums, Barrels and other Containers, Baskets, and other Wares Made Entirely or Mainly of Cane, Rattan, Reed, Bamboo, Willow, Fibres, Leaves and Grass	1.70*	0.875*	1.042*
274	Manufacture of Wooden Industrial Goods n.e.c.	1.204*	1.055*	1.35*
275	Manufacture of Cork and Cork Products	2.252*	4.30*	0.466**
276	Manufacture of Wooden Furniture And Fixtures	0.613**	0.444**	1.51*
279	Manufacture of Products of Wood, Bamboo Cane, Reed and Grass (Incl. Articles Made from Shell etc) n.e.c.	1.175*	2.53*	0.86*
280	Manufacture of Pulp, Paper and Paper Board incl. Manufacture of Newsprint	0.767*	1.14*	1.26*
281	Manufacture of Containers and Boxes of Paper and Paper Board	1.77*	1.49*	1.478*
282	Manufacture of Paper and Paper Board Articles and Pulp Products n.e.c.	1.732*	1.894*	1.62*
283	Manufacture of Paper and Paper Board Articles not elsewhere classified	1.94*	2.12*	1.81*
316	Manufacture of Refined Petroleum Products n.e.c. (Obtained from Products or Residues from Petroleum Refining)	0.0058@	0.326**	0.032@
317	Manufacture of Matches	0.0033@	0.1807*	0.018@
378	Manufacture of Bullock Carts, Push-Carts, Hand Carts *etc.*	0.0022@	0.121***	0.012@
385	Manufacture of Sports and Athletic Goods	1.583*	0.907*	1.266*
386	Manufacture of Musical Instruments (Excl. Toys)	DNF	DNF	DNF
387	Manufacture of Stationery Articles n.e.c.	0.187***	0.0114@	0.027@
Total Wood & Forest Based Industries		1.0107@	1.163*	0.96*

(DNF means data not found for all the years).

**Significant at 1% level of significance*

*** Significant at 5% level of significance*

**** Significant at 10% level of significance*

@ Insignificant

Table (5.2.15): Growth in Total Factor Productivity Indices: U.P.

Industry Code	Description of 3 Digit Asi Industries (Based on NIC 1970 & 1987)	Kendrick Index	Solow Index	Divisia Index
226	Manufacture of Bidi	3.475*	1.34*	2.242*
270	Sawing and Planing of Wood (Other Than Plywood)	0.906*	0.478**	1.398*
271	Manufacture of Veneer Sheets, Plywood and their Products	0.764*	1.2048*	0.927*
272	Manufacture of Structural Wooden Goods such as Beams, Posts, Doors & Windows (Excluding Hewing and Rough Shaping of Poles, Bolts & other Materials)	0.582**	1.0339*	0.855*
273	Manufacture of Wooden and Cane Boxes, Crates, Drums, Barrels and other Containers, Baskets, and other Wares Made Entirely or Mainly of Cane, Rattan, Reed, Bamboo, Willow, Fibres, Leaves and Grass	1.246*	-0.638**	0.763*
274	Manufacture of Wooden Industrial Goods n.e.c.	0.756*	-1.263*	0.846*
275	Manufacture of Cork and Cork Products	1.625*	0.867*	0.337**
276	Manufacture of Wooden Furniture And Fixtures	0.378**	0.274***	0.932*
279	Manufacture of Products of Wood, Bamboo Cane, Reed and Grass (Incl. Articles Made from Shell etc) n.e.c.	1.294*	2.784*	0.944*
280	Manufacture of Pulp, Paper and Paper Board incl. Manufacture of Newsprint	0.794*	1.178*	1.302*
281	Manufacture of Containers and Boxes of Paper and Paper Board	1.73*	1.455*	1.44*
282	Manufacture of Paper and Paper Board Articles and Pulp Products n.e.c.	1.456*	-0.576**	1.36*
283	Manufacture of Paper and Paper Board Articles not elsewhere classified	1.32*	1.421*	1.28*
316	Manufacture of Refined Petroleum Products n.e.c. (Obtained from Products or Residues from Petroleum Refining)	0.0058@	-0.322**	0.032***
317	Manufacture of Matches	0.0032@	0.176***	0.017***
378	Manufacture of Bullock Carts, Push-Carts, Hand Carts *etc.*	DNF	DNF	DNF
385	Manufacture of Sports and Athletic Goods	0.988*	-0.566**	0.789*
386	Manufacture of Musical Instruments (Excl. Toys)	DNF	DNF	DNF
387	Manufacture of Stationery Articles n.e.c.	0.183***	-0.184***	0.026***
Total Wood & Forest Based Industries		0.878*	0.872*	0.708*

(DNF means data not found for all the years).
*Significant at 1% level of significance
** Significant at 5% level of significance
*** Significant at 10% level of significance
@ Insignificant

Table (5.2.16): Growth in Total Factor Productivity Indices: West Bengal

Industry Code	Description of 3 Digit Asi Industries (Based on NIC 1970 & 1987)	Kendrick Index	Solow Index	Divisia Index
226	Manufacture of Bidi	0.888*	1.348*	1.341*
270	Sawing and Planing of Wood (Other Than Plywood)	0.772*	0.0828*	1.191*
271	Manufacture of Veneer Sheets, Plywood and their Products	1.194*	0.977*	1.445*
272	Manufacture of Structural Wooden Goods such as Beams, Posts, Doors & Windows (Excluding Hewing and Rough Shaping of Poles, Bolts & other Materials)	-0.572**	-0.509**	-0.872*
273	Manufacture of Wooden and Cane Boxes, Crates, Drums, Barrels and other Containers, Baskets, and other Wares Made Entirely or Mainly of Cane, Rattan, Reed, Bamboo, Willow, Fibres, Leaves and Grass	1.84*	0.945*	1.13*
274	Manufacture of Wooden Industrial Goods n.e.c.	1.27*	1.115*	1.42*
275	Manufacture of Cork and Cork Products	1.568*	2.996*	0.324**
276	Manufacture of Wooden Furniture And Fixtures	2.458*	-1.029*	1.454*
279	Manufacture of Products of Wood, Bamboo Cane, Reed and Grass (Incl. Articles Made from Shell etc) n.e.c.	0.613**	0.2752***	0.447**
280	Manufacture of Pulp, Paper and Paper Board incl. Manufacture of Newsprint	0.606**	0.4784*	0.995*
281	Manufacture of Containers and Boxes of Paper and Paper Board	1.679*	-0.3004***	0.802*
282	Manufacture of Paper and Paper Board Articles and Pulp Products n.e.c.	1.0294*	0.845*	0.962*
283	Manufacture of Paper and Paper Board Articles not elsewhere classified	1.178*	-1.288*	1.101*
316	Manufacture of Refined Petroleum Products n.e.c. (Obtained from Products or Residues from Petroleum Refining)	0.0061@	-0.6383*	0.034@
317	Manufacture of Matches	DNF	DNF	DNF
378	Manufacture of Bullock Carts, Push-Carts, Hand Carts *etc.*	DNF	DNF	DNF
385	Manufacture of Sports and Athletic Goods	DNF	DNF	DNF
386	Manufacture of Musical Instruments (Excl. Toys)	2.676*	-2.844*	0.029***
387	Manufacture of Stationery Articles n.e.c.	0.108***	-0.7115*	0.015@
Total Wood & Forest Based Industries		1.227*	1.218*	0.989*

(DNF means data not found for all the years).

**Significant at 1% level of significance*

*** Significant at 5% level of significance*

**** Significant at 10% level of significance*

@ Insignificant

Table (5.2.17): Growth in Total Factor Productivity Indices: Eastern Region

Industry Code	Description of 3 Digit Asi Industries (Based on NIC 1970 & 1987)	Kendrick Index	Solow Index	Divisia Index
226	Manufacture of Bidi	0.8647**	1.312*	1.3053*
270	Sawing and Planing of Wood (Other Than Plywood)	-0.037@	0.0808@	0.51**
271	Manufacture of Veneer Sheets, Plywood and their Products	1.1156*	0.5375**	1.3528*
272	Manufacture of Structural Wooden Goods such as Beams, Posts, Doors & Windows (Excluding Hewing and Rough Shaping of Poles, Bolts & other Materials)	-0.074@	0.4014***	0.9333**
273	Manufacture of Wooden and Cane Boxes, Crates, Drums, Barrels and other Containers, Baskets, and other Wares Made Entirely or Mainly of Cane, Rattan, Reed, Bamboo, Willow, Fibres, Leaves and Grass	0.608**	0.5953**	0.943*
274	Manufacture of Wooden Industrial Goods n.e.c.	1 0283*	0.46**	1.1507*
275	Manufacture of Cork and Cork Products	1 8016*	3.4424*	0.372***
276	Manufacture of Wooden Furniture And Fixtures	0.9908*	-0.439**	1.2915*
279	Manufacture of Products Of Wood, Bamboo Cane, Reed and Grass (Incl. Articles Made from Shell etc) n.e.c.	0.487**	0.7861*	0.355***
280	Manufacture of Pulp, Paper and Paper Board incl. Manufacture of Newsprint	0.5923**	0.5115**	0.9723*
281	Manufacture of Containers and Boxes of Paper and Paper Board	1.131*	0.5917**	0.9478*
282	Manufacture of Paper and Paper Board Articles and Pulp Products n.e.c.	1.3462*	1.4022*	1.2578*
283	Manufacture of Paper and Paper Board Articles not elsewhere classified	1.4103*	0.672**	1.3194*
316	Manufacture of Refined Petroleum Products n.e.c. (Obtained from Products or Residues from Petroleum Refining)	0.0052@	-0.342***	0.0287@
317	Manufacture of Matches	0.0129@	0.0105@	0.0144@
378	Manufacture of Bullock Carts, Push-Carts, Hand Carts *etc.*	DNF	DNF	DNF
385	Manufacture of Sports and Athletic Goods	DNF	DNF	DNF
386	Manufacture of Musical Instruments (Excl. Toys)	0.6356**	-0.675*	0.007@
387	Manufacture of Stationery Articles n.e.c.	0.124***	-0.816*	0.018@
Total Wood & Forest Based Industries		0.7051**	0.6879*	0.8963*

(DNF means data not found for all the years).

**Significant at 1% level of significance*

*** Significant at 5% level of significance*

**** Significant at 10% level of significance*

@ Insignificant

Table (5.2.18): Growth in Total Factor Productivity Indices: Western Region

Industry Code	Description of 3 Digit Asi Industries (Based on NIC 1970 & 1987)	Kendrick Index	Solow Index	Divisia Index
226	Manufacture of Bidi	1.465*	1.4823*	1.4743*
270	Sawing and Planing of Wood (Other Than Plywood)	0.6971*	0.1503***	1.0757*
271	Manufacture of Veneer Sheets, Plywood and their Products	1.0298*	0.5978**	1.2493*
272	Manufacture of Structural Wooden Goods such as Beams, Posts, Doors & Windows (Excluding Hewing and Rough Shaping of Poles, Bolts & other Materials)	1.002*	1.0476*	1.4719*
273	Manufacture of Wooden and Cane Boxes, Crates, Drums, Barrels and other Containers, Baskets, and other Wares Made Entirely or Mainly of Cane, Rattan, Reed, Bamboo, Willow, Fibres, Leaves and Grass	1.6285*	0.4458**	0.9968*
274	Manufacture of Wooden Industrial Goods n.e.c.	1.1651*	0.1492***	1.304*
275	Manufacture of Cork and Cork Products	2.6891*	-0.381***	0.4388**
276	Manufacture of Wooden Furniture And Fixtures	0.5708**	-0.215***	1.407*
279	Manufacture of Products Of Wood, Bamboo Cane, Reed and Grass (Incl. Articles Made from Shell etc) n.e.c.	0.9662*	2.4933*	0.844*
280	Manufacture of Pulp, Paper and Paper Board incl. Manufacture of Newsprint	0.761*	0.35**	1.255*
281	Manufacture of Containers and Boxes of Paper and Paper Board	1.5625*	1.314*	1.31*
282	Manufacture of Paper and Paper Board Articles and Pulp Products n.e.c.	1.6019*	1.77*	1.68*
283	Manufacture of Paper and Paper Board Articles not elsewhere classified	1.6471*	1.80*	1.54*
316	Manufacture of Refined Petroleum Products n.e.c. (Obtained from Products or Residues from Petroleum Refining)	1.3847*	-0.403**	1.706*
317	Manufacture of Matches	DNF	DNF	DNF
378	Manufacture of Bullock Carts, Push-Carts, Hand Carts *etc.*	DNF	DNF	DNF
385	Manufacture of Sports and Athletic Goods	1.2758*	0.731*	1.0199*
386	Manufacture of Musical Instruments (Excl. Toys)	2.9633*	0.321***	0.032@
387	Manufacture of Stationery Articles n.e.c.	0.3687**	-0.002@	0.218***
Total Wood & Forest Based Industries		1.295*	0.7375*	1.197*

(DNF means data not found for all the years).

**Significant at 1% level of significance*

*** Significant at 5% level of significance*

**** Significant at 10% level of significance*

@ Insignificant

Table (5.2.19): Growth in Total Factor Productivity Indices: Northern Region

Industry Code	Description of 3 Digit Asi Industries (Based on NIC 1970 & 1987)	Kendrick Index	Solow Index	Divisia Index
226	Manufacture of Bidi	3.2*	2.14*	2.54*
270	Sawing and Planing of Wood (Other Than Plywood)	0.8182*	0.1056@	1.262*
271	Manufacture of Veneer Sheets, Plywood and their Products	0.23***	0.583**	0.536**
272	Manufacture of Structural Wooden Goods such as Beams, Posts, Doors & Windows (Excluding Hewing and Rough Shaping of Poles, Bolts & other Materials)	1.017*	1.808*	1.49*
273	Manufacture of Wooden and Cane Boxes, Crates, Drums, Barrels and other Containers, Baskets, and other Wares Made Entirely or Mainly of Cane, Rattan, Reed, Bamboo, Willow, Fibres, Leaves and Grass	1.814*	0.247***	1.008*
274	Manufacture of Wooden Industrial Goods n.e.c.	1.077*	0.462**	1.206*
275	Manufacture of Cork and Cork Products	1.4732*	2.202*	0.307**
276	Manufacture of Wooden Furniture And Fixtures	0.8365*	0.21***	1.26*
279	Manufacture of Products Of Wood, Bamboo Cane, Reed and Grass (Incl. Articles Made from Shell etc) n.e.c.	1.4988*	2.743*	1.3775
280	Manufacture of Pulp, Paper and Paper Board incl. Manufacture of Newsprint	0.6941*	0.609**	1.14*
281	Manufacture of Containers and Boxes of Paper and Paper Board	1.751*	1.86*	1.461*
282	Manufacture of Paper and Paper Board Articles and Pulp Products n.e.c.	1.562*	1.166*	1.46*
283	Manufacture of Paper and Paper Board Articles not elsewhere classified	1.536*	1.674*	1.45*
316	Manufacture of Refined Petroleum Products n.e.c. (Obtained from Products or Residues from Petroleum Refining)	0.201***	0.109***	0.188***
317	Manufacture of Matches	0.0037@	0.202***	0.0195@
378	Manufacture of Bullock Carts, Push-Carts, Hand Carts *etc.*	DNF	DNF	DNF
385	Manufacture of Sports and Athletic Goods	1.438	0.4463	1.1493
386	Manufacture of Musical Instruments (Excl. Toys)	DNF	DNF	DNF
387	Manufacture of Stationery Articles n.e.c.	0.3283**	-0.02@	0.047**
Total Wood & Forest Based Industries		1.0635*	1.080*	0.975*

(DNF means data not found for all the years).

**Significant at 1% level of significance*

*** Significant at 5% level of significance*

**** Significant at 10% level of significance*

@ Insignificant

Table (5.2.20): Growth in Total Factor Productivity Indices: Southern Region

Industry Code	Description of 3 Digit Asi Industries (Based on NIC 1970 & 1987)	Kendrick Index	Solow Index	Divisia Index
226	Manufacture of Bidi	0.566**	0.859*	0.855*
270	Sawing and Planing of Wood (Other Than Plywood)	0.6648*	0.1046@	1.164*
271	Manufacture of Veneer Sheets, Plywood and their Products	1.0185*	0.2098***	1.4008*
272	Manufacture of Structural Wooden Goods such as Beams, Posts, Doors & Windows (Excluding Hewing and Rough Shaping of Poles, Bolts & other Materials)	2.63*	2.104*	1.457*
273	Manufacture of Wooden and Cane Boxes, Crates, Drums, Barrels and other Containers, Baskets, and other Wares Made Entirely or Mainly of Cane, Rattan, Reed, Bamboo, Willow, Fibres, Leaves and Grass	1.445*	0.5373**	0.894*
274	Manufacture of Wooden Industrial Goods n.e.c.	1.428*	0.696*	1.29*
275	Manufacture of Cork and Cork Products	1.990*	3.801*	0.412**
276	Manufacture of Wooden Furniture And Fixtures	0.889*	-0.41**	1.134*
279	Manufacture of Products Of Wood, Bamboo Cane, Reed and Grass (Incl. Articles Made from Shell etc) n.e.c.	0.9824*	1.307*	0.652**
280	Manufacture of Pulp, Paper and Paper Board incl. Manufacture of Newsprint	0.6862*	0.731*	1.126*
281	Manufacture of Containers and Boxes of Paper and Paper Board	1.809*	0.94*	1.5925*
282	Manufacture of Paper and Paper Board Articles and Pulp Products n.e.c.	1.449*	0.954*	1.2706*
283	Manufacture of Paper and Paper Board Articles not elsewhere classified	0.776*	1.041*	1.2692*
316	Manufacture of Refined Petroleum Products n.e.c. (Obtained from Products or Residues from Petroleum Refining)	0.0058@	0.326***	0.032@
317	Manufacture of Matches	1.33*	0.0292@	0.018@
378	Manufacture of Bullock Carts, Push-Carts, Hand Carts *etc.*	0.0013@	0.0738@	0.0073@
385	Manufacture of Sports and Athletic Goods	1.235*	0.688*	0.9612*
386	Manufacture of Musical Instruments (Excl. Toys)	DNF	DNF	DNF
387	Manufacture of Stationery Articles n.e.c.	0.202***	-0.052@	0.0237@
Total Wood & Forest Based Industries		1.010*	1.163*	0.96*

(DNF means data not found for all the years).

**Significant at 1% level of significance*

*** Significant at 5% level of significance*

**** Significant at 10% level of significance*

@ Insignificant

Table (5.2.21): Growth in Total Factor Productivity Indices: Other States

Industry Code	Description of 3 Digit Asi Industries (Based on NIC 1970 & 1987)	Kendrick Index	Solow Index	Divisia Index
226	Manufacture of Bidi	0.686**	0.997*	0.094*
270	Sawing and Planing of Wood (Other Than Plywood)	1.142*	2.638*	1.039*
271	Manufacture of Veneer Sheets, Plywood and their Products	1.102*	1.749*	1.008*
272	Manufacture of Structural Wooden Goods such as Beams, Posts, Doors & Windows (Excluding Hewing and Rough Shaping of Poles, Bolts & other Materials)	0.8076*	1.225*	1.013*
273	Manufacture of Wooden and Cane Boxes, Crates, Drums, Barrels and other Containers, Baskets, and other Wares Made Entirely or Mainly of Cane, Rattan, Reed, Bamboo, Willow, Fibres, Leaves and Grass	1.1062*	1.712*	0.969*
274	Manufacture of Wooden Industrial Goods n.e.c.	0.8382*	1.975*	0.894*
275	Manufacture of Cork and Cork Products	1.094*	2.232*	1.245*
276	Manufacture of Wooden Furniture And Fixtures	0.5828**	-1.842*	0.929*
279	Manufacture of Products Of Wood, Bamboo Cane, Reed and Grass (Incl. Articles Made from Shell etc) n.e.c.	0.935*	1.080*	0.830*
280	Manufacture of Pulp, Paper and Paper Board incl. Manufacture of Newsprint	0.668**	1.226*	0.638*
281	Manufacture of Containers and Boxes of Paper and Paper Board	0.8945*	0.999*	0.88*
282	Manufacture of Paper and Paper Board Articles and Pulp Products n.e.c.	0.9268*	1.142*	0.911*
283	Manufacture of Paper and Paper Board Articles not elsewhere classified	1.035*	1.126*	0.932*
316	Manufacture of Refined Petroleum Products n.e.c. (Obtained from Products or Residues from Petroleum Refining)	0.0089@	-2.177*	0.038***
317	Manufacture of Matches	0.0072@	2.308*	0.625**
378	Manufacture of Bullock Carts, Push-Carts, Hand Carts *etc.*	0.4851**	0.498**	0.493**
385	Manufacture of Sports and Athletic Goods	0.8547*	0.903*	0.903*
386	Manufacture of Musical Instruments (Excl. Toys)	1.2886*	-0.316**	0.112***
387	Manufacture of Stationery Articles n.e.c.	0.6253**	-0.211***	0.22***
Total Wood & Forest Based Industries		0.5292**	1.240*	1.184*

**Significant at 1% level of significance*
*** Significant at 5% level of significance*
**** Significant at 10% level of significance*
@ Insignificant

Table (5.2.22): Growth in Total Factor Productivity Indices: All India

Industry Code	Description of 3 Digit Asi Industries (Based on NIC 1970 & 1987)	Kendrick Index	Solow Index	Divisia Index
226	Manufacture of Bidi	0.853**	1.295*	1.288*
270	Sawing and Planing of Wood (Other Than Plywood)	0.611**	0.29***	0.943*
271	Manufacture of Veneer Sheets, Plywood and their Products	0.934*	0.843*	1.133*
272	Manufacture of Structural Wooden Goods such as Beams, Posts, Doors & Windows (Excluding Hewing and Rough Shaping of Poles, Bolts & other Materials)	0.9238*	1.64*	1.357*
273	Manufacture of Wooden and Cane Boxes, Crates, Drums, Barrels and other Containers, Baskets, and other Wares Made Entirely or Mainly of Cane, Rattan, Reed, Bamboo, Willow, Fibres, Leaves and Grass	1.521*	0.782**	0.931*
274	Manufacture of Wooden Industrial Goods n.e.c.	0.993*	0.87*	1.112*
275	Manufacture of Cork and Cork Products	1.90*	3.63*	1.393*
276	Manufacture of Wooden Furniture And Fixtures	0.48***	0.348***	1.183*
279	Manufacture of Products Of Wood, Bamboo Cane, Reed and Grass (Incl. Articles Made from Shell etc) n.e.c.	0.92*	1.98*	0.671**
280	Manufacture of Pulp, Paper and Paper Board incl. Manufacture of Newsprint	0.457**	0.678**	0.75*
281	Manufacture of Containers and Boxes of Paper and Paper Board	1.399*	1.176*	1.168*
282	Manufacture of Paper and Paper Board Articles and Pulp Products n.e.c.	1.381*	1.51*	1.29*
283	Manufacture of Paper and Paper Board Articles not elsewhere classified	1.413*	1.545*	1.319*
316	Manufacture of Refined Petroleum Products n.e.c. (Obtained from Products or Residues from Petroleum Refining)	0.0037@	0.205@	0.0205@
317	Manufacture of Matches	0.0022@	0.12***	0.012***
378	Manufacture of Bullock Carts, Push-Carts, Hand Carts *etc.*	0.0013@	0.074***	0.0073@
385	Manufacture of Sports and Athletic Goods	1.276*	0.731*	1.02*
386	Manufacture of Musical Instruments (Excl. Toys)	2.963*	0.322***	0.032***
387	Manufacture of Stationery Articles n.e.c.	0.146@	0.0089@	0.021@
Total Wood & Forest Based Industries		0.523**	0.92*	1.12*

**Significant at 1% level of significance*

*** Significant at 5% level of significance*

**** Significant at 10% level of significance*

@ Insignificant

References

Ahluwalia I.J. (1985): 'Industrial Growth in India: Stagnation since the Mid-Sixties'', Oxford University Press, New Delhi, 1985.

Ahluwalia, I.J. (1991): Productivity and Growth in Indian Manufacturing, Oxford University Press, New Delhi.

Balakrishnan, P and Pushpangadan, K. (1994): "Total Factor Productivity Growth in Manufacturing Industry: A Fresh Look", *Economic and Political Weekly,* 29: 2028-35.

Balakrishnan, P. & Pushpangadan, K. (1998): "What do we know about productivity growth in Indian industry?", *Economic and Political Weekly,* 33: 2241–46.

Balakrishnan, P. (2004): "Measuring Productivity in Manufacturing Sector", *Economic and Political Weekly,* April 3-10.

Baltagi, B.H. & J.M. Griffin. (1988). "A Generalized Error Component Model with Heteroscedastic Disturbances." *International Economic Review* 29, 745-753.

Banerji, A., -"Capital Intensity and Productivity in Indian Industry", Macmillon, Delhi, 1975.

Barro, R. (1991): "Economic Growth in a Cross-Section of Countries", *Quarterly Journal of Economics,* 106: 407-444.

Barro, R., Mankiw, N. & X. Sala-i-Martin, (1995): "Capital Mobility in Neoclassical Models of Growth", *American Economic Review,* 85(1): 103-115.

Brahmananda, P.R. (1982): "Productivity in the Indian Economy: Rising Inputs for Falling Outputs", Himalaya Publishing House, Mumbai.

Cobb, C.W. & P.H. Douglas (1928), "A Theory of Production", in American Economic Review, 18: 139-165. Denison, E. (1985), "Trends in American Economic Growth", Washington: Brookings Institution.

Christiansen L.R., Jorgensen D.W, Lau L.J. (1971): "Conjugate duality and the transcendental logarithmic production function", in *Econometrica,* vol. 39

Christiansen L.R., Jorgensen D.W, Lau L.J. (1973): "Transcendental logarithmic production frontier", in *Review of Economics and Statistics,* vol. 55.

Dholakia, R. H. & Dholakia, B. H. (1994): "Total Factor Productivity Growth in Indian Manufacturing", *Economic and Political Weekly,* 29, 342-344.

Dhillon, S. S., - "Productivity Trends and Factor Substitutability in Manufacturing Sector of Karnataka", Margin: 16 (1); Oct., 1983.

Dhillon, S.S., - "Productivity Trends & Factor Substitutability in Manufacturing Sector in Karnataka", *Margin*: 16 (1), Oct. 1983.

Diewert, W. E. (2000): "The Quadratic Approximation Lemma and Decompositions of Superlative Indexes", Discussion Paper 00-15, Department of Economics, University of British Columbia, Canada

D.M. Wu (1975): "Estimation of the Cobb-Douglas Production Function", *Econometrica,* Vol. 43, No. 4, p.-739-744.

Fabricant, S. (1954): "Economic Progress and Economic Change", 34th Report of the National Bureau of Economic Research, New York: NBER.

Goldar, B. N. (1986): "Productivity Growth in Indian Industry", Allied Publishers, New Delhi.

Hasim, S. R. & Dadi, S. S., -"Capital-Output Relations in Indian Manufacturing (1946-1964)", M. S. University of Baroda, 1973.

Kathuria, V., Raj, R.S.N & Sen, K (2011): "Productivity Measurement in Indian Manufacturing: A Comparison of Alternative Methods", Working Paper No. 31, Institute of Development Policy and Management, School of Environment and Development, University of Manchester, UK.

Kmenta J. (1967): "On Estimation of CES Production Function", in *International Economic Review*.

Krugman, P. (1990): "The Age of Diminished Expectations", MIT Press, Massachusetts, Cambridge.

Kumar, V., - "Productivity Trends in Selected Industries in India.", *Indian Labour Journal*, 24 (1); Jan. 1983.

Lu, Y.C. & L. Fletcher (1968): "A Generalization of the CES Production Function", *Rev. Econ. & Stats*, 50(4).

Metha, S. S., -"Productivity, Production function and Technical change: A Survey of Some India Industries.", Concept Publishing Company, New Delhi, 1980.

Pendse, N.G. & Baghel, LMS (2008): "Technological Change and Productivity Growth in Manufacturing Sector of India", Sarup & Sons, New Delhi.

Rajalakshmi, K., - "Sources of Growth in the Total Manufacturing Sector in Rajasthan and at India Level", *Indian Journal of Industrial Relations*: 19 (1), July, 1983.

Singh, L., - "Productivity Trends and Factor Substitutability in Punjab Industry", *Indian Journal of Economics*: 67 (266), Jan. 1987.

6

Production Function Analysis

6.1: Introduction

Production function is a technological or engineering concept which shows for a given state of technological knowledge and managerial ability, the maximum rates of output that can be obtained from different combinations of the productive factors during a given period of time. In brief, the production function is a catalog of different output possibilities. Hence, production function is the process by which inputs are converted into output. In other words, production function is defined as the technical or engineering relationship between factor inputs and output. The production function formalizes the relationship between the quantity of output yielded by a product process and the quantities of the various input used in that process. Inputs refer to the various factor prices, which are used in the production. Outputs on the contrary refers o the quantity of goods produced by the firm with the help of the various inputs. In the form of mathematical equations, the production function can also be expressed in which output is the dependent variable and inputs are the independent variables. Therefore, relationship of these terms in general form can be stated as: **Q: $(X_1, X_2, X_3, \ldots\ldots. X_n)$**. Where Q is the rate of output of a given commodity and $X_1, X_2, X_3, \ldots\ldots X_n$ are the various factor of production used per unit of the time.

The main purpose of applying production functions is to estimate marginal productivity and elasticity of inputs, returns to scale and elasticity of substitution between the inputs. Among the available literate of production functions, the choice in this study was in favour of Cobb-Douglas, Constant Elasticity of Substitution and Variable Elasticity of Substitution because these are the most popular and widely used form of production functions. These production function involve treatment of elasticity of substitution constant and equal to unity in C–D specification and constant but not necessarily equal to unity in case of C.E.S. specification. While C.E.S. production function major limitation is, it assumes the elasticity of substitution to be constant for all input combinations. In the literature, several functional forms have been suggested and used in empirical studies in which elasticity of substitution is variable. These are called V.E.S. (variable elasticity of substitution) production function. A vast detailed discussion has already been made in the previous chapter (in Chapter-IV) regarding the specification and from of above mentioned production function specification. A time trend variable has been included in production functions to represent technological progress.

However, Production functions for the Wood & Forest based Industries (WFBI) for India and Indian Regions has been estimated with a view to identify the contribution of factors other than capital and labour inputs to output growth. TFPG, which is an indicator of the change in efficiency in factor use, discussed earlier has also been measured for explaining the resource use efficiency. The conceptual issues associated with the problems of aggregation, valuation of capital, under utilization of capital and changing quality of labour in the specification of a production function have been extensively discussed in research methodological chapter.

In case of growth accounting estimates, direct estimation of the production function has an advantage in that, it is not necessary to assume competitive equilibrium in order to derive an estimate of productivity growth. The efficiency parameter, the scale parameter and the extent of substitutability between factors can be obtained by directly estimating the parameters of a suitably production function through regression analysis. However, statistical problems associated with endogeneity of input variables, multi-collinearity among the explanatory variables and autocorrelation of the residuals tend to impinges upon the robustness and other desirable properties of the production function estimates. In the specification of a production function, intermediate inputs can be explicitly incorporated by relating changes in the gross value of output to changes in labour, capital and intermediate inputs. When intermediate inputs are explicitly recognized, it implies that the estimated TFP growth rates are lower than that base on the value added function. The analysis in the present study is based on the value added production function.

6.2: Brief Methodological Frame of Production Functions

In the following section an attempt has been made to present the estimates of production functions for various industries under consideration and total wood and forest based industrial sector using alternative specification outlined above e.g. Cobb-Douglas, Constant Elasticity of Substitution (CES) and Variable Elasticity of Substitution (VES).

Cobb-Douglas functional form of production functions is widely used to represent the relationship of an output to inputs. It was proposed by Knut Wicksell (1851-1926), and tested against statistical evidence by Charles Cobb and Paul Douglas in 1928. In 1928 Charles Cobb and Paul Douglas published a study in which they modeled the growth of the American economy during the period 1899-1922. They considered a simplified view of the economy in which production output is determined by the amount of labor involved and the amount of capital invested. While there are many other factors affecting economic performance, their model proved to be remarkably accurate. In its most standard form for production of a single good with two factors, the C-D function can be written as:

$$V = Ao\, L^{\alpha} K^{\beta} e^{\lambda t} \quad (6.1)$$

A natural logarithmic transformation of the C-D production function yields an equation that is linear in the natural logarithms of output and inputs and in time and is written as:

$$\text{Ln } V = a + \alpha \text{ Ln } L + \beta \text{ Ln } K + \lambda t \qquad (6.2)$$

The OLS (Ordinary Least Square) estimation to this equation yields, α, β and λ estimates. While, λ provides a measure of the TFP growth, the sum of the estimates of α and β is a measure of the degree of homogeneity of the production function. Thus, returns to scale are decreasing, constant or less than one, one, or greater than one.

If the function is transformed so as to have Ln (V/L) as the dependent variable, the estimates of this function provide a direct test of whether the degree of homogeneity is significantly different from unity or that the returns to scale are significantly different from being constant. The ratio form of the production function can be written as:

$$\text{Ln } (V/L) = a + \beta \text{ Ln } (K/L) + (\alpha + \beta - 1) \text{ Ln } L + \lambda t \qquad (6.3)$$

A restrictive form of the C-D production function explicitly assumes constant returns to scale i. e. $(\alpha + \beta = 1)$ and is written as:

$$\text{Ln } (V/L) = a + \beta \text{ Ln } (K/L) + \lambda t \qquad (6.4)$$

As mentioned earlier, the Cobb-Douglas production function assumes that the elasticity of substitution between labour and capital is equal to unity. One way of testing for this assumption is to specify a production function with constant (but not necessarily unitary) elasticity of substitution and test whether the elasticity is significantly different from one. The Constant Elasticity of Substitution production function is written as:

$$V = Yo\, e^{\lambda t} [\delta L^{-\varrho} + (1-\delta) K^{-\varrho}]^{-V/\varrho} \qquad (6.5)$$

Where,

λ is efficiency parameter is the distribution parameter

ϱ is the substitution parameter, and

V is the scale parameter.

The CES production function cannot be transformed into a function that is linear in the parameter and thus requires a non-linear estimation procedure.

Kmenta (1967) provides a method of linearizing the CES function by obtaining an approximation to the function. If the interest is centered on testing for the assumption of unitary elasticity of substitution, it is possible to derive from the CES production function, assuming competitive equilibrium (so that the real wage rate equals the marginal product of labour), a linear relationship between labour productivity, on the one hand, and the real wage rate and the level of employment, on the other.

$$\text{Ln } (V/L) = a + b \text{ Ln } (w) + c \text{ LnL} + d_t \qquad (6.6)$$

Where,

$$b = \frac{v}{(v+p)}, c = \frac{p(v-1)}{(v+p)} \text{ and } \quad d = \frac{\lambda p}{(v+p)}$$

The SMAC form of CES production function can be written as:

$$\text{Ln } V/L = a + b \text{ Ln } w \quad (6.7)$$

If time variable in incorporated the above specification, the equation becomes:

$$\text{Ln } V/L = a + b \text{ Ln } w + dt \quad (6.8)$$

A major limitation of C.E.S. production function is that it assumes the elasticity of substitution to be constant for all input combinations. Several functional forms have been suggested in the literature and used in empirical studies in which elasticity of substitution is variable. These are called V.E.S. (variable elasticity of substitution) production function. We take up for discussion one type of V.E.S. function, which has been suggested by Hildebrand and Liu (1965) and has been applied in a number of studies for Indian industries. The functional form suggested by Hildebrand and Liu is as follows:

$$Y = A\left[(1-\delta) K^{n} + \delta K^{mn} L^{(1-m)n}\right]^{1/n} \quad (6.9)$$

Where Y is output (value added), K capital and L labour. It reduces to the CES form for m = 0. It has the properties of a neoclassical production function, i.e. positive marginal product and downward sloping marginal product curve, over the relevant range of inputs. This function is homogeneous of degree one (i.e. constant returns to scale) and has variable elasticity of substitution.

An important implication of this functional form is that labour productivity becomes a log-linear function of wage rate and capital- labour ratio (under the condition of cost minimization and competitive markets):

$$\text{Ln}(Y/L) = a + b \ln (w) + c \ln (K/L) \quad (6.10)$$

Using the parameters of the above equation, the elasticity of substitution may be obtained as: $\sigma = b / [1-(c/S_k)]$, Where S_k is the share of capital.

6.3: Estimates of Production Functions

Banerji (1975) had made a systematic attempt to test for the assumptions underlying the C-D specification and found empirical support for these assumptions. Goldar (1986) has estimated C-D production function for the large-scale registered manufacturing sector. Attempts have also been made to estimates production functions for individual industries of manufacturing using data on a time series data of firms within the industry, the most notable example of this being the recently published estimates of Translog production function for five industries by Little, Mazumdar & Page (1987).

6.3.1 (I): Estimates of C-D Production Function for Eastern Region

Various estimates of Cobb-Douglas production function for Wood & Forest based Industrial sector as well as individual industry for the Eastern Region of India is presented in Table 6.1.1 to 6.1.4. The estimates of the unrestricted Cobb-Douglas function without time trend along with time trend and ratio form is worked out in following paras. The results of unrestricted Cobb-Douglas production function without time trend are presented in Table 6.1.1 to 6.1.2.

To begin with, let's have a look of Table 6.1.1. The analysis of table reveals that most of the industry group at three digit level shows that the labour and capital taken together explains more than 90% variation in dependent variable along with overall manufacturing sector. The industry group that explains more than 90% variation in dependent variable are Manufacture of Bidi (226); manufacture veneer, plywood and their products (270), manufacture of structural wooden goods (273), manufacture of wooden industrial goods... (274); manufacture of cork and cork products (275); manufacture of wooden furniture & fixtures (276); manufacture of container and boxes of paper and paper board (281); and total WFBI as well. The industry group that explains between 70-90% variations in dependent variable are Manufacture of sawing & plaining of wood (271), manufacture of wood, bamboo and cane products n.e.c. (279); and manufacture of pulp, paper and paper board including newsprint (280). Efficiency parameter was observed positive and significant in most of the industry group at three-digit level. But, in some cases it turns to be negative and significant such as manufacture of Bidi (226); manufacture of cork & cork products (275); manufacture of turpentine, synthetic resin...(316); and manufacture of miscellaneous products n.e.c. (387). Labour coefficient was found positive and significant in almost all cases. However, the negative and significant labour coefficient was found only in 3 cases i.e. Industry code 316, 386 and for 387, whereas, insignificant labour coefficient was not found for any industry. The capital coefficient however showed mixed results. Though the most of the industries have shown negative and significant parameters, only one industry have shown significant positive trend in capital coefficients and industry belong to this category are manufacture of miscellaneous products n.e.c.(387). Whereas, insignificant capital coefficient were found in case of industry group like, 226, 273, 275 and 386.

For Eastern region of India, the results of unrestricted Cobb-Douglas production function with time trend are presented as Model II in Table 6.1.2. Inclusion of time variable has improved the fit. Inclusion of time variable brought reversal in negative sign of capital coefficient. Labour coefficient was found positive and significant in almost all cases. Sign of time coefficient was found negative and significant in most of the cases. This indicates negative growth in total factor productivity in most of the cases. Efficiency parameter was observed positive and significant in most of the industry group at three-digit level in model II. Whereas in some of the industry group cases it turns to be negative and significant such as Industry Code (226, 316, and 387). However, it was insignificant in some of the industry group cases such as Manufacture of structural wooden goods (273) and manufacture of cork and cork products (275). Labour coefficient was found positive and significant in almost all cases. However, the negative and significant labour coefficient was found only in case of Manufacture of sports & athletic goods and play equipments (386) and manufacture of miscellaneous products n.e.c.(387). The capital coefficient however continued to show mixed results as it obvious for Model I of C-D production function. Though, the most of the industries have shown improvement in their magnitude, as well as, reversal in sign of capital coefficient. As compared to negative sign of capital coefficient in model I, Inclusion of time variable in Model II, brought reversal in negative sign of capital coefficient in most of the cases. The positive and significant capital coefficient was observed in case of industry group 272, 273, 279, 280, and 282. Sign of time coefficient was found negative and significant in 14 cases while insignificant in 6 cases and total WFBI as well. This indicates negative growth in total factor productivity in most of the cases. However, time coefficient was not found positive and significant in any group of industries belongs to WFBI.

If the function is transformed so as to have Ln (V/L) as the dependent variable, the estimates of this function provide a direct test of whether the degree of homogeneity is significantly different from unity or that the returns to scale are significantly different from being constant. The estimates of ratio form of Cobb-Douglas production function for the Eastern Region of India are presented as a Model III in Table 6.1.3. The estimates of ratio form of Cobb-Douglas production function suggests the industry groups where the degree of homogeneity is significantly different from unity or that the returns to scale are significantly different from being constant are manufacture of Bidi (226), manufacture of wooden industrial goods...(274), and manufacture of wooden furniture and fixtures (276). Significant positive coefficient time indicates upward trend in total factor productivity. The industries groups belong to this category are 226, 270, 273, 274, 276, and 386. Significant negative time coefficient indicates downward trend in total factor productivity. The industries groups belong to this category are 275, 316, and 387. Negative and Insignificant coefficients of time trend implies that the production function are not subject to any kind of significant shifts in the industry under reference period and but here in eastern region not a single such type of case has been observed. So far as best fit of the industries under study are concern, except 6 industries (such as, industry code: 275, 280, 282, 316, 386 and 387) rest of group of industries are explaining fit of regression very good as evident from highly significant values of Adjusted R^2 and F.

A restrictive form of C-D production function explicitly assumes constant returns to scale. The results of this specification for Eastern Region as Model IV are presented in Table 6.1.4. In this specification the efficiency parameter turns to be negative and significant in most of the cases except in two industries such as Manufacture of cork & cork products (275); Manufacture of miscellaneous products n.e.c. (387) and total WFBI of eastern Region. In this specification the capital intensity parameter was found positive and significant in 6 industries such as industry code 226, 270, 274, 275, 276 and 279 while in other cases, either found insignificant and negative or insignificant. In this specification the significant negative time coefficient was observed only in two industries. The industries groups belong to this category are Manufacture of cork and cork products; and manufacture of turpentine, synthetic resin, *etc* (316), indicates downward trend in total factor productivity. Whereas, insignificant time coefficient was found in most of the cases and total WFBI as well. Only one industry namely manufacture of musical instruments (386) depicted positive and significant time coefficient suggesting upward trend in total factor productivity growth.

6.3.1 (II): Estimates of C.E.S. Production Function for Eastern Region

As discussed earlier in previous Chapter II & IV that Elasticity of substitution (σ) reflecting responsiveness of factor proportion to a change in the relative factor prices is a crucial parameter, particularly in case of developing countries like India.

It has important policy implications for economic growth, resource allocation and relative income distribution. High elasticity of substitution (or easy substitutability) has a favorable impact on the levels of both output and employment. It has been seen that the conflict between high output and high employment (the dilemma faced by most of the underdeveloped and developing economies) essentially arise out from low substitutability of production structure. Apart from these, estimate of elasticity of substitution is necessary to make choice of the form of production function to be applied for empirical analysis. The results of this specification for Eastern Region of India are presented as Model I in Table 6.1.5.

Model I of Constant Elasticity of Substitution (C.E.S.) specification is the original SMAC formulation relating the Ln (V/L) to Ln (W). The coefficient of wage rate (giving estimate of σ) is found to significant even at one per cent level of significance in case of most of the industry groups, where it was found to be different from unity. However, in case of 3 industry groups (such as Industry code 272, 279 and 316) it was found to be insignificant. Incorporation of time variable in model II (presented in Table 6.1.6) resulted in a nominal improvement in overall fit. Though the coefficient of wage rate (giving estimate of σ) is found to be significant in only 3 cases, but in some cases it turns to insignificant, the industry groups belong to this category are 226, 273, 276, 281, 316, and 386 industries of WFBI of Eastern region. The coefficient of wage rate (giving estimate of σ) is found to be positive and significant in case of industry manufacture of veneer, plywood and their products (270); manufacture of wooden and cane boxes...(272); manufacture of wooden industrial goods...(274); manufacture of cork and cork products (275); manufacture of wood, bamboo and cane products n.e.c. (279); manufacture of pulp, paper and paper board including newsprint (280); manufacture of paper products n.e.c. like dolls (282); and total wood & forest based industries of eastern region.

Inclusion of labour variable in Model III (Table 6.1.7) the coefficients of σ (wage rate) were found positive in case of industry code 271, 270, 274, 275, 279, 280, and 282 only and in most of the other cases σ were experiences insignificant or negative trend similarly like labour variable which were also found to be negative or insignificant. The coefficient of time although found significant but not positive.

Model IV of Constant Elasticity of Substitution (C.E.S.) specification is the Kmenta approximation. In this Model IV (presented in Table 6.1.8), the terms Ln(L) and Ln(K) are those of the C-D function while the last term accounts for non-unitary elasticity of substitution. In practices, the closer the elasticity of substitution to unity, the better the approximation. In this specification distribution parameter, scale parameter, and substitution parameter was not found significant in any case of industry groups under study. The time coefficient representing total factor productivity growth was found positive and significant only in 2 cases of industry groups like manufacture of Bidi (226); manufacture of container and boxes of paper and paper board (281) and total WFBI of eastern region. However, the time coefficient was found negative and significant in cases of industry groups like sawing and plaining of wood (271); manufacture of wood, bamboo & cane products (279); manufacture of pulp, paper and paper board including newsprint

(282); manufacture of turpentine, synthetic resin...(316); and manufacture of miscellaneous products n.e.c (387). In rest of the cases of industry groups of WFBI, the time coefficient was found insignificant.

6.3.1 (III): Estimates of V.E.S. Production Function for Eastern Region

To determine the relevant form of production functions, it has always been wise to examine whether elasticity of substitution is variable or constant. However, a major limitation of C.E.S. production function is that it assumes the elasticity of substitution to be constant for all input combinations. Several functional forms have been suggested in the literature and used in empirical studies in which elasticity of substitution is variable. These are called V.E.S. (variable elasticity of substitution) production function. It has the properties of a neoclassical production function, i.e. positive marginal product and downward sloping marginal product curve, over the relevant range of inputs. This function is homogeneous of degree one (i.e. constant returns to scale) and has variable elasticity of substitution. An important implication of this functional form is that labour productivity becomes a log-linear function of wage rate and capital-labour ratio (under the condition of cost minimization and competitive markets). The results of this specification for the Eastern Region of India are presented in Table 6.1.9 & 6.1.10 as Model I and Model II.

To begin with model I, it has been observed that the coefficient of wage rate was found positive and significant in most of the cases of industry groups except in cases of industry groups like manufacture of structural wooden goods (273); manufacture of pulp, paper and paper board (280); manufacture of musical instruments (386), where it was found insignificant. The coefficient of capital-labour ratio (i.e. LnK/L) was found negative and significant in cases of industry groups 273, 280, 282, 316, 386. The coefficient of capital-labour ratio was found insignificant in cases of industry groups like 271, 273, 280, 282, 316, 386 and 387. While for rest of the industry groups (such as industry code 226, 270, 272, 275, 276, 279 281, and total WFBI), it was found positive and significant. It means coefficients of K/L ratio are statistically different from zero in the model which means that the changes in this ratio may influence the value of elasticity of substitution.

The time coefficient in Model II (presented in Table 6.1.10) of VES specification was found positive and significant suggesting growth in total factor productivity in case of only 2 industry groups such as Manufacture of wooden and cane boxes, crates (272); and manufacture of wooden furniture & fixtures (276). Whereas negative and significant time coefficient was found in case of industry groups like 271, 274, 279, 280, and 282. In rest of the cases of industry groups, it was found insignificant which experiencing absence of technological changes in the period under study.

6.3.2 (I): Estimates of C-D Production Function for Western Region

Various estimates of Cobb-Douglas production function for Wood & Forest based Industrial sector as well as individual industry for the Western Region of India is presented in Table 6.2.1 to 6.2.4. The estimates of the unrestricted Cobb-Douglas function without time trend along with time trend and ratio form is worked out.

The results of unrestricted Cobb-Douglas production function without time trend are presented in Table 6.2.1 to 6.2.2. The analysis of Table 6.2.1 reveals that 11 industries of WFBI of the region along with total WFBI at three digit level shows that the labour and capital taken together explains between 80-90% variations in dependent variable. The industry group that explains more than 90% variation in dependent variable is sawing and plaining of wood (271); manufacture wooden and cane boxes (272); manufacture of wooden industrial goods... (274); manufacture of cork and cork products (275); manufacture of wooden furniture & fixtures (276); manufacture of sports and athletic goods and play equipments (385); manufacture of musical instruments (386); and total WFBI as well. Labour coefficient was found positive and significant at 1 per cent level of significance in almost all cases. However, the neither negative nor insignificant labour coefficient was not found for any industry. The capital coefficient however showed mixed results. Though similarly like eastern region, the most of the industries have shown negative and significant parameters, only one industry have shown significant (at 1% level) and positive trend in capital coefficients and industry belong to this category is manufacture of turpentine, synthetic, resin, plastic materials...(316). Whereas, insignificant capital coefficient were found in case of industry groups like, industry code 226, 273, 275 and 386 as it is found for eastern region too as well.

For the Western region of India, the results of unrestricted Cobb-Douglas production function with time trend are presented as Model II in Table 6.2.2. Inclusion of time variable has improved the fit. Inclusion of time variable brought reversal in negative sign of capital coefficient. Labour coefficient was found positive and significant in almost all cases similarly like model I. Sign of time coefficient was found negative and significant in most of the cases. This indicates negative growth in total factor productivity in most of the cases. Efficiency parameter was observed positive and significant in most of the industry group at three-digit level in model II. Whereas in some of the industry group cases it turns to be negative and significant such as Industry Code (226 and 387). Labour coefficient was found positive and significant in almost all cases. However, insignificant labour coefficient was found only in case of Manufacture of turpentine, synthetic, resin, plastic materials...(316). On the other side, capital coefficient however continued to show mixed results as it obvious for Model I of C-D production function. Though, the most of the industries have shown improvement in their magnitude, as well as, reversal in sign of capital coefficient. As compared to negative sign of capital coefficient in model I, Inclusion of time variable in Model II, brought reversal in negative sign of capital coefficient in most of the cases but in most of the cases they were observed insignificant coefficient. The positive and significant capital coefficient was observed in case of industry group 273, 279, 280, 385, and 282. Sign of time coefficient was found negative and significant in 9 cases while insignificant in 8 cases and along with total WFBI as well. This indicates negative growth in total factor productivity in most of the cases. However, time coefficient was not found positive and significant in any group of industries belongs to WFBI, it means these industries are not experiencing technological change.

If the function is transformed so as to have Ln (V/L) as the dependent variable, the estimates of this function provide a direct test of whether the degree of homogeneity is significantly different from unity or that the returns to scale are significantly different from being constant. The estimates of ratio form of Cobb-Douglas production function for the Western Region of India are presented as a Model III in Table 6.2.3. The estimates of ratio form of Cobb-Douglas production function suggests the industry groups where the degree of homogeneity is significantly different from unity or that the returns to scale are significantly different from being constant are manufacture of Bidi (226); manufacture of veneer, plywood and their products (270); Manufacture of wooden and cane boxes, crates, drums, barrels and other wood (272); manufacture of wooden industrial goods... (274); manufacture of cork & cork products (275); manufacture of wooden furniture and fixtures (276); manufacture of wood, bamboo and cane products (279); manufacture of musical instruments (386) and total WFBI. Significant positive coefficient time indicates upward trend in total factor productivity. The industries groups belong to this category are 273, 274, 276, 316, 385, and 386. Significant negative time coefficient indicates downward trend in total factor productivity. The industries groups belong to this category are 275, and 386. Negative and Insignificant coefficients of time trend implies that the production function are not subject to any kind of significant shifts in the industry under reference period and but here in eastern region not a single such type of case has been observed. So far as best fit of the industries under study are concern, except 6 industries (such as, industry code: 270, 275, 280, 282, 316, and 387) rest of group of industries are explaining fit of regression very good as evident from highly significant values of Adjusted R^2 and F.

A restrictive form of C-D production function explicitly assumes constant returns to scale. The results of this specification for Western Region as Model IV are presented in Table 6.2.4. In this specification the efficiency parameter turns to be negative and significant in most of the cases except in one industry such as manufacture of turpentine, synthetic resin, plastic materials...(316); and for total WFBI of Western Region too. In this specification the capital intensity parameter was found positive and significant in 6 industries such as industry code 226, 270, 274, 275, 276, 279, and 386 along with total WFBI, while in other cases, either found insignificant and negative or insignificant. In this specification the significant negative time coefficient was observed only in two industries. The industries groups belong to this category are Manufacture of cork and cork products (275); and manufacture of musical instrument (386), indicates downward trend in total factor productivity. Whereas, insignificant time coefficient was found in most of the cases and total WFBI as well. Only two industries namely manufacture of sports and athletic goods and play equipments (385) and manufacture of turpentine, synthetic resin, plastic materials... (316) depicted positive and significant time coefficient suggesting upward trend in total factor productivity growth.

6.3.2 (II): Estimates of C.E.S. Production Function for Western Region

The results of estimates of C.E.S. production function for Western Region of India are presented in Table 6.2.5 as model I; Table 6.2.6. as Model II; Table

6.2.7 as model III; and Table 6.2.8 as Model IV. Model I (Table 6.2.5) of Constant Elasticity of Substitution (C.E.S.) specification is the original SMAC formulation relating the Ln (V/L) to Ln (W). The coefficient of wage rate (giving estimate of σ) is found to significant even at 1 per cent level of significance in case of most of the industry groups, where it was found to be different from unity. However, in case of 4 industry groups (such as Industry code 272, 279, 385 and 316) it was found to be insignificant. Incorporation of time variable in model II (presented in Table 6.2.6) resulted in a nominal improvement in overall fit. Though the coefficient of wage rate (giving estimate of σ) is found to be significant in only 9 cases, but in some cases it turns to insignificant, the industry groups belong to this category are 226, 273, 276, 281, 316, 385 and 386. The coefficient of wage rate (giving estimate of σ) is found to be positive and significant in case of industry manufacture of veneer, plywood and their products (270); sawing and plaining of wood (271); manufacture of wooden and cane boxes...(272); manufacture of wooden industrial goods...(274); manufacture of cork and cork products (275); manufacture of wood, bamboo and cane products n.e.c. (279); manufacture of pulp, paper and paper board including newsprint (280); manufacture of paper products n.e.c. like dolls (282); manufacture of miscellaneous products n.e.c. (387); and total wood & forest based industries of Western region.

On the other side, inclusion of labour variable in Model III (Table 6.2.7) the coefficients of σ (wage rate) were found positive in case of industry code 270, 271, 274, 275, 279, 280, 282, and total WFBI as well and in rest of the cases, σ were experiences insignificant or negative trend. Similarly, labour variable were also found to be negative or insignificant in majority of groups of industry except industry code 272 and 281. The coefficient of time either found insignificant or significant but not positive. In this model, it has been observed that most of industries are not explaining fit of regression very good as evident from insignificant values of Adjusted R^2 and F in majority of cases.

As previously discuss in the interpretation various models specification of Eastern region, that Model IV of Constant Elasticity of Substitution (C.E.S.) specification is the Kmenta approximation. In this Model IV (presented in Table 6.2.8 for Western Region), the terms Ln(L) and Ln(K) are those of the C-D function while the last term accounts for non-unitary elasticity of substitution. In this specification distribution parameter, scale parameter, and substitution parameter was not found significant in any case of industry groups. The time coefficient representing total factor productivity growth was found positive and significant only in one case of industry groups like manufacture of Bidi (226), while in all other cases, the time coefficient were found either negative and insignificant or insignificant which indicating towards the absence of technological change in the industries in the study period.

6.3.2 (III): Estimates of V.E.S. Production Function for Western Region

The results of this specification for the Western Region of India are presented in Table 6.2.9 & 6.2.10 as Model I and Model II. From Model I of VES production function, it has been observed that the coefficient of wage rate was found positive and significant in most of the cases of industry groups except in cases of industry groups like manufacture of structural wooden goods (273); and manufacture of musical instruments (386), where it was found insignificant. The coefficient of capital-labour ratio (i.e. LnK/L) was found negative and significant in cases of industry groups 273, 280, 282, 387. The coefficient of capital-labour ratio was found insignificant in cases of industry groups like 271, 273, 280, 282, 386 and 387. While for rest of the industry groups (such as industry code 226, 270, 272, 274, 275, 276, 279, 281, 316, 385, and total WFBI), it was found positive and significant which means coefficients of K/L ratio are statistically different from zero in the model which means that the changes in this ratio may influence the value of elasticity of substitution. In this model, it has been observed that majority of industries are explaining fit of regression very good as evident from insignificant values of Adjusted R^2 and F in majority of cases. The time coefficient in Model II (presented in Table 6.2.10) of VES specification was found positive and significant suggesting growth in total factor productivity in case of only total wood & forest based industries of Western Region. Whereas negative and significant time coefficient was found in case of industry groups like 271, 274, 279, 280, and 282. In rest of the cases of industry groups, it was found insignificant; it means that these groups of industries are experiencing absence of technological changes in the period under study.

6.3.3 (I): Estimates of C-D Production Function for Northern Region

Various estimates of Cobb-Douglas production function for Wood & Forest based Industrial sector as well as individual industry for the Northern Region of India is presented in Table 6.3.1 to 6.3.4. The estimates of the unrestricted Cobb-Douglas function without time trend along with time trend and ratio form is worked out in following paras. The results of unrestricted Cobb-Douglas production function without time trend are presented in Table 6.3.1 to 6.3.2.

The analysis of Table 6.3.1 reveals that 10 industries of WFBI of the region along with total WFBI at three digit level shows that the labour and capital taken together explains between 80-90% variations in dependent variable. However, the neither negative nor insignificant labour coefficient was not found for any industry. The capital coefficient however showed mixed results. Though similarly like eastern region, the most of the industries have shown negative and significant parameters, only one industry have shown significant (at 1% level) and positive trend in capital coefficients and industry belong to this category is manufacture of turpentine, synthetic, resin, plastic materials...(316). Whereas, insignificant capital coefficient were found in case of industry groups like, industry code 226, 273, 275, and 279.

For the Northern region of India, the results of unrestricted Cobb-Douglas production function with time trend are presented as Model II in Table 6.3.2. Inclusion of time variable has improved the fit. Inclusion of time variable brought reversal in negative sign of capital coefficient in most of cases. Labour coefficient was found positive and significant in almost all cases similarly like model I except industry code 316. Sign of time coefficient was observed either negative and significant or insignificant for the industries under study. This indicates negative growth in total factor productivity experiencing in majority of the cases. Efficiency parameter was observed positive and significant in most of the industry group at three-digit level in model II. Whereas in some of the industry group cases it turns to be negative and significant such as Industry Code (226 and 275). Labour coefficient was found positive and significant in almost all cases. However, insignificant labour coefficient was found only in case of Manufacture of turpentine, synthetic, resin, plastic materials...(316). On the other side, capital coefficient however continued to show mixed results. Though, the most of the industries have shown improvement in their magnitude, as well as, reversal in sign of capital coefficient. As compared to negative sign of capital coefficient in model I, Inclusion of time variable in Model II, brought reversal in negative sign of capital coefficient in most of the cases but in majority of the cases, they were observed insignificant coefficient. The positive and significant capital coefficient was observed in case of industry group 273, 279, 280, 282, and 385 similar like previous region. Sign of time coefficient was found negative and significant in 8 cases and insignificant in 7 cases and along with total WFBI as well. This indicates negative growth in TFP in most of the cases. However, time coefficient was not found positive and significant in any group of industries belongs to WFBI under northern region of India, it means these industries are not experiencing technological change during the period under study.

The estimates of ratio form of Cobb-Douglas production function for the Northern Region of India are presented as a Model III in Table 6.3.3. The estimates of ratio form of C-D production function suggests the industry groups where the degree of homogeneity is significantly different from unity or that the returns to scale are significantly different from being constant are manufacture of Bidi (226); manufacture of veneer, plywood and their products (270); Manufacture of wooden and cane boxes, crates, drums, barrels and other wood (272); manufacture of wooden industrial goods...(274); manufacture of wooden furniture and fixtures (276); and manufacture of wood, bamboo and cane products (279). Significant positive coefficient time indicates upward trend in total factor productivity. The industries groups found positive time coefficients belong to this category are 273, 274, 276, 279, 316, 385, and 387. Significant negative time coefficient indicates downward trend in total factor productivity. The industries groups belong to this category is manufacture of cork and cork products (275). Negative and Insignificant coefficients of time trend implies that the production function are not subject to any kind of significant shifts in the industry under reference period and but similarly like eastern region, here in northern region too not a single such type of case has been observed. So far as best fit of the industries under study are concern, except 6 industries (such as, industry code: 226, 271, 275, 279, 280, 282, 316, and 387) rest of group of industries are explaining fit of regression very good as evident from highly significant values of Adjusted R^2 and F-statistic.

A restrictive form of C-D production function explicitly assumes constant returns to scale. The results of this specification for Northern Region as Model IV are presented in Table 6.3.4. In this specification the efficiency parameter turns to be negative and significant in most of the cases except total WFBI of Northern

Region. In this specification the capital intensity parameter was found positive and significant in 6 industries such as industry code 226, 270, 274, 275, 276, 279,281 and 386 along with total WFBI, while in other cases, either found insignificant and negative or insignificant. In this specification the significant negative time coefficient was observed only in one industry *i.e* manufacture of cork and cork products (275), indicates downward trend in total factor productivity. Whereas, insignificant time coefficient was found in most of the cases and total WFBI as well. Only one industry manufacture of turpentine, synthetic resin, plastic materials... (316) depicted positive and significant time coefficient suggesting upward trend in total factor productivity growth.

6.3.3 (II): Estimates of C.E.S. Production Function for Northern Region

The results of estimates of C.E.S. production function for Northern Region of India are presented in Table 6.3.5 as model I; Table 6.3.6. as Model II; Table 6.3.7 as model III; and Table 6.3.8 as Model IV. Model I (Table 6.3.5) of Constant Elasticity of Substitution (C.E.S.) specification is the original SMAC formulation relating the Ln (V/L) to Ln (W). The coefficient of wage rate (giving estimate of σ) is found to significant even at 1 per cent level of significance in case of majority of the industry groups, where it was found to be different from unity. However, in case of 3 industry groups (such as Industry code 279, 385 and 316) it was found to be insignificant. Incorporation of time variable in model II (presented in Table 6.3.6) resulted in a nominal improvement in overall fit. Though the coefficient of wage rate (giving estimate of σ) is found to be significant in 10 cases, but in some cases it turns to insignificant, the industry groups belong to this category are 226, 273, 281, and 316. The coefficient of wage rate (giving estimate of σ) is found to be positive and significant in case of industry manufacture of veneer, plywood and their products (270); sawing and plaining of wood (271); manufacture of wooden and cane boxes...(272); manufacture of wooden industrial goods...(274); manufacture of cork and cork products (275); manufacture of wooden furniture and fixtures (276); manufacture of wood, bamboo and cane products n.e.c. (279); manufacture of pulp, paper and paper board including newsprint (280); manufacture of paper products n.e.c. like dolls (282); manufacture of sports and athletic goods and play equipments (385); manufacture of miscellaneous products n.e.c. (387); and total wood & forest based industries. In most of the cases industries are not explaining fit of regression very good as evident from highly significant values of Adjusted R^2 and F.

On the other side, inclusion of labour variable in Model III (Table 6.3.7) the coefficients of σ (wage rate) were found positive in case of industry code 270, 271, 274, 275, 276, 279, 280, 282, 385 and total WFBI as well and in rest of the cases, σ were experiences insignificant or negative trend. Similarly, labour variable were also found to be negative or insignificant in majority of groups of industry except industry code 272 and 281. The coefficient of time either found insignificant or significant but not positive. Only one industry experiencing positive and significant time coefficient and industry is manufacture of sports and athletic goods and play equipments (385). In this model, it has been observed that most of industries are not explaining fit of regression very good as evident from insignificant values of Adjusted R^2 and F in majority of cases.

Model IV of Constant Elasticity of Substitution (C.E.S.) specification is the Kmenta approximation. In this Model IV (presented in Table 6.3.8 for Northern Region), the terms Ln(L) and Ln(K) are those of the C-D function while the last term accounts for non-unitary elasticity of substitution. In this specification distribution parameter, scale parameter, and substitution parameter was not found significant in any case of industry groups. The time coefficient representing total factor productivity growth was found positive and significant only in two cases of industry groups like manufacture of Bidi (226) and manufacture of sports and athletic goods and play equipments (385), while in all other cases, the time coefficient were found either negative and insignificant or insignificant which indicating towards the absence of technological change in the various category of WFB industries. Similarly like Model III of CES specification, here in Model IV too as well industries are not observing fit of regression significantly as it is obvious from the insignificant values of Adjusted R^2 and F in majority of cases.

6.3.3 (III): Estimates of V.E.S. Production Function for Northern Region

The results of this specification for the Northern Region of India are presented in Table 6.3.9 & 6.3.10 as Model I and Model II. From Model I of VES production function, it has been observed that the coefficient of wage rate was found positive and significant in most of the cases of industry groups except in cases of industry groups like manufacture of structural wooden goods (273); and manufacture of cork and cork products (275), where it was found insignificant. The coefficient of capital-labour ratio was found insignificant in cases of industry groups like 271, 273, 280, 282,316, 385 and 387. While for rest of the industry groups (such as industry code 226, 270, 272, 274, 275, 276, 279, 281, and total WFBI), it was found positive and significant which means coefficients of K/L ratio are statistically different from zero in the model which means that the changes in this ratio may influence the value of elasticity of substitution. In this model, it has been observed that some of the industries (such as industry code 270, 271, 276, 385 and total WFBI) are explaining fit of regression very good as evident from insignificant values of Adjusted R^2 and F in majority of cases. The time coefficient in Model II (presented in Table 6.3.10) of VES specification was found positive and significant suggesting growth in total factor productivity in case of industry code 272, only in Northern Region. Whereas negative and significant time coefficient was found in case of industry groups like 271, 274, 275, 279, 280, and 282. In rest of the cases of industry groups, it was found insignificant; it means that these groups of industries are experiencing absence of technological changes in the period under study.

6.3.4 (I): Estimates of C-D Production Function for Southern Region

Various estimates of Cobb-Douglas production function for Wood & Forest based Industrial sector as well as individual industry for the Southern Region of India is presented in Table 6.4.1 to 6.4.4. The estimates of the unrestricted Cobb-Douglas function without time trend along with time trend and ratio form is worked out in following paras. The results of unrestricted Cobb-Douglas production function without time trend are presented in Table 6.4.1 to 6.4.2.

The analysis of Table 6.4.1 reveals that 11 industries of WFBI of the region along with total WFBI at three digit level shows that the labour and capital taken together explains between 80-90% variations in dependent variable. However, for all most all the industries, labour coefficient was found positive and significant except industry code 316 where it was observed significant but negative coefficient. Similarly like other regions, the capital coefficient however showed mixed results. Though the most of the industries have shown significant but negative parameters, only one industry have shown insignificant but positive trend in capital coefficients and industry belong to this category is manufacture of bullock carts, push carts hand carts *etc* (378). Whereas, insignificant capital coefficient were found in case of industry groups like, industry code 226, 273, 279, and 378.

For the southern region of India, the results of unrestricted Cobb-Douglas production function with time trend are presented as Model II in Table 6.4.2. Inclusion of time variable has improved the fit. Inclusion of time variable brought reversal in negative sign of capital coefficient in most of cases. Labour coefficient was found positive and significant in almost all cases similarly like model I except industry code 316. Capital coefficients either observed negative or insignificant in majority of the cases. Sign of time coefficient was observed either negative and significant or insignificant for the all industries under study. This indicates negative growth in total factor productivity experiencing in majority of the cases. Efficiency parameter was observed positive and significant in most of the industry group at three-digit level in Model II. Whereas in some of the industry group cases it turns to be negative and significant such as Industry Code (226, 275 and 316). Labour coefficient was found positive and significant in almost all cases. On the other side, capital coefficient however continued to show mixed results. Though, the most of the industries have shown improvement in their magnitude, as well as, reversal in sign of capital coefficient. As compared to negative sign of capital coefficient in model I, Inclusion of time variable in Model II, brought reversal in negative sign of capital coefficient in most of the cases but in majority of the cases, they were observed insignificant coefficient. The positive and significant capital coefficient was observed in case of industry group similar like previous region. Sign of time coefficient was found negative and significant in 11 cases and insignificant in 5 cases and along with total WFBI as well. This indicates negative growth in TFP in most of the cases. However, time coefficient was not found positive and significant for any industries belong to WFBI under Southern region of India; it means these industries are not experiencing technological change during the period under study.

The estimates of ratio form of Cobb-Douglas production function for the Southern Region of India are presented as a Model III in Table 6.4.3. The estimates of ratio form of C-D production function suggests the industry groups where the degree of homogeneity is significantly different from unity or that the returns to scale are significantly different from being constant are manufacture of veneer, plywood and their products (270); Manufacture of wooden and cane boxes, crates, drums, barrels and other wood (272); manufacture of wooden industrial goods... (274); manufacture of wooden furniture and fixtures (276); and manufacture of wood, bamboo and cane products (279). The industries groups found positive time coefficients belong to this category are 226, 270, 273, 274, 276, 378, 387 and total WFBI. Significant negative time coefficient was found for industry code 275 and 316. Negative and Insignificant coefficients of time trend implies that the production function are not subject to any kind of significant shifts in the industry under reference period and but similarly like eastern region, here in Southern region too not a single such type of case has been observed. So far as best fit of the industries under study are concern, except 6 industries (such as, industry code: 275, 280, 282, 316, 378 and 387) rest of group of industries are explaining fit of regression very good as evident from highly significant values of Adjusted R^2 and F-Statistic.

A restrictive form of C-D production function explicitly assumes constant returns to scale. The results of this specification for Southern Region as Model IV are presented in Table 6.4.4. In this specification the efficiency parameter turns to be negative and significant in most of the cases except total WFBI of Southern Region. In this specification the capital intensity parameter was found positive and significant in 6 industries such as industry code 226, 270, 274, 275, 276, and 279 along with total WFBI, while in other cases, either found insignificant and negative or insignificant. In this specification the significant negative time coefficient was observed only in two industries *i.e* industry code 275 and 316, indicates downward trend in total factor productivity. Whereas, insignificant time coefficient was found in most of the cases and total WFBI as well.

6.3.4 (II): Estimates of C.E.S. Production Function for Southern Region

The results of estimates of C.E.S. production function for Southern Region of India are presented in Table 6.4.5 as model I; Table 6.4.6 as Model II; Table 6.4.7 as model III; and Table 6.4.8 as Model IV. Model I (Table 6.4.5) of Constant Elasticity of Substitution (C.E.S.) specification is the original SMAC formulation relating the Ln (V/L) to Ln (W). The coefficient of wage rate (giving estimate of σ) is found to significant even at 1 per cent level of significance in majority of the industry groups, where it was found to be different from unity. However, in case of 3 industry groups (such as Industry code 272, 316 and 385) it was found to be insignificant. Incorporation of time variable in model II (presented in Table 6.4.6) resulted in a nominal improvement in overall fit. Though the coefficient of wage rate (giving estimate of σ) is found to be significant in 11 industries along with total WFBI, but in some cases it turns to insignificant, the industry groups belong to this category are 226, 273, 276, 281, 316 and 385. The coefficient of wage rate (giving estimate of σ) is found to be positive and significant in case of industry manufacture of veneer, plywood and their products (270); sawing and plaining of wood (271); manufacture of wooden and cane boxes...(272); manufacture of wooden industrial goods...(274); manufacture of cork and cork products (275); manufacture of wood, bamboo and cane products n.e.c. (279); manufacture of pulp, paper and paper board including newsprint (280); manufacture of paper products n.e.c. like dolls (282); manufacture of bullock carts, push carts, hand carts *etc.* (378); manufacture of miscellaneous products n.e.c. (387); and total wood & forest based industries. For industry code 272, 273, 275, 282, 316, 378, and 387, industries are not explaining fit of regression very good while for the rest of the industries explaining fit of regression as evident from highly significant values of Adjusted R^2 and F.

On the other side, inclusion of labour variable in Model III (Table 6.4.7) the coefficients of σ (wage rate) were found positive in case of industry code 270, 271, 274, 275, 279, 280, 378, and total WFBI as well and in rest of the cases, σ were experiences insignificant or negative trend. Similarly, labour variable were also found to be negative or insignificant in majority of groups of industry except industry code 272 and 281. The coefficient of time either found insignificant or significant but not positive. In this model, it has been observed that most of industries are not explaining fit of regression very good as evident from insignificant values of Adjusted R^2 and F in majority of cases.

However, Model IV of Constant Elasticity of Substitution (C.E.S.) specification is the Kmenta approximation. In this Model IV (presented in Table 6.4.8 for Southern Region), Labour coefficients for the most of the industry group experiences insignificant, while capital coefficient for specification were found either insignificant or negative in most of the cases. The time coefficient representing total factor productivity growth was found positive and significant only in four cases of industry groups like manufacture of Bidi (226); manufacture of cork and cork products (275); manufacture of sports and athletic goods and play equipments (385); and total WFBI, while in all other cases, the time coefficient were found either negative and insignificant or insignificant which indicating towards the absence of technological change in the various category of WFB industries. Similarly like Model III of CES specification, here in Model IV too as well only some industries are observing fit of regression significantly as it is obvious from the insignificant values of Adjusted R^2 and **F** in majority of cases.

6.3.4 (III): Estimates of V.E.S. Production Function for Southern Region

The results of this specification for the Southern Region of India are presented in Table 6.4.9 & 6.4.10 as Model I and Model II. From Model I of VES production function, it has been observed that the coefficient of wage rate was found positive and significant in most of the cases of industry groups except in cases of industry groups like manufacture of structural wooden goods (273); and manufacture of pulp, paper and paper board including newsprint (280), where it was found insignificant. The coefficient of capital-labour ratio was found insignificant in cases of industry groups like 273 and 280. While for rest of the industry groups (such as industry code 226, 270, 272, 274, 275, 276, 279, 281, and total WFBI), it was found positive and significant which means coefficients of K/L ratio are statistically different from zero in the model which means that the changes in this ratio may influence the value of elasticity of substitution. In this model, it has been also observed that some of the industries (such as industry code 270, 271, 274, 276, 279, 385 and total WFBI) are explaining fit of regression very good as evident from insignificant values of Adjusted R^2 and F in majority of cases. The time coefficient in Model II (presented in Table 6.4.10) of VES specification was found positive and significant suggesting growth in total factor productivity in case of industry code 272 and 273 only in Southern Region. Whereas negative and significant time coefficient was found in case of industry groups like 271, 274, 275, 279, 280, 282 and 378. In rest of the cases of industry groups, it was found insignificant; it means that these groups of industries are experiencing absence of technological changes in the period under study.

6.3.5 (I): Estimates of C-D Production Function for Other States

Various estimates of Cobb-Douglas production function for Wood & Forest based Industrial sector as well as individual industry for the Other States of India is presented in Table 6.5.1 to 6.5.4. The estimates of the unrestricted Cobb-Douglas function without time trend along with time trend and ratio form has been worked out and the results of unrestricted Cobb-Douglas production function without time trend are presented in Table 6.5.1 to 6.5.2.

The analysis of Table 6.5.1 reveals that 12 industries of WFBI of the Other States along with total WFBI at three digit level shows that the labour and capital taken together explains between 80-90% variations in dependent variable. However, except one industry namely manufacture of structural wooden goods (273), the not a single other industry have experiences neither negative nor insignificant labour coefficient. The capital coefficient however showed mixed results of zigzag pattern. Though the most of the industries have shown negative and significant parameters, only one industry have shown significant (at 1% level) and positive trend in capital coefficients and industry belong to this category is manufacture of turpentine, synthetic, resin, plastic materials...(316). Whereas, insignificant capital coefficient were found in case of industry groups like, industry code 226, 273, 275, 279 and 378. In this specification, majority of industries are experiencing fit of regression significantly as it is obvious from the insignificant values of Adjusted R^2 and F-Value.

For the Other States, the results of unrestricted Cobb-Douglas production function with time trend are presented as Model II in Table 6.5.2. Inclusion of time variable here in this model too has improved the fit. Inclusion of time variable brought reversal in negative sign of capital coefficient in most of cases. Labour coefficient was found positive and significant in almost all cases similarly like model I except industry code 316. Sign of time coefficient was observed either negative and significant or insignificant for the industries under study. This indicates negative growth in total factor productivity experiencing in majority of the cases. Efficiency parameter was observed positive and significant in most of the industry group at three-digit level in model II. Whereas in some of the industry group cases it turns to be negative and significant such as Industry Code (226, 275 and 386). Labour coefficient was found positive and significant in each and every industry under study except industry code 316, where it was found insignificant but positive. On the other side, capital coefficient however continued to show mixed results. Though, the most of the industries have shown improvement in their magnitude, as well as, reversal in sign of capital coefficient. As compared to negative sign of capital coefficient in model I, Inclusion of time variable in Model II, brought reversal in negative sign of capital coefficient in most of the cases but in majority of the cases, they were observed either insignificant or negative coefficient. The positive and significant capital coefficient was observed in industry group 273, 279, 280, 282, and 385 similar like previous region. Sign of time coefficient was found negative and significant in 8 cases and insignificant in 10 cases and along with total WFBI as well. This indicates negative growth in TFP in most of the cases. However, time coefficient was not found positive and significant in any group of industries belongs to WFBI under Other States of India, it means these industries are not experiencing technological change during the period under study. In spite of this, all industries under study in Other States have experiences fit of regression significantly as it is obvious from the insignificant values of Adjusted R^2 and F-Value.

The estimates of ratio form of Cobb-Douglas production function for the Other States of India are presented as a Model III in Table 6.5.3. The estimates of ratio form of C-D production function suggests the industry groups where the degree of homogeneity is significantly different from unity or that the returns to scale are significantly different from being constant are manufacture of Bidi (226); manufacture of veneer, plywood and their products (270); manufacture of sawing and palining of wood (271); Manufacture of wooden and cane boxes, crates, drums, barrels and other wood (272); manufacture of wooden industrial goods...(274); manufacture of wooden furniture and fixtures (276); and manufacture of wood, bamboo and cane products (279). Significant positive coefficient time indicates upward trend in total factor productivity. The industries groups found positive time coefficients belong to this category are 273, 274, 276, 316, 378, 385, and 387. Significant negative time coefficient indicates downward trend in total factor productivity. The industries groups belong to this category are manufacture of cork and cork products (275). Negative and Insignificant coefficients of time trend implies that the production function are not subject to any kind of significant shifts in the industry under reference period and but similarly like other region, here in other States too not a single such type of case has been observed. So far as best fit of the industries under study are concern, except 9 industries (such as, industry code: 270, 271, 273, 275, 280, 282, 316, 378 and 386) all the industries are explaining fit of regression very good as evident from highly significant values of Adjusted R^2 and F-Value.

A restrictive form of C-D production function explicitly assumes constant returns to scale. The results of this specification for Other States as Model IV are presented in Table 6.5.4. In this specification the efficiency parameter turns to be negative and significant in most of the cases except industry code 316 and total WFBI of other States. In this specification the capital intensity parameter was found positive and significant in 6 industries such as industry code 226, 270, 272, 274, 275, 276, 279, and 386 along with total WFBI, while in other cases, either found insignificant and negative or insignificant. In this specification the significant and positive time coefficient was found for manufacture of sports and athletic goods and play equipments (385) which indicates upward trend in TFP. While negative but significant time coefficient was observed for manufacture of cork and cork products (275), indicates downward trend in total factor productivity. Whereas, insignificant time coefficient was found in most of the cases and total WFBI as well.

6.3.5 (II): Estimates of C.E.S. Production Function for Other States

The results of estimates of C.E.S. production function for other States of India are presented in Table 6.5.5 as model I; Table 6.5.6 as Model II; Table 6.5.7 as model

III; and Table 6.5.8 as Model IV. Model I (Table 6.5.5) of Constant Elasticity of Substitution (C.E.S.) specification is the original SMAC formulation relating the Ln (V/L) to Ln (W). The coefficient of wage rate (giving estimate of σ) is found to significant even at 1 per cent level of significance in case of majority of the industry groups, where it was found to be different from unity. However, in case of 3 industry groups (such as Industry code 279, 316, and 385) it was found to be insignificant. Incorporation of time variable in model II (presented in Table 6.5.6) resulted in a nominal improvement in overall fit. Though, the coefficient of wage rate (giving estimate of σ) is found to be significant in 11 cases. The coefficient of wage rate (giving estimate of σ) is found to be positive and significant in case of industry manufacture of veneer, plywood and their products (270); sawing and plaining of wood (271); manufacture of wooden and cane boxes...(272); manufacture of wooden industrial goods...(274); manufacture of cork and cork products (275); manufacture of wood, bamboo and cane products n.e.c. (279); manufacture of pulp, paper and paper board including newsprint (280); manufacture of paper products n.e.c. like dolls (282); manufacture of bullock carts, push carts hand carts *etc* (378); manufacture of miscellaneous products n.e.c. (387); and total wood & forest based industries. 11 industries in this model are not explaining fit of regression, while industry code 276, 272, 279, 281, 387, and total WFBI are found very good as evident from highly significant values of Adjusted R^2 and F-Statistic.

On the other side, inclusion of labour variable in Model III (Table 6.5.7) the coefficients of σ (wage rate) were found positive in case of industry code 270, 271, 274, 275, 279, 280, 378, and total WFBI as well and in rest of the cases, σ were experiences insignificant or negative trend. Similarly, labour coefficient was also found to be negative or insignificant in majority of groups of industry except industry code 272 and 281. The coefficient of time either found insignificant or significant but not positive. Only one industry experiencing positive and significant time coefficient for the industry code (316). In this model, it has been observed that 6 industries such as industry code 270, 271, 275, 282, 316, and 386 are not explaining fit of regression while others found best in fit of regression as evident from insignificant values of Adjusted R^2 and F.

Model IV of Constant Elasticity of Substitution (C.E.S.) specification is the Kmenta approximation. In this Model IV (presented in Table 6.5.8 for Other States), the terms Ln(L) and Ln(K) are those of the C-D function while the last term accounts for non-unitary elasticity of substitution. In this specification distribution parameter, scale parameter, and substitution parameter was not found significant in any case of industry groups. The time coefficient representing total factor productivity growth was found positive and significant only in two cases of industry groups like manufacture of container and boxes of paper and paper board (281); manufacture of sports and athletic goods and play equipments (385) and total WFBI, while in all other cases, the time coefficient were found either negative and insignificant or insignificant which indicating towards the absence of technological change in the various category of WFB industries. In this Model IV, 9 industries are not observing fit of regression significantly as it is obvious from the insignificant values of Adjusted R^2 and F in majority of cases.

6.3.5 (III): Estimates of V.E.S. Production Function for other States

The results of this specification for the Other States of India are presented in Table 6.5.9 & 6.5.10 as Model I and Model II. From Model I of VES production function, it has been observed that the coefficient of wage rate was found positive and significant in most of the cases of industry groups except in cases of industry groups like manufacture of structural wooden goods (273); and manufacture of musical instruments (386), where it was found insignificant. The coefficient of K/L ratio was found insignificant in cases of industry groups like 271, 273, 280, 282,316, 385, 386, 387 and total WFBI as well. While for rest of the industry groups (such as industry code 226, 270, 272, 274, 275, 276, 279, 281, and 378), it was found positive and significant which means coefficients of K/L ratio are statistically different from zero in the model which means that the changes in this ratio may influence the value of elasticity of substitution. In this model, it has been observed that some of the industries (such as industry code 226, 272, 276, 279, 281, 385, 387 and total WFBI) are explaining fit of regression very good as evident from insignificant values of Adjusted R^2 and F in majority of cases. The time coefficient in Model II (presented in Table 6.5.10) of VES specification was not found positive and significant in group of industries. Whereas negative and significant time coefficient was found in case of industry groups such as 274, 275, 279, 280, and 282. In rest of the industry groups, it was found insignificant; it means that these groups of industries are experiencing absence of technological changes in the period under study.

6.3.6 (I): Estimates of C-D Production Function for All India

Various estimates of Cobb-Douglas production function for Wood & Forest based Industrial sector as well as individual industry for the All India is presented in Table 6.6.1 to 6.6.4. The estimates of the unrestricted Cobb-Douglas function without time trend along with time trend and ratio form has been worked out and he results of unrestricted Cobb-Douglas production function without time trend for All India presented are in Table 6.6.1 to 6.6.2.

The analysis of Table 6.6.1 reveals that most of the industry group at 3-digit level shows that the independent variables such as labour and capital taken together explains more than 90% variation in dependent variable along with overall manufacturing sector. The industry group that explains more than 90% variation in dependent variable in majority of industries along with total wood and forest based industries of India except the only three industries such as industry code 279, 316 and 378. Labour coefficient was found positive and significant in almost all cases. However, neither negative and significant nor insignificant labour coefficients were observed for any groups of industry. For all India, capital coefficient showed somewhat mixed results similarly like the regions under study. Though the most of the industries have shown negative and significant parameters, only one industry have shown significant positive trend in capital coefficients and industry belong to this category are manufacture of turpentine, synthetic resin, plastic materials and synthetic fibres (316). Whereas, insignificant capital coefficient was found for the industry group code: 226, 273, 275, 378 and 386.

For All India, the results of unrestricted Cobb-Douglas production function with time trend are presented as Model II in Table 6.1.2. Inclusion of time variable has improved the fit. Inclusion of time variable brought reversal in negative sign of capital coefficient. Labour coefficient was found positive and significant in almost all cases. Sign of time coefficient was found either negative and significant or insignificant in majority of the cases. This indicates negative growth in total factor productivity in most of the industries of All India for the period under study. Efficiency parameter was observed positive and significant in most of the industry group at three-digit level in Model II. Whereas in three of the industry group cases it turns to be negative and significant such as Industry Code (226, 275, and 386). However, it was insignificant in some of the industry group cases such as Manufacture of structural wooden goods (273); and manufacture of cork and cork products (275). Labour coefficient for All India was found positive and significant in almost all industry group cases. However, the insignificant labour coefficient was found only in case of industry group (316). On the other side, the capital coefficient however continued to show mixed results as it obvious for Model I of C-D production function. Though, here in Model II, the most of the industries have shown improvement in their magnitude, as well as, reversal in sign of capital coefficient. As compared to negative sign of capital coefficient in Model I, Inclusion of time variable in Model II, brought reversal in negative sign of capital coefficient in most of the cases. The positive and significant capital coefficient was observed in case of industry group 273, 279, 280, 282, and 378. Sign of time coefficient was found negative and significant in 6 cases, insignificant time trend only in 3 cases, while total WFBI of All India have observed significant negative time trend. This indicates negative growth in total factor productivity in most of the cases. However, time coefficient was not found positive and significant in any group of industries belongs to All India WFBI. All most all industries are explaining fit of regression very good as evident from insignificant values of Adjusted R^2 and F-Value.

If the function is transformed so as to have Ln (V/L) as the dependent variable, the estimates of this function provide a direct test of whether the degree of homogeneity is significantly different from unity or that the returns to scale are significantly different from being constant. The estimates of ratio form of Cobb-Douglas production function for All India are presented as a Model III in Table 6.6.3. The estimates of ratio form of Cobb-Douglas production function suggests the industry groups where the degree of homogeneity is significantly different from unity or that the returns to scale are significantly different from being constant are manufacture of Bidi (226), manufacture of veneer, plywood and their products (270); manufacture of wooden industrial goods (274); manufacture of wooden furniture and fixtures (276); and manufacture of wood, bamboo and cane products n.e.c. (279). Significant and positive coefficient time indicates upward trend in total factor productivity. The industries groups belong to this category are 226, 270, 273, 274, 276, 316, 378,375 387 and total all India as well. Significant with negative time coefficient indicates downward trend in total factor productivity and only one industry groups belong to this category *i.e* manufacture of cork and cork products (275). Negative and Insignificant coefficients of time trend implies that the production function are not subject to any kind of significant shifts in the industry under reference period and but here in case of All India, such type of case has been observed only for manufacture of musical instruments (386). So far as best fit of the industries under study are concern, except some industries (such as, industry code: 273, 275, 280, 282, 316, and 386) rest of group of industries are explaining fit of regression very good as evident from highly significant values of Adjusted R^2 and F.

A restrictive form of C-D production function explicitly assumes constant returns to scale. The results of this specification for All India as Model IV are presented in Table 6.6.4. In this specification the efficiency parameter turns to be negative and significant in most of the cases except in two industries such as Manufacture of turpentine, synthetic resin, plastic materials *etc* (316) and total WFBI of eastern Region. Efficiency parameter found insignificant for industries 275, 378 & 386. In this specification the capital intensity parameter was found positive and significant in 9 industries such as industry code 226, 270, 272, 274, 275, 276 and 279, 386, and for total All India; while in other cases, either found insignificant and negative or insignificant. Except manufacture of paper products n.e.c. (282), Labour coefficient either insignificant or negative. On the other side, significant negative time coefficient was observed for 4 industries (i.e. industry code 271, 275, 280 and 386), indicates downward trend in total factor productivity. Whereas, insignificant time coefficient was found in rest of the cases and total WFBI as well. Only one industry namely manufacture of veneer, plywood and their products (270); manufacture of turpentine, synthetic resin, plastic materials *etc* (316); manufacture of bullock carts, push carts, hand carts *etc* (378); and manufacture of sports and athletic goods and play equipments (385) depicted positive and significant time coefficient suggesting upward trend in total factor productivity growth.

6.3.6 (II): Estimates of C.E.S. Production Function for All India

The Elasticity of substitution (σ) reflecting responsiveness of factor proportion to a change in the relative factor prices is a crucial parameter, particularly in case of developing countries like India. It has important policy implications for economic growth, resource allocation and relative income distribution. High elasticity of substitution (or easy substitutability) has a favorable impact on the levels of both output and employment. It has been seen that the conflict between high output and high employment (the dilemma faced by most of the underdeveloped and developing economies) essentially arise out from low substitutability of production structure. Apart from these, estimate of elasticity of substitution is necessary to make choice of the form of production function to be applied for empirical analysis. The results of this specification for All India are presented as Model I in Table 6.6.5.

Model I of Constant Elasticity of Substitution (C.E.S.) specification is the original SMAC formulation relating the Ln (V/L) to Ln (W). The coefficient of wage rate (giving estimate of σ) is found to be significant even at one per cent level of significance in case of majority of the industry groups, where it was found to be different from unity. However, in case of 4 industry groups (such as Industry code 272, 279, 316, and 385) it was found to be insignificant. Incorporation of time variable in model II (presented in Table 6.6.6) resulted in a nominal improvement in overall fit. Though the coefficient of wage rate (giving estimate of σ) is found to be significant in only majority of the cases, but in some cases it turns to insignificant, the industry groups belong to this category are 226, 273, 276, 281, 316, 385 and 386 industries of WFBI of All India. The coefficient of wage rate (giving estimate of σ) is found to be positive and significant in case of industry manufacture of veneer, plywood and their products (270); sawing & plaining of wood (271); manufacture of wooden and cane boxes…(272); manufacture of wooden industrial goods…(274); manufacture of cork and cork products (275); manufacture of wood, bamboo and cane products n.e.c. (279); manufacture of pulp, paper and paper board including newsprint (280); manufacture of paper products n.e.c. like dolls (282); manufacture of bullock carts, push carts, hand carts *etc* (378); manufacture of miscellaneous products n.e.c. like pencils (387); and total wood & forest based industries of All India. So far as best fit of the industries under study are concern, except some industries (such as, industry code: 275, 282, 316, and 386) rest of group of industries are explaining fit of regression very good as evident from highly significant values of Adjusted R^2 and F-Value.

Inclusion of labour variable in Model III (Table 6.6.7) the coefficients of σ (wage rate) were found positive in case of industry code 270, 271, 273, 274, 275, 279, 280, 378 and for total WFBI of All India, while for rest of cases, wage rate (σ) were experiences insignificant or negative trend similarly like labour variable which were also found to be negative or insignificant. The coefficient of time, either found insignificant or negative.

Model IV of Constant Elasticity of Substitution (C.E.S.) specification is the Kmenta approximation. In this Model IV (presented in Table 6.6.8), the terms Ln (L) and Ln (K) are those of the C-D function while the last term accounts for non-unitary elasticity of substitution. In practices, the closer the elasticity of substitution to unity, the better the approximation. In this specification distribution parameter, scale parameter, and substitution parameter was not found significant in any case of industry groups under study. The time coefficient representing total factor productivity growth was found positive and significant only for manufacture of Bidi (226). In rest of the cases of industry groups of WFBI, the time coefficient was found insignificant.

6.3.6 (III): Estimates of V.E.S. Production Function for All India

To determine the relevant form of production functions, it has always been wise to examine whether elasticity of substitution is variable or constant. However, a major limitation of C.E.S. production function is that it assumes the elasticity of substitution to be constant for all input combinations. Several functional forms have been suggested in the literature and used in empirical studies in which elasticity of substitution is variable. These are called V.E.S. (variable elasticity of substitution) production function. It has the properties of a neoclassical production function, i.e. positive marginal product and downward sloping marginal product curve, over the relevant range of inputs. This function is homogeneous of degree one (i.e. constant returns to scale) and has variable elasticity of substitution. An important implication of this functional form is that labour productivity becomes a log-linear function of wage rate and K/L ratio (under the condition of cost minimization and competitive markets). The results of this specification for the All India are presented in Table 6.6.9 & 6.6.10 as Model I and Model II.

It is obvious from Model I that the coefficient of wage rate was found positive and significant in most of the cases of industry groups except in cases of industry groups like manufacture of structural wooden goods (273); and manufacture of musical instruments (386), where it was found insignificant. The coefficient of capital-labour ratio (i.e. Ln K/L) was found insignificant in cases of industry groups like 271, 273, 280, 282, 316, 378, 385, 386 and 387. While for the industry groups, such as industry code 226, 270, 272, 275, 276, 279 281, and total WFBI, it was found positive and significant. It means coefficients of K/L ratio are statistically different from zero in the model which means that the changes in this ratio may influence the value of elasticity of substitution. The time coefficient in Model II (presented in Table 6.6.10) of VES specification was found positive and significant suggesting growth in total factor productivity in case of three industry groups such as Manufacture of wooden and cane boxes, crates (272); manufacture of wooden furniture & fixtures (276); and manufacture of turpentine, synthetic resin, plastic materials *etc* (316). Whereas negative and significant time coefficient was found in case of industry groups like 271, 274, 279, 280, 282, and 378. In rest of the cases of industry groups, it was found insignificant which experiencing absence of technological changes in the period under study. So far as best fit of the industries under study are concern, except some industries (such as, industry code: 275, 282, 316, and 386) rest of group of industries are explaining fit of regression very good as evident from highly significant values of Adjusted R^2 and F-Value.

6.4: The Overall Analysis of Production Function Estimates

In the overall analysis, the various specifications of production function for the Eastern Region of India, the industry groups which have shown negative and significant growth in total factor productivity are 270, 271, 272, 276, 316, 387 and total wood & forest based industries of the region. In case of Western region, the industry groups which have shown negative and significant growth in total factor productivity are 270, 271, 272, 276, 282, 386, and 387. While for the Northern Region of India, the industry groups which have shown negative and significant growth in total factor productivity are 270, 271, 272, 276, 385, 387 and total wood & forest based industries of the region. On the other side, in case of Southern Region of India, the industry groups which have shown negative and significant growth in total factor productivity are 270, 271, 272, 274, 316, 387 and total wood & forest based industries of the region. For the Other States, the industry groups which have shown negative and significant growth in total factor productivity are 270, 271, 272, 274, 276, 279, 280, 282, 386, 387 and total wood & forest based industries of the region. At last, for All India, overall analysis of the various specifications of production function, the industry groups which have shown negative and significant growth in total factor productivity are 226, 270, 271, 272, 274, 275, 276, 279, 280, 386, 387 and total wood & forest based industries. It means, majority of wood and forest based industries of all India are not experiencing technological change, thus, suggesting technological retrogation. The overall analysis the various specifications of production function showing insignificant time coefficient in the remaining industry groups, thus indicating neither technological progress nor technological retrogation (i.e. technological stagnation).

6.5: Comparision with earlier estimates

It is useful at this stage to highlight main findings of this study and compare this with the results of some earlier studies. So far as the estimates of σ are concerned, no general conclusions can be drawn on the basis of existing studies. Diwan & Gujarati (1968) on the basis of time series data of 1946-58 find σ to be significantly less than unity in 26 out of 28 industries studied. Bhasin & Seth (1980) also find less than unity estimates of σ in most of the industries studied for the period of 1950-65. Similar results regarding jute industry for the period 1958-78, Karnataka manufacturing for the period 1968-78 and Punjab manufacturing for the period 1967-80 are bought out in the respective studies of Verma (1985), Dhillon (1983) and Singh (1987). On the other hand, estimates of σ are not found to be significantly different from unity in 21 out of 28 industries studied by Venkataswami (1968). Similarly, Banerji (1975) finds estimates of σ not significantly different from unity with regard to Indian manufacturing (1946-64) and five industries [i.e. cotton textiles, jute textiles (1946-64), sugar (1946-63), Paper and bicycles (1946-58)]. Similar estimates of σ were found in an inter-state study on sugar industry by Subramaniyan (1979), Gupta (1985) also finds evidence in favour of unitary σ in Maharastra manufacturing for the period of 1969-77. Large inter-industry variations in the estimates of σ were found in a study by Mehta (1980) for 1953-65.

In econometrics estimation of production functions, technical change can be estimated using the single time trend (TT) approach (i.e. inclusion of a deterministic time trend in the estimation of a production function) (Solow, 1957; Tinbergen, 1942; Christensen et.al., 1973) or alternatively by the general index (GI) approach (Baltagi & Griffin, 1988). An intermediate approach is multiple time trend approach where multiplicity of trends has been introduced to capture structural changes such as pre- and post-economic reform periods.. With the TT approach, the trend may be linear or non-linear, and certain specifications such as flexible functional forms may allow interactions between time and other explanatory variables. This allows the rate of technical change to be non- constant and non-neutral (Gollop

& Jorgenson, 1980, Jorgenson & Fraumenti, 1981, Gollop & Roberts, 1983). The derivative of the production function with respect to time provides measure of the rate of technical change.

Production functions for the industrial sector as a whole as well as for seven important industries in India are worked out based on cross-section data relating to individual firms for the two years 1951 and 1952 (V. N. Murti & V. K. Sastry, 1957, De-Min Wu, 1975), derives the exact distribution of the indirect least squares estimator of the coefficients of the C-D function within the context of a stochastic production model of Marschak-Andrews type. The stochastic term in C-D type models is either specified to be additive or multiplicative (Stephen M. Goldfeld & Richard E. Quandt, 1976). They developed a model in which a C-D type function is coupled with simultaneous multiplicative and additive errors. This specification is a natural generalization of the "pure" models in which either additive or multiplicative stochastic terms are introduced. A C-D type function with both multiplicative and additive errors has been proposed by Goldfeld & Quandt (1976). They suggested a maximum likelihood approach to the estimation of a C-D type model when the model includes both multiplicative and additive disturbance terms.

In a study, Goldar (1986) finds σ to be quite close to unity for aggregate CMI (1951-65) and large scale sector (ASI sample sector) for 1960-78, estimate of σ is found to be less one. Conflicting evidences are also available regarding the returns to scale. Findings of many earlier studies have gone in favour of constant returns to scale. Datta (1955) on the basis of cross-section data of 1946 and 1947, Murti and Sastry (1957) on the basis of time series data of 1951, Datta Mazumdar (1966) on the basis of time series data of 1951 to 1961 and Narasimhan & Fabrycy (1974) on the basis of pooled cross-section time series data of 1946-58 have shown evidences in favour of constant returns to scale. Banerji (1975) in his study finds constant returns operating with regard to Indian manufacturing and jute textile industry and cotton textile industry. However, contrary to this, the remaining three industries (namely sugar, paper and bicycles) have experienced increasing returns to scale. Mehta (1980) has found evidence in favour of constant returns to scale in 23 out of 27 industries studied. Hasim & Dadi (1973) find Indian manufacturing sector experiencing returns to scale during 1946-64. However, for the period 1953 to 64 evidence is in favour of constant returns to scale. Goldar (1986) on the basis of bot the C-D and CES production function find categorical evidence in favour of constant returns to scale in aggregate CMI (for 1951-65) and also in large scale (1959-79) and small -scale (for 1960-78) sectors. Almost similar conclusions in respect of total manufacturing sector for all-India level and Rajasthan are reached by Rajlakshmi (1983). In another study of Rajlakshmi (1981), she has found evidences in favour of constant returns to scale in non-metallic mineral products and non-ferrous basic metals of Rajasthan. Analyzing jute industry at all-India level for the period of 1950-78, Verma (1985) has also reached the similar conclusions.

In a study on five industries (namely food products, cotton textiles, basic metals and alloys, transport equipment, and electricity industry) of Punjab for the period of 1967-82, Singh and Singhal (1986) through application of C-D production function have found the first industry operating under decreasing and rest of the industries under constant returns to scale. However, on the basis of CES production function, basic metals and alloys is found to be experiencing increasing returns to scale, where food products and transport equipment have operated under decreasing returns to scale. The remaining two industries have experienced constant returns to scale. Categorical evidence in favour of increasing returns to scale has been brought out in the studies of Yeh, Yeong-Her (1966), Diwan (1968) and Sankar (1970) for the period 1953-58. Yeh in his study notices increasing returns to scale in 17 industries, where as constant and decreasing returns have been under operation in two and ten industries, respectively. Diwan & Gujarati (1968) observe significant economies of scale on the basis of CES production function for 29 industries for the period of 1946-58.

Including time trend in the production function to capture technological change often gives unrealistic estimates. Quite frequently time trend turns out to be insignificant; leading to the conclusion that neutral technological progress is not a major source of output growth so far as Indian industries are concerned. In a study, Venkataswami (1975) notices that new industries have demonstrated much more technological progress as compared to old ones. His study relates to manufacturing sector for the period of 1946-67. Sankar (1970) observes positive neutral shifts in 6 industries and negative in two for the period of 1953-58. Narasimham & Fabrycy (1974) have not observed any significant shift in the production function for 1946-58, Banerji (1975) finds evidence of a shift in production function in sugar, paper and bicycles, whereas, cotton and jute have not experienced significant technical progress, Goldar (1986), however, finds significant shifts in production function in aggregate CMI, large scale sector and to some extent in small sector also, conversely, according to Subramaniyan (1979), sugar industry has not experienced away significant technical progress during the period of 1953-69.

The present study also supports the findings of earlier studies. Neutral technological progress does not seem to have played any important role in output growth. The present study seems to be a distinction and identifies labour as a factor most crucial to output growth. In the above mentioned production function studies, coefficient of capital is generally found to be insignificant owing to mainly measurement problems, multi-collinearity, over-capitalization and under-utilization of the installed capacity. The findings of the earlier ones, differs mainly because of differences in the choice of the period and the coverage of real units. Insignificant labour coefficient in some cases paints a disquieting picture so far as the role of labour in the output growth is concerned.

Table (6.1.1): C-D Production Function for Eastern Region

Dependent Variable: Ln (V) Model: I							
Industry Code	**Constant**	**Ln L**	**Ln K**	**R^2**	**Adjusted R^2**	**F-Statistic**	**D.F.**
226	-0.734*	2.1536*	-0.012@	0.9346	0.9851	264.37	37
	(-0.269)	(0.1591)	(0.1224)				
270	46.078*	2.4796*	-5.579*	0.9452	0.9963	327.72	37
	(0.7852)	(0.4959)	(1.2811)				
271	5.2575*	1.0563*	-0.588*	0.8833	0.931	140.03	37
	(0.1921)	(0.096)	(0.156)				
272	4.8536*	3.6215*	-0.784**	0.9092	0.9583	185.24	37
	(0.7467)	(0.336)	(0.2987)				
273	2.1112*	2.1298*	-0.355@	0.9421	0.993	301.05	37
	(0.355)	(0.1495)	(0.2802)				
274	1.5937*	1.4523*	-0.296*	0.9314	0.9817	251.18	37
	(0.1799)	(0.0514)	(0.0771)				
275	-0.425@	2.1113*	-0.088@	0.9345	0.985	263.94	37
	(0.2932)	(0.1026)	(0.1173)				
276	5.0132*	1.3885*	-0.658*	0.9247	0.9747	227.18	37
	(0.2046)	(0.1169)	(0.19)				
279	0.8225@	1.7246*	0.0796@	0.665	0.7009	36.724	37
	(0.6103)	(0.3184)	(0.0796)				
280	0.9132**	1.0158*	-0.195*	0.8641	0.9108	117.63	37
	(0.4002)	(0.0513)	(0.0718)				
281	5.9799*	1.5297*	-0.643*	0.9304	0.9807	247.3	37
	(0.3303)	(0.1391)	(0.2086)				
282	1.5479*	1.6681*	-0.301**	-0.816	-0.86	-8.313	37
	(0.4208)	(0.0751)	(0.1353)				
316	-3.144*	-0.465*	-0.217***	0.7316	0.7711	50.427	37
	(0.579)	(0.145)	(0.114)				
386	22.086@	-64.17*	2.0836@	-0.917	-0.967	-8.85	37
	(30.84)	(6.251)	(9.585)				
387	-1.016*	-0.319*	0.1177*	-0.298	-0.314	-4.247	37
	(0.05)	(0.024)	(0.03)				
Total	18.868*	1.1792*	-1.209*	0.9449	0.996	317.25	37
WFBI	(0.4329)	(0.1149)	(0.3433)				

(Figures within brackets are the standard error)
**Significant at 1% level of significance*
*** Significant at 5% level of significance*
**** Significant at 10% level of significance*
@ Insignificant

Table (6.1.2): C-D Production Function for Eastern Region

Dependent Variable: Ln (V) Model: II								
Industry Code	**Constant**	**Ln (L)**	**Ln (K)**	**T**	**R^2**	**Adjusted R^2**	**F-Statistic**	**D.F.**
226	-0.56**	1.774*	0.196@	-0.09**	0.9204	0.9971	138.69	36
	(0.245)	(0.233)	(0.147)	(0.037)				
270	36.57*	2.273*	-3.8**	-0.21***	0.9021	0.9773	110.57	36
	(0.744)	(0.496)	(1.57)	(0.124)				
271	4.765*	1.044*	-0.5	-0.01@	0.9129	0.9889	125.71	36
	(0.192)	(0.108)	(0.444)	(0.06)				
272	4.219*	2.24*	0.747*	-0.34*	0.9205	0.9972	138.88	36
	(0.635)	(0.523)	(0.56)	(0.112)				
273	0.299@	1.476*	0.617***	-0.17*	0.9024	0.9776	110.91	36
	(0.28)	(0.206)	(0.318)	(0.037)				
274	1.761*	1.272*	-0.17@	-0.04***	0.9203	0.997	138.6	36
	(0.167)	(0.141)	(0.129)	(0.026)				
275	-0.43@	2.111*	-0.09@	0.0007@	0.9214	0.9982	140.75	36
	(0.293)	(0.381)	(0.176)	(0.044)				
276	2.382*	1.257*	-0.06@	-0.07@	0.9119	0.9879	124.21	36
	(0.205)	(0.146)	(0.468)	(0.059)				
279	5.041*	0.425**	0.902*	-0.35*	0.9209	0.9976	139.71	36
	(0.425)	(0.186)	(0.265)	(0.053)				
280	0.544**	0.38*	0.626*	-0.15*	0.8971	0.9719	104.62	36
	(0.246)	(0.103)	(0.144)	(0.021)				
281	1.408*	1.339*	0.33@	-0.17***	0.9205	0.9972	138.92	36
	(0.313)	(0.157)	(0.504)	(0.087)				
282	1.623*	0.751*	0.586**	-0.23**	-0.826	-0.895	-5.428	36
	(0.316)	(0.225)	(0.241)	(0.06)				
316	-3.61*	-0.34@	-0.25@	-0.02@	0.8215	0.89	55.227	36
	(0.59)	(0.61)	(0.19)	(0.07)				
386	64.59**	-77.5*	6.251@	-2.5@	0.9196	0.9962	137.25	36
	(31.3)	(24.6)	(12.1)	(3.75)				
387	-0.78*	-0.29*	0.057@	0.01@	-0.298	-0.323	-2.755	36
	(0.05)	(0.04)	(0.13)	(0.02)				
Total	12.18*	1.179*	-0.66@	-0.08@	0.9214	0.9982	140.67	36
WFBI	(0.448)	(0.119)	(1.717)	(0.224)				

(Figures within brackets are the standard error)

**Significant at 1% level of significance*

*** Significant at 5% level of significance*

**** Significant at 10% level of significance*

@ Insignificant

Table (6.1.3): Estimates of C-D Production Function for Eastern Region

Dependent Variable Ln (V/L) Model III							
Industry Code	Const.	Ln (K/L)	T	R^2	Adjusted R^2	F-Statistic	D.F.
226	-1.96*	0.451*	0.016**	0.7748	0.8167	63.649	37
	(0.176)	(0.166)	(0.006)				
270	-10.4*	1.036*	0.082*	0.9353	0.9859	267.44	37
	(0.521)	(0.298)	(0.018)				
271	-2.38*	0.296@	0.01@	0.6683	0.7044	37.273	37
	(0.201)	(0.204)	(0.014)				
272	-7.55*	1.372*	0.046@	0.9109	0.9601	189.13	37
	(0.731)	(0.435)	(0.044)				
273	-7.2*	-0.36@	0.069*	0.8294	0.8742	89.941	37
	(0.395)	(0.422)	(0.023)				
274	-2.34*	0.485*	0.018*	0.7098	0.7482	45.249	37
	(0.225)	(0.146)	(0.008)				
275	1.039**	1.355*	-0.07*	0.4289	0.4521	13.894	37
	(0.454)	(0.391)	(0.022)				
276	-3.2*	0.455***	0.03***	0.9315	0.9819	251.57	37
	(0.222)	(0.241)	(0.015)				
279	-4.97*	1.098*	0.011@	0.7309	0.7704	50.248	37
	(1.08)	(0.232)	(0.038)				
280	-2.25*	-0.06@	0.033@	0.4233	0.4462	13.579	37
	(0.267)	(0.303)	(0.02)				
281	-3.41*	0.338@	0.023@	0.8131	0.8571	80.483	37
	(0.263)	(0.216)	(0.015)				
282	-3.76*	-0.03@	0.014@	-0.038	-0.04	-0.677	37
	(0.416)	(0.171)	(0.017)				
316	1.029***	-0.13@	-0.04***	0.3813	0.4019	11.401	37
	(0.58)	(0.13)	(0.02)				
386	-43.9@	-38.1**	2.254***	0.2063	0.2175	4.8086	37
	(34.4)	(17)	(1.34)				
387	1.065*	0.045@	-0.02***	-0.251	-0.265	-3.712	37
	(0.07)	(0.07)	(0.01)				
Total	-1.51***	1.167*	0.006@	0.9436	0.9946	309.51	37
WFBI	(0.078)	(0.168)	(0.012)				

(Figures within brackets are the standard error)

**Significant at 1% level of significance*

*** Significant at 5% level of significance*

**** Significant at 10% level of significance*

@ Insignificant

Table (6.1.4): Estimates of C-D Production Function for Eastern Region

Dependent Variable ln(V/L) Model IV								
Industry Code	**Const.**	**Ln (K/L)**	**Ln (L)**	**T**	**R^2**	**Adjusted R^2**	**F-Statistic**	**D.F.**
226	-2.05*	0.571*	0.13@	-0.01@	0.7954	0.8617	46.651	36
	(0.176)	(0.204)	(0.131)	(0.026)				
270	-15.2*	1.017*	0.769@	0.025@	0.9204	0.9971	138.81	36
	(0.525)	(0.302)	(0.946)	(0.074)				
271	-3.76*	0.27@	0.209@	-0.01@	0.6723	0.7283	24.619	36
	(0.201)	(0.217)	(0.486)	(0.054)				
272	-11.9*	0.975@	0.743@	-0.01@	0.923	0.9999	143.86	36
	(0.732)	(0.582)	(0.724)	(0.071)				
273	-9.48*	-0.43@	0.417@	0.041@	0.8582	0.9297	72.626	36
	(0.396)	(0.432)	(0.463)	(0.037)				
274	-3.8*	0.378**	0.233@	0.008@	0.7511	0.8137	36.212	36
	(0.217)	(0.154)	(0.141)	(0.01)				
275	0.562*	1.333*	0.119@	-0.07*	0.4401	0.4768	9.4324	36
	(0.462)	(0.4)	(0.221)	(0.025)				
276	-6.63*	0.428***	0.612@	-0.02@	0.921	0.9977	139.9	36
	(0.219)	(0.238)	(0.478)	(0.044)				
279	-8.5*	0.971*	1.072@	-0.11@	0.7777	0.8425	41.981	36
	(1.043)	(0.236)	(0.661)	(0.085)				
280	-7.35*	-0.05@	0.631@	-0.05@	0.4412	0.478	9.4746	36
	(0.269)	(0.307)	(0.798)	(0.101)				
281	-5.14*	0.296@	0.271@	-0.01@	0.8218	0.8903	55.34	36
	(0.268)	(0.23)	(0.445)	(0.057)				
282	-6.66*	-0.22@	0.452**	-0.01@	-0.198	-0.215	-1.983	36
	(0.38)	(0.179)	(0.198)	(0.018)				
316	-1.5***	-0.25@	0.439@	-0.09***	0.426	0.4615	8.9059	36
	(0.58)	(0.18)	(0.44)	(0.05)				
386	0.208@	-35.9**	-7.13@	2.459***	0.2212	0.2396	3.4083	36
	(34.9)	(17.6)	(12)	(1.38)				
387	0.695*	0.027@	0.059@	-0.03@	-0.251	-0.272	-2.408	36
	(0.07)	(0.08)	(0.16)	(0.02)				
Total	2.842*	1.148*	-0.36@	0.049@	0.9074	0.983	117.59	36
WFBI	(0.076)	(0.166)	(0.282)	(0.036)				

(Figures within brackets are the standard error)

**Significant at 1% level of significance*

*** Significant at 5% level of significance*

**** Significant at 10% level of significance*

@ Insignificant

Table (6.1.5): Estimates of C.E.S. Production Function for Eastern Region

Dependent Variable: Ln(V/L) Model: I						
Industry Code	Constant	Ln (W)	R^2	Adjusted R^2	F-Statistic	D.F.
226	0.2765@ (0.509)	0.8529* (0.1468)	0.7831	0.8037	68.598	38
270	-6.831* (0.6736)	1.3431* (0.2893)	0.7157	0.7345	47.831	38
271	-1.913* (0.2353)	0.5738* (0.0648)	0.9107	0.9347	193.77	38
272	-9.412* (0.7766)	0.5787@ (0.5003)	0.6146	0.6308	30.299	38
273	-2.771* (0.2877)	0.8669* (0.1495)	0.9336	0.9582	267.14	38
274	-1.55* (0.446)	0.4691** (0.2313)	0.2017	0.207	4.8006	38
275	-2.186* (0.2258)	0.7214* (0.1408)	0.9377	0.9624	285.98	38
276	-2.134* (0.5729)	0.7936* (0.1549)	0.7794	0.7999	67.129	38
279	-2.839* (0.6182)	0.5068@ (0.3131)	0.608	0.624	29.469	38
280	-1.508* (0.157)	0.3478* (0.0657)	0.6351	0.6518	33.069	38
281	-3.312* (0.4554)	0.2903*** (0.1686)	0.1876	0.1925	4.3875	38
282	-0.983* (0.1788)	0.8867* (0.0721)	-0.797	-0.818	-8.427	38
316	1.5657* (0.444)	-0.207@ (0.365)	0.0462	0.0474	0.9203	38
386	44.881* (7.501)	-18.54* (3.334)	0.9466	0.9715	336.81	38
387	0.1786* (0.072)	-0.232* (0.026)	-0.248	-0.255	-3.776	38
Total WFBI	-1.246* (0.1254)	0.7986* (0.0328)	0.9409	0.9657	302.49	38

(Figures within brackets are the standard error)
**Significant at 1% level of significance*
*** Significant at 5% level of significance*
**** Significant at 10% level of significance*
@ Insignificant

Table (6.1.6): Estimates of C.E.S. Production Function for Eastern Region

Dependent Variable: Ln(V/L) Model: II							
Industry Code	**Const.**	**Ln (W)**	**T**	**R^2**	**Adjusted R^2**	**F-Statistic**	**D.F.**
226	-5.0*	0.009@	0.026@	0.624	0.6577	30.702	37
	(0.204)	(0.277)	(0.028)				
270	-3.6*	2.496***	-0.15@	0.854	0.9002	108.21	37
	(0.595)	(1.236)	(0.136)				
271	1.486*	1.282**	-0.09***	0.7386	0.7785	52.273	37
	(0.186)	(0.517)	(0.047)				
272	-5.42*	1.923***	-0.03@	0.8486	0.8945	103.69	37
	(0.818)	(1.012)	(0.108)				
273	-4.08*	0.45@	0.006@	0.8228	0.8673	85.902	37
	(0.396)	(0.615)	(0.065)				
274	-0.4**	0.943*	-0.06*	0.838	0.8833	95.698	37
	(0.181)	(0.175)	(0.019)				
275	0.308@	1.048***	-0.09**	0.1976	0.2083	4.5558	37
	(0.516)	(0.509)	(0.043)				
276	-4.25*	0.203@	0.032*	0.8933	0.9416	154.88	37
	(0.24)	(0.99)	(0.113)				
279	9.17*	4.702*	-0.34*	0.7915	0.8343	70.229	37
	(1.00)	(0.86)	(0.088)				
280	7.051*	2.911*	-0.26*	0.8814	0.929	137.49	37
	(0.13)	(0.349)	(0.035)				
281	-6.07*	-0.49@	0.089@	0.7687	0.8103	61.483	37
	(0.275)	(0.772)	(0.071)				
282	0.863**	1.366**	-0.09**	-0.323	-0.34	-4.517	37
	(0.364)	(0.525)	(0.039)				
316	1.545*	0.159@	-0.06@	0.3363	0.3545	9.374	37
	(0.59)	(1.34)	(0.13)				
386	-5.5@	-30.7@	2.625@	0.0878	0.0925	1.7806	37
	(37.1)	(28.2)	(2.5)				
387	-0.52*	-0.4**	0.026@	-0.26	-0.274	-3.817	37
	(0.06)	(0.17)	(0.02)				
Total	-0.75*	0.949***	-0.02@	0.9209	0.9707	215.38	37
WFBI	(0.128)	(0.488)	(0.052)				

(Figures within brackets are the standard error)
**Significant at 1% level of significance*
*** Significant at 5% level of significance*
**** Significant at 10% level of significance*
@ Insignificant

Table (6.1.7): Estimates of C.E.S. Production Function for Eastern Region

Dependent Variable: Ln (V/L) Model: III								
Industry Code	Const.	Ln (W)	Ln (L)	T	R^2	Adjusted R^2	F-Statistic	D.F.
226	-3.45*	0.16@	-0.12@	0.032@	0.6522	0.7066	22.503	36
	(0.206)	(0.334)	(0.147)	(0.028)				
270	-1.53**	2.64***	-0.23@	-0.15@	0.8804	0.9538	88.334	36
	(0.608)	(1.508)	(1.306)	(0.136)				
271	3.814*	1.46**	-0.27@	-0.07@	0.7936	0.8597	46.14	36
	(0.189)	(0.627)	(0.513)	(0.055)				
272	-16.3*	0.47@	1.37***	-0.05@	0.9204	0.9971	138.75	36
	(0.777)	(1.236)	(0.736)	(0.105)				
273	-5.65*	0.269@	0.181@	0.011@	0.6349	0.6878	20.868	36
	(0.405)	(0.93)	(0.688)	(0.071)				
274	3.054*	1.414*	-0.35***	-0.09*	0.9207	0.9974	139.29	36
	(0.171)	(0.301)	(0.184)	(0.023)				
275	0.224@	1.037***	0.015@	-0.09**	0.1972	0.2136	2.9477	36
	(0.528)	(0.557)	(0.27)	(0.044)				
276	-7.4*	0.396@	0.716@	-0.05*	0.9145	0.9907	128.35	36
	(0.234)	(0.978)	(0.515)	(0.126)				
279	13.11*	5.288*	-0.59@	-0.32*	0.8434	0.9137	64.628	36
	(1.011)	(1.21)	(0.841)	(0.093)				
280	12.94*	3.123*	-0.64@	-0.21*	0.9188	0.9953	135.71	36
	(0.126)	(0.363)	(0.403)	(0.049)				
281	-23.8*	-2.78**	1.665**	0.077@	0.8692	0.9416	79.743	36
	(0.242)	(1.078)	(0.608)	(0.063)				
282	2.744*	1.727***	-0.13@	-0.1***	0.396	0.429	7.8675	36
	(0.371)	(1.032)	(0.328)	(0.056)				
316	1.537**	0.159@	0.001@	-0.06@	0.2976	0.3224	5.0843	36
	(0.61)	(1.37)	(0.33)	(0.14)				
386	-52.4@	-40@	4.417@	3.167@	0.0922	0.0999	1.2188	36
	(38)	(81.2)	(36.1)	(5.21)				
Total	-3.5*	-0.34*	0.264**	-0.02***	0.9209	0.9976	139.71	36
WFBI	(0.03)	(0.11)	(0.11)	(0.01)				

(Figures within brackets are the standard error)
**Significant at 1% level of significance*
*** Significant at 5% level of significance*
**** Significant at 10% level of significance*
@ Insignificant

Table (6.1.8): Estimates of C.E.S. Production Function for Eastern Region

Dependent Variable: Ln (V/L) Model: IV									
Industry Code	**Const.**	**Ln (L)**	**Ln (K)**	**(lnL-lnK)²**	**T**	**R²**	**Adjusted R²**	**F-Statistic**	**D.F.**
226	-5.81*	0.455@	-0.39@	-0.12@	0.044***	0.7843	0.8739	31.816	35
	(0.184)	(0.662)	(0.814)	(0.11)	(0.0256)				
270	-22.7*	0.31@	0.963**	0.0660@	0.086@	0.8914	0.9933	71.828	35
	(0.566)	(1.05)	(0.36)	(0.277)	(0.08)				
271	-6.75*	0.873@	-0.34@	0.993@	-0.024@	0.7049	0.7855	20.901	35
	(0.203)	(0.647)	(0.346)	(0.603)	(0.055)				
272	-16.9*	0.052@	1.008@	0.538@	0.048@	0.889	0.9906	70.092	35
	(0.754)	(1.184)	(0.691)	(1.482)	(0.086)				
273	-6.57*	0.31@	-0.14@	0.157@	0.017@	0.8679	0.9671	57.488	35
	(0.404)	(0.76)	(0.622)	(0.314)	(0.037)				
274	-6.0*	0.702***	-0.32@	-0.24@	0.016@	0.7157	0.7975	22.027	35
	(0.234)	(0.353)	(0.343)	(0.157)	(0.015)				
275	-3.67*	-0.01@	0.125@	-0.6@	0.007@	0.1467	0.1635	1.5043	35
	(0.556)	(0.648)	(0.78)	(0.805)	(0.026)				
276	-7.77*	0.531@	-0.04@	-0.28@	0.038@	0.8933	0.9954	73.271	35
	(0.216)	(0.68)	(0.51)	(0.241)	(0.048)				
279	-13.7*	0.602@	1.027**	-0.12@	-0.093@	0.71	0.7911	21.422	35
	(1.162)	(0.82)	(0.496)	(0.486)	(0.111)				
280	-4.5*	1.103@	-0.72@	0.146@	-0.035@	0.513	0.5716	9.2171	35
	(0.258)	(1.241)	(1.013)	(0.524)	(0.093)				
281	-6.39*	-0.23@	0.537@	-0.37@	0.012***	0.8502	0.9474	49.661	35
	(0.266)	(0.666)	(0.382)	(0.595)	(0.060)				
282	-5.8*	0.476@	-0.06@	-0.25@	-0.017	-0.336	-0.374	-2.201	35
	(0.379)	(0.377)	(0.328)	(0.329)	(0.018)				
316	2.037*	1.524@	-1.62@	0.534@	-0.041@	0.7342	0.8181	24.169	35
	(0.53)	(2.88)	(3.1)	(0.8)	(0.1)				
386	175.7*	20.21@	-43.9@	-6.79@	0.458@	0.2903	0.3235	3.5792	35
	(36.4)	(49.1)	(62.2)	(31.3)	(1.46)				
387	0.942*	-0.12@	0.109@	0.073@	-0.013@	-0.284	-0.316	-1.935	35
	(0.07)	(0.22)	(0.09)	(-0.12)	(-0.02)				
Total	-2.00*	1.367@	-1.49***	1.008**	0.106**	0.8946	0.9968	74.267	35
WFBI	(0.097)	(0.891)	(0.882)	(0.36)	(0.043)				

(Figures within brackets are the standard error)
**Significant at 1% level of significance*
*** Significant at 5% level of significance*
**** Significant at 10% level of significance*
@ Insignificant

Table (6.1.9): Estimates of V.E.S. Production Function for Eastern Region

Dependent Variable: Ln (V/L) Model: I							
Industry Code	Const.	Ln (W)	Ln (K/L)	R^2	Adjusted R^2	F-Statistic	D.F.
226	-1.11*	0.628*	0.47*	0.7716	0.8133	62.498	37
	(0.177)	(0.135)	(0.162)				
270	-7.28*	1.756*	0.963*	0.9355	0.9861	268.28	37
	(0.502)	(0.236)	(0.293)				
271	-1.99*	0.39*	0.19@	0.6925	0.7299	41.663	37
	(0.196)	(0.084)	(0.203)				
272	-5.95*	1.784*	1.25**	0.9118	0.9611	191.25	37
	(0.73)	(0.201)	(0.526)				
273	-4.71*	0.289@	-0.37@	0.8515	0.8975	106.08	37
	(0.39)	(0.25)	(0.41)				
274	-1.91*	0.597*	0.338**	0.7935	0.8364	71.088	37
	(0.198)	(0.103)	(0.139)				
275	-1.77*	0.535***	1.033**	0.2276	0.2399	5.4513	37
	(0.509)	(0.265)	(0.47)				
276	-2.18*	0.719*	0.453***	0.9272	0.9773	235.62	37
	(0.224)	(0.14)	(0.255)				
279	-3.88*	1.44*	0.886*	0.7621	0.8033	59.264	37
	(1.039)	(0.281)	(0.256)				
280	-1.1*	0.196@	-0.31@	0.5436	0.573	22.035	37
	(0.239)	(0.121)	(0.257)				
281	-2.56*	0.589*	0.377***	0.8017	0.845	74.793	37
	(0.266)	(0.111)	(0.215)				
282	-2.86*	0.251***	-0.18@	-0.174	-0.183	-2.742	37
	(0.393)	(0.146)	(0.176)				
316	-0.05@	-0.52**	-0.12@	0.368	0.3879	10.772	37
	(0.58)	(0.19)	(0.13)				
386	27.95@	-14.7@	-28.8@	0.1924	0.2028	4.4074	37
	(35.9)	(13.1)	(17.4)				
387	0.362*	-0.15*	0.05@	-0.288	-0.304	-4.137	37
	(0.06)	(0.03)	(0.06)				
Total	-1.32***	1.199*	1.102*	0.9467	0.9979	328.59	37
WFBI	(0.077)	(0.067)	(0.175)				

(Figures within brackets are the standard error)
**Significant at 1% level of significance*
*** Significant at 5% level of significance*
**** Significant at 10% level of significance*
@ Insignificant

Table (6.1.10): Estimates of V.E.S. Production Function for Eastern Region

Dependent Variable: Ln (V/L) Model: II								
Industry Code	Const.	Ln (W)	Ln (K/L)	T	R^2	Adjusted R^2	F-Statistic	D.F.
226	-1.71*	0.`506@	0.453*	0.011@	0.7758	0.8405	41.524	36
	(0.181)	(0.308)	(0.17)	(0.025)				
270	-3.59*	2.593**	0.923*	-0.100@	0.8431	0.9133	64.469	36
	(0.507)	(1.05)	(0.298)	(0.116)				
271	1.748*	1.382**	0.214@	-0.09***	0.7872	0.8528	44.391	36
	(0.185)	(0.523)	(0.192)	(0.047)				
272	-6.57*	1.627***	1.271**	0.018***	0.9127	0.9887	125.43	36
	(0.747)	(0.934)	(0.55)	(0.101)				
273	-5.37*	0.118@	-0.39@	0.017@	0.9187	0.9953	135.6	36
	(0.398)	(0.716)	(0.428)	(0.067)				
274	-0.12@	1.024*	0.249**	-0.05*	0.8942	0.9687	101.45	36
	(0.17)	(0.168)	(0.123)	(0.018)				
275	2.273*	1.713*	1.176*	-0.100@	0.572	0.6197	16.037	36
	(0.451)	(0.504)	(0.421)	(0.038)				
276	-5.22*	-0.07@	0.517***	0.093*	0.9197	0.9964	137.49	36
	(0.226)	(0.943)	(0.267)	(0.11)				
279	5.776*	3.868*	0.629**	-0.26*	0.9052	0.9806	114.53	36
	(0.88)	(0.816)	(0.232)	(0.084)				
280	7.153*	2.965*	0.045@	-0.27*	0.8817	0.9552	89.437	36
	(0.134)	(0.4)	(0.153)	(0.038)				
281	-5.15*	-0.17@	0.34@	0.07@	0.8995	0.9745	107.4	36
	(0.267)	(0.777)	(0.219)	(0.071)				
282	1.246*	1.461*	-0.22@	-0.09**	0.497	0.5384	11.857	36
	(0.357)	(0.519)	(0.161)	(0.039)				
316	2.165*	0.293**	-0.14	-0.08@	-0.285	-0.309	-2.661	36
	(0.6)	(1.35)	(0.13)	(0.13)				
386	-90.7**	-57***	-35.7***	3.834@	-0.355	-0.385	-3.144	36
	(34.7)	(29.3)	(17.4)	(2.39)				
387	-0.4*	-0.36***	0.028@	0.023@	-0.29	-0.314	-2.698	36
	(0.04)	(0.2)	(0.03)	(0.02)				
Total WFBI	-0.56*	1.433*	1.108*	-0.02@	0.9188	0.9954	135.8	36
	(0.078)	(0.304)	(0.177)	(0.031)				

(Figures within brackets are the standard error)

**Significant at 1% level of significance*

*** Significant at 5% level of significance*

**** Significant at 10% level of significance*

@ Insignificant

Table (6.2.1): Estimates of C-D Production Function for Western Region

Dependent Variable: Ln (V) Model: I							
Industry Code	Constant	Ln (L)	Ln (K)	R^2	Adjusted R^2	F-Statistic	D.F.
226	-0.578**	1.696*	-0.0101@	0.7671	0.8085	39.516	37
	(-0.212)	(0.125)	(0.096)				
270	13.265*	0.7138*	-1.606*	0.8696	0.9166	80.02	37
	(0.226)	(0.143)	(0.368)				
271	5.4932*	1.1037*	-0.615*	0.9405	0.9914	189.79	37
	(0.2007)	(0.1003)	(0.163)				
272	1.7379*	1.2967*	-0.281**	0.9435	0.9945	200.45	37
	(0.267)	(0.120)	(0.107)				
273	1.4754*	1.4885*	-0.248@	0.8204	0.8647	54.807	37
	(0.248)	(0.1045)	(0.196)				
274	1.6064*	1.4639*	-0.298*	0.9314	0.9817	162.81	37
	(0.181)	(0.052)	(0.077)				
275	-0.757@	3.7568*	-0.157@	0.9082	0.9573	118.67	37
	(0.522)	(0.1826)	(0.209)				
276	5.2406*	1.4515*	-0.688*	0.9306	0.9809	160.93	37
	(0.214)	(0.122)	(0.198)				
279	0.3112@	0.6525*	0.0301*	0.665	0.7009	23.817	37
	(0.231)	(0.121)	(0.030)				
280	0.8024**	0.8926*	-0.171*	0.7593	0.8003	37.853	37
	(0.352)	(0.045)	(0.063)				
281	3.8206*	0.9774*	-0.411*	0.8243	0.8689	56.301	37
	(0.211)	(0.089)	(0.133)				
282	1.2291*	1.3245*	-0.239**	0.0834	0.0879	1.0913	37
	(0.3341)	(0.056)	(0.107)				
316	0.3213*	0.0476*	0.0222*	0.8267	0.8714	57.247	37
	(0.0592)	(0.015)	(0.012)				
385	1.9113*	1.0022*	-0.267*	0.9462	0.9974	211.12	37
	(0.208)	(0.116)	(0.089)				
386	-0.845@	2.4546*	-0.08@	0.9166	0.9661	131.88	37
	(1.179)	(0.239)	(0.366)				
387	1.0217*	0.3214*	-0.118*	0.3002	0.3164	5.1472	37
	(0.051)	(0.024)	(0.0304)				
Total WFBI	13.375	0.836*	-0.857*	0.9254	0.9755	148.93	37
	(0.307)	(0.081)	(0.243)				

(Figures within brackets are the standard error)

**Significant at 1% level of significance*

*** Significant at 5% level of significance*

**** Significant at 10% level of significance*

@ Insignificant

Table (6.2.2): Estimates of C-D Production Function for Western Region

Dependent Variable: Ln (V) Model: II								
Industry Code	**Constant**	**Ln (L)**	**Ln (K)**	**T**	**R^2**	**Adjusted R^2**	**F-Statistic**	**D.F.**
226	-0.44**	1.397*	0.154@	-0.07**	0.7686	0.8327	39.858	36
	(0.193)	(0.183)	(0.116)	(0.029)				
270	10.53*	0.654*	-1.09**	-0.06***	0.9038	0.9791	112.74	36
	(0.214)	(0.143)	(0.452)	(0.036)				
271	4.979*	1.091*	-0.53@	-0.01@	0.9125	0.9885	125.12	36
	(0.201)	(0.113)	(0.464)	(0.063)				
272	1.511*	0.802*	0.267@	-0.12*	0.9038	0.9791	112.73	36
	(0.227)	(0.187)	(0.201)	(0.04)				
273	0.209@	1.032*	0.431**	-0.12*	0.8941	0.9686	101.35	36
	(0.196)	(0.144)	(0.222)	(0.026)				
274	1.775*	1.283*	-0.17@	-0.04@	0.9042	0.9795	113.22	36
	(0.168)	(0.143)	(0.13)	(0.026)				
275	-0.76@	3.757*	-0.16@	0.0006@	0.9144	0.9906	128.17	36
	(0.522)	(0.678)	(0.313)	(0.078)				
276	2.49*	1.314*	-0.06@	-0.08@	0.9131	0.9892	126.15	36
	(0.214)	(0.153)	(0.489)	(0.061)				
279	1.907*	0.161*	0.341*	-0.13*	0.9029	0.9781	111.58	36
	(0.161)	(0.07)	(0.1)	(0.02)				
280	0.478**	0.334*	0.55*	-0.14*	0.7883	0.854	44.684	36
	(0.216)	(0.09)	(0.126)	(0.018)				
281	0.9*	0.855*	0.211@	-0.11***	0.8204	0.8888	54.815	36
	(0.2)	(0.1)	(0.322)	(0.056)				
282	1.289*	0.597*	0.465**	-0.18*	0.0844	0.0914	1.1062	36
	(0.251)	(0.179)	(0.191)	(0.048)				
316	0.369*	0.035@	0.026@	0.002@	0.8215	0.89	55.227	36
	(0.061)	(0.062)	(0.019)	(0.007)				
385	1.367*	0.565*	0.267***	-0.12*	0.9186	0.9952	135.43	36
	(0.158)	(0.139)	(0.158)	(0.03)				
386	-2.47**	2.965*	-0.24@	0.096@	0.9196	0.9962	137.25	36
	(1.195)	(0.94)	(0.462)	(0.144)				
387	0.788*	0.291*	-0.06@	-0.01@	0.3001	0.3251	5.1453	36
	(0.051)	(0.041)	(0.129)	(0.02)				
Total	8.635*	0.836*	-0.47@	-0.05@	0.9084	0.9841	118.96	36
WFBI	(0.318)	(0.085)	(1.217)	(0.159)				

(Figures within brackets are the standard error)
**Significant at 1% level of significance*
*** Significant at 5% level of significance*
**** Significant at 10% level of significance*
@ Insignificant

Table (6.2.3): Estimates of C-D Production Function for Western Region

Dependent Variable: Ln (V/L) Model: III							
Industry Code	Const.	Ln (K/L)	T	R^2	Adjusted R^2	F-Statistic	D.F.
226	-1.55*	0.355*	0.013@	0.6102	0.6431	18.782	37
	(0.139)	(0.131)	(0.005)				
270	-3.00*	0.298*	0.024@	0.4107	0.4329	8.3639	37
	(0.15)	(0.086)	(0.005)				
271	-2.49*	0.31@	0.011@	0.6968	0.7345	27.582	37
	(0.21)	(0.214)	(0.015)				
272	-2.7*	0.491*	0.016@	0.9175	0.9671	133.49	37
	(0.262)	(0.156)	(0.016)				
273	-5.03*	-0.25@	0.048*	0.6377	0.6721	21.12	37
	(0.276)	(0.295)	(0.016)				
274	-2.36*	0.489*	0.019***	0.7173	0.7561	30.449	37
	(0.227)	(0.147)	(0.008)				
275	1.848*	2.411*	-0.12*	0.4289	0.4521	9.0136	37
	(0.807)	(0.696)	(0.039)				
276	-3.35*	0.476***	0.032**	0.8998	0.9484	107.72	37
	(0.232)	(0.252)	(0.015)				
279	-1.88*	0.416*	0.004@	0.7309	0.7704	32.596	37
	(0.409)	(0.088)	(0.014)				
280	-1.98*	-0.05@	0.029@	0.372	0.3921	7.1084	37
	(0.234)	(0.267)	(0.018)				
281	-2.18*	0.216@	0.015@	0.6456	0.6805	21.86	37
	(0.168)	(0.138)	(0.01)				
282	-2.98*	-0.02@	0.011@	0.0039	0.0041	0.0465	37
	(0.33)	(0.136)	(0.014)				
316	-0.11**	0.013@	0.004***	0.3813	0.4019	7.3945	37
	(0.059)	(0.01)	(0.002)				
385	-2.88*	0.022@	0.05*	0.9165	0.966	131.64	37
	(0.208)	(0.252)	(0.016)				
386	1.679@	1.455**	-0.09***	0.2063	0.2175	3.1199	37
	(1.315)	(0.65)	(0.051)				
387	-1.07***	-0.05@	0.022*	0.2525	0.2661	4.0525	37
	(0.066)	(0.068)	(0.005)				
Total	-1.07***	0.828*	0.004@	0.9436	0.9947	200.93	37
WFBI	(0.055)	(0.119)	(0.008)				

(Figures within brackets are the standard error)
**Significant at 1% level of significance*
*** Significant at 5% level of significance*
**** Significant at 10% level of significance*
@ Insignificant

Table (6.2.4): Estimates of C-D Production Function for Western Region

	Dependent Variable ln(V/L) Model IV							
Industry Code	Const.	Ln (K/L)	Ln (L)	T	R^2	Adjusted R^2	F-Statistic	D.F.
226	-1.611*	0.45*	0.1021@	-0.007@	0.6264	0.6786	20.12	36
	(0.1388)	(0.1609)	(0.1031)	(0.0202)				
270	-4.386*	0.2927*	0.2213@	0.0071@	0.4214	0.4565	8.7397	36
	(0.1511)	(0.0868)	(0.2724)	(0.0214)				
271	-3.928*	0.2822@	0.2182@	-0.013@	0.701	0.7594	28.134	36
	(0.2132)	(0.227)	(0.5079)	(0.0564)				
272	-4.274*	0.3489@	0.266@	-0.004@	0.913	0.9891	125.88	36
	(0.262)	(0.2085)	(0.2593)	(0.0254)				
273	-6.626*	-0.3@	0.2912@	0.0287@	0.6599	0.7149	23.284	36
	(0.2768)	(0.3016)	(0.3238)	(0.0261)				
274	-3.827*	0.3809**	0.2345@	0.0078@	0.7591	0.8224	37.813	36
	(0.2189)	(0.1555)	(0.1425)	(0.0104)				
275	0.9992@	2.3715*	0.2113@	-0.128*	0.4401	0.4768	9.4324	36
	(0.8218)	(0.7122)	(0.3939)	(0.0444)				
276	-6.933*	0.4477***	0.6402@	-0.024@	0.9172	0.9936	132.93	36
	(0.2292)	(0.249)	(0.4996)	(0.0458)				
279	-3.214*	0.3674*	0.4055***	-0.042@	0.7777	0.8425	41.981	36
	(0.3945)	(0.0893)	(0.2499)	(0.0321)				
280	-6.459*	-0.042@	0.5545@	-0.04@	0.3877	0.42	7.5982	36
	(0.2362)	(0.2696)	(0.7015)	(0.0884)				
281	-3.284*	0.1888@	0.1733@	-0.007@	0.6525	0.7069	22.532	36
	(0.171)	(0.1466)	(0.2843)	(0.0367)				
282	-5.287*	-0.177@	0.3592**	-0.006@	0.0202	0.0219	0.2474	36
	(0.3019)	(0.142)	(0.1575)	(0.0143)				
316	0.1534*	0.0258@	-0.045@	0.009***	0.426	0.4615	8.9059	36
	(0.0593)	(0.0185)	(0.0453)	(0.0054)				
385	-3.119*	0.0099@	0.0495*	0.044***	0.9173	0.9937	133.1	36
	(0.2129)	(0.2614)	(0.1832)	(0.0267)				
386	-0.008@	1.3724***	0.2726@	-0.09***	0.2212	0.2396	3.4083	36
	(1.334)	(0.6742)	(0.4606)	(0.052)				
387	-0.699*	-0.027@	-0.06@	0.03@	0.2529	0.274	4.0621	36
	(0.067)	(0.0842)	(0.1627)	(0.019)				
Total	2.0148*	0.8137*	-0.252@	0.035@	0.9211	0.9978	140.02	36
WFBI	(0.054)	(0.117)	(0.200)	(0.025)				

(Figures within brackets are the standard error)
**Significant at 1% level of significance*
*** Significant at 5% level of significance*
**** Significant at 10% level of significance*
@ Insignificant

Table (6.2.5): Estimates of C.E.S. Production Function for Western Region

Dependent Variable: Ln (V/L) Model: I						
Industry Code	Constant	Ln (W)	R^2	Adjusted R^2	F-Statistic	D.F.
226	0.2178@ (0.4009)	0.6716* (0.1156)	0.6167	0.6329	30.57	38
270	-1.967* (0.1939)	0.3866* (0.0833)	0.9075	0.9314	186.41	38
271	-1.999* (0.2458)	0.5995* (0.0677)	0.9398	0.9645	296.61	38
272	-3.37* (0.2781)	0.2072@ (0.1791)	0.619	0.6353	30.869	38
273	-1.936* (0.2011)	0.6058* (0.1045)	0.9179	0.9421	212.43	38
274	-1.562* (0.4495)	0.4728** (0.2332)	0.2039	0.2093	4.8663	38
275	-3.89* (0.4018)	1.2836* (0.2505)	0.9478	0.9727	344.98	38
276	-2.231* (0.5989)	0.8296* (0.162)	0.7528	0.7726	57.861	38
279	-1.074* (0.2339)	0.1917@ (0.1184)	0.608	0.624	29.469	38
280	-1.325* (0.1379)	0.3056* (0.0577)	0.5581	0.5728	23.996	38
281	-2.116* (0.291)	0.1855*** (0.1077)	0.149	0.1529	3.3267	38
282	-0.78* (0.142)	0.704* (0.0573)	0.0815	0.0836	1.6859	38
316	-0.16* (0.0453)	0.0211@ (0.0373)	0.0462	0.0474	0.9203	38
385	-0.664@ (0.8526)	0.3506@ (0.3119)	0.1341	0.1376	2.9425	38
386	-1.717* (0.2869)	0.7093* (0.1275)	0.9466	0.9715	336.81	38
387	-0.18** (0.0724)	0.2338* (0.0257)	0.2498	0.2564	6.3266	38
Total WFBI	-0.884* (0.0889)	0.5661* (0.0233)	0.9381	0.9628	287.95	38

(Figures within brackets are the standard error)
**Significant at 1% level of significance*
*** Significant at 5% level of significance*
**** Significant at 10% level of significance*
@ Insignificant

Table (6.2.6): Estimates of C.E.S. Production Function Western Region

		Dependent Variable ln(V/L) Model II					
Industry Code	**Const.**	**Ln (W)**	**T**	**R^2**	**Adjusted R^2**	**F-Statistic**	**D.F.**
226	-3.93*	0.007@	0.02@	0.4914	0.518	11.596	37
	(0.161)	(0.218)	(0.022)				
270	-1.04*	0.719**	-0.04@	0.1892	0.1995	2.8007	37
	(0.171)	(0.356)	(0.039)				
271	1.553*	1.339**	-0.09***	0.7701	0.8117	40.191	37
	(0.194)	(0.541)	(0.049)				
272	-1.94*	0.689***	-0.01@	0.8547	0.9009	70.576	37
	(0.293)	(0.362)	(0.039)				
273	-2.85*	0.315@	0.004@	0.6326	0.6668	20.661	37
	(0.277)	(0.43)	(0.046)				
274	-0.4**	0.951*	-0.06*	0.8469	0.8927	66.376	37
	(0.183)	(0.176)	(0.019)				
275	0.548@	1.865**	-0.15***	0.1976	0.2083	2.9555	37
	(0.918)	(0.905)	(0.076)				
276	-4.45*	0.212@	0.034@	0.8628	0.9094	75.468	37
	(0.251)	(1.034)	(0.118)				
279	3.469*	1.779*	-0.13*	0.7915	0.8343	45.55	37
	(0.378)	(0.325)	(0.033)				
280	6.196*	2.558*	-0.23*	0.7745	0.8163	41.211	37
	(0.115)	(0.307)	(0.031)				
281	-3.88*	-0.31@	0.057@	0.6104	0.6433	18.797	37
	(0.176)	(0.493)	(0.046)				
282	0.685**	1.085*	-0.07@	0.033	0.0348	0.4098	37
	(0.289)	(0.416)	(0.031)				
316	-0.16**	-0.02@	0.006@	0.3363	0.3545	6.0801	37
	(0.061)	(0.137)	(0.013)				
385	-1.98*	0.281@	0.025@	0.9185	0.9681	135.16	37
	(0.208)	(0.642)	(0.06)				
386	0.21@	1.175@	-0.1@	0.0878	0.0925	1.1545	37
	(1.419)	(1.078)	(0.096)				
387	0.524*	0.401**	-0.03@	0.2617	0.2758	4.2529	37
	(0.06)	(0.174)	(0.02)				
Total	-0.53*	0.673***	-0.01@	0.9209	0.9707	139.72	37
WFBI	(0.091)	(0.346)	(0.037)				

(Figures within brackets are the standard error)
**Significant at 1% level of significance*
*** Significant at 5% level of significance*
**** Significant at 10% level of significance*
@ Insignificant

Table (6.2.7): Estimates of C.E.S. Production Function for Western Region

Dependent Variable: Ln (V/L) Model: III								
Industry Code	Const.	Ln (W)	Ln (L)	T	R^2	Adjusted R^2	F-Statistic	D.F.
226	-2.72*	0.126@	-0.10@	0.025@	0.5069	0.5491	12.336	36
	(0.162)	(0.263)	(0.116)	(0.022)				
270	-0.44**	0.76***	-0.07@	-0.04@	0.1892	0.205	2.8002	36
	(0.175)	(0.434)	(0.376)	(0.039)				
271	3.985*	1.525**	-0.28@	-0.08@	0.7761	0.8408	41.595	36
	(0.197)	(0.655)	(0.536)	(0.058)				
272	-5.83*	0.168@	0.491***	-0.02@	0.8992	0.9741	107.05	36
	(0.278)	(0.443)	(0.263)	(0.037)				
273	-3.95*	0.188@	0.127@	0.008@	0.6351	0.688	20.886	36
	(0.283)	(0.65)	(0.481)	(0.05)				
274	3.078*	1.425*	-0.35***	-0.09*	0.8816	0.9551	89.351	36
	(0.172)	(0.303)	(0.185)	(0.023)				
275	0.399@	1.845***	0.026@	-0.15***	0.1976	0.2141	2.9551	36
	(0.939)	(0.991)	(0.48)	(0.078)				
276	-7.74*	0.414@	0.749@	-0.06@	0.8855	0.9593	92.803	36
	(0.245)	(1.022)	(0.538)	(0.131)				
279	4.958*	2.001*	-0.22@	-0.12*	0.7995	0.8661	47.85	36
	(0.382)	(0.458)	(0.318)	(0.035)				
280	11.37*	2.745*	-0.56@	-0.18*	0.788	0.8537	44.604	36
	(0.111)	(0.319)	(0.354)	(0.043)				
281	-15.2*	-1.78**	1.064**	0.049@	0.7203	0.7803	30.903	36
	(0.154)	(0.689)	(0.389)	(0.04)				
282	2.179*	1.371***	-0.11@	-0.08***	0.0338	0.0366	0.4198	36
	(0.295)	(0.82)	(0.26)	(0.044)				
316	-0.16**	-0.02@	0.0001@	0.006@	0.3363	0.3643	6.0805	36
	(0.062)	(0.141)	(0.034)	(0.014)				
385	-2.13*	0.253@	0.014@	0.026@	0.9185	0.995	135.24	36
	(0.212)	(0.776)	(0.212)	(0.062)				
386	2.004@	1.53@	-0.17@	-0.12@	0.0895	0.097	1.1796	36
	(1.452)	(3.107)	(1.38)	(0.199)				
Total	3.52*	0.338*	-0.27**	0.015@	0.9211	0.9978	140.02	36
WFBI	(0.026)	(0.111)	(0.107)	(0.013)				

(Figures within brackets are the standard error)
**Significant at 1% level of significance*
*** Significant at 5% level of significance*
**** Significant at 10% level of significance*
@ Insignificant

Table (6.2.8): Estimates of C.E.S. Production Function for Western Region

Dependent Variable ln(V/L) Model: IV									
Industry Code	Const.	Ln (L)	Ln (K)	(LnL-LnK)^2	T	R^2	Adj. R^2	F-Statistic	D.F.
226	-4.57*	0.359@	-0.3@	-0.100@	0.035***	0.6177	0.6883	14.138	35
	(0.145)	(0.521)	(0.64)	(0.087)	(0.02)				
270	-6.53*	0.089@	0.277**	0.019**	0.025@	0.1405	0.1566	1.4303	35
	(0.163)	(0.302)	(0.104)	(0.08)	(0.025)				
271	-7.05*	0.912@	-0.36@	1.037@	-0.03@	0.7349	0.8189	24.256	35
	(0.212)	(0.676)	(0.361)	(0.63)	(0.058)				
272	-6.06*	0.019@	0.361@	0.193@	0.017@	0.8925	0.9945	72.645	35
	(0.27)	(0.424)	(0.247)	(0.531)	(0.031)				
273	-4.6*	0.217@	-0.1@	0.11@	0.012@	0.6673	0.7436	17.55	35
	(0.282)	(0.531)	(0.435)	(0.219)	(0.026)				
274	-6.05*	0.71***	-0.32@	-0.24@	0.017@	0.7233	0.806	22.873	35
	(0.236)	(0.356)	(0.346)	(0.158)	(0.016)				
275	-6.53*	-0.02@	0.222@	-1.07@	0.013@	0.1467	0.1635	1.5043	35
	(0.989)	(1.153)	(1.388)	(1.432)	(0.047)				
276	-8.12*	0.555@	-0.05@	-0.3@	0.04@	0.889	0.9906	70.102	35
	(0.226)	(0.711)	(0.533)	(0.252)	(0.05)				
279	-5.16*	0.228@	0.389**	-0.05@	-0.04@	0.71	0.7911	21.422	35
	(0.44)	(0.31)	(0.188)	(0.184)	(0.042)				
280	-3.95*	0.969@	-0.63@	0.128@	-0.03@	0.4508	0.5023	7.1823	35
	(0.226)	(1.09)	(0.89)	(0.461)	(0.082)				
281	-4.08*	-0.14@	0.343@	-0.24@	0.008@	0.6751	0.7523	18.181	35
	(0.17)	(0.425)	(0.244)	(0.38)	(0.039)				
282	-4.6*	0.378@	-0.04@	-0.20@	-0.01@	0.0344	0.0383	0.3117	35
	(0.301)	(0.3)	(0.26)	(0.261)	(0.014)				
316	-0.21*	-0.16@	0.166@	-0.06@	0.004@	0.7342	0.8181	24.169	35
	(0.054)	(0.29)	(0.317)	(0.081)	(0.01)				
385	-3.21*	0.134@	-0.08@	-0.05@	0.044@	0.8917	0.9936	72.066	35
	(0.218)	(0.55)	(0.47)	(0.345)	(0.026)				
386	-6.72*	-0.77@	1.678@	0.26@	-0.02@	0.2903	0.3235	3.5792	35
	(1.392)	(1.88)	(2.378)	(1.199)	(0.056)				
387	-0.95*	0.123@	-0.11@	-0.07@	0.013@	0.2859	0.3186	3.5032	35
	(0.066)	(0.22)	(0.089)	(0.118)	(0.022)				
Total	-1.42**	0.97@	-1.06	0.714*	0.075@	0.8909	0.9927	71.452	35
WFBI	(0.069)	(0.63)	(0.625)	(0.255)	(0.031)				

(Figures within brackets are the standard error)

**Significant at 1% level of significance*

*** Significant at 5% level of significance*

**** Significant at 10% level of significance*

@ Insignificant

Table (6.2.9): Estimates of V.E.S. Production Function for Western Region

Dependent Variable: Ln (V/L) Model: I							
Industry Code	Const.	Ln (W)	Ln (K/L)	R^2	Adjusted R^2	F-Statistic	D.F.
226	-0.87*	0.494*	0.37*	0.6077	0.6405	18.586	37
	(0.139)	(0.106)	(0.128)				
270	-2.1*	0.505*	0.277*	0.2179	0.2297	3.3433	37
	(0.144)	(0.068)	(0.084)				
271	-2.07*	0.408*	0.198@	0.7221	0.7611	31.174	37
	(0.205)	(0.088)	(0.212)				
272	-2.13*	0.639*	0.448**	0.9183	0.968	134.95	37
	(0.262)	(0.072)	(0.188)				
273	-3.29*	0.202@	-0.26@	0.6547	0.6901	22.749	37
	(0.272)	(0.175)	(0.287)				
274	-1.92*	0.602*	0.34**	0.8019	0.8452	48.572	37
	(0.199)	(0.104)	(0.14)				
275	-3.15*	0.952**	1.839**	0.2276	0.2399	3.5353	37
	(0.905)	(0.471)	(0.837)				
276	-2.28*	0.752*	0.473***	0.8956	0.944	102.91	37
	(0.235)	(0.146)	(0.266)				
279	-1.47*	0.545*	0.335*	0.7621	0.8033	38.446	37
	(0.393)	(0.106)	(0.097)				
280	-0.97*	0.172***	-0.27@	0.4777	0.5035	10.974	37
	(0.21)	(0.106)	(0.226)				
281	-1.63*	0.376*	0.241***	0.6366	0.671	21.018	37
	(0.17)	(0.071)	(0.138)				
282	-2.27*	0.200***	-0.14@	0.0178	0.0188	0.2178	37
	(0.312)	(0.116)	(0.104)				
316	0.005@	0.053*	0.013*	0.368	0.3878	6.986	37
	(0.06)	(0.019)	(0.033)				
385	-0.94*	0.571*	0.069*	0.9176	0.9672	133.69	37
	(0.208)	(0.11)	(0.236)				
386	-1.07@	0.564@	1.1@	0.1924	0.2028	2.859	37
	(1.372)	(0.502)	(0.666)				
387	-0.36*	0.151*	-0.05@	0.2896	0.3052	4.8916	37
	(0.061)	(0.027)	(0.06)				
Total	-0.94*	0.85***	0.781*	0.9411	0.9919	191.64	37
WFBI	(0.055)	(0.047)	(0.124)				

(Figures within brackets are the standard error)
**Significant at 1% level of significance*
*** Significant at 5% level of significance*
**** Significant at 10% level of significance*
@ Insignificant

Table (6.2.10): Estimates of V.E.S. Production Function for Western Region

Dependent Variable: Ln (V/L) Model: II								
Industry Code	Const.	Ln (W)	Ln (K/L)	T	R^2	Adjusted R^2	F-Statistic	D.F.
226	-1.35*	0.398@	0.357**	0.009@	0.6109	0.6618	18.84	
	(0.142)	(0.242)	(0.134)	(0.02)				36
270	-1.03*	0.746**	0.266*	-0.03@	0.4529	0.4906	9.9338	36
	(0.146)	(0.302)	(0.086)	(0.033)				
271	1.827*	1.444**	0.224@	-0.09***	0.7953	0.8616	46.622	36
	(0.193)	(0.546)	(0.20)	(0.049)				
272	-2.35*	0.583***	0.455**	0.006@	0.9187	0.9953	135.6	36
	(0.267)	(0.334)	(0.197)	(0.036)				
273	-3.76*	0.083@	-0.27@	0.012@	0.6564	0.7111	22.924	36
	(0.278)	(0.5)	(0.299)	(0.047)				
274	-0.12@	1.032*	0.251**	-0.06*	0.8857	0.9595	92.987	36
	(0.171)	(0.169)	(0.124)	(0.018)				
275	4.045*	3.049*	2.093*	-0.18**	0.473	0.5124	10.77	36
	(0.802)	(0.897)	(0.748)	(0.068)				
276	-5.45*	-0.07@	0.54***	0.097@	0.9034	0.9787	112.22	36
	(0.236)	(0.986)	(0.279)	(0.115)				
279	2.185*	1.463*	0.238*	-0.1*	0.8836	0.9572	91.093	36
	(0.333)	(0.309)	(0.088)	(0.032)				
280	6.286*	2.605*	0.04@	-0.24*	0.7748	0.8394	41.286	36
	(0.117)	(0.352)	(0.134)	(0.033)				
281	-3.29*	-0.11@	0.217@	0.045@	0.6538	0.7083	22.662	36
	(0.17)	(0.496)	(0.14)	(0.045)				
282	0.989*	1.16*	-0.18@	-0.07**	0.0406	0.044	0.5078	36
	(0.283)	(0.412)	(0.127)	(0.031)				
316	-0.22*	-0.03@	0.014@	0.008@	0.386	0.4182	7.544	36
	(0.061)	(0.138)	(0.014)	(0.014)				
385	-1.78*	0.335@	0.039@	0.021@	0.9189	0.9955	135.97	36
	(0.212)	(0.748)	(0.259)	(0.066)				
386	3.468*	2.181***	1.365**	-0.15@	0.3555	0.3851	6.6191	36
	(1.326)	(1.119)	(0.665)	(0.092)				
387	0.399*	0.363***	-0.03@	-0.02@	0.2922	0.3166	4.9539	36
	(0.061)	(0.196)	(0.063)	(0.021)				
Total	-0.94*	0.85*	0.781*	0.714*	0.9102	0.9861	121.69	36
WFBI	(0.055)	(0.047)	(0.124)	(0.255)				

(Figures within brackets are the standard error)
**Significant at 1% level of significance*
*** Significant at 5% level of significance*
**** Significant at 10% level of significance*
@ Insignificant

Table (6.3.1): Estimates of C-D Production Function for Northern Region

Dependent Variable: Ln (V) Model: I							
Industry Code	Constant	Ln (L)	Ln (K)	R^2	Adjusted R^2	F-Statistic	D.F.
226	-0.362*	1.0626*	-0.006@	0.4806	0.5066	17.118	37
	(-0.133)	(0.0785)	(0.0604)				
270	11.669*	0.6279*	-1.413*	0.8362	0.8814	94.443	37
	(0.1989)	(0.1256)	(0.3244)				
271	11.735*	2.3577*	-1.313*	0.9045	0.9534	175.22	37
	(0.4287)	(0.2143)	(0.3483)				
272	1.4165*	1.0569*	-0.229*	0.8769	0.9243	131.78	37
	(0.2179)	(0.0981)	(0.0872)				
273	1.4759*	1.489*	-0.248@	0.9405	0.9913	292.42	37
	(0.2482)	(0.1045)	(0.1959)				
274	1.5323*	1.3964*	-0.284*	0.9412	0.9921	296.13	37
	(0.173)	(0.0494)	(0.0741)				
275	-0.522*	2.5913*	-0.108@	0.9425	0.9934	303.24	37
	(0.3599)	(0.126)	(0.144)				
276	4.0079*	1.1101*	-0.526*	0.8661	0.9129	119.66	37
	(0.1636)	(0.0935)	(0.1519)				
279	0.2383@	0.4997*	0.0231@	0.4456	0.4697	14.869	37
	(0.1768)	(0.0922)	(0.0231)				
280	0.7762**	0.8635*	-0.166*	0.7345	0.7742	51.18	37
	(0.3402)	(0.0436)	(0.0611)				
281	3.1462*	0.8049*	-0.338*	0.8468	0.8926	102.26	37
	(0.1738)	(0.0732)	(0.1098)				
282	1.2625*	1.3606*	-0.245**	0.4507	0.4751	15.179	37
	(0.3432)	(0.0613)	(0.1103)				
316	1.7375*	0.2572*	0.12***	0.8267	0.8714	88.251	37
	(0.3201)	(0.08)	(0.0629)				
385	1.777*	0.9318*	-0.249*	0.9462	0.9973	325.37	37
	(0.1934)	(0.1077)	(0.0829)				
387	1.926*	0.606*	-0.223*	0.566	0.5966	24.127	37
	(0.096)	(0.0446)	(0.0574)				
Total WFBI	18.077*	1.1298*	-1.158*	0.9422	0.9931	301.57	37
	(0.4148)	(0.1101)	(0.3289)				

(Figures within brackets are the standard error)

**Significant at 1% level of significance*

*** Significant at 5% level of significance*

**** Significant at 10% level of significance*

@ Insignificant

Table (6.3.2): Estimates of C-D Production Function for Northern Region

Dependent Variable: Ln (V) Model: II								
Industry Code	Constant	Ln (L)	Ln (K)	T	R^2	Adjusted R^2	F-Statistic	D.F.
226	-0.278**	0.8754*	0.0966*	-0.042**	0.4815	0.5216	11.144	36
	(0.1207)	(0.1147)	(0.0724)	(0.0181)				
270	9.2622*	0.5756*	-0.963**	-0.052***	0.8381	0.9079	62.12	36
	(0.1884)	(0.1256)	(0.3977)	(0.0314)				
271	10.636*	2.3309*	-1.125@	-0.027@	0.8989	0.9738	106.69	36
	(0.4287)	(0.2411)	(0.9913)	(0.134)				
272	1.2313*	0.6538*	0.2179@	-0.098*	0.8862	0.9601	93.448	36
	(0.1852)	(0.1525)	(0.1634)	(0.0327)				
273	0.209@	1.0318*	0.431***	-0.118*	0.8748	0.9477	83.854	36
	(0.1959)	(0.1437)	(0.222)	(0.0261)				
274	1.693*	1.2234*	-0.161@	-0.037@	0.8942	0.9687	101.39	36
	(0.1606)	(0.1359)	(0.1236)	(0.0247)				
275	-0.522@	2.5913*	-0.108@	0.0006@	0.8894	0.9635	96.518	36
	(0.3599)	(0.468)	(0.2159)	(0.054)				
276	1.9046*	1.0049*	-0.047@	-0.058@	0.8668	0.939	78.09	36
	(0.1636)	(0.12)	(0.3739)	(0.0467)				
279	1.4606*	0.123**	0.2614*	-0.1*	0.6226	0.6745	19.797	36
	(0.123)	(0.054)	(0.0769)	(0.0154)				
280	0.4623**	0.3227*	0.532*	-0.131*	0.7626	0.8262	38.548	36
	(0.2093)	(0.087)	(0.1221)	(0.0174)				
281	0.7408*	0.7042*	0.1738@	-0.091***	0.8427	0.9129	64.287	36
	(0.1646)	(0.0823)	(0.2652)	(0.0457)				
282	1.3238*	0.6129*	0.478***	-0.184*	0.4563	0.4943	10.071	36
	(0.2574)	(0.1839)	(0.1961)	(0.049)				
316	1.9975*	0.1865@	0.1397@	0.0085@	0.8215	0.89	55.227	36
	(0.3274)	(0.3357)	(0.1046)	(0.0379)				
385	1.2706*	0.5248*	0.2486***	-0.11@	0.8949	0.9694	102.13	36
	(0.1473)	(0.1289)	(0.1473)	(0.0276)				
387	1.4862*	0.5485*	-0.108@	-0.019@	0.5658	0.613	15.637	36
	(0.0957)	(0.0765)	(0.2424)	(0.0383)				
Total	11.67*	1.1298**	-0.629@	-0.072@	0.9142	0.9904	127.83	36
WFBI	(0.4291)	(0.1144)	(1.6447)	(0.2145)				

(Figures within brackets are the standard error)

**Significant at 1% level of significance*

*** Significant at 5% level of significance*

**** Significant at 10% level of significance*

@ Insignificant

Table (6.3.3): Estimates of C-D Production Function for Northern Region

Dependent Variable: Ln (V/L) Model: III							
Industry Code	Const.	Ln (K/L)	T	R^2	Adjusted R^2	F-Statistic	D.F.
226	-0.969*	0.2223*	0.008@	0.3823	0.403	11.45	37
	(0.087)	(0.0817)	(0.0031)				
270	-2.642*	0.2624*	0.0207*	0.8138	0.8578	80.856	37
	(0.132)	(0.0756)	(0.0045)				
271	-5.318*	0.6612@	0.023@	0.647	0.682	33.908	37
	(0.4482)	(0.456)	(0.0316)				
272	-2.204*	0.4003*	0.0133@	0.8527	0.8988	107.09	37
	(0.2135)	(0.1268)	(0.0129)				
273	-5.03*	-0.248@	0.0482*	0.6099	0.6429	28.924	37
	(0.2762)	(0.2952)	(0.0159)				
274	-2.249*	0.4664*	0.0177**	0.7173	0.7561	46.94	37
	(0.2165)	(0.1403)	(0.0078)				
275	1.2749*	1.6627*	-0.08*	0.4148	0.4372	13.113	37
	(0.5568)	(0.4801)	(0.027)				
276	-2.559*	0.364***	0.0242**	0.8374	0.8827	95.276	37
	(0.1777)	(0.193)	(0.0118)				
279	-1.439*	0.3182*	0.0032*	0.4898	0.5163	17.76	37
	(0.313)	(0.0672)	(0.011)				
280	-1.914*	-0.053@	0.0279@	0.3599	0.3794	10.402	37
	(0.2267)	(0.2579)	(0.0171)				
281	-1.793*	0.1776@	0.012@	0.6632	0.699	36.429	37
	(0.1385)	(0.1136)	(0.0079)				
282	-3.064*	-0.023@	0.0112@	0.0209	0.022	0.3949	37
	(0.339)	(0.1396)	(0.0141)				
316	-0.57***	0.0704@	0.0221**	0.3813	0.4019	11.401	37
	(0.32)	(0.0712)	(0.0091)				
385	-2.67*	0.0204@	0.0461*	0.9052	0.9541	176.65	37
	(0.194)	(0.234)	(0.0146)				
387	-2.02*	-0.085@	0.042*	0.476	0.5017	16.805	37
	(0.124)	(0.1274)	(0.0097)				
Total	-1.45**	1.1184*	0.0056@	0.9402	0.991	290.86	37
WFBI	(0.074)	(0.1608)	(0.011)				

(Figures within brackets are the standard error)
**Significant at 1% level of significance*
*** Significant at 5% level of significance*
**** Significant at 10% level of significance*
@ Insignificant

Table (6.3.4): Estimates of C-D Production Function for Northern Region

Dependent Variable: Ln (V/L) Model: IV								
Industry Code	Const.	Ln (K/L)	Ln (L)	T	R^2	Adjusted R^2	F-Statistic	D.F.
226	-1.01@	0.2819*	0.064@	-0.004@	0.3924	0.4251	7.7498	36
	(0.087)	(0.1008)	(0.0646)	(0.0127)				
270	-3.859*	0.2575*	0.1947@	0.0063@	0.8205	0.8889	54.852	36
	(0.1329)	(0.0764)	(0.2397)	(0.0188)				
271	-8.391*	0.603@	0.4662@	-0.027@	0.6508	0.705	22.364	36
	(0.4555)	(0.485)	(1.0851)	(0.1206)				
272	-3.483*	0.284@	0.2168@	-0.003@	0.8641	0.9361	76.3	36
	(0.2136)	(0.17)	(0.2114)	(0.021)				
273	-6.629*	-0.3@	0.2913@	0.0287@	0.6311	0.6837	20.529	36
	(0.2769)	(0.3017)	(0.3239)	(0.0261)				
274	-3.65*	0.3633**	0.2237@	0.0074@	0.7591	0.8224	37.813	36
	(0.2088)	(0.1483)	(0.1359)	(0.0099)				
275	0.6892@	1.636*	0.1458@	-0.088*	0.4256	0.4611	8.8914	36
	(0.5668)	(0.4913)	(0.2717)	(0.0306)				
276	-5.303*	0.3424***	0.4896@	-0.019@	0.8536	0.9247	69.967	36
	(0.175)	(0.1905)	(0.3821)	(0.035)				
279	-2.461*	0.2813*	0.3106@	-0.032@	0.5212	0.5646	13.063	36
	(0.302)	(0.0684)	(0.1914)	(0.0246)				
280	-6.248*	-0.041@	0.5364@	-0.038@	0.375	0.4063	7.2	36
	(0.2285)	(0.2608)	(0.6786)	(0.0855)				
281	-2.704*	0.1555*	0.1427@	-0.005@	0.6703	0.7262	24.397	36
	(0.1408)	(0.021)	(0.2341)	(0.0302)				
282	-5.431*	-0.181@	0.369**	-0.006@	0.1093	0.1184	1.4725	36
	(0.3101)	(0.1459)	(0.1618)	(0.0147)				
316	-0.8293*	0.1395@	-0.242@	0.051***	0.426	0.4615	8.9059	36
	(0.3206)	(0.1)	(0.2452)	(0.0297)				
385	-2.901*	0.0092@	0.046@	0.040@	0.906	0.9815	115.66	36
	(0.198)	(0.243)	(0.1703)	(0.0249)				
387	-1.318*	-0.051@	-0.112@	0.0542@	0.4768	0.5165	10.936	36
	(0.126)	(0.1588)	(0.3068)	(0.0351)				
Total	2.723*	1.099*	-0.34@	0.0472@	0.9124	0.9884	125	36
WFBI	(0.073)	(0.1587)	(0.2703)	(0.0343)				

(Figures within brackets are the standard error)
Significant at 1% level of significance
*** Significant at 5% level of significance*
**** Significant at 10% level of significance*
@ Insignificant

Table (6.3.5): Estimates of C.E.S. Production Function for Northern Region

Dependent Variable: Ln (V/L) Model: I						
Industry Code	Constant	Ln (W)	R^2	Adjusted R^2	F-Statistic	D.F.
226	0.1364@	0.4208*	0.3864	0.3966	11.965	38
	(0.2512)	(0.0724)				
270	-1.73*	0.3401*	0.6227	0.6391	31.358	38
	(0.1706)	(0.0733)				
271	-4.271*	1.2806*	0.8816	0.9048	141.47	38
	(0.5251)	(0.1447)				
272	-2.747*	0.1689*	0.5753	0.5904	25.737	38
	(0.2266)	(0.046)				
273	-1.937*	0.606*	0.8779	0.901	136.61	38
	(0.2011)	(0.1045)				
274	-1.49*	0.451**	0.2039	0.2093	4.8663	38
	(0.4288)	(0.2224)				
275	-2.683*	0.8854*	0.9263	0.9507	238.8	38
	(0.2771)	(0.1728)				
276	-1.706*	0.6345*	0.7006	0.719	44.46	38
	(0.458)	(0.1239)				
279	-0.823*	0.1468@	0.4074	0.4181	13.062	38
	(0.1791)	(0.0907)				
280	-1.282*	0.2957*	0.5399	0.5541	22.295	38
	(0.1334)	(0.0558)				
281	-1.742*	0.153***	0.1531	0.1571	3.4348	38
	(0.2396)	(0.0887)				
282	-0.802*	0.7232*	0.4404	0.452	14.953	38
	(0.1459)	(0.0588)				
316	-0.865*	0.1143@	0.0462	0.0474	0.9203	38
	(0.2452)	(0.2018)				
385	-0.618@	0.3259@	0.1324	0.1359	2.8995	38
	(0.7928)	(0.29)				
387	-0.339**	0.4408*	0.4711	0.4835	16.924	38
	(0.1365)	(0.0485)				
Total	-1.194*	0.7651*	0.9171	0.9412	210.19	38
WFBI	(0.1201)	(0.0315)				

(Figures within brackets are the standard error)
*Significant at 1% level of significance
** Significant at 5% level of significance
*** Significant at 10% level of significance
@ Insignificant

Table (6.3.6): Estimates of C.E.S. Production Function for Northern Region

Dependent Variable: Ln (V/L) Model: II							
Industry Code	**Const.**	**Ln (W)**	**T**	**R^2**	**Adjusted R^2**	**F-Statistic**	**D.F.**
226	-2.464*	0.0042@	0.0127@	0.3079	0.3246	5.3388	37
	(0.1008)	(0.1364)	(0.0139)				
270	-0.913*	0.6321**	-0.039@	0.7431	0.7832	34.706	37
	(0.1507)	(0.3129)	(0.0345)				
271	3.3168*	2.8613**	-0.19***	0.715	0.7536	30.099	37
	(0.4153)	(1.1547)	(0.1045)				
272	-1.581*	0.561***	-0.01@	0.7943	0.8373	46.343	37
	(0.2386)	(0.2953)	(0.0316)				
273	-2.853*	0.3148@	0.0039@	0.605	0.6377	18.381	37
	(0.2769)	(0.4297)	(0.0457)				
274	-0.381**	0.907*	-0.061*	0.8469	0.8927	66.376	37
	(0.1742)	(0.1681)	(0.0185)				
275	0.3779@	1.2866**	-0.106***	0.1911	0.2014	2.8353	37
	(0.6334)	(0.6244)	(0.0522)				
276	-3.399*	0.1624**	0.0257@	0.803	0.8464	48.9	37
	(0.1916)	(0.0711)	(0.090)				
279	2.6567*	1.3622*	-0.099*	0.5304	0.5591	13.554	37
	(0.2898)	(0.2491)	(0.0254)				
280	5.9937*	2.4744*	-0.224*	0.7492	0.7897	35.849	37
	(0.1108)	(0.2965)	(0.0297)				
281	-3.196*	-0.259@	0.0466@	0.627	0.6609	20.17	37
	(0.1445)	(0.4061)	(0.0375)				
282	0.7036**	1.1142**	-0.071**	0.1785	0.1882	2.6081	37
	(0.2966)	(0.4278)	(0.0319)				
316	-0.854**	-0.088@	0.0326@	0.3363	0.3545	6.0801	37
	(0.3275)	(0.7419)	(0.073)				
385	-1.836*	0.2615*	0.023@	0.9072	0.9562	117.32	37
	(0.1934)	(0.0967)	(0.056)				
387	0.988*	0.7558**	-0.05@	0.4934	0.52	11.685	37
	(0.1123)	(0.3272)	(0.037)				
Total	-0.715*	0.9096**	-0.016@	0.9171	0.9667	132.77	37
WFBI	(0.123)	(0.437)	(0.0501)				

(Figures within brackets are the standard error)

**Significant at 1% level of significance*

*** Significant at 5% level of significance*

**** Significant at 10% level of significance*

@ Insignificant

Table (6.3.7): Estimates of C.E.S. Production Function for Northern Region

Dependent Variable: Ln (V/L) Model: III								
Industry Code	Const.	Ln (W)	Ln (L)	T	R^2	Adjusted R^2	F-Statistic	D.F.
226	-1.703*	0.0791@	-0.06@	0.0157@	0.3176	0.3441	5.585	36
	(0.1014)	(0.1648)	(0.0724)	(0.0139)				
270	-0.388@	0.6688***	-0.059@	-0.038@	0.7431	0.805	34.711	36
	(0.1538)	(0.382)	(0.3307)	(0.0345)				
271	8.5117*	3.258**	-0.6@	-0.161@	0.7206	0.7807	30.949	36
	(0.4206)	(1.3985)	(1.144)	(0.1232)				
272	-4.754*	0.1373@	0.4***	-0.014@	0.8357	0.9053	61.037	36
	(0.2266)	(0.3607)	(0.2147)	(0.0305)				
273	-3.947*	0.1881@	0.1267@	0.0078@	0.6074	0.658	18.565	36
	(0.2834)	(0.6504)	(0.4807)	(0.0496)				
274	2.9361*	1.3593*	-0.332***	-0.088*	0.8816	0.9551	89.351	36
	(0.1644)	(0.2892)	(0.1767)	(0.0222)				
275	0.2753@	1.272***	0.018@	-0.106***	0.1911	0.207	2.835	36
	(0.6478)	(0.6838)	(0.3311)	(0.054)				
276	-5.918*	0.3167**	0.5726@	-0.043@	0.8241	0.8928	56.221	36
	(0.187)	(0.147)	(0.4113)	(0.1005)				
279	3.7967*	1.532*	-0.17@	-0.094*	0.5358	0.5805	13.851	36
	(0.2929)	(0.3505)	(0.2437)	(0.0269)				
280	10.998*	2.6549*	-0.546@	-0.176*	0.7623	0.8258	38.484	36
	(0.1073)	(0.3088)	(0.3428)	(0.0419)				
281	-12.53*	-1.464**	0.8762*	0.0402@	0.7399	0.8016	34.136	36
	(0.1271)	(0.5671)	(0.3201)	(0.0329)				
282	2.2382*	1.408***	-0.109@	-0.085***	0.1825	0.1977	2.6789	36
	(0.3028)	(0.8421)	(0.2672)	(0.0454)				
316	-0.849**	-0.088@	-0.001@	0.0326@	0.3363	0.3643	6.0805	36
	(0.3349)	(0.7596)	(0.182)	(0.0772)				
385	-1.98*	0.2348*	0.0129@	0.0239*	0.9072	0.9828	117.31	36
	(0.197)	(0.019)	(0.197)	(0.058)				
Total	6.6374*	0.6366*	-0.501**	0.0281@	0.9127	0.9888	125.45	36
WFBI	(0.0491)	(0.2092)	(0.2016)	(0.0242)				

(Figures within brackets are the standard error)
**Significant at 1% level of significance*
*** Significant at 5% level of significance*
**** Significant at 10% level of significance*
@ Insignificant

Table (6.3.8): Estimates of C.E.S. Production Function for Northern Region

Dependent Variable ln(V/L) Model: IV									
Industry Code	Const.	Ln (L)	Ln (K)	(LnL-LnK)²	T	R²	Adj. R²	F-Statistic	D.F.
226	-2.865*	0.225@	-0.19@	-0.061@	0.02***	0.387	0.4312	5.5241	35
	(0.0906)	(0.33)	(0.401)	(0.0543)	(0.012)				
270	-5.748*	0.078@	0.244**	0.0167@	0.022@	0.7954	0.8863	34.016	35
	(0.1434)	(0.266)	(0.0911)	(0.0701)	(0.022)				
271	-15.06*	1.95@	-0.766@	2.216***	-0.054@	0.6823	0.7603	18.792	35
	(0.4528)	(1.44)	(0.7716)	(1.345)	(0.1232)				
272	-4.936*	0.015@	0.2942@	0.1569@	0.014@	0.8597	0.958	53.616	35
	(0.2201)	(0.345)	(0.2016)	(0.4326)	(0.0251)				
273	-4.596*	0.217@	-0.099@	0.1097@	0.012@	0.6382	0.7111	15.435	35
	(0.2821)	(0.532)	(0.4349)	(0.2194)	(0.0261)				
274	-5.772*	0.68***	-0.308@	-0.232@	0.016@	0.7233	0.806	22.873	35
	(0.2249)	(0.34)	(0.3299)	(0.1508)	(0.0148)				
275	-4.501*	-0.016@	0.153@	-0.738@	0.009@	0.1419	0.1581	1.4469	35
	(0.682)	(0.795)	(0.9573)	(0.988)	(0.0324)				
276	-6.212*	0.424@	-0.035@	-0.226@	0.030@	0.8673	0.9664	57.188	35
	(0.1729)	(0.543)	(0.4078)	(0.192)	(0.0386)				
279	-3.955*	0.174@	0.297**	-0.035@	-0.027@	0.4758	0.5302	7.9421	35
	(0.3367)	(0.237)	(0.1437)	(0.1407)	(0.0323)				
280	-3.825*	0.937@	-0.611@	0.123@	-0.03@	0.4361	0.4859	6.7669	35
	(0.2189)	(1.054)	(0.8609)	(0.446)	(0.0794)				
281	-3.359*	-0.119@	0.282@	-0.194@	0.006***	0.6935	0.7728	19.798	35
	(0.1399)	(0.350)	(0.2012)	(0.313)	(0.032)				
282	-4.727*	0.388@	-0.045@	-0.202@	-0.013@	0.1858	0.207	1.9967	35
	(0.3089)	(0.3077)	(0.2672)	(0.268)	(0.0147)				
316	-1.126*	-0.84@	0.897@	-0.295@	0.03@	0.7342	0.8181	24.169	35
	(0.2909)	(1.594)	(1.714)	(0.439)	(0.055)				
385	-2.985*	0.124@	-0.07@	-0.048@	0.04***	0.8905	0.9922	71.123	35
	(0.2026)	(0.5073)	(0.4374)	(0.320)	(0.024)				
387	-1.787*	0.231@	-0.207@	-0.138@	0.025@	0.539	0.6006	10.23	35
	(0.125)	(0.409)	(0.168)	(0.222)	(0.0421)				
Total	-1.915*	1.31@	-1.43***	0.965**	0.101@	0.8935	0.9956	73.425	35
WFBI	(0.093)	(0.8538)	(0.8452)	(0.3447)	(0.0415)				

(Figures within brackets are the standard error)
**Significant at 1% level of significance*
*** Significant at 5% level of significance*
**** Significant at 10% level of significance*
@ Insignificant

Table (6.3.9): Estimates of V.E.S. Production Function for Northern Region

Dependent Variable: Ln (V/L) Model: I							
Industry Code	Const.	Ln (W)	Ln (K/L)	R^2	Adjusted R^2	F-Statistic	D.F.
226	-0.546*	0.3097*	0.2319*	0.3807	0.4013	7.3774	37
	(0.0873)	(0.0664)	(0.0801)				
270	-1.844*	0.4446*	0.2437*	0.8309	0.8758	58.955	37
	(0.1271)	(0.0599)	(0.0741)				
271	-4.43*	0.8713*	0.4233@	0.6704	0.7066	24.405	37
	(0.437)	(0.1883)	(0.4522)				
272	-1.736*	0.5205*	0.3649**	0.8535	0.8996	69.903	37
	(0.2131)	(0.0585)	(0.1534)				
273	-3.292*	0.2019@	-0.259@	0.6261	0.66	20.097	37
	(0.2723)	(0.1749)	(0.2867)				
274	-1.832*	0.5738*	0.3245**	0.8019	0.8452	48.572	37
	(0.1899)	(0.0987)	(0.134)				
275	-2.171*	0.656@	1.268**	0.2201	0.232	3.3861	37
	(0.6242)	(0.525)	(0.5771)				
276	-1.742*	0.575*	0.362***	0.8334	0.8785	60.05	37
	(0.1794)	(0.1118)	(0.2034)				
279	-1.123*	0.4171*	0.2566*	0.5107	0.5383	12.526	37
	(0.301)	(0.0814)	(0.0741)				
280	-0.934*	0.17***	-0.26@	0.4621	0.4871	10.308	37
	(0.2028)	(0.1028)	(0.2188)				
281	-1.345*	0.31*	0.198***	0.6539	0.6892	22.671	37
	(0.1401)	(0.0582)	(0.1132)				
282	-2.334*	0.205***	-0.146@	0.0964	0.1016	1.2798	37
	(0.3208)	(0.1191)	(0.144)				
316	0.0282@	0.29*	0.067@	0.368	0.3878	6.986	37
	(0.3225)	(0.1047)	(0.072)				
385	-0.871*	0.531*	0.0645@	0.9064	0.9554	116.2	37
	(0.1933)	(0.1023)	(0.2196)				
387	-0.687*	0.284*	-0.095@	0.546	0.5755	14.432	37
	(0.115)	(0.051)	(0.1125)				
Total	-1.266*	1.149**	1.056*	0.9405	0.9913	189.57	37
WFBI	(0.0737)	(0.0641)	(0.1676)				

(Figures within brackets are the standard error)
*Significant at 1% level of significance
** Significant at 5% level of significance
*** Significant at 10% level of significance
@ Insignificant

Table (6.3.10): Estimates of V.E.S. Production Function for Northern Region

Dependent Variable: Ln (V/L) Model: II								
Industry Code	Const.	Ln (W)	Ln (K/L)	T	R^2	Adjusted R^2	F-Statistic	D.F.
226	-0.843*	0.249@	0.2235*	0.0055@	0.3828	0.4147	7.4426	36
	(0.0891)	(0.152)	(0.084)	(0.0124)				
270	-0.909*	0.66*	0.2337*	-0.024@	0.8376	0.9074	61.892	36
	(0.1283)	(0.266)	(0.0756)	(0.0293)				
271	3.9025*	3.0854*	0.478@	-0.202***	0.7384	0.7999	33.872	36
	(0.4127)	(1.167)	(0.428)	(0.105)				
272	-1.916*	0.475***	0.371**	0.005***	0.8538	0.925	70.079	36
	(0.2179)	(0.272)	(0.161)	(0.029)				
273	-3.756*	0.0827@	-0.274@	0.012@	0.6278	0.6801	20.241	36
	(0.2782)	(0.500)	(0.299)	(0.047)				
274	-0.118@	0.984*	0.2396**	-0.052*	0.8857	0.9595	92.987	36
	(0.1631)	(0.1615)	(0.1185)	(0.017)				
275	2.79*	2.103*	1.444*	-0.123**	0.4574	0.4955	10.116	36
	(0.553)	(0.619)	(0.516)	(0.0466)				
276	-4.17*	-0.056@	0.413***	0.0744@	0.8407	0.9108	63.33	36
	(0.1805)	(0.754)	(0.214)	(0.088)				
279	1.6734*	1.1207*	0.182*	-0.075*	0.5921	0.6414	17.419	36
	(0.255)	(0.236)	(0.067)	(0.0242)				
280	6.0806*	2.520*	0.0383@	-0.227*	0.7495	0.812	35.904	36
	(0.1135)	(0.340)	(0.1297)	(0.0324)				
281	-2.708*	-0.089@	0.179@	0.0367@	0.6716	0.7276	24.541	36
	(0.1402)	(0.408)	(0.115)	(0.0372)				
282	1.0159*	1.192*	-0.183@	-0.076**	0.2195	0.2378	3.3748	36
	(0.291)	(0.423)	(0.131)	(0.031)				
316	-1.197*	-0.162@	0.0749@	0.045@	0.386	0.4182	7.544	36
	(0.3314)	(0.745)	(0.074)	(0.073)				
385	-1.651*	0.3112@	0.0363@	0.0196@	0.9077	0.9833	118.01	36
	(0.1974)	(0.695)	(0.241)	(0.0614)				
387	0.7517*	0.685***	-0.052@	-0.044@	0.551	0.5969	14.726	36
	(0.1145)	(0.37)	(0.119)	(0.0402)				
Total	-0.532	1.3727*	1.062*	-0.023@	0.9203	0.997	138.64	36
WFBI	(0.074)	(0.2916)	(0.167)	(0.03)				

(Figures within brackets are the standard error)
**Significant at 1% level of significance*
*** Significant at 5% level of significance*
**** Significant at 10% level of significance*
@ Insignificant

Table (6.4.1): Estimates of C-D Production Function for Southern Region

Dependent Variable Ln (V) Model I							
Industry Code	Constant	Ln (L)	Ln (K)	R^2	Adjusted R^2	F-Statistic	D.F.
226	-1.121*	3.2889*	-0.019@	0.875	0.9223	129.5	37
	(-0.411)	(0.2429)	(0.1869)				
270	13.194*	0.71*	-1.598*	0.9455	0.9966	320.95	37
	(0.2248)	(0.142)	(0.3668)				
271	6.0111*	1.2077*	-0.672*	0.9344	0.9849	263.51	37
	(0.2196)	(0.1098)	(0.1784)				
272	0.988)*	0.7376*	-0.16**	0.612	0.6451	29.18	37
	(0.1521)	(0.0684)	(0.0608)				
273	1.5743*	1.5882*	-0.265@	0.9405	0.9913	292.42	37
	(0.2647)	(0.1115)	(0.209)				
274	1.2318*	1.1225*	-0.228*	0.9412	0.9921	296.13	37
	(0.1391)	(0.0397)	(0.0596)				
275	-0.335@	1.6635*	-0.069@	0.9425	0.9934	303.24	37
	(0.231)	(0.0809)	(0.0924)				
276	5.7253*	1.5857*	-0.751*	0.9395	0.9903	287.29	37
	(0.2337)	(0.1335)	(0.217)				
279	0.4553@	0.9547*	0.0441@	0.8515	0.8975	106.08	37
	(0.3378)	(0.1763)	(0.044)				
280	0.7452**	0.829*	-0.159*	0.7052	0.7433	44.254	37
	(0.3266)	(0.0419)	(0.0586)				
281	3.6802*	0.9414*	-0.396*	0.945	0.9961	317.86	37
	(0.2033)	(0.0856	(0.1284)				
282	1.6881*	1.819**	-0.328**	-0.909	-0.958	-8.809	37
	(0.4589)	(0.0819)	(0.1475)				
316	-14.27*	-2.112*	-0.986***	0.8267	0.8714	88.251	37
	(-2.628)	(-0.657)	(-0.516)				
378	0.2216@	0.4646*	0.0214@	0.4144	0.4368	13.092	37
	(0.1644)	(0.0915)	(0.021)				
385	2.862*	1.5007*	-0.4*	0.9346	0.9851	264.37	37
	(0.3114)	(0.1735)	(0.1335)				
387	2.2685*	0.7136*	-0.263*	0.6665	0.7025	36.972	37
	(0.1127)	(0.0526)	(0.0676)				
Total	16.607*	1.038*	-1.064*	0.9042	0.9531	174.65	37
WFBI	(0.381)	(0.1012)	(0.3022)				

(Figures within brackets are the standard error)

**Significant at 1% level of significance*

*** Significant at 5% level of significance*

**** Significant at 10% level of significance*

@ Insignificant

Table (6.4.2): Estimates of C-D Production Function for Southern Region

			Dependent Variable Ln (V) Model II					
Industry Code	**Constant**	**Ln (L)**	**Ln (K)**	**T**	**R^2**	**Adjusted R^2**	**F-Statistic**	**D.F.**
226	-0.86**	2.7096*	0.299@	-0.131**	0.9123	0.9884	124.9	36
	(0.3737)	(0.3551)	(0.224)	(0.056)				
270	10.473*	0.6508*	-1.089**	-0.059***	0.9476	0.9988	334.55	36
	(0.213)	(0.142)	(0.4497)	(0.0355)				
271	5.4485*	1.194*	-0.576@	-0.014@	0.9286	0.9788	240.6	36
	(0.2196)	(0.1235)	(0.5078)	(0.0686)				
272	0.8593*	0.4563*	0.1521@	-0.068*	0.6185	0.6519	29.993	36
	(0.1293)	(0.1065)	(0.1141)	(0.0228)				
273	0.2229@	1.1006*	0.46***	-0.125*	0.9481	0.9993	337.95	36
	(0.209)	(0.1532)	(0.2368)	(0.0279)				
274	1.3609*	0.9834*	-0.129@	-0.03@	0.9417	0.9926	298.82	36
	(0.1291)	(0.1093)	(0.0993)	(0.0199)				
275	-0.335@	1.6635*	-0.069@	0.0007@	0.9419	0.9928	299.92	36
	(0.231)	(0.3004)	(0.1386)	(0.0347)				
276	2.7208*	1.4355*	-0.067@	-0.083@	0.9403	0.9911	291.38	36
	(0.2337)	(0.1669)	(0.5341)	(0.0668)				
279	2.7908*	0.235**	0.4994*	-0.191*	0.9469	0.9981	329.9	36
	(0.235)	(0.1028)	(0.1469)	(0.0294)				
280	0.4438**	0.3098*	0.5108*	-0.126*	0.7321	0.7717	50.556	36
	(0.201)	(0.0837)	(0.1172)	(0.0167)				
281	0.8666*	0.8238*	0.2033@	-0.107***	0.9405	0.9913	292.42	36
	(0.1926)	(0.0963)	(0.3102)	(0.0535)				
282	1.77*	0.8195*	0.6392*	-0.246*	-0.921	-0.971	-8.87	36
	(0.3442)	(0.2458)	(0.2622)	(0.066)				
316	-16.4*	-1.532@	-1.147@	-0.07**	0.8215	0.8659	85.141	36
	(2.689)	(2.756)	(0.859)	(0.311)				
378	0.529*	0.286*	0.1644*	-0.036*	0.5633	0.5937	23.863	36
	(0.1215)	(0.0715)	(0.0357)	(0.0071)				
385	2.0464*	0.8452*	0.40***	-0.178*	0.9486	0.9999	341.42	36
	(0.2373)	(0.2076)	(0.2373)	(0.0445)				
387	1.7502*	0.646*	-0.128@	-0.023@	0.6663	0.7023	36.939	36
	(0.1127)	(0.090)	(0.285)	(0.045)				
Total	10.721*	1.038*	-0.578@	-0.066@	0.9418	0.9927	299.37	36
WFBI	(0.3942)	(0.1051)	(1.511)	(0.197)				

(Figures within brackets are the standard error)
**Significant at 1% level of significance*
*** Significant at 5% level of significance*
**** Significant at 10% level of significance*
@ Insignificant

Table (6.4.3): Estimates of C-D Production Function for Southern Region

Dependent Variable Ln (V/L) Model III							
Industry Code	Const.	Ln(K/L)	T	R^2	Adjusted R^2	F-Statistic	D.F.
226	-3.000*	0.6881*	0.0249*	0.9398	0.9906	288.81	37
	(0.2693)	(0.253)	(0.0095)				
270	-2.988*	0.2967*	0.0234*	0.9202	0.9699	213.33	37
	(0.1492)	(0.0854)	(0.0051)				
271	-2.724*	0.3387@	0.0118@	0.6683	0.7044	37.273	37
	(0.2296)	(0.2336)	(0.0162)				
272	-1.538*	0.2794*	0.0093@	0.5951	0.6273	27.19	37
	(0.149)	(0.0885)	(0.009)				
273	-5.365*	-0.265@	0.0514*	0.6099	0.6429	28.924	37
	(0.2947)	(0.32)	(0.017)				
274	-1.808*	0.375*	0.0142**	0.7173	0.7561	46.94	37
	(0.174)	(0.113)	(0.0063)				
275	0.8185**	1.067*	-0.052*	0.4148	0.4372	13.113	37
	(0.3574)	(0.308)	(0.0173)				
276	-3.656*	0.52***	0.0346**	0.9084	0.9575	183.47	37
	(0.2539)	(0.275)	(0.0169)				
279	-2.749*	0.608*	0.0062@	0.9359	0.9865	270.11	37
	(0.5981)	(0.1284)	(0.021)				
280	-1.838*	-0.05@	0.0268@	0.3455	0.3642	9.7659	37
	(0.2176)	(0.25)	(0.0164)				
281	-2.097*	0.2078@	0.014@	0.7401	0.7801	52.681	37
	(0.162)	(0.133)	(0.0092)				
282	-4.097*	-0.031@	0.0149@	-0.042	-0.044	-0.746	37
	(0.4533)	(0.1867)	(0.0188)				
316	4.67***	-0.578@	-0.182**	0.3813	0.4019	11.401	37
	(2.63)	(0.584)	(0.075)				
378	-2.32*	0.0739@	0.0285*	0.2717	0.2864	6.9016	37
	(0.294)	(0.0761)	(0.0081)				
385	-4.306*	0.0329@	0.0743@	0.9052	0.9541	176.65	37
	(0.312)	(0.3768)	(0.0236)				
387	-2.38*	-0.1@	0.0494*	0.5605	0.5908	23.593	37
	(0.146)	(0.1501)	(0.0114)				
Total	-1.33***	1.0274*	0.0051*	0.9402	0.991	290.86	37
WFBI	(0.068)	(0.1477)	(0.0101)				

(Figures within brackets are the standard error)

**Significant at 1% level of significance*

*** Significant at 5% level of significance*

**** Significant at 10% level of significance*

@ Insignificant

Table (6.4.4): Estimates of C-D Production Function for Southern Region

			Dependent Variable ln(V/L) Model IV					
Industry Code	Const.	Ln (K/L)	Ln (L)	T	R^2	Adjusted R^2	F-Statistic	D.F.
226	-3.124*	0.8727*	0.1981@	-0.013@	0.8937	0.9958	73.533	35
	(0.2691)	(0.3121)	(0.200)	(0.039)				
270	-4.363*	0.2911*	0.2201@	0.0071@	0.9277	0.9778	237.38	35
	(0.1503)	(0.0864)	(0.271)	(0.021)				
271	-4.298*	0.3088@	0.2388@	-0.014@	0.6723	0.7086	37.954	35
	(0.2333)	(0.2484)	(0.5558)	(0.062)				
272	-2.431*	0.1985@	0.1513@	-0.002@	0.603	0.6356	28.099	35
	(0.149)	(0.1186)	(0.1475)	(0.014)				
273	-7.07*	-0.32@	0.3107@	0.0306@	0.6311	0.6652	31.649	35
	(0.2954)	(0.3218)	(0.3455)	(0.028)				
274	-2.934*	0.292**	0.1798***	0.006@	0.7591	0.8001	58.295	35
	(0.1679)	(0.119)	(0.1093)	(0.008)				
275	0.4424@	1.050*	0.0936@	-0.057*	0.4256	0.4486	13.708	35
	(0.3639)	(0.3154)	(0.1744)	(0.019)				
276	-7.575*	0.489***	0.6994@	-0.027@	0.926	0.9761	231.5	35
	(0.2504)	(0.272)	(0.5458)	(0.050)				
279	-4.703*	0.5376*	0.5934@	-0.062@	0.9325	0.9829	255.57	35
	(0.577)	(0.1307)	(0.3657)	(0.047)				
280	-5.999*	-0.039@	0.515@	-0.037@	0.3601	0.3796	10.411	35
	(0.219)	(0.2504)	(0.6515)	(0.082)				
281	-3.163*	0.1819@	0.1669@	-0.006@	0.748	0.7884	54.913	35
	(0.1648)	(0.1412)	(0.2739)	(0.035)				
282	-7.262*	-0.243@	0.4933**	-0.008@	-0.221	-0.233	-3.348	35
	(0.4146)	(0.195)	(0.2163)	(0.02)				
316	-6.81*	-1.145@	1.9898@	-0.41***	0.426	0.449	13.73	35
	(2.633)	(0.821)	(2.013)	(0.244)				
378	0.0186@	0.0107@	-0.246**	0.0778*	0.5418	0.5711	21.875	35
	(0.1501)	(0.0629)	(0.1129)	(0.024)				
385	-4.671*	0.0148@	0.0741@	0.065@	0.906	0.955	178.31	35
	(0.3188)	(0.3915)	(0.2743)	(0.04)				
387	-1.553*	-0.06@	-0.132@	0.0638@	0.5615	0.5919	23.689	35
	(0.1487)	(0.187)	(0.361)	(0.041)				
Total	2.5016*	1.0104*	-0.313@	0.0434@	0.9409	0.9918	294.53	35
WFBI	(0.067)	(0.1458)	(0.2483)	(0.031)				

(Figures within brackets are the standard error)

**Significant at 1% level of significance*

*** Significant at 5% level of significance*

**** Significant at 10% level of significance*

@ Insignificant

Table (6.4.5): Estimates of C.E.S. Production Function for Southern Region

Dependent Variable ln(V/L) Model I						
Industry Code	Constant	Ln (W)	R^2	Adjusted R^2	F-Statistic	D.F.
226	0.4223@ (0.7774)	1.3025* (0.2242)	0.9046	0.9284	180.16	38
270	-1.956* (0.1929)	0.3846* (0.0828)	0.7041	0.7226	45.211	38
271	-2.188* (0.269)	0.656* (0.0741)	0.9107	0.9347	193.77	38
272	-1.917* (0.1582)	0.1179@ (0.1019)	0.4015	0.4121	12.746	38
273	-2.066* (0.2145)	0.6464* (0.1115)	0.8779	0.901	136.61	38
274	-1.198* (0.3447)	0.3626** (0.1788)	0.2039	0.2093	4.8663	38
275	-1.722* (0.1779)	0.5684* (0.1109)	0.9263	0.9507	238.8	38
276	-2.437* (0.6543)	0.9064* (0.1769)	0.7601	0.7801	60.2	38
279	-1.572* (0.3422)	0.2805*** (0.1733)	0.7785	0.799	66.779	38
280	-1.231* (0.1281)	0.2839* (0.0536)	0.5183	0.5319	20.444	38
281	-2.038* (0.2803)	0.1787*** (0.1038)	0.1708	0.1753	3.9137	38
282	-1.072* (0.195)	0.967* (0.0787)	-0.889	-0.912	-8.942	38
316	7.1052* (2.013)	-0.939@ (1.657)	0.0462	0.0474	0.9203	38
378	-0.995* (0.0986)	0.296* (0.033)	0.6083	0.6243	29.507	38
385	-0.995@ (1.2768)	0.525@ (0.467)	0.1324	0.1359	2.8995	38
387	-0.399** (0.1608)	0.519* (0.057)	0.5548	0.5694	23.677	38
Total WFBI	-1.097* (0.1104)	0.703** (0.0289)	0.9171	0.9412	210.19	38

(Figures within brackets are the standard error)
**Significant at 1% level of significance*
*** Significant at 5% level of significance*
**** Significant at 10% level of significance*
@ Insignificant

Table (6.4.6): Estimates of C.E.S. Production Function for Southern Region

		Dependent Variable ln(V/L) Model II					
Industry Code	**Const.**	**Ln (W)**	**T**	**R^2**	**Adjusted R^2**	**F-Statistic**	**D.F.**
226	-7.628*	0.0131@	0.039@	0.9469	0.9981	329.9	37
	(0.312)	(0.4223)	(0.043)				
270	-1.032*	0.7147**	-0.044@	0.8402	0.8856	97.27	37
	(0.1704)	(0.3538)	(0.0391)				
271	1.699*	1.466**	-0.099***	0.7386	0.7785	52.273	37
	(0.2127)	(0.5915)	(0.0535)				
272	-1.103*	0.392***	-0.007@	0.5544	0.5844	23.017	37
	(0.1665)	(0.2061)	(0.0221)				
273	-3.043*	0.3358@	0.0042@	0.605	0.6377	28.335	37
	(0.2954)	(0.4584)	(0.0488)				
274	-0.306**	0.7291*	-0.049*	0.8469	0.8927	102.34	37
	(0.1401)	(0.1351)	(0.0149)				
275	0.2426@	0.826**	-0.068**	0.1911	0.2014	4.3706	37
	(0.4066)	(0.4009)	(0.0335)				
276	-4.856*	0.232@	0.0367*	0.8711	0.9182	125.02	37
	(0.2737)	(1.13)	(0.1285)				
279	5.0763*	2.6028*	-0.189*	0.9448	0.9959	316.64	37
	(0.5538)	(0.476)	(0.0485)				
280	5.754*	2.3755*	-0.215*	0.7193	0.7582	47.407	37
	(0.1063)	(0.285)	(0.0285)				
281	-3.738*	-0.303@	0.0546@	0.6997	0.7375	43.105	37
	(0.169)	(0.475)	(0.0439)				
282	0.9407**	1.4898**	-0.095**	-0.36	-0.379	-4.897	37
	(0.3966)	(0.572)	(0.0426)				
316	7.0113*	0.7227@	-0.268@	0.3363	0.3545	9.374	37
	(2.689)	(6.092)	(0.596)				
378	-0.558*	0.4839*	-0.025**	0.5147	0.5425	19.621	37
	(0.198)	(0.090)	(0.0114)				
385	-2.96*	0.4211@	0.0371*	0.9072	0.9562	180.85	37
	(0.3114)	(0.961)	(0.0905)				
387	1.1636*	0.8901**	-0.059@	0.581	0.6124	25.653	37
	(0.132)	(0.3854)	(0.044)				
Total	-0.66*	0.84***	-0.014@	0.9171	0.9667	204.66	37
WFBI	(0.113)	(0.43)	(0.046)				

(Figures within brackets are the standard error)

**Significant at 1% level of significance*

*** Significant at 5% level of significance*

**** Significant at 10% level of significance*

@ Insignificant

Table (6.4.7): Estimates of C.E.S. Production Function for Southern Region

			Dependent Variable ln(V/L) Model III					
Industry Code	Const.	Ln (W)	Ln (L)	T	R^2	Adjusted R^2	F-Statistic	D.F.
226	-5.27*	0.245@	-0.187@	0.048@	0.9135	0.9896	126.73	36
	(0.314)	(0.510)	(0.224)	(0.043)				
270	-0.439*	0.756***	-0.066@	-0.043@	0.8402	0.9102	63.094	36
	(0.174)	(0.432)	(0.374)	(0.039)				
271	4.3601*	1.6688**	-0.307@	-0.082@	0.7444	0.8064	34.948	36
	(0.2155)	(0.7164)	(0.586)	(0.063)				
272	-3.318*	0.0958@	0.28***	-0.01@	0.5833	0.6319	16.798	36
	(0.1582)	(0.2517)	(0.149)	(0.021)				
273	-4.21*	0.2006@	0.135@	0.008@	0.6074	0.658	18.565	36
	(0.3023)	(0.6938)	(0.51)	(0.053)				
274	2.3602*	1.0927*	-0.267***	-0.071*	0.8816	0.9551	89.351	36
	(0.1321)	(0.2324)	(0.142)	(0.018)				
275	0.1767@	0.82***	0.0116@	-0.07**	0.1911	0.207	2.835	36
	(0.4159)	(0.44)	(0.21)	(0.034)				
276	-8.454*	0.452@	0.82@	-0.062@	0.894	0.9685	101.21	36
	(0.2671)	(1.12)	(0.59)	(0.144)				
279	7.2546*	2.927*	-0.325@	-0.18*	0.9209	0.9976	139.61	36
	(0.5596)	(0.67)	(0.47)	(0.051)				
280	10.559*	2.55*	-0.524@	-0.17*	0.7318	0.7928	32.743	36
	(0.103)	(0.296)	(0.329)	(0.040)				
281	-14.65*	-1.713*	1.025*	0.047@	0.8258	0.8946	56.886	36
	(0.1487)	(0.663)	(0.374)	(0.038)				
282	2.9927*	1.883@	-0.146@	-0.13***	-0.368	-0.399	-3.228	36
	(0.4048)	(1.126)	(0.357)	(0.061)				
316	6.9738*	0.7227@	0.0047@	-0.27@	0.3363	0.3643	6.0805	36
	(2.75)	(6.237)	(1.50)	(0.63)				
378	-0.21@	0.5576*	-0.11***	-0.013@	0.5433	0.5886	14.275	36
	(0.187)	(0.0944)	(0.056)	(0.013)				
385	-3.188*	0.378@	0.021@	0.038@	0.9072	0.9828	117.31	36
	(0.317)	(1.163)	(0.317)	(0.093)				
Total	11.5*	1.1*	-0.865	0.048@	0.9025	0.9777	111.08	36
WFBI	(0.085)	(0.36)	(0.348)	(0.042)				

(Figures within brackets are the standard error)
*Significant at 1% level of significance
** Significant at 5% level of significance
*** Significant at 10% level of significance
@ Insignificant

Table (6.4.8): Estimates of C.E.S. Production Function for Southern Region

Dependent Variable ln(V/L) Model IV									
Industry Code	Const.	Ln (L)	Ln (K)	(LnL-LnK)^2	T	R^2	Adj. R^2	F-Statistic	D.F.
226	-8.867*	0.695@	-0.589@	-0.189@	0.07***	0.8763	0.9764	61.986	35
	(0.280)	(1.01)	(1.243)	(0.168)	(0.04)				
270	-6.499*	0.088@	0.275**	0.0189@	0.024@	0.8893	0.9909	70.292	35
	(0.162)	(0.30)	(0.103)	(0.0793)	(0.025)				
271	-7.716*	0.997@	-0.393@	1.135@	-0.027@	0.7049	0.7855	20.901	35
	(0.232)	(0.739)	(0.395)	(0.689)	(0.063)				
272	-3.445*	0.010@	0.205@	0.1095@	0.01@	0.6	0.6686	13.125	35
	(0.154)	(0.241)	(0.140)	(0.302)	(0.017)				
273	-4.903*	0.231@	-0.106@	0.117@	0.012@	0.6382	0.7111	15.435	35
	(0.301)	(0.567)	(0.464)	(0.234)	(0.028)				
274	-4.64*	0.542**	-0.247@	-0.187@	0.013@	0.7233	0.806	22.873	35
	(0.181)	(0.273)	(0.2652)	(0.121)	(0.012)				
275	-2.889*	-0.01@	0.0982@	-0.474@	0.006*	0.1419	0.1581	1.4469	35
	(0.44)	(0.510)	(0.61)	(0.634)	(0.020)				
276	-8.87*	0.606@	-0.05@	-0.322@	0.043@	0.8909	0.9927	71.452	35
	(0.247)	(0.776)	(0.58)	(0.275)	(0.055)				
279	-7.56*	0.333@	0.568**	-0.066@	-0.051@	0.8791	0.9796	63.619	35
	(0.64)	(0.454)	(0.27)	(0.27)	(0.062)				
280	-3.67*	0.900@	-0.586@	0.12@	-0.028@	0.4187	0.4666	6.3025	35
	(0.21)	(1.012)	(0.826)	(0.428)	(0.076)				
281	-3.93*	-0.14@	0.330@	-0.227@	0.007***	0.7739	0.8623	29.95	35
	(0.164)	(0.409)	(0.235)	(0.366)	(0.037)				
282	-6.32*	0.519@	-0.061@	-0.27@	-0.018@	-0.375	-0.418	-2.386	35
	(0.413)	(0.411)	(0.357)	(0.359)	(0.019)				
316	9.2452*	6.917@	-7.36@	2.42@	-0.188@	0.7342	0.8181	24.169	35
	(2.4)	(13.09)	(14.07)	(3.61)	(0.45)				
378	-2.44*	0.187@	-0.144@	-0.051@	0.014@	0.3059	0.3409	3.8563	35
	(0.296)	(0.129)	(0.096)	(0.04)	(0.018)				
385	-4.808*	0.200@	-0.113@	-0.077@	0.07***	0.8806	0.9812	64.533	35
	(0.326)	(0.817)	(0.704)	(0.516)	(0.038)				
387	-2.104*	0.272@	-0.243@	-0.163@	0.029@	0.6347	0.7072	15.203	35
	(0.147)	(0.482)	(0.198)	(0.261)	(0.05)				
Total WFBI	-1.759*	1.203@	-1.31@	0.887*	0.093**	0.8935	0.9956	73.425	35
	(0.085)	(0.784)	(0.77)	(0.32)	(0.038)				

(Figures within brackets are the standard error)
**Significant at 1% level of significance*
*** Significant at 5% level of significance*
**** Significant at 10% level of significance*
@ Insignificant

Table (6.4.9): Estimates of V.E.S. Production Function for Southern Region

		Dependent Variable: Ln (V/L) Model: I					
Industry Code	Const.	Ln (W)	Ln (K/L)	R^2	Adjusted R^2	F-Statistic	D.F.
226	-1.69*	0.96*	0.7178*	0.948	0.9992	337.27	37
	(0.270)	(0.2054)	(0.248)				
270	-2.085*	0.503*	0.2756*	0.9395	0.9903	287.29	37
	(0.144)	(0.067)	(0.084)				
271	-2.2698	0.446*	0.2168@	0.6925	0.7299	41.663	37
	(0.224)	(0.096)	(0.232)				
272	-1.212*	0.363*	0.2547**	0.5956	0.6278	27.247	37
	(0.148)	(0.040)	(0.107)				
273	-3.512*	0.215@	-0.276@	0.6261	0.6599	30.978	37
	(0.290)	(0.186)	(0.306)				
274	-1.473*	0.461*	0.2609**	0.8019	0.8452	74.887	37
	(0.153)	(0.079)	(0.107)				
275	-1.394*	0.421**	0.8142**	0.2201	0.232	5.221	37
	(0.400)	(0.208)	(0.370)				
276	-2.488*	0.821*	0.517***	0.9042	0.9531	174.61	37
	(0.256)	(0.159)	(0.290)				
279	-2.145*	0.797*	0.490*	0.9473	0.9985	332.54	37
	(0.575)	(0.155)	(0.142)				
280	-0.896*	0.16@	-0.249@	0.4436	0.4676	14.749	37
	(0.195)	(0.098)	(0.210)				
281	-1.573*	0.362*	0.23***	0.7298	0.7692	49.968	37
	(0.164)	(0.068)	(0.13)				
282	-3.121*	0.274***	-0.196@	-0.194	-0.204	-3.006	37
	(0.43)	(0.159)	(0.192)				
316	-0.231@	-2.37**	-0.553@	0.368	0.3879	10.772	37
	(2.65)	(0.86)	(0.59)				
378	-1.137*	0.379*	0.0713@	0.4886	0.515	17.675	37
	(0.210)	(0.069)	(0.054)				
385	-1.403*	0.855*	0.104@	0.9064	0.9554	179.15	37
	(0.311)	(0.165)	(0.354)				
387	-0.809*	0.334*	-0.112@	0.643	0.6778	33.321	37
	(0.135)	(0.06)	(0.132)				
Total WFBI	-1.163**	1.06***	0.97*	0.9405	0.9913	292.42	37
	(0.067)	(0.058)	(0.154)				

(Figures within brackets are the standard error)
**Significant at 1% level of significance*
*** Significant at 5% level of significance*
**** Significant at 10% level of significance*
@ Insignificant

Table (6.4.10): Estimates of V.E.S. Production Function for Southern Region

Dependent Variable: Ln (V/L) Model: II								
Industry Code	**Const.**	**Ln(W)**	**Ln(K/L)**	**T**	**R^2**	**Adjusted R^2**	**F-Statistic**	**D.F.**
226	-2.609*	0.772***	0.692**	0.017@	0.8931	0.9675	100.26	36
	(0.276)	(0.47)	(0.259)	(0.038)				
270	-1.028*	0.7425**	0.264*	-0.027@	0.9125	0.9885	125.1	36
	(0.145)	(0.300)	(0.085)	(0.033)				
271	1.999*	1.5805*	0.245@	-0.103***	0.7628	0.8264	38.59	36
	(0.211)	(0.597)	(0.219)	(0.054)				
272	-1.34*	0.331***	0.26**	0.004***	0.5959	0.6456	17.696	36
	(0.152)	(0.19)	(0.112)	(0.020)				
273	-4.006*	0.088@	-0.292@	0.012**	0.6278	0.6801	20.241	36
	(0.296)	(0.534)	(0.319)	(0.05)				
274	-0.095@	0.791*	0.193**	-0.042*	0.8857	0.9595	92.987	36
	(0.131)	(0.13)	(0.095)	(0.014)				
275	1.791*	1.35*	0.93*	-0.079*	0.4574	0.4955	10.116	36
	(0.355)	(0.397)	(0.331)	(0.03)				
276	-5.957*	-0.08@	0.590***	0.106@	0.9121	0.9881	124.52	36
	(0.258)	(1.077)	(0.305)	(0.126)				
279	3.1974*	2.141*	0.348@	-0.144*	0.9225	0.9994	142.8	36
	(0.487)	(0.452)	(0.128)	(0.046)				
280	5.84*	2.42*	0.0368@	-0.218*	0.7195	0.7795	30.781	36
	(0.11)	(0.327)	(0.124)	(0.031)				
281	-3.168*	-0.104@	0.209@	0.043@	0.7496	0.8121	35.923	36
	(0.164)	(0.478)	(0.134)	(0.043)				
282	1.3584*	1.59*	-0.244@	-0.102**	-0.443	-0.48	-3.684	36
	(0.389)	(0.566)	(0.175)	(0.042)				
316	9.825*	1.33@	-0.615@	-0.37@	0.386	0.4182	7.544	36
	(2.721)	(6.12)	(0.608)	(0.60)				
378	-0.33***	0.56@	0.0745@	-0.026**	0.5337	0.5782	13.735	36
	(0.19)	(0.101)	(0.05)	(0.011)				
385	-2.66*	0.501@	0.0584@	0.032@	0.9077	0.9833	118.01	36
	(0.318)	(1.119)	(0.388)	(0.098)				
387	0.885*	0.81***	-0.061@	-0.052@	0.6489	0.703	22.178	36
	(0.135)	(0.44)	(0.14)	(0.047)				
Total WFBI	-0.488*	1.261*	0.975*	-0.022@	0.9201	0.9967	138.14	36
	(0.068)	(0.268)	(0.156)	(0.027)				

(Figures within brackets are the standard error)
**Significant at 1% level of significance*
*** Significant at 5% level of significance*
**** Significant at 10% level of significance*
@ Insignificant

Table (6.5.1): Cobb-Douglas Production Function for Other States

Dependent Variable: Ln (V) Model: I							
Industry Code	Const.	Ln (L)	Ln (K)	R^2	Adjusted R^2	F- Statistic	D.F.
226	-0.856*	2.513*	-0.0143@	0.948	0.9992	337.27	37
	(-0.314)	(0.1856)	(0.1427)				
270	5.119*	0.275*	-0.619*	0.9419	0.9928	299.92	37
	(0.087)	(0.055)	(0.1423)				
271	3.662*	(0.7358*	-0.4097*	0.67	0.7062	37.561	37
	(0.133)	(0.0668)	(0.1087)				
272	1.715*	1.279*	-0.277*	0.926	0.9761	231.5	37
	(0.2638)	(0.1187)	(0.105)				
273	1.091*	1.1006@	-0.1834@	0.765	0.8064	60.223	37
	(0.183)	(0.0772)	(0.1448)				
274	0.998*	0.910*	-0.185*	0.699	0.7368	42.962	37
	(0.1127)	(0.032)	(0.0483)				
275	-0.325@	1.614*	-0.067@	0.925	0.975	228.17	37
	(0.2241)	(0.078)	(0.0896)				
276	10.91*	3.02*	-1.432*	0.94	0.9908	289.83	37
	(0.445)	(0.254)	(0.4136)				
279	0.456@	0.956*	0.0441@	0.8528	0.8989	107.18	37
	(0.338)	(0.176)	(0.044)				
280	0.699**	0.778*	-0.149*	0.6623	0.6981	36.282	37
	(0.3067)	(0.039)	(0.055)				
281	5.78*	1.481*	-0.622*	0.9361	0.9867	271.01	37
	(0.319)	(0.134)	(0.2019)				
282	2.002*	2.157*	-0.388**	0.921	0.9708	215.68	37
	(0.544)	(0.097)	(0.175)				
316	5.178*	0.766*	0.358***	0.802	0.8454	74.934	37
	(0.953)	(0.238)	(0.187)				
378	0.0028@	0.0058*	0.0003@	0.522	0.5502	20.203	37
	(0.0021)	(0.0012)	(0.00027)				
385	0.006*	0.0032*	-0.0008**	0.521	0.5492	20.122	37
	(0.0006)	(0.0004)	(0.0003)				
386	-0.050@	0.1457*	-0.0047**	0.939	0.9898	284.78	37
	(0.067)	(0.0142)	(0.022)				
387	3.914*	1.231*	-0.454*	0.941	0.9919	295.06	37
	(0.1943)	(0.091)	(0.117)				
Total	46.282*	2.892*	-2.966*	0.937	0.9876	275.15	37
WFBI	(1.062)	(0.282)	(0.086)				

(Figures within brackets are the standard error)

**Significant at 1% level of significance*

*** Significant at 5% level of significance*

**** Significant at 10% level of significance*

@ Insignificant

Table (6.5.2): Cobb-Douglas Production Function for Other States

Dependent Variable: Ln (V) Model: II								
Industry Code	**Constant**	**Ln (L)**	**Ln (K)**	**T**	**R^2**	**Adjusted R^2**	**F-Statistic**	**D.F.**
226	-0.657** (0.285)	2.0702* (0.2713)	0.2284@ (0.1713)	-0.10** (0.043)	0.904	0.98	113.3	36
270	4.0634* (0.082)	0.2525* (0.0551)	-0.422@ (0.1745)	-0.023@ (0.014)	0.915	0.991	128.9	36
271	3.3195* (0.134)	0.7275* (0.0753)	-0.351@ (0.3094)	-0.008@ (0.042)	0.891	0.965	98.18	36
272	1.4908* (0.224)	0.7916* (0.1847)	0.2639@ (0.1979)	-0.119* (0.039)	0.921	0.998	139.9	36
273	0.1545@ (0.1448)	0.7627* (0.1062)	0.32*** (0.1641)	-0.087* (0.019)	0.923	1	143.8	36
274	1.1033* (0.1047)	0.7973* (0.0886)	-0.105@ (0.0805)	-0.024@ (0.016)	0.918	0.994	133.6	36
275	-0.325@ (0.2242)	1.6141* (0.2914)	-0.067@ (0.1345)	0.03@ (0.034)	0.914	0.99	127.5	36
276	5.186* (0.445)	2.7361* (0.3182)	-0.127@ (1.0181)	-0.16@ (0.127)	0.894	0.969	101.3	36
279	2.795* (0.2354)	0.2354** (0.103)	0.5002* (0.1471)	-0.191* (0.029)	0.913	0.989	125.9	36
280	0.4168** (0.1887)	0.291* (0.0786)	0.4797* (0.1101)	-0.118* (0.016)	0.915	0.991	128.7	36
281	1.363* (0.303)	1.2957* (0.1514)	0.3197@ (0.488)	-0.17** (0.084)	0.918	0.995	134.9	36
282	2.099* (0.408)	0.9719* (0.2916)	0.758** (0.311)	-0.29* (0.078)	0.913	0.989	125.6	36
316	5.9524* (0.976)	0.5559@ (1.0003)	0.4163@ (0.3118)	0.0249@ (0.113)	0.843	0.913	64.48	36
378	0.0067* (0.0015)	0.0036* (0.0009)	0.0021* (0.0005)	-0.0005* (0.00004)	0.903	0.978	111.2	36
385	0.0044* (0.0005)	0.0018* (0.0004)	0.001*** (0.0005)	-0.0001@ (0.006)	0.921	0.998	140.5	36
386	-0.147** (0.0709)	0.1759* (0.0558)	-0.014@ (0.0274)	0.006@ (0.0085)	0.921	0.998	139.5	36
387	3.0195* (0.1944)	1.1145* (0.1555)	-0.22@ (0.4924)	-0.039@ (0.0778)	0.918	0.995	135	36
Total WFBI	29.878* (1.0985)	2.8926* (0.293)	-1.611* (0.298)	-0.183@ (0.5492)	0.902	0.977	109.8	36

(Figures within brackets are the standard error)
**Significant at 1% level of significance*
*** Significant at 5% level of significance*
**** Significant at 10% level of significance*
@ Insignificant

Table (6.5.3): Cobb-Douglas Production Function for Other States

Dependent Variable Ln (V/L) Model III							
Industry Code	Const.	Ln (K/L)	T	R^2	Adjusted R^2	F-Statistic	D.F.
226	-1.99*	0.457*	0.016**	0.904	0.9529	174.21	37
	(0.179)	(0.168)	(0.006)				
270	-3.13*	0.311*	0.025*	0.357	0.3763	10.271	37
	(0.156)	(0.09)	(0.005)				
271	-2.46*	0.306*	0.011@	0.4805	0.5065	17.111	37
	(0.207)	(0.211)	(0.015)				
272	-2.43*	0.441*	0.015@	0.945	0.9961	317.86	37
	(0.235)	(0.14)	(0.014)				
273	-4.78*	-0.24@	0.046*	0.4963	0.5231	18.228	37
	(0.262)	(0.28)	(0.015)				
274	-2.07*	0.43*	0.016**	0.5327	0.5615	21.089	37
	(0.2)	(0.129)	(0.007)				
275	0.857**	1.118*	-0.05*	0.3966	0.418	12.16	37
	(0.374)	(0.323)	(0.018)				
276	-2.93*	0.417***	0.028***	0.931	0.9813	249.62	37
	(0.204)	(0.221)	(0.014)				
279	-2.26*	0.501*	0.005@	0.9375	0.9882	277.5	37
	(0.493)	(0.106)	(0.017)				
280	-2.48*	-0.07@	0.036@	0.3245	0.342	8.8871	37
	(0.294)	(0.334)	(0.022)				
281	-2.43*	0.241@	0.016@	0.931	0.9813	249.62	37
	(0.188)	(0.154)	(0.011)				
282	-3.6*	-0.03@	0.013@	0.0612	0.0645	1.206	37
	(0.398)	(0.164)	(0.017)				
316	-1.23***	0.153@	0.048**	0.4634	0.4884	15.976	37
	(0.695)	(0.154)	(0.02)				
378	-3.79*	0.121@	0.047*	0.0034	0.0036	0.0631	37
	(0.482)	(0.125)	(0.013)				
385	-3.48*	0.027@	0.06*	0.0025	0.0026	0.0464	37
	(0.252)	(0.305)	(0.019)				
386	1.833@	1.589**	-0.09@	0.0116	0.0122	0.2171	37
	(1.435)	(0.709)	(0.056)				
387	-3.56*	-0.15@	0.074*	0.9271	0.9772	235.27	37
	(0.219)	(0.225)	(0.017)				
Total	-1.35*	1.048*	0.005@	0.943	0.994	306.06	37
WFBI	(0.07)	(0.151)	(0.01)				

(Figures within brackets are the standard error)

**Significant at 1% level of significance*

*** Significant at 5% level of significance*

**** Significant at 10% level of significance*

@ Insignificant

Table (6.5.4): Cobb-Douglas Production Function for Other States

			Dependent Variable: Ln (V/L) Model: IV					
Industry Code	Const.	Ln (K/L)	Ln (L)	T	R^2	Adjusted R^2	F-Statistic	D.F.
226	-2.387*	0.6667*	0.1513@	-0.01@	0.8928	0.9672	99.94	36
	(0.2056)	(0.2384)	(0.153)	(0.03)				
270	-1.693*	0.1129*	0.0854@	0.0028@	0.36	0.39	6.75	36
	(0.0583)	(0.0335)	(0.105)	(0.008)				
271	-2.619*	0.1881@	0.1455@	-0.008@	0.4833	0.5236	11.224	36
	(0.142)	(0.1513)	(0.338)	(0.037)				
272	-4.218*	0.345***	0.2625@	-0.004@	0.9048	0.9802	114.05	36
	(0.258)	(0.206)	(0.256)	(0.025)				
273	-4.9*	-0.222@	0.2153@	0.0212@	0.5136	0.5564	12.671	36
	(0.2047)	(0.223)	(0.239)	(0.019)				
274	-2.379*	0.2368**	0.146***	0.0048@	0.5637	0.6107	15.504	36
	(0.136)	(0.0966)	(0.088)	(0.006)				
275	0.4293@	1.0189*	0.0908@	-0.055*	0.4069	0.4408	8.2327	36
	(0.353)	(0.306)	(0.169)	(0.019)				
276	-14.44*	0.932***	1.333@	-0.051@	0.9184	0.9949	135.04	36
	(0.477)	(0.5186)	(1.040)	(0.095)				
279	-4.711*	0.5385*	0.5944@	-0.062@	0.9178	0.9943	134.07	36
	(0.578)	(0.1309)	(0.366)	(0.047)				
280	-5.634*	-0.037@	0.4836@	-0.035@	0.3381	0.3663	6.1296	36
	(0.206)	(0.235)	(0.612)	(0.077)				
281	-4.976*	0.2861@	0.2625@	-0.01@	0.9147	0.9909	128.65	36
	(0.259)	(0.222)	(0.431)	(0.055)				
282	-8.613*	-0.288@	0.5851**	-0.01@	0.3201	0.3468	5.6497	36
	(0.492)	(0.231)	(0.256)	(0.023)				
316	2.4713*	0.4156@	-0.722@	0.147@	0.5178	0.561	12.886	36
	(0.955)	(0.298)	(0.730)	(0.088)				
378	0.0029@	0.0016@	-0.038**	0.012*	0.0832	0.0901	1.089	36
	(0.023)	(0.0097)	(0.017)	(0.004)				
385	-2.272*	0.0072@	0.0361*	0.032***	0.5627	0.6096	15.441	36
	(0.155)	(0.190)	(0.133)	(0.019)				
386	-0.0004@	0.0706**	0.014@	-0.005@	0.0107	0.0116	0.1298	36
	(0.0686)	(0.0347)	(0.024)	(0.003)				
387	-2.679*	-0.104@	-0.228@	0.110@	0.9039	0.9792	112.83	36
	(0.256)	(0.322)	(0.623)	(0.071)				
Total	2.1363*	0.8628*	-0.267**	0.037@	0.8332	0.9026	59.942	36
WFBI	(0.057)	(0.1245)	(0.120)	(0.027)				

(Figures within brackets are the standard error)

**Significant at 1% level of significance*

*** Significant at 5% level of significance*

**** Significant at 10% level of significance*

@ Insignificant

Table (6.5.5): Estimates of C.E.S. Production Function for Other States

Dependent Variable ln(V/L) Model I						
Industry Code	Constant	Ln (W)	R^2	Adjusted R^2	F-Statistic	D.F.
226	0.3227@	0.9951*	0.9137	0.9377	201.16	38
	(0.594)	(0.171)				
270	-0.759*	0.1492*	0.2732	0.2804	7.142	38
	(0.075)	(0.032)				
271	-1.333*	0.3997*	0.6547	0.6719	36.025	38
	(0.164)	(0.045)				
272	-3.326*	0.2045@	0.9482	0.9732	347.8	38
	(0.274)	(0.177)				
273	-1.432*	0.448*	0.7144	0.7332	47.527	38
	(0.148)	(0.077)				
274	-0.97*	0.2939**	0.1514	0.1554	3.3898	38
	(0.279)	(0.145)				
275	-1.67*	0.5515*	0.8855	0.9088	146.94	38
	(0.172)	(0.107)				
276	-4.645*	1.7276*	0.941	0.9658	303.03	38
	(1.247)	(0.337)				
279	-1.574*	0.281@	0.7798	0.8003	67.285	38
	(0.343)	(0.173)				
280	-1.156*	0.2666*	0.4868	0.4996	18.023	38
	(0.120)	(0.050)				
281	-3.206*	0.281@	0.948	0.9729	346.38	38
	(0.441)	(0.163)				
282	-1.271*	1.15***	0.9486	0.9736	350.65	38
	(0.231)	(0.093)				
316	-2.579*	0.3406@	0.0415	0.0426	0.8226	38
	(0.731)	(0.601)				
378	-0.013*	0.0038*	0.7736	0.794	64.922	38
	(0.0013)	(0.0004)				
385	-0.006@	0.0031@	0.1059	0.1087	2.2504	38
	(0.0075)	(0.0027)				
386	-0.005*	0.0021*	0.8802	0.9034	139.6	38
	(0.0009)	(0.0004)				
387	-0.688*	0.8955*	0.9431	0.9679	314.92	38
	(0.277)	(0.098)				
Total	-3.057*	1.96**	0.9428	0.9676	313.17	38
WFBI	(0.307)	(0.080)				

(Figures within brackets are the standard error)
**Significant at 1% level of significance*
*** Significant at 5% level of significance*
**** Significant at 10% level of significance*
@ Insignificant

Table (6.5.6): Estimates of C.E.S. Production Function for Other States

Dependent Variable ln(V/L) Model II							
Industry Code	**Const.**	**Ln (W)**	**T**	**R^2**	**Adjusted R^2**	**F-Statistic**	**D.F.**
226	-5.828*	0.01@	0.03@	0.7281	0.7675	49.54	37
	(0.238)	(0.323)	(0.033)				
270	-0.4*	0.2773**	-0.017@	0.326	0.3436	8.9481	37
	(0.066)	(0.137)	(0.0152)				
271	1.035*	0.893**	-0.06***	0.531	0.5597	20.946	37
	(0.129)	(0.360)	(0.033)				
272	-1.914*	0.6794***	-0.012@	0.9418	0.9927	299.37	37
	(0.289)	(0.357)	(0.038)				
273	-2.109*	0.2327@	0.0029@	0.4924	0.519	17.946	37
	(0.205)	(0.317)	(0.034)				
274	-0.248**	0.5911*	-0.039*	0.629	0.663	31.365	37
	(0.113)	(0.109)	(0.012)				
275	0.2354@	0.8014**	-0.066**	0.1827	0.1926	4.1355	37
	(0.395)	(0.389)	(0.032)				
276	-9.255*	0.4422@	0.07*	0.941	0.9919	295.06	37
	(0.522)	(2.154)	(0.245)				
279	5.0846*	2.607*	-0.19*	0.942	0.9929	300.47	37
	(0.555)	(0.477)	(0.048)				
280	5.404*	2.231*	-0.202*	0.6755	0.712	38.511	37
	(0.10)	(0.267)	(0.026)				
281	-5.879*	-0.476@	0.0858@	0.9395	0.9903	287.29	37
	(0.266)	(0.747)	(0.069)				
282	0.8379**	1.33**	-0.085**	0.3927	0.4139	11.963	37
	(0.353)	(0.509)	(0.038)				
316	-1.878**	-0.194@	0.0716@	0.3017	0.318	7.9929	37
	(0.720)	(1.632)	(0.159)				
378	-0.71*	0.6155*	-0.032**	0.6545	0.6899	35.046	37
	(0.252)	(0.114)	(0.0145)				
385	-1.852*	0.2638@	0.0232@	0.7254	0.7646	48.871	37
	(0.195)	(0.602)	(0.056)				
386	0.2072@	1.1567@	-0.099@	0.0816	0.086	1.6437	37
	(1.397)	(1.061)	(0.094)				
387	2.0074*	1.5357**	-0.101@	0.9145	0.9639	197.87	37
	(0.228)	(0.665)	(0.075)				
Total	-0.686*	0.8724***	-0.015	0.9428	0.9938	304.93	37
WFBI	(0.118)	(0.45)	(0.021)				

(Figures within brackets are the standard error)

**Significant at 1% level of significance*

*** Significant at 5% level of significance*

**** Significant at 10% level of significance*

@ Insignificant

Table (6.5.7): Estimates of C.E.S. Production Function for Other States

Dependent Variable: Ln (V/L) Model: III								
Industry Code	Const.	Ln (W)	Ln (L)	T	R^2	Adjusted R^2	F-Statistic	D.F.
226	-4.03*	0.187@	-0.14@	0.037@	0.751	0.8136	36.193	36
	(0.24)	(0.39)	(0.171)	(0.033)				
270	-0.17**	0.293***	-0.03@	-0.02@	0.326	0.3532	5.8042	36
	(0.068)	(0.168)	(0.145)	(0.015)				
271	2.657*	1.017**	-0.19@	-0.05@	0.5351	0.5797	13.812	36
	(0.131)	(0.437)	(0.357)	(0.039)				
272	-5.76*	0.166@	0.48***	-0.02@	0.9148	0.991	128.8	36
	(0.274)	(0.437)	(0.26)	(0.037)				
273	-2.92*	0.139@	0.094@	0.006@	0.9188	0.9954	135.86	36
	(0.21)	(0.481)	(0.355)	(0.037)				
274	1.913*	0.886*	-0.22@	-0.06*	0.6547	0.7093	22.752	36
	(0.107)	(0.188)	(0.115)	(0.015)				
275	0.172@	0.793***	0.011@	-0.07**	0.1827	0.1979	2.6825	36
	(0.404)	(0.426)	(0.206)	(0.034)				
276	-16.1*	0.862@	1.559@	-0.12@	0.9147	0.9909	128.68	36
	(0.509)	(2.129)	(1.12)	(0.274)				
279	7.266*	2.932*	-0.33@	-0.18*	0.8936	0.9681	100.78	36
	(0.561)	(0.671)	(0.466)	(0.052)				
280	9.916*	2.394*	-0.49@	-0.16*	0.6873	0.7446	26.375	36
	(0.097)	(0.278)	(0.309)	(0.038)				
281	-23.1*	-2.69**	1.612*	0.074@	0.945	1.0238	206.18	36
	(0.234)	(1.043)	(0.589)	(0.061)				
282	2.666*	1.67@	-0.13@	-0.1***	0.4014	0.4349	8.0468	36
	(0.361)	(1.003)	(0.318)	(0.054)				
316	-1.87**	-0.19@	-0.002@	0.072*	0.3017	0.3268	5.1846	36
	(0.737)	(1.671)	(0.402)	(0.17)				
378	-0.27@	0.709*	-0.14***	-0.02@	0.6909	0.7485	26.822	36
	(0.238)	(0.12)	(0.072)	(0.016)				
385	-2.0@	0.237@	0.013@	0.024@	0.7254	0.7859	31.7	36
	(0.199)	(0.728)	(0.199)	(0.059)				
386	1.97@	1.507@	-0.17@	-0.12@	0.0832	0.0901	1.089	36
	(1.43)	(3.059)	(1.359)	(0.196)				
Total	13.49*	1.293*	-1.02**	0.057@	0.9198	0.9965	137.71	36
WFBI	(0.1)	(0.425)	(0.41)	(0.049)				

(Figures within brackets are the standard error)
**Significant at 1% level of significance*
*** Significant at 5% level of significance*
**** Significant at 10% level of significance*
@ Insignificant

Table (6.5.8): Estimates of C.E.S. Production Function for Other States

				Dependent Variable: Ln (V/L) Model: IV					
Industry Code	Const.	Ln (L)	Ln (K)	$(\ln L - \ln K)^2$	T	R^2	Adjusted R^2	F-Statistic	D.F.
226	-6.774*	0.531@	-0.45@	-0.144@	0.035@	0.892	0.9934	71.91	35
	(0.2142)	(0.7724)	(0.949)	(0.1285)	(0.021)				
270	-2.522*	0.0344@	0.107*	0.0073@	0.019@	0.349	0.3888	4.6888	35
	(0.063)	(0.116)	(0.039)	(0.030)	(0.019)				
271	-4.701*	0.608@	-0.239@	0.691@	-0.02@	0.507	0.5646	8.9877	35
	(0.141)	(0.450)	(0.240)	(0.419)	(0.05)				
272	-5.976*	0.018@	0.356@	0.19@	0.009@	0.89	0.9921	71.044	35
	(0.266)	(0.42)	(0.244)	(0.523)	(0.017)				
273	-3.397*	0.160@	-0.073@	0.0811@	0.010@	0.519	0.5788	9.4564	35
	(0.208)	(0.393)	(0.321)	(0.162)	(0.023)				
274	-3.762*	0.44***	-0.201@	-0.151@	0.0032@	0.537	0.5985	10.153	35
	(0.146)	(0.221)	(0.215)	(0.098)	(0.003)				
275	-2.803*	-0.01@	0.095@	-0.46@	0.0058@	0.136	0.1511	1.3726	35
	(0.425)	(0.495)	(0.596)	(0.615)	(0.021)				
276	-16.91*	1.155@	-0.095@	-0.614@	0.028@	0.894	0.9957	73.487	35
	(0.471)	(1.5)	(1.110)	(0.525)	(0.035)				
279	-7.569*	0.334@	0.57***	-0.066@	-0.027@	0.89	0.9918	70.846	35
	(0.644)	(0.454)	(0.275)	(0.269)	(0.032)				
280	-3.448*	0.845@	-0.55@	0.111@	-0.027@	0.393	0.4381	5.6699	35
	(0.197)	(0.95)	(0.776)	(0.401)	(0.072)				
281	-6.18*	-0.22@	0.52***	-0.357@	0.002**	0.892	0.9937	72.104	35
	(0.257)	(0.64)	(0.370)	(0.575)	(0.008)				
282	-5.629*	0.463@	-0.054@	-0.241@	-0.015@	0.409	0.4554	6.0479	35
	(0.368)	(0.366)	(0.318)	(0.319)	(0.016)				
316	-2.476*	-1.85@	1.97@	-0.649@	0.047@	0.659	0.734	16.887	35
	(0.639)	(3.51)	(3.769)	(0.966)	(0.004)				
378	-3.106*	0.238@	-0.184@	-0.065@	0.022@	0.389	0.4336	5.5731	35
	(0.377)	(0.164)	(0.123)	(0.0491)	(0.030)				
385	-3.011*	0.125@	-0.071@	-0.048@	0.007***	0.725	0.8073	23.01	35
	(0.2043)	(0.512)	(0.441)	(0.3232)	(0.004)				
386	-6.615*	-0.76@	1.653@	0.256@	-0.019@	0.27	0.3009	3.2363	35
	(1.370)	(1.85)	(2.34)	(1.18)	(0.059)				
387	-3.63*	0.469@	-0.42@	-0.281@	0.036@	0.873	0.973	60.256	35
	(0.254)	(0.83)	(0.342)	(0.45)	(0.062)				
Total	-1.837**	1.256@	-1.37*	0.926*	0.092**	0.889	0.9911	70.371	35
WFBI	(0.089)	(0.82)	(0.42)	(0.330)	(0.04)				

Figures within brackets are the standard error)
**Significant at 1% level of significance*
*** Significant at 5% level of significance*
**** Significant at 10% level of significance*
@ Insignificant

Table (6.5.9): Estimates of V.E.S. Production Function for Other States

Dependent Variable: Ln (V/L) Model: I							
Industry Code	Const.	Ln (W)	Ln (K/L)	R^2	Adjusted R^2	F-Statistic	D.F.
226	-1.291* (0.2064)	0.7324* (0.156)	0.548* (0.189)	0.9003	0.949	167.06	37
270	-0.809* (0.0557)	0.195* (0.026)	0.1069* (0.032)	0.3645	0.3842	10.611	37
271	-1.383* (0.1364)	0.2719* (0.058)	0.132@ (0.141)	0.4978	0.5247	18.338	37
272	-2.102* (0.2581)	0.6302* (0.070)	0.442** (0.185)	0.937	0.9876	275.15	37
273	-2.434* (0.2013)	0.1493@ (0.129)	-0.191@ (0.212)	0.5096	0.5371	19.224	37
274	-1.194* (0.1238)	0.3739* (0.064)	0.211** (0.087)	0.5955	0.6277	27.235	37
275	-1.352* (0.388)	0.42** (0.202)	0.79** (0.359)	0.2104	0.2218	4.9296	37
276	-4.742* (0.4884)	1.57* (0.304)	0.985*** (0.554)	0.9454	0.9965	320.33	37
279	-2.149* (0.576)	0.798* (0.156)	0.491* (0.142)	0.9475	0.9987	333.88	37
280	-0.842* (0.1828)	0.1503* (0.093)	-0.234@ (0.197)	0.4166	0.4391	13.211	37
281	-2.474* (0.26)	0.570* (0.107)	0.365*** (0.208)	0.876	0.9234	130.69	37
282	-2.78* (0.382)	0.244*** (0.142)	-0.174@ (0.171)	0.212	0.2235	4.9772	37
316	0.062@ (0.709)	0.635* (0.230)	0.148@ (0.158)	0.3301	0.3479	9.1161	37
378	-1.446* (0.267)	0.482* (0.088)	0.0907* (0.069)	0.6215	0.6551	30.377	37
385	-0.88* (0.195)	0.535* (0.103)	0.065@ (0.221)	0.7247	0.7639	48.699	37
386	-1.052@ (1.35)	0.555@ (0.494)	1.083@ (0.656)	0.1789	0.1886	4.0308	37
387	-1.4* (0.234)	0.58* (0.103)	-0.193@ (0.228)	0.9421	0.993	301.02	37
Total WFBI	-1.214* (0.0706)	1.102*** (0.062)	1.013@ (0.074)	0.946	0.9971	324.09	37

(Figures within brackets are the standard error)
**Significant at 1% level of significance*
*** Significant at 5% level of significance*
**** Significant at 10% level of significance*
@ Insignificant

Table (6.5.10): Estimates of V.E.S. Production Function for Other States

			Dependent Variable: Ln (V/L) Model: II					
Industry Code	**Const.**	**Ln (W)**	**Ln (K/L)**	**T**	**R^2**	**Adjusted R^2**	**F-Statistic**	**D.F.**
226	-1.99*	0.59@	0.529**	0.013@	0.9052	0.9806	114.58	36
	(0.211)	(0.359)	(0.198)	(0.029)				
270	-0.4*	0.288**	0.103*	-0.01@	0.3674	0.398	6.9693	36
	(0.056)	(0.117)	(0.033)	(0.013)				
271	1.218*	0.963**	0.149@	-0.06@	0.5484	0.5941	14.572	36
	(0.129)	(0.364)	(0.134)	(0.033)				
272	-2.32*	0.575***	0.449**	0.006@	0.8944	0.9689	101.58	36
	(0.264)	(0.33)	(0.195)	(0.036)				
273	-2.78*	0.061@	-0.2@	0.009@	0.5109	0.5535	12.535	36
	(0.206)	(0.37)	(0.221)	(0.035)				
274	-0.08@	0.641*	0.156**	-0.03*	0.6578	0.7126	23.067	36
	(0.106)	(0.105)	(0.077)	(0.011)				
275	1.738*	1.31*	0.899*	-0.08*	0.4373	0.4737	9.3258	36
	(0.345)	(0.386)	(0.322)	(0.029)				
276	-11.4*	-0.15@	1.125***	0.203@	0.8893	0.9634	96.412	36
	(0.492)	(2.053)	(0.582)	(0.24)				
279	3.203*	2.145*	0.35*	-0.14*	0.8928	0.9672	99.919	36
	(0.488)	(0.453)	(0.129)	(0.046)				
280	5.482*	2.272*	0.035@	-0.21*	0.6757	0.732	25.003	36
	(0.102)	(0.307)	(0.117)	(0.029)				
281	-4.98*	-0.16@	0.329@	0.068@	0.845	0.9154	65.419	36
	(0.258)	(0.752)	(0.212)	(0.069)				
282	1.21*	1.419*	-0.22@	-0.09**	0.4827	0.5229	11.197	36
	(0.347)	(0.504)	(0.156)	(0.038)				
316	-2.63*	-0.36@	0.165@	0.099@	0.3463	0.3752	6.357	36
	(0.729)	(1.639)	(0.163)	(0.162)				
378	-0.42***	0.711*	0.095@	-0.03**	0.6788	0.7354	25.36	36
	(0.245)	(0.13)	(0.063)	(0.014)				
385	-1.67*	0.314@	0.037@	0.02@	0.7258	0.7863	31.764	36
	(0.199)	(0.701)	(0.243)	(0.062)				
386	3.415*	2.148**	1.344**	-0.14@	0.3305	0.358	5.9238	36
	(1.305)	(1.102)	(0.655)	(0.09)				
387	1.527*	1.39***	-0.11@	-0.09@	0.9148	0.991	128.85	36
	(0.233)	(0.752)	(0.241)	(0.082)				
Total	-0.51*	1.317*	1.018@	-0.02@	0.9184	0.9949	135.02	36
WFBI	(0.071)	(0.28)	(0.064)	(0.029)				

(Figures within brackets are the standard error)

**Significant at 1% level of significance*

*** Significant at 5% level of significance*

**** Significant at 10% level of significance*

@ Insignificant

Table (6.6.1): Estimates of C-D Production Function for All India

		Dependent Variable: Ln (V) Model: I					
Industry Code	Constant	Ln L	Ln K	R^2	Adjusted R^2	F-Statistic	D.F.
226	-0.744* (0.2728)	2.1824* (0.1612)	-0.0124@ (0.124)	0.947	0.998	329.9	37
270	13.826* (0.2356)	0.744* (0.1488)	-1.674* (0.3844)	0.939	0.99	293	37
271	5.4312* (0.1984)	1.0912* (0.0992)	-0.6076* (0.1612)	0.94	0.99	295.6	37
272	1.56* (0.24)	1.164* (0.108)	-0.252** (0.096)	0.947	0.999	341.5	37
273	1.4012* (0.2356)	1.4136* (0.0992)	-0.2356@ (0.186)	0.948	1.000	348.5	37
274	1.4136* (0.1596)	1.2882* (0.0456)	-0.2622* (0.0684)	0.93	0.98	250.5	37
275	-0.350@ (0.242)	1.7424* (0.0847)	-0.0726@ (0.0968)	0.913	0.962	198.4	37
276	4.5962* (0.1876)	1.273* (0.1072)	-0.603* (0.1742)	0.94	0.991	299.3	37
279	0.375@ (0.2783)	0.7865* (0.1452)	0.0363@ (0.036)	0.701	0.739	44.63	37
280	1.0057* (0.4407)	1.1187* (0.0565)	-0.2147* (0.0791)	0.945	0.996	327.7	37
281	4.266* (0.2356)	1.0912* (0.0992)	-0.4588* (0.1488)	0.942	0.993	308	37
282	1.4832* (0.4032)	1.5984* (0.072)	-0.288** (0.1296)	0.931	0.981	255.6	37
316	3.7696* (0.6944)	0.558 (0.1736)	0.2604*** (0.1364)	0.732	0.771	51.79	37
378	0.3627@ (0.2691)	0.7605* (0.1497)	0.0351@ (0.035)	0.678	0.715	40.04	37
385	2.316* (0.252)	1.2144* (0.1404)	-0.324* (0.108)	0.948	1.000	347.8	37
386	-0.9222@ (1.2876)	2.6796* (0.261)	-0.087@ (0.4002)	0.945	0.996	325.8	37
387	3.3945* (0.1686)	1.0678* (0.07868)	-0.3934* (0.1012)	0.94	0.991	296.1	37
Total WFBI	16.937* (0.3886)	1.0586* (0.10318)	-1.0854* (0.3082)	0.925	0.975	234.3	37

(Figures within brackets are the standard error)
**Significant at 1% level of significance*
*** Significant at 5% level of significance*
**** Significant at 10% level of significance*
@ Insignificant

Table (6.6.2): Estimates of C-D Production Function for All India

Dependent Variable: Ln (V) Model: II								
Industry Code	**Constant**	**Ln (L)**	**Ln (K)**	**T**	**R^2**	**Adjusted R^2**	**F-Statistic**	**D.F.**
226	-0.5704** (0.248)	1.798* (0.2356)	0.1984@ (0.1488)	-0.087** (0.037)	0.9189	0.9955	135.97	36
270	10.974* (0.2232)	0.682* (0.1488)	-1.1408* (0.4712)	-0.062@ (0.037)	0.9193	0.995	136.7	36
271	4.9228* (0.1984)	1.0788* (0.1116)	-0.5208@ (0.4588)	-0.012@ (0.062)	0.9198	0.9965	137.63	36
272	1.356* (0.204)	0.72* (0.168)	0.24@ (0.18)	-0.108* (0.036)	0.9076	0.9832	117.87	36
273	0.1984@ (0.186)	0.9796* (0.1364)	0.4092*** (0.2108)	-0.1116* (0.0248)	0.9191	0.9957	136.33	36
274	1.5618* (0.1482)	1.1286* (0.1254)	-0.1482@ (0.114)	-0.034@ (0.0228)	0.9129	0.989	125.77	36
275	-0.3509@ (0.242)	1.7424* (0.3146)	-0.0726@ (0.1452)	0.001@ (0.0363)	0.8972	0.972	104.73	36
276	2.1842* (0.1876)	1.1524* (0.134)	-0.0536@ (0.4288)	-0.067@ (0.053)	0.9094	0.9852	120.45	36
279	2.299* (0.1936)	0.1936** (0.0847)	0.4114* (0.121)	-0.157@ (0.0242)	0.918	0.9945	134.34	36
280	0.5989** (0.2712)	0.4181* (0.113)	0.6893* (0.1582)	-0.169* (0.0226)	0.9199	0.9965	137.78	36
281	1.0044* (0.2232)	0.9548* (0.1116)	0.2356@ (0.3596)	-0.12*** (0.062)	0.9195	0.9961	137.07	36
282	1.5552* (0.3024)	0.72* (0.216)	0.5616** (0.2304)	-0.216 (0.057)	0.9119	0.9879	124.21	36
316	4.3337* (0.7104)	0.4047@ (0.7282)	0.3031@ (0.22704)	0.018@ (0.082)	0.727	0.7876	31.95	36
378	0.8658* (0.1989)	0.468* (0.117)	0.2691* (0.0585)	-0.058* (0.0117)	0.8722	0.9449	81.89	36
385	1.656* (0.192)	0.684* (0.168)	0.324@ (0.192)	-0.144* (0.036)	0.908	0.9837	118.43	36
386	-2.697* (1.305)	3.2364* (1.0266)	-0.261@ (0.5046)	0.104@ (0.156)	0.9148	0.991	128.85	36
387	2.6189* (0.1686)	0.96664* (0.13488)	-0.1910@ (0.4271)	-0.034@ (0.067)	0.9193	0.9959	136.7	36
Total WFBI	10.9344* (0.402)	1.0586* (0.1072)	-0.5896@ (1.541)	-0.067* (0.201)	0.9198	0.9965	137.63	36

(Figures within brackets are the standard error)

**Significant at 1% level of significance*

*** Significant at 5% level of significance*

**** Significant at 10% level of significance*

@ Insignificant

Table (6.6.3): Estimates of C-D Production Function for All India

Dependent Variable Ln (V/L) Model III							
Industry Code	Const.	Ln (K/L)	T	R^2	Adjusted R^2	F-Statistic	D.F.
226	-1.99*	0.457*	0.017*	0.7852	0.8276	67.627	37
	(0.179)	(0.168)	(0.006)				
270	-3.13*	0.311*	0.025*	0.9464	0.9976	326.78	37
	(0.156)	(0.09)	(0.005)				
271	-2.46*	0.306@	0.011@	0.7125	0.751	45.848	37
	(0.208)	(0.211)	(0.015)				
272	-2.43*	0.441*	0.0165@	0.9391	0.9899	285.28	37
	(0.235)	(0.14)	(0.014)				
273	-4.78*	-0.24@	0.046*	0.6375	0.672	32.534	37
	(0.262)	(0.28)	(0.015)				
274	-2.07*	0.43*	0.016***	0.7541	0.7949	56.734	37
	(0.2)	(0.129)	(0.007)				
275	0.857**	1.118*	-0.05*	0.4281	0.4512	13.848	37
	(0.374)	(0.323)	(0.018)				
276	-2.94*	0.417***	0.028***	0.9485	0.9998	340.72	37
	(0.204)	(0.221)	(0.014)				
279	-2.26*	0.501*	0.005@	0.771	0.8127	62.286	37
	(0.493)	(0.106)	(0.017)				
280	-2.48*	-0.07@	0.036@	0.4662	0.4914	16.157	37
	(0.294)	(0.334)	(0.022)				
281	-2.43*	0.241@	0.016@	0.7791	0.8212	65.248	37
	(0.188)	(0.154)	(0.011)				
282	-3.6*	-0.03@	0.013@	0.0453	0.0477	0.8778	37
	(0.398)	(0.164)	(0.017)				
316	-1.23***	0.153@	0.048*	0.3374	0.3556	9.4203	37
	(0.695)	(0.154)	(0.02)				
378	-3.79*	0.121@	0.047*	0.4447	0.4687	14.815	37
	(0.482)	(0.125)	(0.013)				
385	-3.49*	0.027@	0.06*	0.9352	0.9858	266.99	37
	(0.253)	(0.305)	(0.019)				
386	1.833@	1.589*	-0.09@	0.2127	0.2242	4.998	37
	(1.435)	(0.709)	(0.056)				
387	-3.56*	-0.15@	0.074*	0.8388	0.8841	96.264	37
	(0.219)	(0.225)	(0.017)				
Total WFBI	-1.36*	1.048*	0.005*	0.9458	0.9969	322.83	37
	(0.07)	(0.151)	(0.01)				

(Figures within brackets are the standard error)
**Significant at 1% level of significance*
*** Significant at 5% level of significance*
**** Significant at 10% level of significance*
@ Insignificant

Table (6.6.4): Estimates of Cobb-Douglas Production Function for All India

Dependent Variable: Ln (V/L) Model: IV								
Industry Code	Const.	Ln (K/L)	Ln (L)	T	R^2	Adjusted R^2	F-Statistic	D.F.
226	-2.07*	0.579*	0.131@	-0.01@	0.806	0.8732	49.856	36
	(0.179)	(0.207)	(0.133)	(0.026)				
270	-4.57*	0.305*	0.231@	0.007*	0.9172	0.9937	132.96	36
	(0.158)	(0.091)	(0.284)	(0.022)				
271	-3.88*	0.279@	0.216@	-0.01***	0.7167	0.7764	30.361	36
	(0.211)	(0.224)	(0.502)	(0.056)				
272	-3.84*	0.313*	0.239@	-0.004@	0.9052	0.9806	114.53	36
	(0.235)	(0.187)	(0.233)	(0.023)				
273	-6.29*	-0.29@	0.277@	0.027@	0.6597	0.7147	23.261	36
	(0.263)	(0.286)	(0.308)	(0.025)				
274	-3.37*	0.335**	0.206@	0.007@	0.798	0.8645	47.406	36
	(0.193)	(0.137)	(0.125)	(0.009)				
275	0.463@	1.100*	0.098@	-0.06*	0.4392	0.4758	9.3991	36
	(0.381)	(0.33)	(0.183)	(0.021)				
276	-6.08*	0.393***	0.562@	-0.02@	0.9179	0.9944	134.14	36
	(0.201)	(0.218)	(0.438)	(0.04)				
279	-3.87*	0.443*	0.489@	-0.05@	0.8204	0.8887	54.808	36
	(0.476)	(0.108)	(0.301)	(0.039)				
280	-8.1*	-0.05@	0.695@	-0.05*	0.4859	0.5264	11.342	36
	(0.296)	(0.338)	(0.879)	(0.111)				
281	-3.67*	0.211@	0.193@	-0.01@	0.7874	0.853	44.444	36
	(0.191)	(0.164)	(0.317)	(0.041)				
282	-6.38*	-0.21@	0.433**	-0.01@	0.2371	0.2569	3.7298	36
	(0.364)	(0.171)	(0.19)	(0.017)				
316	1.799**	0.303@	-0.53@	0.109***	0.377	0.4084	7.2604	36
	(0.696)	(0.217)	(0.532)	(0.065)				
378	0.03@	0.018@	-0.4**	0.128*	0.8869	0.9608	94.063	36
	(0.246)	(0.103)	(0.185)	(0.04)				
385	-3.78*	0.012@	0.06@	0.053***	0.9124	0.9884	124.92	36
	(0.258)	(0.317)	(0.222)	(0.032)				
386	-0.01@	1.498**	0.298@	-0.101***	0.228	0.247	3.544	36
	(1.456)	(0.736)	(0.503)	(0.057)				
387	-2.32*	-0.09@	-0.2@	0.096@	0.8402	0.9102	63.081	36
	(0.223)	(0.28)	(0.541)	(0.062)				
Total	2.551*	1.031*	-0.32@	0.044@	0.9195	0.9961	137.08	36
WFBI	(0.068)	(0.149)	(0.253)	(0.032)				

(Figures within brackets are the standard error)

**Significant at 1% level of significance*

*** Significant at 5% level of significance*

**** Significant at 10% level of significance*

@ Insignificant

Table (6.6.5): Estimates of C.E.S. Production Function for All India

Dependent Variable ln(V/L) Model I						
Industry Code	Constant	Ln (W)	R^2	Adjusted R^2	F-Statistic	D.F.
226	0.2802@ (0.5158)	0.8643* (0.1488)	0.7936	0.8145	73.054	38
270	-2.05* (0.2021)	0.403* (0.0868)	0.7378	0.7572	53.464	38
271	-1.977* (0.243)	0.5927* (0.067)	0.9471	0.972	340.17	38
272	-3.025* (0.2496)	0.186@ (0.1608)	0.6336	0.6503	32.856	38
273	-1.839* (0.191)	0.5754* (0.0992)	0.9176	0.9417	211.58	38
274	-1.375* (0.3956)	0.4161** (0.2052)	0.2143	0.2199	5.1823	38
275	-1.804* (0.1863)	0.5953* (0.1162)	0.9156	0.9397	206.12	38
276	-1.956* (0.5253)	0.7276* (0.142)	0.8034	0.8245	77.643	38
279	-1.295* (0.2819)	0.2311@ (0.1428)	0.6413	0.6582	33.969	38
280	-1.661* (0.1729)	0.3831* (0.0723)	0.6995	0.7179	44.228	38
281	-2.362* (0.3249)	0.2071*** (0.1203)	0.1798	0.1845	4.1651	38
282	-0.942* (0.1714)	0.8496* (0.0691)	0.9555	0.9806	407.97	38
316	-1.877* (0.532)	0.248@ (0.4377)	0.0409	0.042	0.8102	38
378	-1.629* (0.1615)	0.4844* (0.0538)	0.948	0.9729	346.38	38
385	-0.805@ (1.0332)	0.4248@ (0.378)	0.1368	0.1404	3.0111	38
386	-1.874* (0.3132)	0.7743* (0.1392)	0.9258	0.9502	237.06	38
387	-0.597** (0.2405)	0.7767* (0.0854)	0.8301	0.8519	92.83	38
Total WFBI	-1.119* (0.1126)	0.7169* (0.0295)	0.937	0.9617	282.59	38

Table (6.6.6): Estimates of C.E.S. Production Function for All India

Dependent Variable ln(V/L) Model II							
Industry Code	**Const.**	**Ln (W)**	**T**	**R^2**	**Adjusted R^2**	**F-Statistic**	**D.F.**
226	-5.06*	0.009@	0.026@	0.6324	0.6666	31.826	37
	(0.207)	(0.28)	(0.029)				
270	-1.08*	0.749***	-0.05@	0.8804	0.928	136.18	37
	(0.179)	(0.371)	(0.041)				
271	1.535*	1.324*	-0.09***	0.7874	0.83	68.518	37
	(0.192)	(0.534)	(0.048)				
272	-1.74*	0.618***	-0.01@	0.8748	0.9221	129.26	37
	(0.263)	(0.325)	(0.035)				
273	-2.71*	0.299@	0.004@	0.6324	0.6666	31.826	37
	(0.263)	(0.408)	(0.043)				
274	-0.35*	0.837*	-0.06*	0.8903	0.9385	150.2	37
	(0.161)	(0.155)	(0.017)				
275	0.254@	0.865**	-0.07***	0.1972	0.2079	4.5452	37
	(0.426)	(0.42)	(0.035)				
276	-3.9*	0.186@	0.03@	0.9208	0.9706	215.12	37
	(0.22)	(0.907)	(0.103)				
279	4.182*	2.144*	-0.16*	0.8349	0.88	93.553	37
	(0.456)	(0.392)	(0.04)				
280	7.765*	3.206*	-0.29*	0.9481	0.9993	337.73	37
	(0.144)	(0.384)	(0.038)				
281	-4.33*	-0.35@	0.063@	0.7366	0.7764	51.725	37
	(0.196)	(0.551)	(0.051)				
282	0.827**	1.309*	-0.08***	0.3874	0.4083	11.697	37
	(0.349)	(0.503)	(0.037)				
316	-1.85**	-0.19@	0.071*	0.2976	0.3137	7.8383	37
	(0.711)	(1.61)	(0.158)				
378	-0.91*	0.792*	-0.04***	0.8424	0.8879	98.886	37
	(0.324)	(0.147)	(0.019)				
385	-2.39*	0.341@	0.03@	0.9372	0.9879	276.09	37
	(0.252)	(0.778)	(0.073)				
386	0.23@	1.282@	-0.11@	0.0905	0.0954	1.8404	37
	(1.549)	(1.176)	(0.104)				
387	1.741*	1.332**	-0.09@	0.8694	0.9164	123.14	37
	(0.198)	(0.577)	(0.065)				
Total	-0.67*	0.852***	-0.02@	0.6324	0.6666	31.826	37
WFBI	(0.115)	(0.438)	(0.047)				

(Figures within brackets are the standard error)
**Significant at 1% level of significance*
***Significant at 5% level of significance*
**** Significant at 10% level of significance*
@ Insignificant

Table (6.6.7): Estimates of C.E.S. Production Function for All India

Dependent Variable Ln (V/L) Model III								
Industry Code	Const.	Ln (W)	Ln (L)	T	R^2	Adjusted R^2	F-Statistic	D.F.
226	-3.5*	0.162@	-0.12@	0.032@	0.6522	0.7066	22.507	36
	(0.208)	(0.339)	(0.149)	(0.029)				
270	-0.46**	0.792***	-0.07@	-0.05@	0.8804	0.9538	88.334	36
	(0.182)	(0.453)	(0.392)	(0.041)				
271	3.94*	1.508**	-0.28@	-0.07@	0.7936	0.8597	46.14	36
	(0.195)	(0.647)	(0.53)	(0.057)				
272	-5.24*	0.151@	0.44***	-0.02@	0.9204	0.9971	138.75	36
	(0.25)	(0.397)	(0.236)	(0.034)				
273	-3.75*	0.179***	0.12@	0.007@	0.6349	0.6878	20.866	36
	(0.269)	(0.618)	(0.456)	(0.047)				
274	2.709*	1.254*	-0.31***	-0.08*	0.9027	0.9779	111.31	36
	(0.152)	(0.267)	(0.163)	(0.021)				
275	0.185@	0.856***	0.012@	-0.07***	0.1972	0.2137	2.9482	36
	(0.436)	(0.46)	(0.223)	(0.036)				
276	-6.79*	0.363@	0.657@	-0.05*	0.9186	0.9951	135.36	36
	(0.214)	(0.897)	(0.472)	(0.115)				
279	5.976*	2.412*	-0.27@	-0.15@	0.8843	0.958	91.75	36
	(0.461)	(0.552)	(0.384)	(0.042)				
280	14.25*	3.44*	-0.71@	-0.23*	0.9188	0.9953	135.71	36
	(0.139)	(0.4)	(0.444)	(0.054)				
281	-17.0*	-1.99**	1.188*	0.055@	0.8692	0.9417	79.771	36
	(0.172)	(0.769)	(0.434)	(0.045)				
282	2.63*	1.655@	-0.13@	-0.1***	0.396	0.429	7.8675	36
	(0.356)	(0.989)	(0.314)	(0.053)				
316	-1.84**	-0.19@	-0.01@	0.071@	0.2976	0.3224	5.0843	36
	(0.727)	(1.648)	(0.397)	(0.167)				
378	-0.34@	0.913*	-0.17***	-0.02@	0.8892	0.9633	96.303	36
	(0.307)	(0.154)	(0.092)	(0.021)				
385	-2.58*	0.306@	0.017@	0.031@	0.9072	0.9828	117.31	36
	(0.257)	(0.941)	(0.257)	(0.076)				
386	2.187@	1.67@	-0.18@	-0.13@	0.0922	0.0999	1.2191	36
	(1.585)	(3.391)	(1.507)	(0.218)				
Total	11.7*	1.122*	-0.88**	0.05@	0.9096	0.9854	120.79	36
WFBI	(0.087)	(0.369)	(0.355)	(0.043)				

(Figures within brackets are the standard error)
**Significant at 1% level of significance*
*** Significant at 5% level of significance*
**** Significant at 10% level of significance*
@ Insignificant

Table (6.6.8): Estimates of C.E.S. Production Function for All India

Dependent Variable ln(V/L) Model IV									
Industry Code	**Const.**	**Ln (L)**	**Ln (K)**	**(lnL-lnK)²**	**T**	**R^2**	**Adj. R^2**	**F-Statistic**	**D.F.**
226	-5.88*	0.461@	-0.39@	-0.13@	0.045***	0.7948	0.8857	33.9	35
	(0.186)	(0.671)	(0.825)	(0.112)	(0.026)				
270	-6.81*	0.093@	0.289*	0.02@	0.026@	0.8804	0.981	64.425	35
	(0.17)	(0.315)	(0.108)	(0.083)	(0.026)				
271	-6.97*	0.902@	-0.36@	1.026@	-0.03@	0.7514	0.8373	26.453	35
	(0.21)	(0.668)	(0.357)	(0.623)	(0.057)				
272	-5.44*	0.017@	0.324@	0.173@	0.016@	0.8468	0.9436	48.365	35
	(0.242)	(0.38)	(0.222)	(0.476)	(0.028)				
273	-4.36*	0.206@	-0.09@	0.104@	0.011@	0.6671	0.7434	17.536	35
	(0.268)	(0.505)	(0.413)	(0.208)	(0.025)				
274	-5.33*	0.62***	-0.28@	-0.21@	0.015@	0.7604	0.8473	27.766	35
	(0.208)	(0.314)	(0.304)	(0.139)	(0.014)				
275	-3.03*	-0.01@	0.103@	-0.5@	0.006@	0.1464	0.1631	1.5008	35
	(0.459)	(0.535)	(0.644)	(0.664)	(0.022)				
276	-7.12*	0.486@	-0.04@	-0.26@	0.035@	0.8946	0.9969	74.283	35
	(0.198)	(0.623)	(0.468)	(0.221)	(0.044)				
279	-6.23*	0.275@	0.468**	-0.05@	-0.04@	0.749	0.8346	26.109	35
	(0.53)	(0.374)	(0.226)	(0.221)	(0.051)				
280	-4.96*	1.215@	-0.79@	0.161@	-0.04@	0.565	0.6296	11.365	35
	(0.284)	(1.366)	(1.115)	(0.577)	(0.103)				
281	-4.56*	-0.16@	0.383@	-0.26@	0.009@	0.8147	0.9078	38.466	35
	(0.19)	(0.475)	(0.273)	(0.424)	(0.043)				
282	-5.55*	0.457@	-0.05@	-0.24@	-0.02@	0.4032	0.4493	5.9115	35
	(0.363)	(0.361)	(0.314)	(0.315)	(0.017)				
316	-2.44**	-1.83@	1.947@	-0.64@	0.05@	0.6498	0.724	16.233	35
	(0.631)	(3.458)	(3.719)	(0.954)	(0.119)				
378	-4.0*	0.307@	-0.24@	-0.08@	0.022@	0.5008	0.558	8.7766	35
	(0.486)	(0.212)	(0.158)	(0.063)	(0.03)				
385	-3.89*	0.162@	-0.09@	-0.06@	0.053@	0.8794	0.9799	63.78	35
	(0.264)	(0.661)	(0.57)	(0.418)	(0.031)				
386	-7.33*	-0.84@	1.832@	0.284@	-0.02@	0.2993	0.3335	3.7372	35
	(1.519)	(2.052)	(2.596)	(1.309)	(0.061)				
387	-3.15*	0.407@	-0.36@	-0.24@	0.044@	0.9498	1.0583	165.48	35
	(0.22)	(0.722)	(0.297)	(0.391)	(0.074)				
Total	-1.79***	1.227@	-1.34***	0.905*	0.095@	0.8902	0.9919	70.904	35
WFBI	(0.087)	(0.8)	(0.792)	(0.323)	(0.039)				

(Figures within brackets are the standard error)

**Significant at 1% level of significance*

*** Significant at 5% level of significance*

**** Significant at 10% level of significance*

@ Insignificant

Table (6.6.9): Estimates of V.E.S. Production Function for All India

Dependent Variable: Ln (V/L) Model: I							
Industry Code	Const.	Ln (W)	Ln (K/L)	R^2	Adjusted R^2	F-Statistic	D.F.
226	-1.121* (0.1793)	0.6361* (0.1363)	0.4763* (0.1645)	0.7819	0.8242	66.341	37
270	-2.185* (0.1505)	0.5268* (0.0709)	0.2888* (0.0878)	0.9384	0.9892	282.04	37
271	-2.05* (0.2022)	0.4032* (0.0872)	0.1959@ (0.2093)	0.7383	0.7782	52.191	37
272	-1.912* (0.2347)	0.5732* (0.0644)	0.4019** (0.169)	0.94	0.9908	289.63	37
273	-3.126* (0.2585)	0.1917@ (0.166)	-0.246@ (0.2722)	0.6545	0.6898	35.041	37
274	-1.691* (0.1752)	0.5293* (0.0911)	0.2994** (0.1236)	0.843	0.8886	99.357	37
275	-1.46* (0.4197)	0.4413** (0.2183)	0.8528* (0.388)	0.2271	0.2394	5.4364	37
276	-1.997* (0.2057)	0.6594* (0.1282)	0.415*** (0.233)	0.9256	0.9756	230.08	37
279	-1.767* (0.4738)	0.6565* (0.1281)	0.4039* (0.116)	0.8039	0.8474	75.851	37
280	-1.209* (0.2627)	0.2159*** (0.1332)	-0.337@ (0.283)	0.5987	0.631	27.597	37
281	-1.823* (0.19)	0.4201* (0.0789)	0.2687*** (0.153)	0.7682	0.8097	61.303	37
282	-2.742* (0.3768)	0.2408*** (0.14)	-0.172@ (0.168)	0.2091	0.2204	4.8907	37
316	0.0611@ (0.6997)	0.6267* (0.2272)	0.146@ (0.156)	0.3256	0.3432	8.9328	37
378	-1.862* (0.3442)	0.6207* (0.1138)	0.1168@ (0.089)	0.7998	0.843	73.913	37
385	-1.136* (0.2519)	0.6919* (0.1333)	0.0841@ (0.286)	0.9364	0.987	272.2	37
386	-1.167@ (1.4978)	0.6156@ (0.5483)	1.2003@ (0.727)	0.1984	0.2091	4.5777	37
387	-1.21* (0.2027)	0.5001* (0.0898)	-0.167@ (0.1982)	0.9214	0.9712	216.99	37
Total WFBI	-1.186* (0.069)	1.0767* (0.06)	0.9895* (0.157)	0.9368	0.9874	274.2	37

(Figures within brackets are the standard error)

**Significant at 1% level of significance*

*** Significant at 5% level of significance*

**** Significant at 10% level of significance*

@ Insignificant

Table (6.6.10): Estimates of V.E.S. Production Function for All India

Dependent Variable: Ln (V/L) Model: II								
Industry Code	Const.	Ln (W)	Ln (K/L)	T	R^2	Adjusted R^2	F-Statistic	D.F.
226	-1.73*	0.512@	0.459**	0.011@	0.7862	0.8517	44.117	36
	(0.183)	(0.312)	(0.172)	(0.025)				
270	-1.08*	0.778**	0.277*	-0.03@	0.9192	0.9958	136.58	36
	(0.152)	(0.315)	(0.09)	(0.035)				
271	1.806*	1.428**	0.221@	-0.09***	0.8132	0.881	52.237	36
	(0.191)	(0.54)	(0.198)	(0.049)				
272	-2.11*	0.523***	0.408**	0.006***	0.9032	0.9785	111.97	36
	(0.24)	(0.3)	(0.177)	(0.033)				
273	-3.57*	0.079@	-0.26@	0.011@	0.6562	0.7109	22.905	36
	(0.264)	(0.475)	(0.284)	(0.045)				
274	-0.11@	0.908*	0.221**	-0.05*	0.9115	0.9875	123.62	36
	(0.151)	(0.149)	(0.109)	(0.016)				
275	1.876*	1.414*	0.971*	-0.08**	0.472	0.5114	10.728	36
	(0.372)	(0.416)	(0.347)	(0.031)				
276	-4.78*	-0.06@	0.474***	0.085*	0.9064	0.9819	116.22	36
	(0.207)	(0.865)	(0.245)	(0.101)				
279	2.634*	1.764*	0.287*	-0.12@	0.9206	0.9973	139.19	36
	(0.401)	(0.372)	(0.106)	(0.038)				
280	7.878*	3.265*	0.05@	-0.29@	0.9101	0.9859	121.47	36
	(0.147)	(0.441)	(0.168)	(0.042)				
281	-3.67*	-0.12@	0.242@	0.05@	0.789	0.8548	44.875	36
	(0.19)	(0.554)	(0.156)	(0.051)				
282	1.194	1.4*	-0.21@	-0.09**	0.4762	0.5159	10.91	36
	(0.342)	(0.497)	(0.154)	(0.037)				
316	-2.6*	-0.35@	0.163@	0.098*	0.3416	0.3701	6.2266	36
	(0.719)	(1.617)	(0.161)	(0.16)				
378	-0.54***	0.915*	0.122	-0.04**	0.8736	0.9464	82.966	36
	(0.315)	(0.166)	(0.082)	(0.018)				
385	-2.15*	0.406@	0.047@	0.026@	0.9038	0.9791	112.7	36
	(0.257)	(0.906)	(0.314)	(0.08)				
386	3.786*	2.381***	1.49**	-0.16@	0.3664	0.397	6.9407	36
	(1.447)	(1.222)	(0.726)	(0.1)				
387	1.325*	1.207***	-0.09@	-0.08@	0.9091	0.9849	120.03	36
	(0.202)	(0.652)	(0.209)	(0.071)				
Total	-0.5**	1.286*	0.995*	-0.02@	0.9096	0.9854	120.79	36
WFBI	(0.07)	(0.273)	(0.159)	(0.028)				

(Figures within brackets are the standard error)

**Significant at 1% level of significance*

*** Significant at 5% level of significance*

**** Significant at 10% level of significance*

@ Insignificant

References

Arrow, K., Chenery, H., Minhas, B. & Solow, R. (1961): 'Capital-labor substitution and economic efficiency', *Review of Economics and Statistics* 43(3), 225–250.

Banerji, A., -"Capital Intensity and Productivity in Indian Industry", Macmillon, Delhi, 1975.

Bhasin, V. K. & Seth, V. K., -"Estimation of Production Function for Indian Manufacturing Industries", *Indian Journal IR*; 15(3), Jan. 1980.

Bhasin V.K. and V.K. Seth – "Estimation of Production Function for Indian Manufacturing Sector: Pooling of Cross-section and Time-series Data, *Artha Vijnana*, 19(2), June, 113-28, 1977.

Bhasin V.K. and V.K. Seth – "Estimation of Production Function for Indian Manufacturing Industries", *Indian Journal of Industrial Relations*, 15(3), Jan., 395-409, 1980.

Bramananda, P. R. –"Productivity in the Indian Economy: Rising Inputs for Falling Output.", Himalaya Publishing House, Bombay, 1982.

Cobb, C.W. & P.H. Douglas (1928), "A Theory of Production", in American Economic Review, 18: 139-165. Denison, E. (1985), "Trends in American Economic Growth", Washington: Brookings Institution.

Christiansen L.R., Jorgensen D.W, Lau L.J. (1971): "Conjugate duality and the transcendental logarithmic production function", in *Econometrica*, vol. 39

Christiansen L.R., Jorgensen D.W, Lau L.J. (1973): "Transcendental logarithmic production frontier", in *Review of Economics and Statistics*, vol. 55.

Datta, M. M. –"The Production Function for Indian Manufacturing", Shankhya: 15 (4), 1955.

Datta, Majumdar, D., -"Productivity of Labour and Capital in Indian Manufacturing during 1951-1961", *Arthaniti*, 1966.

Dhillon, S. S., - "Productivity Trends and Factor Substitutability in Manufacturing Sector of Karnataka", *Margin*: 16 (1); Oct., 1983.

Dhillon, S.S., - "Productivity Trends & Factor Substitutability in Manufacturing Sector in Karnataka", *Margin*: 16 (1), Oct. 1983.

Dholakia, R.H., - "An Inter-State Analysis of Capital and Output in the Registered Manufacturing Sector", *Indian Journal of Industrial Relations*: 15 (1), 1979.

Dholkia, B. H., - "Measurement of Capital Input and Estimation of Time Series Production Function in India Manufacturing", *Indian Economic Journal*; 24(3): pp. 333-355, Jan-Mar. 1977.

D.M. Wu (1975): "Estimation of the Cobb-Douglas Production Function", *Econometrica*, Vol. 43, No. 4, p.-739-744.

Dutt, M.M. – "Production Function for Indian Manufacturers", *Sankhya*, 15(pt. 4), 1955.

Ferguson, C.F. -"Time-Series Production Functions and Technological Progress in American Manufacture in Industry", *Journal of Political Economy* Vol. 72-3, pp. 135, 1964-64.

Goldar, B.N., - "Unit Size and Economic Efficiency in Small Scale Washing Soap Industry in India", *Artha Vijnana;* 27(1): pp. 21-40, Mar. 1985.

Goldar, B. N., - "Productivity Growth in Indian Industry", Allied Publishers, New Delhi 1986, p. 15.

Goldar, B. N., -"Employment Growth in Indian Industry", *Indian Journal of Industrial Relations,* 22(3), Jan., 271-85, 1987.

Goldar, B. N., -"Productivity Trends in Indian Manufacturing Industry: 1951-1978", *Indian Economic Review.*: 18(1): Jan-June 1983; pp. 77.

Goldfeld S. M. & Quandt R. E. (1970): "The Estimation of Cobb-Douglas Type Functions with Multiplicative and Additive," International Economic Review, Vol. 11, No. 2, pp. 251-257.

Goldfield S. M. & Quandt R. E. (1976): "Nonlinear Methods of Econometrics", North-Holland Publication Company, Amsterdam, New York.

Hasim, S. R. & Dadi, S. S., -"Capital-Output Relations in Indian Manufacturing (1946-1964)", M. S. University of Baroda, 1973.

Hilderbrand, G.H. and T.C. Liu- "Manufacturing Production Functions in the U. S., 1957: An Inter Industry and Interstate comparison of Productivity, Cornell University Press, New York, 1965.

Jogenson, D. W. & Griliches, z. – "The Explanation of Productivity Growth", *Review of Economic & Studies.,* 34., July, 1967.

Kmenta, J. – "On Estimation of CES Production Function", *International Economic Review,* 8(3), June, 180-9, 1967.

Kendrick, J. W. – "Productivity Trends Capital and Labour.", *Review of Economic & Statistics,* Vol. 38 Aug. 1956, pp. 248-57.

Kendrick, J. W. –"Post-War Productivity Trends in the United States, 1948-1969", *National Beauro of Economic Research,* New York, 1973.

Kendrick, J. W. –"Productivity Trends in the United States", Princton University Press for National Beauro of Economic Research, 1961.

Kumar, V., - "Productivity Trends in Selected Industries in India.", *Indian Labour Journal,* 24 (1); Jan. 1983.

Lu, Y and L. Fletcher – "A Generalisation of CES Production Function", *Review of Economic & Statistics* 1968.

Metha, S. S., -"Productivity, Production function and Technical change: A Survey of Some India Industries.", Concept Publishing company, New Delhi, 1980.

Mukerji, G. P., - "On Some Indices of Productivity". *Indian Economic Journal*: 10 (1) p. 143. July 1962.

Murty, V. N. & Sastry, V. K., -"Production Function for Indian Industry", *Econometrica*: 25(21), pp. 205-21, April, 1957.

Narsimham, G. V. L. and M. Z.,Fabrycy -"Relative Efficiencies of Organised Industries in India; 1949-58", *The Journal of Development Studies,* 1974.

Rajalakshmi, K., - "Sources of Growth in the Total Manufacturing Sector in Rajasthan and at India Level", *Indian Journal of Industrial Relations*: 19 (1), July, 1983.

Sankar, U., -"Elasticities of Substitution and Returns to Scale in Indian Manufacturing Industries", *International Economic Review.*; 11(3), Oct. 1970.

Singh, L., - "Productivity Trends and Factor Substitutability in Punjab Industry", *Indian Journal of Economics*: 67 (266), Jan. 1987.

Subramaiyan, G., -"Measurement of Productivity and Production function in Indian Sugar Industry: A Regional Analysis," Indian Journal of Regional Science: 11(2), 1979.

Solow, R. W. (1957): "Technical Change and Aggregate Production Function", *Rev. Econ. Stat.*, Vol. 39: pp.312 – 320.

Tinbergen, J. (1942): "Zur Theorie des Langfristigen Wirtschaftsentwicklung", reprinted in Tinbergen, J. (1959) Selected Papers, Amsterdam. Zaman, G., Goschin, Z., Pârþachi, I and C. Herþeliu (2007), "The Contribution of Labour and Capital to Romania's and Moldova's Economic Growth", *Journal of Applied Quantitative Methods*, 2(1): 176-185.

Venkataswami, T. S., -"Production Function an Technical Change in Indian Manufacturing Industries", *Indian Journal of Economics*; LV(219): pp 435-64, April 1975.

Yeong-Yeh –"Economics of Scale for Indian Manufacturing Industries", *Indian Economic Journal*; 14(2): pp. 275-91. 1966.

7

Summary and Conclusion

Historically, forests have played a major role to influence patterns of economic development of an economy, supporting livelihoods not only for human but also for the habitat living in and around the forest, helping structure economic change, and promoting sustainable growth. For millennia before the industrial revolution, forests, woodland, and trees were the source of land for cultivation and settlement, of construction materials, of fuel and energy, and indeed of food and nutrition as well. Historical evidence also indicates that the early societies comprising mostly hunting-gathering households depended on forests for nearly all of their livelihood needs. The agricultural transition translated into deforestation for cultivable land at a scale that was limited only by available technologies, by the labor available to cultivate land and grow crops and by the presence or absence of markets. Forest provided fodder, firewood, and subsistence timber–goods for which they are still the major source for most poor households in the developing world. The extended use and exploitation of forest resources even before the industrial revolution had led to efforts to conserve forested areas and plant new trees in specific regions of the world. In Europe, France and Germany were leaders in developing policies in the 17th and 18th centuries to regulate the use of and to protect forests.

Since last sixty years, changing views about the importance of forests and their relationship to humans have been associated with corresponding changes in emphases in how forests are managed. Earlier efforts to manage forests for sustained yields of commercial timber were in many countries replaced in the 1970s and 1980s by the watchwords of sustainable forest management and ecosystem management as the multi-functional nature of forests and woodlands became obvious. Increasing levels of deforestation despite efforts to reduce and reverse loss forests led many governments and international agencies to attempt policy experiments that would have a positive influence on net deforestation. So at last, for this introductory section of the final drafting of UGC sponsored major research project, it can be said that Forests perform a number of productive, protective and regulative functions. They are important components of the economic and ecological systems. The role of forestry is to regulate the hydrological and atmospheric cycles, to provide a habitat for biodiversity and to prevent soil erosion, siltation and flooding. The economic returns are derived from the large variety of forest products. Forests also provide recreational, aesthetic and cultural functions and are an important source of livelihood for local communities.

A *forest dependent industry* or *forest based industries* is any industry that uses raw materials from the forested portion of the ecosystem, including amenities and services, such as oxygen, water, electricity, recreation, and migratory animals. A forest based industry also includes any industry that uses extractive good such as minerals wood fiber, forage for livestock, resident fish and game animals, and pelts from fur-bearing mammals. Forest dependent industries based on amenities and services are extractive if the products either enter and/or leave the forest under their own volition. Such industries includes both the sport and commercial fisher who catches migratory salmon and steel hood, the farmer who uses water to irrigate crops, the person who markets those crops, the electrical company that uses water converted to electricity and the municipal water company itself.

Forest based industries that are based on extractive products, on the other hand, physically remove raw materials from the forest and, for the most part, make them available for further refinement and use that goes beyond the initial extraction. Such industries include timber companies that cut trees, ranchers who graze their livestock in forested allotments, miners who extract ore, and hunters, fishers and trappers who kill and remove forest-dwelling wildlife. Forest dependent industries that refine the extracted products include jewelers, Carpenters, boat builders, artisan woodworkers and furriers. Finally, these forest-based industries are all interwoven because each industry uses one or more of the other's produced, such as water, electricity, wood fiber, red meat, vegetables, and so on.

Wood and forest-based industries are essentially those industries which depend on raw materials drawn from forests; save for some exceptional cases. Development of forest based industries is the story of development of civilization. Today our access to major forest products is through industry only. Special features of forest-based industries become active when they furnish a very wide range of products, both consumable goods and intermediate goods flowing into many sectors of the economy. In the context of country like India, the forest-based industries do play a significant role in the economic development, putting forest products to more meaningful use. Therefore, *wood & forest-based* Industry contributes significantly to a nation's economic development for following reasons: (1) Wood & forest based industries are a primary method of transforming raw –material of forests and/or products into finished products for various uses of human life, (2) Wood and forest based industries often constitute the majority of a developing nation's manufacturing sector, (3) Wood and Forest based industrial products are the major exports from a developing nation; and (4) The Wood and forest based industries provide diverse products, employment and income.

India is a densely populated country with high pressure on land and forest resources. The total geographic area of the country is 32,87,240 km^2 (327.8 million ha) and the population is 1,210 million (Census 2011) which gives population density of 382 persons per km^2 and 0.06 ha. per capita forest. It is the 7th largest country of the World's geographic area, and 1.8 per cent of its forests, but supports almost 18 percent of its population. At present, India's population is almost equal to be combined population of USA, Brazil, Indonesia, Pakistan, Bangladesh and Japan. Comparing with USA, India's geographic area is almost one-third of the area of USA whereas the population is 4 times, thus population density is 12 times of that of USA. As per Census 2011, about 68.8 percent of India's population (833 million) lives in rural areas, and most of them have land based economy which uses forest resources one way or the other. It is estimated that about 200 million people live in and around forests, and fully depend for their livelihood on forest resources. In addition, the livestock population of India is about 530 million as per the 18th Livestock Census, 2007 (MoA 2010) and has registered a growth of 9.2 per cent during 2003 to 2007. About 38% of livestock depend on fodder derived from forests by direct grazing or by harvesting, causing additional burden on the forests, and is responsible for forest degradation. With only 2.4% of the land area, India accounts for 7 to 8 percent of the recorded species of the world (Gokhale, 2010). This biodiversity is of immense economic, ecological, social, and cultural value. Approximately 275 million people in India (27% of the total population) are known to live in the forest fringes and earn bulk of their livelihood from forests (World Bank, 2001; 2006; Poffenberger, 2000, Sinha *et al.*, 2010).

Economic development necessarily depends upon productivity of the factors of production. Productivity is the result of a large number of social and managerial factors. In the most general sense, productivity usually denotes the possession or use of power to create or to make. To many, it also can note the quality or state of being productive. Such knowledge, as possession of power to bring forth goods and services; or the quality or state of being productive, do not precisely explain what productivity in economic terms actually means. Productivity is a key and major factor in the victory of any socio-economic system because of its direct relationship with economic welfare and also to achieve various goals of economic growth. The concept of productivity has come into greater prominence during the recent years and has assumed great importance and significance in the context of industrial development. If the same production units can be used more efficiently, the net addition to the total national product will be much higher which will result in the process of industrial growth.

Productivity increase is, thus, an indispensable and powerful stimulus to and the end-result of a complex socio-economic process of economic development. There would be no gain saying the fact that mere possession and magnitude of rich diversified endowment of natural resources and technical progress, automation and complex machanisation of production, are by no means the sole factors exerting a decisive influence on productivity growth or on the progress of a nation. Productivity cannot, however, be increased in any country or under any social system by simple decree.

Thus the economic development of any nation depends upon two distinct but related factors. The Efficiency and competitive capacity of industry or the quality of the goods produced, these can only be achieved by introducing innovations and technological changes as soon as they become available, by making the best possible available resources such as equipment's, materials, and man-power, by the introduction of productivity services. In case of productivity analysis, a similar interest is discernible in various countries to gauge the relative performance of individual industrial enterprises. To be sure, the productivity studies are still in the embryonic stage. In earlier accounts of the history of economic ideas, the term "productivity" was hardly used; instead the classical economist spoke of 'production' and the "rate of production', which essentially meant more or less what is now referred to as 'productivity'. The interest of economists in productivity growth is venerable. John Stuart Mill, like Karl Marx, was a growth theorist. Alfred Marshall was much interested in long-run economic change. Since the mid 1950s, considerable research has proceeded closely guided by the neoclassical formulation. Some of this work has been theoretical. Various forms of the production function have been invented. Models have been developed which assume that technological advance must be embodied in new capital. Technological advance has been made indigenous to the theory through linking it to and increasing R & D capital stock.

'Productivity' may be defined as the ratio between the productions of a given commodity measured by volume, and one or more of the corresponding inputs factors also measured by volume. It is evident that this concept includes many different kinds of productivity according to whether one or another of several different factors is considered. It is the measurement of the economic soundness of the means. It is the combine effect of a number of factors like scientific management, technological development, proper allocation and scientific utilisation of available resources, human or labour incentive etc

In terms of economics, thus 'productivity' can be expressed as the ratio of input in the form of production resources to output in form of production of goods and services obtained from consumption of such resources. Industrial productivity, in the aforesaid broad and fundamental economic sense, has also come to mean a ratio between output of wealth and the input of resources used up in the process of production. In a modern manufacturing industry, the effectiveness of the expenditure of all other resources are also equally important, because the concept of productive efficiency is a function of many independent and inter-dependent variables.

Thus shorn of all technical jargon, the concept of productivity implies the rate of proper and efficient utilization of power of all the associated factors of production. The object is to minimize losses and waste of every type and in every sphere, so that goods and services of kind and quality, most wanted by consumers, may be produced at lower and lower costs. If productivity is increasing in an economy, it means that its factors of production and commodity inputs are manifesting increases in their output efficiency. The notion of higher productivity is, therefore, consequently related to the optimum utilization of the inputs of all the available potential production factors so as to increase output, both qualitatively and quantitatively. If only one production factor is required in a production process, the productivity is usually called simple productivity. 'Single Factor' or Partial Factor Productivity: When a number of factors are involved in a production process, but the output is related to any single factor unit, productivity, thus measured, is called single factor or partial factor productivity, such as labour productivity, capital productivity, raw- materials productivity, machine productivity, capital intensity or productivity of organization, etc. Thus, ratio of any factor of production to output is known as partial factor productivity.

Composite or Overall or Total Factor Productivity: It is an indicator of the change in efficiency in factor use in estimated simultaneously with the returns to scale which is another parameter which has a bearing on efficiency. In fact, total productivity measures only that component of technological change, which increases the overall efficiency of factor. Thus, when the output is related to the entire input complex (labour, capital, raw- materials, power, organization, etc. combine together), the relation between output and input is multifactor (or overall, or composite, or total) productivity.

The economists have always recognized the central role that technological change plays in economic development. The importance attached to the problems of economic growth has varied considerably during the history of economic thought. According to David Ricardo the most productive areas of land would be occupied first and the quality of the land to be occupied subsequently would gradually decline. As a consequence diminishing marginal productivity would be encountered in expanding the land under cultivation until no further increase in output could be obtained.

Like Marx, Marshall also was very clear as far as the importance of technical progress is concerned. In his 'Principles of Economics' he conceived the importance of 'knowledge' as the chief engine of economic progress. In the same way, Schumpeter argued that the fundamental impulse that sets and keeps the capitalist engine in motion comes from the new consumers' goods, the new methods of production or transportation, the new market, the new form of industrial organization that capitalist enterprises creates. The problem of economic growth occupied a central place in the thinking of classical economist although it was very different from those it was analyzed in ways, which are very different from those prevalent now days. Furthermore, technical change, though not thoroughly formalized, played an important part in their thinking.

Technical change or technological change or technological progress is a shorthand expression for any kind of upward shift in the production function indicating improvements. It is the pool of knowledge relating to art of production and signifies creation of new set of production alternatives. Due to technological progress, productivity rises and the produced output is generally in the shape of new products rather than old products. We may identify it by one of the following three ways: (a) Output increases for the same inputs. (b) Same output is realised with less input. (c) Output may not rise but quality and durability of good as well as working condition improve.

Technological change has been defined directly or indirectly in terms of its effects on productivities of inputs in terms of Q/L and Q/K ratios, when technological progress takes place, both the ratios may increase and one of them relatively rises. Mansfield offers its direct definition as follows: Technological change is the advance of technology, such advance often taking the form of new methods of producing existing products, new designs which enable the production of products with important new characteristics and new techniques of organization, marketing and management.

Testing empirically, with the tool of production, various economists found technological change to be the most important source of output growth. But what greatly proved to be decisive were the findings of Fabricant, Abramowitz and Solow. They find out that 80 to 90 percent of output growth per head in the U. S. economy could not be accounted for by increase in capital per head, thereby, giving this credit to technological change. Many other studies have also reached a similar conclusion.

However, it would be wrong to ascribe this large contribution to technological change alone, as there are some other factors also contribution to growth. These are technological progress in the narrow sense, economies of scale, external economies, improvement in health, education and the skill of the labour force, better management and the movement of labour force form low productivity to high productivity sectors. This is the reason why the names given to this group have ranged from output per unit of input, efficiency index, total factor productivity, change in productive efficiency, technical change, and the measure of our ignorance. Attempts were made, therefore, in the subsequent researches to firstly, disaggregate the residual factor measuring the factor inputs in conventional manner; and secondly, labour and capital inputs were adjusted for changes in quality and composition so that much more measured could be attributed to increase in factor inputs.

Earlier, in the works of Solow, Abramovitz and others, the technological change was taken as a unexplained residual meaning, thereby, that it is *disembodied*. But, further researches led by Solow attempted to search for a solution were technological change is embodied in the latest vintage of capital. Analogously, technological change can be said to be embodied in the latest vintage of labour. New–Classical economists viewed technological progress as exogenous variables. But it was subsequently that 'technical progress' does not occur by accident but through deliberate diversion of resources to activities, which generate progress in pursuit of fame and profit, or both. In the modern growth models, Kaldor and Mirless followed by others have tried to incorporate technological change as an endogenous variable into the economic system.

Generally, while discussing technological change, it is assumed that there is an aggregate economic system and are only two types of inputs (L & K) which are homogeneous within themselves. Analogous to the concept of unitary elasticity of demand, economists make use of the notion of "neutrality" to serve as a criterion to measure bias of technological progress. Neutral technical change means, those innovations which or neither capital neither saving nor labour saving i. e. they are neutral in their effect in the use of the two factors at the margin. It is posited that the firm can choose between 'neutral' technological changes and 'non - neutral' or 'biased' technological changes. However, the interest in classifying technological change as labour saving, capital saving and neutral stems historically from a concern for its effect upon the distribution of income.

The Hicks classification is biased upon the comparison of points at which the capital-labour ratio is constant. Technical progress is said to be Hicks-neutral if, at any constant value of the capital-labour ratio, the ratio of relative shares remains constant. Technical progress is said to be labour saving in Hick's sense if, any constant value of the capital-labour ratio, the ratio of relative shares is increasing. Hicks classification of technological change is based on short-run where input supply is fixed, if we concentrate on the immediate effects of technological change, but Harrod classification of technological change is related to long run, where supplies are flexible in the long run impact of technical change.

Among the types of technical progress that we have obtained in this way are several, which are well known in the literature: Hicks, Harrod, Solow, and factor augmenting technical progress. We have defined these concepts of bias for disembodied technical change. Consider now that technical change is embodied. At the aggregate level, technological change may be embodied or disembodied. Generally, embodied and endogenous technological change goes together. The new machines produced are superior which incorporate innovations and various improvements and new internal (endogenous) to the economic system. Arrow's conception of learning by doing and Kaldor's technological progress functions are the classic examples of endogenous – embodied technological change. Embodied technical change refers to the introduction of new technology in the physical capital inputs

The changes in technology itself are brought about by the combination of activities of research, invention (or knowledge creating activities), development and innovations (or the activities applying knowledge to production process), and the spread of technology depends upon the rate of its adoption and diffusion. Research and development spending may be considered as the monetary equivalent of inputs to the process producing technological advances. R & D spending is an input to both invention and innovation. However, to the extent that R & D tends to be directed more towards development than research, we might argue that an analysis of R & D is concerned more with innovation than with invention.

The impact of R&D activity on growth is difficult to measure; many times falling to create anything new but when successful, spectacular result has been achieved. The development of organised R&D in the modern firm has brought about a dramatic change in the capacity of industry to generate a stream of new and modified products, and to use variations in this capacity as a competitive weapon between firms and between national economies. The institutionalization of R&D has changed but it has not solved the problem of how to innovate. Research and development and technological innovations continuously create new knowledge and uncertainty. Schumpeter defined three phases in the process of technological change: invention, innovation and diffusion.

Schumpeter made a sharp distinction between invention, innovation and imitation, with the greater emphasis being laid on the innovation and the charismatic figure of the entrepreneur who against all odds takes the bold step to innovate. Technological innovation thus, is a vital component in the performance of firms and of other organisations. That is not to say that the only effect of innovation is to make firms (or industry) perform better. The greater efficiency that firms achieve as a consequence of various types of innovation is only an intermediate step towards the achievement of economic growth and economic development. The innovation process is sometime referred to as additions to the economy's book of blue prints; however to confuse matters further, this blueprint terminology is sometimes used to refer to invention, and then only certain blueprints are selected for development to the innovation stage.

The relationship between technological change and economic development is intimate and complex. Many contemporary events have contributed to this complexity. In fact, economic development is a way to bridge the technological gap between industrialized and underdeveloped countries. Technological choice and adoption are important aspects of economic engineering. Each factor has gained so much specialization that one unit of ant category of these factors is not easily substitutible for the other. Technology as a factor of production has made such an important contribution that the emerging industrial estate is radically different from the old one.

Technological invention is different form technological innovations. The former is a scientific fact, whereas the latter is the result of entrepreneurship of entrepreneurial ingenuity. The former is abundantly available in industrialized countries; it incubates in research laboratories and installations of those countries. In developing countries, the task has become complicated due to three factors, viz., the problems relating to transfer of technology, the mixed nature of the economy, and concentration of technological knowledge mostly in advanced countries.

The transfer of technology from industrialized countries to developing areas is beset with many technical, economic educational, sociological and political problems. Having imported technology, its duplication, extension and development pose another type of problem. The establishment of specialized research centres in industrialised countries is no solution of these problems. Encouragement given by the planning authorities in a developing country could provide resources for the establishment of research installations, but the main source of inventions and innovations does not follow merely from such establishments. The history of the industrial revolution in the west shows that the process of rapid technological transformation has not been merely the result of new improved technical processes.

The review of literature serves as a background for any research and scientific investigation which helps in understanding it in proper perspective. An exhaustive review of literature help any researcher to identify variables relevant for research, decide tools and techniques to be adopted, delineate a new area of study and relate the present study with the previous ones. Productivity and Technological change assumes a greater significance for developing nations like India, which are severely constrained by shortage of capital still in the era of 21st century. It therefore, become more imperative for a nation to make use of this scarce resource which a greater efficiency and in such a manner which could ensure maximum possible exploitation of existing factor endowments. A significant of literature is available, thus, an attempt has been made to review the literature available on the subject. Keeping this in view, all the available literatures concerning the problem is being reviewed under the following heads/sub-heads: (I) General Theoretical Frame related to present study; (II) Overview of India's Manufacturing Sector; (III) Methodological Issues related to Productivity, Production Functions & Technological Change aspects: and (IV) Wood & Forest based related aspects. During the time of review of literature, it has been observed that the existing literature on Wood & Forest based Industries sources at the India's provincial/ regional/state level is very poor and it does not provide sufficient understanding of the subject into the consideration. Very few studies attempt to investigate productivity growth, technological change and production functions in general or specific to WFBI sectors of the Indian economy.

The development of any area of economy has always been influenced by a large number of factors operating in that economy separately or altogether. The choices of these factors operating may be tangible, while some other might be intangible. But past studies observed that the choice of these factors has been subjected to data to data availability. Keeping these views in mind, and hence to understand and measure the contribution of individual inputs, and the factor other than the inputs like technical change or batter utilization of existing resources, it has becomes essential to procure data on output and inputs particularly those concerning capital and labour, the data on wages paid to the labour is also needed, besides appropriate price deflators to be used for making the time series data comparable. Therefore, the main purpose of this study was devoted to the productivity trends and statistical estimation of production functions and technical change in the wood and forest-based industries for Indian states and India as a whole. With the overall objective of the proposed study was to analyze the growth of technological change and productivity growth in wood and forest-based industries of Indian states as well as India as a whole.

The data for the purpose of the present study was related to period 1973-74 to 2012-13 (last year of study depends upon the availability of the data till aforesaid year). Per 1973-74 has been not considered because of the introduction of National Industrial Classification (NIC) in 1973-74 and comparability with the later period. For the purpose of the proposed study, the regions and states which have been taken are shown in Annexure II. While the Wood and Forest–based Industries which has been selected from the Annual Survey of Industries classification, C.S.O., New Delhi.

The ASI definition of variables which is used in the present study are as: Fixed Capital; Working Capital; Worker: Persons other than Workers; Total Employees; Total Emoluments; Total Inputs; Depreciation; Value Added; Fuels & Materials; Apart from the above variables, the data on Price indices for deflation purposes were taken from various sources. As a measure of output, a choice arises between gross output and value added. However, value added figure is generally preferred, as it is felt that gross output varies widely with the changes in the stages of productive process of an industry. Further, for making comparison between firms, industries and even countries, value added figure is found to be more meaningful.

Once the concept of value added is finally selected the question of making choice between net value added (NVA) and gross value added (GVA) arises. In case of productivity analysis, the choice boils down to gross value added as a measure of output. Generally gross value added at current prices in any industry is defined as the value of the output (at current prices) minus the value of the raw materials (at current prices). This is the value of service produced within the industry, by the factors of production, in processing the raw materials. In this study gross value added is obtained by combining the figures of net value added and depreciation as reported in the data sources.

Of the entire variable in the study the most difficult to define conceptually as well as measure in the "capital input "capital play an important role in production and distribution brings it at a centre-stage of every sphere of economic activity. Gross fixed capital stock at constant prices has been taken in the present study as a measure of capital input. However, what is reported in ASI represents the depreciated book value of capital assets. In other words, ASI fixed capital data for any particular year represents a simple aggregation of expenditure on capital assets, valued depreciation change. Therefore, for any meaningful economic analysis, these figures need to be adjusted simultaneously for both price change as well as arbitrary depreciation charges.

For estimation purposes of capital input series, we have made use of *Perpetual Inventory Method*. The perpetual inventory method, which is based on the relationship between the capital stock at a point of time and investments up to that point, has been used almost universally for this purpose. The price index number used for deflating the capital input series was the Price index of Machinery and Machine Tools of India.

One of the measure causes of technological change and also a source of future income is heterogeneity of labour. This, however, does not prevent our labour friends from merrily, aggregating man-hours among industries and overtime. However, the presence of a well-developed and reliable second hand market in case of labour makes problems less severe. Apart from these, there are other factors, which make measurement of labour input less complex in comparison to capital input. The problem of converting value into constant terms, as far as labour is converted, does not arise due to the existence of physical measures like employment and man-hours. Besides, the problem of depreciation need not be taken into account unless one thinks of efficiency deteriorating seriously with age.

For the measurement of labour input, there are three alternatives available: (i) non-hours worked; (ii) number of worker; and (iii) number of employees. In the present study total employees including both workers and persons other than workers have been taken as the measure of labour input, for the purpose of production function the labour input used was total emoluments to employees. The main reason for disfavoring the first two is that these measures do not take into account the services of persons other than workers who are as much important for getting the work done as the workers who operate the machines. A serious limitation of our measure of labour input is that it treats workers and persons other than workers as perfect substitutes. Also the chosen measure of labour input does not take into account the changing age, sex, skill and educational composition of the industrial labour force. Besides, these adjustments one said to embody labour and arbitrary elements of choice of them.

The present study has cover trends in SAGR, partial factor productivity ratios (SFPs) and total factor productivity (TFPG) indices, and estimate parameters of Cobb-Douglas, CES, and VES production function. In this study an attempt has made to provide an in-depth look at the trends in Productivity growth in various industry groups of manufacturing sector of India. This can be done in terms of the conventional measures of productivity such as partial or single factor productivity, but these measures have well-know limitations. In situations where capital intensity is increasing over time, partial productivity measures such as labour productivity may show an increase but this is more a reflection of rising capital-labour ratios rather than pure productivity increases. These problems can be resolved by analysing total factor productivity growth, which identifies the contribution an increase in output of influences other than increases in the factor inputs. TFP, which is an indicator of the change in efficiency in factor use, is estimated simultaneously with the returns to scale, which is another parameter, which has a bearing on efficiency. Total factor productivity growth encompasses the effect not only of technical progress but also of better utilisation of capacities, learning-by-doing improved skills of labour, etc. It is therefore a composite measure of technological change and changes in the efficiency with which known technology is applied to production. It certainly provides a useful supplement to the conventional analysis of the trend in partial productivity measures.

Thus, the measure of Single factor productivity used in this study as Labour productivity (LP), Capital productivity (CP), Capital intensity (IC), Raw-material productivity, Fuel Productivity (RMP), Enterprise productivity (FP), and Emoluments per worker (EW) while total factor productivity used in this study has been derived from a Kendrick, Solow and Divisia indices. Kendrick's index of TFP is based on a linear production function, which assumes infinite elasticity of substitution between the factors of production. But Solow assumes a more general neoclassical production where the elasticity of substitution need not be a constant, but technical change is of a Hicks-neutral type. Increase in the partial factor productivity ratio signifies efficiency in the use of particular input over time. Labour and capital productivities indicate efficiency in the use of these inputs, the changes occurring in these are generated by various economic forces interacting simultaneously, whereas, the capital-labour ratio gives an idea of capital deepening over the period and its relationship with labour and capital productivities. While, emoluments per worker gives an idea about the economic status, living standard of workers and their economic viability over the period of time. The purpose of measuring these ratios at constant price was to study the real change.

Partial productivity ratios are the reflectors of value added by input factors separately, that is to say, of the extent of the effective utilization of materials, labour capital, machines, etc. and the cause for their low or high performance. This is a simplest measure of productivity, ratio of any factor of production to output is known as partial factor productivity of that factor; in other words per unit output of any factor of production is known as partial factor productivity of that factor.

The more conventional measure of productivity is labour productivity and capital productivity. Since trends in the partial factor productivities are dominantly affected by the trends in factor intensity, i. e. the capital-labour ratio. If the capital-labour ratio were increasing over time, the analysis of partial productivity changes would overstate the increase in labpur productivity and understate the increase in capital productivity. The trends in the partial productivity indices with respect to individual factors include the effect of capital deepening in the context of capital accumulation in a growing economy. Thus, increases in labour productivity may be due to improved efficiency and technological progress as well as more capital per unit of labour, and the two sets of influences does not operate independently of each other. More specifically, labour productivity may increase because of factors such as learning-by-doing experience, improved skills, etc., and because there are better and more machines to work with. It can be also said that when capital intensity is increasing, the measured increase in partial labour productivity will be larger than the increase in the pure productivity component since part of the measured increase is due to the rising K/L ratio.

Similarly, when analyzing the change in capital productivity, the observed increase in capital productivity will understate the increase in the pure productivity of capital. It is, therefore, reasonable to argue that in a situation where capital intensity is increasing over time, the analysis of partial productivity changes would overstate the increase in labour productivity and understate the increase in capital productivity. In order to measure of single factor productivity, the various types of single factor productivity ratios have been calculated. Labour Productivity has been calculated as the ratio of GVA, at constant prices to the number of workers. Capital Productivity has been measured as ratio of product to capital, at constant price. Capital Intensity has been measured as the ratio of gross fixed capital stock at constant prices to the total number of employees.

Total factor productivity, which is an indicator of the change in efficiency in factor use, is estimated simultaneously with the returns to scale, which is another parameter, which has a bearing on efficiency. Total productivity in fact measures only that component of technological change, which increases the overall efficiency of factors. The total productivity often referred to as the "residual" or the index of "technical progress is defined as output per unit of labour and capital combined. The TFP indices are based on the assumptions of competitive equilibrium and constant returns to scale. Thus, TFP may be defined as the ratio of output to weighted combination of inputs. Various TFP indices suggested differ from one another with regard to the weighting scheme involved. Thus, there are many ways of measuring total factor productivity, but the two indices most often used in empirical research are Kendrick's arithmetic measure and Solow's geometric index. Thus, the Kendrick, Solow and Divisia index of total factor productivity has been used in the present study.

Kendrick index of total factor productivity is an arithmetic measure of rate of technological change, and is based on linear production function, which assumes infinite elasticity of substitution between the factors of production. The index is defined as the ratio of value added in production to a weighted average of the two factors of production. In this index two factors of production, labour and capital have used by Kendrick for the measurement purpose. He further assumed perfect competition in market and factors are paid according to their marginal product. Technological change is Hicks-neutral type and condition of constant returns to scale prevails. Thus, the Kendrick index may be interpreted as the ratio of actual output to the output, which would have resulted from increased inputs in the absence of technology change. Solow Index provides a geometric measure of rate of technological change, which is developed by the Solow. Solow assumes a more general neoclassical production function where the elasticity of substitution need not be a constant, but technological change is of a Hick-neutral type. Assuming that the factors are paid the value of their marginal products, the growth of total factor productivity is measured as the difference between the rate of growth of value added and the rate of growth of total factor inputs. Under the assumptions of competitive equilibrium and constant returns to scale the following equation is obtained. The need for using the Divisia index has been noted by Solow, Jorgenson, Griliches, Christensen, Cummings, and Gollop have used approximations to the Divisia index in their studies. Christensen, Jorgenson, Griliches and others have discussed the properties of the Divisia index, which makes its application highly desirable. It has been pointed out that the rate of growth of the Divisia index of prices and quantities add up to the rate of growth of the value and such indexes are symmetric in different directions of time. Divisia index also have the reproductive property that "a Divisia index of the components."

A production function is a complex analytical tool, which describe the maximum output that can be obtained from a given set of inputs in the existing state of technological knowledge. The production function expresses the technological or engineering relationship between various inputs and output. Output of an industry in general, depends on land, labour capital, organization and technical progress. By estimating the parameters of a production function, one can measure all the elements of technological change, such as, the efficiency of the technology, the capital intensity of technology, the elasticity of substitution and the economies of scale. However, economist concept of production function is much wider and deals with entire production activity of a firm or economy.

The most popular and widely used form of production functions are Cobb-Douglas, Constant Elasticity of Substitution, and Variable Elasticity of Substitution. These production function involve treatment of elasticity of substitution constant and equal to unity in C–D specification and constant but not necessarily equal to unity in case of C.E.S. specification. A major limitation of C.E.S. production function is that it assumes the elasticity of substitution to be constant for all input combinations. Several functional forms have been suggested in the literature and used in empirical studies in which elasticity of substitution is variable. These are called V.E.S. production function. In the specification of a production function, intermediate inputs can be explicitly incorporated by relating changes in the gross value of output to changes in labour, capital and intermediate inputs. Alternatively, the latter can be noted out by relating changes in value added to changes in labour and capital. When intermediate inputs are explicitly recognized, it implies that the estimated TFP growth rate is lower than that base on the value added function. In fact they are lower, exactly, by an amount proportional to the share of intermediate inputs in the gross value of output. The analysis in the present study is based on the value added production function.

For the purpose of analysis, The present study covers 20 industries group of wood & forest based industries and total as well; 16 States, and 5 regions along with All India as whole. The data used for the analytical purpose are gathered through secondary sources for the period 1973-74 to 2012-13. Thus, the validity and reliability of the results will depend upon the validity and reliability of the data. Moreover, the implications of the present study confined to coverage of the industries.

In the present study, Inter-state comparison of SFPs can be summarized on the basis of overall discussion of Table 5.1.1 to 5.1.16, and thus vast interpretation of these tables in terms of high and low growth of SFPs among the states for overall wood & forest based industries can be outlined as follows:

- High Labour Productivity (LP) observed for the state Delhi with 1.66% per annum.
- Low Labour Productivity (LP) observed for the state Assam with 1.25% per annum.
- High Capital Productivity (CP) observed for the state Andhra Pradesh with 5.6% per annum.
- Low Capital Productivity (CP) observed for the state Uttar Pradesh with 1.01% per annum.
- High Capital Intensity (IC) observed for the state Assam with 4.6% per annum.
- Low Capital Intensity (IC) observed for the state Uttar Pradesh with 0.86% per annum.
- High Raw-Material Productivity (RMP) observed for the state Assam with 1.17% per annum.
- Low Raw-Material Productivity (RMP) observed for the state Bihar with 0.78% per annum.
- High Fuel Productivity (FP) observed for the state Tamil Nadu with 1.67% per annum.
- Low Fuel Productivity (FP) observed for the state Assam with 1.14 % per annum.
- High Enterprise Productivity (EP) observed for the state Haryana with 6.56% per annum.
- Low Enterprise Productivity (EP) observed for the state Assam with 2.72% per annum.
- High Emoluments per worker (EW) observed for the state Delhi with 1.6% per annum.

- Low Emoluments per worker observed for the state Assam with 1.078% per annum.

On the other side, Overall discussion of Table 5.1.1 to 5.1.16 can be summarized as high and low growth of SFPs estimation among the industries within the selected states for various 3-digit minor group industries of wood & forest based industries are as follows:

- High Labour Productivity (LP) has been observed for industry code (282) of state Delhi and it was 2.57% per annum.
- Low Labour Productivity (LP) has been observed for industry code (316) of state Delhi and it was only 0.31% per annum.
- High Capital Productivity (CP) has been observed for industry code (273) of state Delhi and it was quite high 11.06% per annum.
- Low Capital Productivity (CP) has been observed for industry code (271) of state UP and it was 0.75% per annum.
- High Capital Intensity (IC) has been observed for industry code (385) of state Karnataka and it was 9.12% per annum.
- Low Capital Intensity (IC) has been observed for industry code (316) of state Punjab and it was very low only 0.37% per annum.
- High Raw-Material Productivity (RMP) has been observed for industry code (271) of state Delhi and it was 2.26% per annum.
- Low Raw-Material (RMP) has been observed for industry code (316) of state Punjab and it was also quite low i.e. only 0.15% per annum.
- High Fuel Productivity (FP) has been observed for industry code (274) of state Tamil Nadu and it was 3.11% per annum.
- Low Fuel Productivity (FP) has been observed for industry code (316) of state Punjab and it was also quite low i.e. only 0.23% per annum.
- High Enterprise Productivity (EP) has been observed for industry code (316) of state Punjab and it was 11.91% per annum.
- Low Enterprise Productivity (EP) has been observed for industry code (279) of state Assam and it was only 0.57% per annum.
- High Emoluments per worker (EW) has been observed for industry code (282) of state Delhi and it was 2.35 per cent per annum.
- Low Emoluments per worker (EW) has been observed for industry code (281) of state Assam and it was only 0.35% per annum.

However, on the basis of overall discussion of Table 5.1.17 to 5.1.21, Inter-state comparison of SFPs can be summarized and thus vast interpretation of these tables in terms of high and low growth of SFPs among the states for overall wood & forest based industries are outlined as follows:

- High Labour Productivity (LP) observed for the Northern Region with 1.52% per annum.

- Low Labour Productivity (LP) observed for the Other States with 0.73% per annum.
- High Capital Productivity (CP) observed for the Southern region with 5.73% per annum.
- Low Capital Productivity (CP) observed for the Other States with 2.48 % per annum.
- High Capital Intensity (IC) observed for the Northern region with 3.17% per annum.
- Low Capital Intensity (IC) observed for the Southern Region with 1.387% per annum.
- High Raw-Material Productivity observed for the Other States with 1.46% per annum.
- Low Raw-Material Productivity observed for the Western Region with 0.88% per annum.
- High Fuel Productivity (FP) observed for the Other States with 3.52% per annum.
- Low Fuel Productivity (FP) observed for the Eastern Region with 1.35% per annum.
- High Enterprise Productivity observed for the Northern Region with 5.09% per annum.
- Low Enterprise Productivity (EP) observed for the Other States with 1.85% per annum.
- High Emoluments per worker observed for the Northern Region with 1.44% per annum.
- Low Emoluments per worker observed for the Southern Region with 1.113% per annum.

On the other side, Overall discussion of Table 5.1.1 to 5.1.16 can be summarized as high and low growth of SFPs estimation among the industries within the selected states for various 3-digit minor group industries of wood & forest based industries are as follows:

- High Labour Productivity (LP) has been observed for industry code (378) of Northern Region and it was 5.15% per annum.
- Low Labour Productivity (LP) has been observed for industry code (276) of Other States and it was only 0.13% per annum.
- High Capital Productivity (CP) has been observed for industry code (385) of Northern Region and it was quite high 27.31% per annum.
- Low Capital Productivity (CP) has been observed for industry code (387) of Other States and it was 0.59% per annum.
- High Capital Intensity (IC) has been observed for industry code (385) of Northern Region and it was 18.65% per annum.

- ◈ Low Capital Intensity (IC) has been observed for industry code (387) of Other States and it was very low only 0.18% per annum.
- ◈ High Raw-Material Productivity (RMP) has been observed for industry code (385) of Northern Region and it was 3.23 % per annum.
- ◈ Low Raw-Material (RMP) has been observed for industry code (387) of Eastern Region and it was also quite low i.e. only 0.225% per annum.
- ◈ High Fuel Productivity (FP) has been observed for industry code (385) of Northern Region and it was 5.2% per annum.
- ◈ Low Fuel Productivity (FP) has been observed for industry code (387) of Eastern Region and it was also quite low i.e. only 0.36% per annum.
- ◈ High Enterprise Productivity (EP) has been observed for industry code (385) of Northern Region and it was 24.51% per annum.
- ◈ Low Enterprise Productivity (EP) has been observed for industry code (270) of Other States and it was only 0.77% per annum.
- ◈ High Emoluments per worker (EW) has been observed for industry code (385) of Northern Region and it was 4.61% per annum.
- ◈ Low Emoluments per worker (EW) has been observed for industry code (387) of Eastern States and it was only 0.32% per annum.

Overall conclusion of partial factor productivity ratios can be summarized that in majority of the cases labour productivity have increased at a rate of about 1.0% per annum. The capital intensity has shown similar trends ranging from 0 % to 2.0% per annum. However, Capital productivity have shown impressive trends and growth rates were increasing between 0.0% to more than 11 per cent per annum for most of the industries of WFBI. Thus, it looks that increase in capital is resulting in technological progress in most of the industries of WFBI. This is quite natural since most of the economic reform measure aims at promoting use of capital for most of the WFBI of India. The growth observed in case of these productivity ratios can't be treated in isolation with their relationship with other ratios. As discussed earlier high growth rate of capital productivity in most of cases can't be attributed nearly as an improvement in quality and efficiency of capital factor but quality and improvement in human labour and the technological specification may also be responsible for this growth. Most of the industries are labour intensive and employ large number of work force, therefore, role of latent potential of labour can't be ruled out.

The measurement of TFP is beset with source well-know conceptual and empirical problem and also involves several restrictive assumptions, such as static competitive equilibrium and constant returns to scale etc. Of late years, the methodology of TFP measurement has advanced remarkably by incorporating the recent advances in production and cost theory and in index number theory and practice. With the time, new developments a rigorous theoretical frame work has evolved and new indexes of total factor productivity have been devised which require less rigid assumptions about the production technology and the nature of technical progress. But, still there many important questions remain regarding the methodology of its measurement still to be resolved adequately. Notwithstanding its measurement problems, the concept of total factor productivity has proved extremely useful in empirical work and a consequence has found favour widely (Pendse & Baghel, 2008).

Studies on total factor productivity acquire great significance in the context of growth in developing economies like India. These economies are characterized by acute, scarcity of resources (particularly capital and must use the available resources as best as they can. Also, generation of surplus, which plays a pivotal role in their growth, depends crucially on the efficiency with which resources are used. Evidently, the various studies B.N. Goldar, P.R. Bramhananda, I.J. Ahluwalia, Rajlakshmi, Mehta, Subramaniyan, Hasim & Dadi, Balakrishnan & Pushpangadan, Dholakia & Dholakia, Raj R.S.N., Unel B., Krishna & Mitra, Kumar, Pendse & Baghel, Kathuria & Sen etc. on productivity trends in Indian manufacturing industry would be of much interest, as it brought out how effective the investments made during the plans (in machinery, human capital and infrastructure) were in raising in overall efficiency in resource use.

Total factor productivity index or the index of technical progress as it often referred, gives us an idea of overall efficiency in the factor use. This index measures the output per unit of capital and labour combined. For empirical purposes, we estimated Kendrick, Solow and Divisia indices of total factor productivity for manufacturing sector of India for the period under study. Studies on total factor productivity acquire great significance in context of growth in developing economics. These economies are characterized by acute scarcity of resources (particularly capital) and must use the available resources as best as they can. Also, generation of surplus, which plays a pivotal role in their growth, depends crucially on the efficiency with which resources are used. Evidently, the study (Goldar, 1983) on productivity trends in Indian manufacturing industry would be of much interest, as it brought out how effective the investments made during the plans (in machinery, human capital and infrastructure) were in raising in overall efficiency in resource use. Estimates of all three TFPG indices Kendrick, Solow and Divisia index) has been used in this study and is computed with the help of procedures and formulae as already discussed in detailed in the Research Methodological Chapter separately (i.e. Chapter-IV).

Overall analysis of TFPG indices reveals that there was significant gain in TFP so far as the 3 industries and overall wood and forest based industries of Andhra Pradesh are concerned. In case of Assam, there was not significant gain in TFP so far as the most of industries and overall wood and forest based industries. In the state Bihar, there was significant gain in TFP in most of industries and total wood and forest based industries. Similarly like the state Andhra Pradesh, and Bihar, the state Delhi has also experiences significant gain in TFP in most of industries and total wood and forest based industries. TFPG growth trend of state Haryana state also observed quite impressive and positive growth performance so far as the industries of WFBI are concerned. There was significant gain in TFP so far as the most of industries and overall wood and forest based industries of Karnataka state are concerned. In the state Kerala, there was significant gain in TFP in most of industries and total wood and forest based industries. Similarly like the state Andhra Pradesh, Bihar and Delhi, Gujarat, Haryana and MP, state Maharashtra have also experiences significant gain in TFP in most of industries and total wood and forest based industries. While, the State Orissa have experiences significant gain in TFP in majority of industries and total wood and forest based industries. TFPG growth trend of Punjab state also observed positive growth performance so far as the industries of WFBI are concerned. In the state Rajasthan, although, there was significant gain in TFP in most of industries and total wood and forest based industries too as well but not with high magnitude and impressive in the perspective of technological change. In the state of Tamil Nadu, 9 industries are experiencing more than one per cent per annum growth and gained considerably in all TFPG indices during the period under study. TFPG growth trend reveals that state Uttar Pradesh although experiencing positive trend in TFPG growth in most of cases but not accelerated growth considerably. TFPG growth trend of West Bengal state observed positive growth performances so far as the industries of Wood and Forest based Industries are concerned.

Overall observation of Chapter-V reveals somewhat zigzag growth pattern of TFPG so far as the 16 states are concerned. Although some states are experiencing technological change with positive growth in TFPG but can't be associated in the level of high magnitude. After analyzing TFP growth trend in various states, an attempt has also been made for an assessment of the contribution of various categories of WFBI to the interstate variations in the rates of TFP growth. In the process it also analyses the differences within the industry and regions too as well. Thus, inter-state variations and trends at regional levels in the total factor productivity growth have been studied as the earlier not a single studies have made such exercises in India and very few studies have studied for few industries comes under WFBI and even only for a short period.

So far as the regions of India are concern, brief analysis of TFPG findings reveals that in the Eastern Region, there was significant gain in TFP so far as the most of eastern region WFB industries and total WFBI are concerned. TFPG growth trend of Western region found quite impressive and positive growth performance so far as the industries of WFBI are concerned. TFPG growth trend reveals that Northern region although experiencing positive trend in TFPG growth in most of industries and total Wood & Forest based Industries as well. In southern region, most of the industries are experiencing either positive trend in growth of TFP but below the one per cent per annum, or found insignificant trend growth. Although, TFP growth for total wood and forest based industries have observed positive and significant with one per cent per annum growth, but not with high magnitude. In the Other States, although, there was significant gain in TFP in most of industries and total wood and forest based industries too as well but not with high magnitude and impressive in the perspective of technological change. TFPG growth trend of All India, observed positive growth performance so far as the industries of WFBI under study are concerned.

Production function is a technological or engineering concept which shows for a given state of technological knowledge and managerial ability, the maximum rates of output that can be obtained from different combinations of the productive factors during a given period of time. In brief, the production function is a catalog of different output possibilities. Hence, production function is the process by which inputs are converted into output. In other words, production function is defined as the technical or engineering relationship between factor inputs and output. The production function formalizes the relationship between the quantity of output yielded by a product process and the quantities of the various input used in that process. Inputs refer to the various factor prices, which are used in the production. Outputs on the contrary refers o the quantity of goods produced by the firm with the help of the various inputs. In the form of mathematical equations, the production function can also be expressed in which output is the dependent variable and inputs are the independent variables. Therefore, relationship of these terms in general form can be stated as: Q: (X_1, X_2, X_3, ... X_n). Where Q is the rate of output of a given commodity and X_1, X_2, X_3, ... X_n are the various factor of production used per unit of the time.

The main purpose of applying production functions is to estimate marginal productivity and elasticity of inputs, returns to scale and elasticity of substitution between the inputs. Among the available literate of production functions, the choice in this study was in favour of Cobb-Douglas, Constant Elasticity of Substitution and Variable Elasticity of Substitution because these are the most popular and widely used form of production functions. These production function involve treatment of elasticity of substitution constant and equal to unity in C–D specification and constant but not necessarily equal to unity in case of C.E.S. specification. While C.E.S. production function major limitation is, it assumes the elasticity of substitution to be constant for all input combinations. In the literature, several functional forms have been suggested and used in empirical studies in which elasticity of substitution is variable. These are called V.E.S. (variable elasticity of substitution) production function. A vast detailed discussion has already been made in the previous chapter (in Chapter-IV) regarding the specification and from of above mentioned production function specification. A time trend variable has been included in production functions to represent technological progress.

However, Production functions for the Wood & Forest based Industries (WFBI) for India and Indian Regions has been estimated with a view to identify the contribution of factors other than capital and labour inputs to output growth. TFPG, which is an indicator of the change in efficiency in factor use, discussed earlier has also been measured for explaining the resource use efficiency. The conceptual issues associated with the problems of aggregation, valuation of capital, under utilization of capital and changing quality of labour in the specification of a production function have been extensively discussed in research methodological chapter.

In case of growth accounting estimates, direct estimation of the production function has an advantage in that, it is not necessary to assume competitive equilibrium in order to derive an estimate of productivity growth. The efficiency parameter, the scale parameter and the extent of substitutability between factors can be obtained by directly estimating the parameters of a suitably production function through regression analysis. However, statistical problems associated with endogeneity of input variables, multi-collinearity among the explanatory variables and autocorrelation of the residuals tend to impinges upon the robustness and other desirable properties of the production function estimates. In the specification of a production function, intermediate inputs can be explicitly incorporated by relating changes in the gross value of output to changes in labour, capital and intermediate inputs. When intermediate inputs are explicitly recognized, it implies that the estimated TFP growth rates are lower than that base on the value added function. The analysis in the present study is based on the value added production function.

In the overall analysis, the various specifications of production function for the Eastern Region of India, the industry groups which have shown negative and significant growth in total factor productivity are 270, 271, 272, 276, 316, 387 and total wood & forest based industries of the region. In case of Western region, the industry groups which have shown negative and significant growth in total factor productivity are 270, 271, 272, 276, 282, 386, and 387. While for the Northern Region of India, the industry groups which have shown negative and significant growth in total factor productivity are 270, 271, 272, 276, 385, 387 and total wood & forest based industries of the region. On the other side, in case of Southern Region of India, the industry groups which have shown negative and significant growth in total factor productivity are 270, 271, 272, 274, 316, 387 and total wood & forest based industries of the region. For the Other States, the industry groups which have shown negative and significant growth in total factor productivity are 270, 271, 272, 274, 276, 279, 280, 282, 386, 387 and total wood & forest based industries of the region. At last, for All India, overall analysis of the various specifications of production function, the industry groups which have shown negative and significant growth in total factor productivity are 226, 270, 271, 272, 274, 275, 276, 279, 280, 386, 387 and total wood & forest based industries. It means, majority of wood and forest based industries of all India are not experiencing technological change, thus, suggesting technological retrogation. The overall analysis the various specifications of production function showing insignificant time coefficient in the remaining industry groups, thus indicating neither technological progress nor technological retrogation (i.e. technological stagnation).

The present study also supports the findings of earlier studies. Neutral technological progress does not seem to have played any important role in output growth. The present study seems to be a distinction and identifies labour as a factor most crucial to output growth. In the above mentioned production function studies, coefficient of capital is generally found to be insignificant owing to mainly measurement problems, multi-collinearity, over-capitalization and under-utilization of the installed capacity. The findings of the earlier ones, differs mainly because of differences in the choice of the period and the coverage of real units. Insignificant labour coefficient in some cases paints a disquieting picture so far as the role of labour in the output growth is concerned.

At, last, on the basis of overall analysis and in-depth elaboration of issues taken here in project which has already been discussed in Chapter-I to Chapter VI, we once again certainly want to explore that Wood and forest based industries can play a vital role in the development process of a country like India because most of the rural population living in and around the forests. Rural population and their dependency on agriculture and forest are still heavy. Economic prosperity of agrarian economy will not only depend upon the integration of agriculture with industry only but also upon the integration of forest with industry. Hence, for the development of rural areas, Wood and forest based industries have a special significance. In the midst of all these problems a special importance can be attached to the question of balance between forest and industry in development. The role of forest in its producer, producer-consumer and consumer aspect can be best utilized for framing policies for development. The analysis of technological change and productivity in wood and forest based industrial segment of economy is very much useful from development aspect of economic growth in a country like India. Technological improvements in wood and forest based industrial sector will enhance output of the economy on one hand and will uplift economic status of bulk of labour force, which is engaged in operation on the other. Balanced economic development can be ensured if wood and forest-based industrial units are established and spread over various part of this county. From overall analysis it becomes clear that diagnosis of this industrial sector requires no explanation and its relevance to society and country is well established.

By 2030, India is likely to have GDP of USD 4 trillion dollars and a population of 1.5 billions. To achieve sustainable growth of our economy, it is essential to develop forest sector and its related activities like forest based industrial development, to achieve double digit per annum growth of GDP. The study like has not been done earlier in India by any academicians, researchers, and also not by any institutions too. Therefore, there is an essential need to work out on the result finally interpreted here in this work for the economic growth and sustainable development of manufacturing sector like Wood and Forest based Industries.

This study has been conducted with the aims at finding out the extent of technological change in wood and Forest Industries and productivity growth. The partial factor productivity ratios have enabled us to quantify the unit factor requirements on saving in the use of factor of production. Thus, inter-state difference in SFP found useful in explaining the gap between developed and less developed states. The total factor productivity (TFP) have also throws light on overall efficiency, which would be helpful for the policy makers in framing policy for efficiency improvement in less developed states. The estimates of production function are giving an idea of marginal product of factor, returns to scale elasticity of substitution among the regions of India and India as a whole as well. The variation in all the measured parameters among the states has given the idea of extent of disparity existing among the states. At last, it can be said that this study would be helpful in framing the policies for complete overall of wood and forest -based industrial factor of India and balanced economic development of country. The observed trends in the various parameters would require identification of courses responsible for the phenomenon, and care of these causes responsible for in policy framing for administrators, scholars, and academicians. Finally, it can be said that this study will prove a mile stone in developmental literature of this country.

www.ingramcontent.com/pod-product-compliance
Ingram Content Group UK Ltd.
Pitfield, Milton Keynes, MK11 3LW, UK
UKHW021530300726
14060UKWH00011B/177